Contemporary Design and Manufacturing Technology

Edited by
Taiyong Wang
Hun Guo
Dunwen Zuo
Ji Xu

Contemporary Design and Manufacturing Technology

Special topic volume with invited peer reviewed papers only.

Edited by

Taiyong Wang, Hun Guo, Dunwen Zuo and Ji Xu

Trans Tech Publications Ltd
Kreuzstrasse 10
CH-8635 Durnten-Zurich
Switzerland
http://www.ttp.net

Volume 819 of
Advanced Materials Research
ISSN print 1022-6680
ISSN cd 1022-6680
ISSN web 1662-8985

Full text available online at *http://www.scientific.net*

***Distributed** worldwide by*

Trans Tech Publications Ltd
Kreuzstrasse 10
CH-8635 Durnten-Zurich
Switzerland

Fax: +41 (44) 922 10 33
e-mail: sales@ttp.net

and in the Americas by

Trans Tech Publications Inc.
PO Box 699, May Street
Enfield, NH 03748
USA

Phone: +1 (603) 632-7377
Fax: +1 (603) 632-5611
e-mail: sales-usa@ttp.net

printed in Germany

Preface

The special volumes are to communicate the latest progress and research results of new theory, new technology, method, equipment and so on in Engineering Technology, and to grasp the updated technological and research trends in international, which will drive international communication and cooperation of production, education and research in this field.

The major topics covered by the special volumes include Advanced Materials and Manufacturing Technologies, Control, Automation and Detection Systems, Advanced Design Technology, Optimization and Modelling and so on.

Table of Contents

Chapter 2: Control, Automation and Detection Systems

Chapter 3: Advanced Design Technology, Optimization and Modelling

CHAPTER 1:

Advanced Materials and Manufacturing Technologies

Advanced Materials Research Vol. 819 (2013) pp 3-6
© (2013) Trans Tech Publications, Switzerland
doi:10.4028/www.scientific.net/AMR.819.3

After grinding NC grinding of large curved surface

HongYu Han[1, a] , Lu Fu[1,b] and Yingliang YU[1,c]

[1]Luohe Vocational and Technical College，Luohe Henan 462002，China

[a] hhy@open.ha.cn, [b] xcwlm@163.com, [c] lzyjdxyyl@163.com

Keywords: CNC grinding Large circular surface Processing Technology The instantaneous automatic compensation for wheel wear

Abstract. Additional power head with common milling machine.Two and a half axis CNC system are used for CNC grinding circular surface curve trajectory. The advantages of this processing method are: We can avoid the influence on processing of lead screw clearance because of changing the parts clamping position,effectively ensure the integrity of the great circle arc curve trajectory. Because of using digital control technology in grinding process is instantaneous automatic compensation of grinding wheel wear, the machining accuracy of parts in outline is improved, the machining error is decreased. it will effectively ensure the consistency of the large circle curve precision.This method has the effect of infer other things from one fact for CNC grinding of large circular arc curve parts. Broaden the scope of application of CNC technology and practical.

Introduction

The radial tolerance of the parts shown in Figure 1 is 0.12mm. Surface roughness $\overset{0.8}{\triangledown}$. Hardness of HRC45～53. Requirements-type surface must have a high-surface precision. Conventional grinding is difficult to guarantee such a high processing requirements.

Additional power head with common milling machine, CNC system is used in two and a half axis, screw feeding respectively control the longitudinal cross perpendicular to the three direction, the grinding of large curved surface.See Figure 2. You can get the desired results. The following illustrates the practical application of research for CNC grinding.

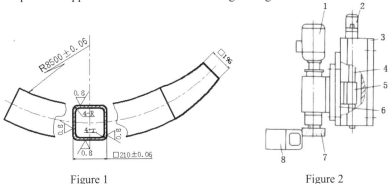

Figure 1   Figure 2

1 - motor speed 2- Servo motor   3 - dovetail bearing
4 - Vertical screw   5 - wire mother Tower
6 - dovetail on the slide   7 - grinding wheel 8- workpiece

Part grinding difficulty analyzing

For parts of the large circular surface processing, The processing difficulty lies circular surface grinding. Because ordinary grinder can not be arc trajectory grinding.

If you rely on in the ordinary grinder mode processing, Constrained by templates manufacturing precision cam wear factor and Bound to affect surface machining precision parts, Another even more important is, Grinding wheel in grinding instantaneous wear, Its the grinding size also instantaneous dynamic changes. So, the result is not satisfactory precision large circular surface grinding.

Loop NC system circular surface grinding, due to the detection point is difficult to be fixed, according to the structure and shape of the part. So it can not be used to solve the automatic compensation for wheel wear.

Part grinding fixture positioning analysis and processing

The parts CNC grinding, The surface under the part of the preceding grinding as positioning reference, The position fixture positioning shown in Figure 3. The biggest advantage of such a fixture: CNC grinding running arc trajectory you can avoid the Y to the pros and cons of the screw to run , Eliminating the gap to the lead screw parts processing, Machining accuracy and makes effective control and assurance. Arc trajectory starting point in the quadrant point is also conducive to the advance processing trajectory coordinate calculation and cut programming.

Grinding wheel wear compensation solutions

In the grinding process of the arc-shaped trajectory, Set in a period of normal wear and tear, The wear of the grinding wheel is carried out in a linear relationship, And with the length of the grinding time was proportional uniform wear. Its circular arc trajectory shown in Figure 4 when Grinding outer circular surface, Reciprocity theory grinding trajectory and the actual grinding trajectory.

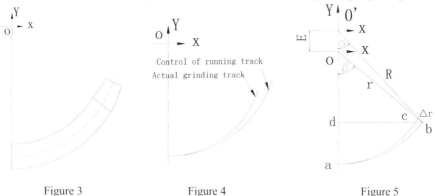

Figure 3   Figure 4   Figure 5

We can see by the the processing trajectory shown in Figure 4: Due to the wear of the grinding wheel, So that the circular arc surface trajectory error generated in the grinding process. This error is proportional to the processing arc length of the arc length and grinding time. It seriously affects the machining accuracy.

We know that: In grinding wheel, according to the circular arc surface part of the material, and the use of hardness. Actual measurement of a given arc length of grinding wheel wear radius Δr. In the grinding of an internal circular arc surface, Should parts processing center coordinates radius mobile X distance along the y axis, See Figure 5. If O is the new center coordinates, R is the new radius, When the the grinding the ab arc trajectory run to the end of b when, Error Δr is produced by dry grinding wheel wear. The actual grinding trajectory for ac arc Curve trajectory ac arc is the coordinate of center O and r is the radius of the theoretical arc trajectory, Making the actual grinding track with theory grinding trajectory coincide ,Thereby achieving the the instantaneous automatic compensation of wheel wear in the grinding process.

From Figure 5: $\triangle$Obd Lak $\triangle$Obd

$$R\cos \beta = (r + \Delta r)\cos \alpha + E$$
$$R\sin \alpha = (r + \Delta r)\sin \alpha \tag{1}$$

Rearranging the equation can be obtained:

$$E_{inner} = \frac{\Delta r(r + \Delta r/2)}{r - (r + \Delta r)\sin\alpha} \qquad (2)$$

Formula: r - requirements of the radius of the arc, mm

$\triangle$r - radius of the given arc length grinding wheel wear, mm

α - require the angle between the cambered beginning and end points and the center of the circle

This is the grinding of an internal arcuate surface center is shifted along the Y axis the amount X is calculated. Thus can be obtained to calculate the the the arcuate surface grinding, Corresponding to the grinding arc radius R:

$$R_{within} = r + x \qquad (3)$$

Similarly, can be deduced the grinding of the outer arcuate surface of the center of the circle along the Y axial direction of the offset X formula. With the corresponding optional grinding circular arc radius R of the formula:

$$E_{outside} = \frac{\Delta r(r + \Delta r/2)}{r + \Delta r - r\cos\alpha} \qquad (4)$$

$$R_{within} = r - x \qquad (5)$$

Same processing large circular surface, Parts flatness there is also the wear of the grinding wheel in grinding. Proportional to the ΔX amount of wear and grinding length and grinding time.

In trial grinding May be determined according to the measured end face of the grinding wheel wear amount ΔX. Well, If you run a length in the plane(In this case, it is assumed that the end face of the grinding wheel wear amount ΔX equal control power wheelhead vertical screw unit displacement accuracy), So that power Wheelhead vertical downward run step, Continue to surface grinding.It can eliminate the error produced in a fixed plane grinding plane. The cycle is repeated indefinitely until the plane grinding.

Unit displacement precision plane on the maximum flatness error is the control of vertical screw motion servo motor after grinding.

Power vertical grinding head screw pitch t＝1. 5mm， Optional servo motor connected thereto. Because servo motor need 1200 pulses per revolution Whereby this direction of the unit displacement accuracy of 1.5/1200=0.00125mm 。 For such a tiny error, to ensure the precision parts processing requirements.

Using the straight line in the vertical direction (Z direction) on the processing instructions to compensate for the wear of the grinding wheel, And X-, Y up (straight line processing instructions) processing slash trajectory to compensate for wheel wear, Control of its movement is the same, Because slash trajectory control, the length of the displacement in the direction of each movement is required to achieve a pulse equivalent , The servo motor was running step. Therefore, the arc-shaped parts when grinding plane can using the straight-line processing instruction to compensate for wheel wear. Or less into to and no feed grinding wheel wear compensation.

The curve machining accuracy of CNC grinding circular surface

Application of the method shown in Figure 3 for parts of the fixture positioning, Automatic wheel wear compensation grinding using special programming calculations, CNC grinding circular surface curve. The advantage of this processing method is: The part clamping position change is in the processing of simple, Since the control is a single direction in the processing movement run, So to avoid a the leadscrew backlash of processing trajectory accuracy, Effectively ensure the integrity of a large circular surface curve trajectory. Wheel wear the realization of the automatic compensation, The calculation is simple, easy programming and debugging, Part contour of the machining accuracy, the machining error is reduced, Ensure consistency of large the arc Powerphones curve accuracy.

Obviously, this processing methods used in a large circular surface curve CNC grinding, Compared with conventional grinding, has incomparable advantages, Satisfactory processing results are obtained in the practical application.

Large the arc Powerphones curve error analysis and the measures taken

CNC machining parts arc length programming. Programming by arc length conversion formula:
$M_z = 2\pi RM / ti$ Shows that, π is an irrational number, So, on the numerator value is also irrational number, The arc length of the theoretical value of the programmed value ranging, Thus making the length of the arc length displacement exist calculation and motion error Δ L. In the process, by increasing the programming of the large circular arc surface trajectory run length, That is slightly larger than the actual grinding trajectory programmed trajectory, equivalent to running track grinding to extend the cut-out, you can eliminate the error.

Conclusion

The actual processing to prove: The machining grinding large circular surface, Due to elimination of the screw clearance processing, effectively ensure the integrity of a large circular surface curve trajectory. Using CNC technology, instantaneous automatic compensation for wheel wear in grinding process, effectively ensure the accuracy of the consistency of the large circular surface curve. Compared with the conventional (including CNC) grinding, Part trajectory control and to improve machining accuracy, the machining error is reduced, Grinding similar large arc curve parts CNC grinding, giving top priority to the role. General business conditions of the lack of high-end CNC systems while avoiding restrictions , Broaden and extend the scope of application of the numerical control technology, has a strong practical.

References

[1] Lin Chun editor of computer-controlled mechanical system design of Shanghai: Shanghai Science and Technology Press, 1991.

[2] Zhang Xin-yi editor of economical NC machine tool system design Beijing: Mechanical Industry Press, 1994.

[3] YU Ying-liang book CNC design and examples of transformation of Beijing: Mechanical Industry Press, 1998.

Advanced Materials Research Vol. 819 (2013) pp 7-12
© *(2013) Trans Tech Publications, Switzerland*
doi:10.4028/www.scientific.net/AMR.819.7

An Adaptive Tool Path Generation for Fused Deposition Modeling

Yu'An Jin[a], Yong He and Jianzhong Fu

The State Key Lab of Fluid Power and Mechatronic Systems, Zhejiang University, Hangzhou, China

[a]Corresponding to: jinyuan7094@163.com

Keywords: FDM. Adaptive slicing. Tool path generation. Tool path adjustment

Abstract: This paper presents an adaptive tool path generation method for Fused Deposition Modeling (FDM), which provides an effective approach for dieless forming and reverse engineering. The proposed method consists of three steps. First, an adaptive slicing considering both surface quality and building time is proposed for following tool path generation. Second, a hybrid tool path strategy is introduced to improve the boundary contour's accuracy and reduce the time for interior filling. This step is an adaptive process to choose an appropriate proportional relationship between the two types of tool paths: contour parallel path and direction parallel path according to the specific fabrication requirements. For further improvement on fabrication quality, a tool path adjustment is employed on the original tool paths. A case study of a sliced layer is used to verify the feasibility of the proposed method and demonstrate it can accomplish the fabrication of models with high surface quality and short building time.

Introduction

Fused Deposition Modeling (FDM) represents a rapid manufacturing technology, which builds a part by gradually adding materials layer by layer [1]. With FDM technology, arbitrary physical parts can be manufactured within a comparatively short time directly from 3D CAD models. However, poor surface quality and low fabrication efficiency are bottlenecks in this technology [2]. Hence, there is an urgent desire of some approaches for surface accuracy improvement and build time reduction.

Tool path in FDM is the trajectory of the nozzle/print head during manufacturing process to fill the interior of each layer [3]. Generally, there are two main tool path trajectories for FDM processes: contour parallel path and direction parallel path as shown in Fig. 1. The former focus on the surface quality at the cost of efficiency, and the latter considers the build time reduction while neglects the surface accuracy. Besides the requirements of tool path of general material removal machining, some special requirements also need to be met in the tool path generation of FDM. The forming principle of FDM is displayed in Fig. 2. The path of the nozzle/print head is covered with forming material. The cross section of material in the vertical direction is an approximate semiellipse which is determined by the property of the forming material. The solid boundary of the forming material in the figure is the final surface of one layer. h is the height of the layer and w is the width of the strip of forming material, and a proportional relationship exists between them. If the distance between tool paths is too large, a gap between two adjacent strips of forming material may appear. However, a too small distance may result in a ridge between strips. Thus, some methods, which can be used to solve fabrication problems in general material removal machining, are not available in FDM.

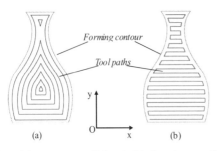

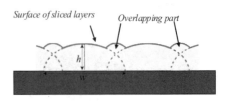

Fig. 1 (a) contour parallel path (b) direction parallel path   Fig. 2 forming principle of RP

Furthermore, in a given FDM machine, the volume of extruded material is constant within a certain period. Hence, the dimension of the semiellipse (h and w) can be represented in Eq. 1.

$$\pi h(w/2)L = kt \tag{1}$$

where the left of the equation is the volume of forming material; L is the length of the tool path; k is the extruding speed of forming material. From Eq. 1, the following equations can be shown:

$$\pi h(w/2)L = kt/L = k/v \tag{2}$$

because of a proportional relationship between h and w, the equation can be shown as follows:

$$Kw2 = k/v \tag{3}$$

where v is the velocity of the head; K is a constant coefficient determined by the forming material property. With Eq. 2，the relationship between v and h can be calculated:

$$K'h^2 = k/v \tag{4}$$

where K' is a constant coefficient determined by the forming material property.

Slicing process of FDM technologies inherently brings a staircase effect, which affects the surface smoothness [4]. Its influence is determined by the layer thickness and the local part geometry. If the layer thickness of all the layers is uniform without consideration of the local geometry, it may result in time-consuming if the thickness is too small while poor surface quality if the thickness is too large. Adaptive slicing modifies the layer thickness to take into account the normal curvature of the surface of the solid model to alleviate the staircase effect, and to decrease the number of layers [5]. The tool path generation proposed in this paper is based on the adaptive slicing.

In this paper, an adaptive slicing algorithm is introduced for further tool path generation firstly; then, a hybrid tool path with the consideration of both fabrication time and surface quality is developed. Meanwhile, for high fabrication quality, some improvements and modification are employed in the original tool paths. Besides, an implementation of the method is presented to verify the feasibility of the proposed approach. Some conclusions are drawn in the end of the paper.

Methodology

In this study, tool path generation from geometric information consists of two steps: obtaining sliced layers with slicing algorithm and tool path generation of each layer.

Adaptive slicing. Adaptive slicing can alleviate the staircase effect by the alteration of layer thickness according to the geometric characteristic of models. FDM processes have permissible layer thickness extents according to the hardware conditions [5]:

$$h_{min} \leqslant h \leqslant h_{max} \tag{5}$$

where h is the permissible layer thickness depending on the process.

The thickness of each sliced layer is mainly determined by the curvature of the surface. The cusp height c should be limited to a certain value. The cusp height c of one sliced layer is defined as the maximum valve among the cusp height in the boundary of the layer. The cusp height c_i of each point is the distance between the vertex of a layer (O) and the actual surface of part model in the normal direction, and the normal direction is obtained by the normal vectors of the adjacent cross sections. Within a specified cusp height tolerance c_{max}, the thickness of each layer (h) can be estimated using the normal vectors of the current slice. At the same time, a judgment of h using Eq.5 should be taken into account.

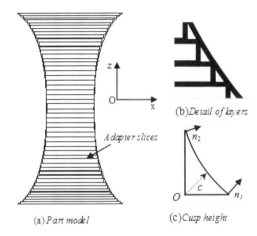

(b)*Detail of layers*

Adapter slices

(a)*Part model*   (c)*Cusp height*

Fig. 3 schema of adaptive slicing

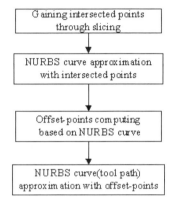

Fig. 4 the flow of generating tool path of boundary contouring

Tool path generation. Once the thickness (h) of each sliced layer has been determined, the offset of the tool path are determined at the same time according to Eqs. 3,4 and the choosing overlapping rate. To meet both the accuracy and efficiency requirements, a hybrid tool path is proposed with the consideration of the characteristic of FDM technologies. The tool paths of one sliced layer contain two parts: contour parallel path along the boundary of the layer, direction parallel path in the interior area of the layer. The former can guarantee the surface quality of the model and the latter enables a fast fabrication in the interior-filling.

Tool path generation of boundary contouring. After a slicing process from a 3D model represented by STL, a sliced layer is represented by a set of intersected points on the boundary. To maintain the accuracy of the original model, the offset of the nozzle/print head has been taken into consideration. A direct point-offset approach can be employed in the intersected points to generate the offset of the point-set boundary. This can be performed by three steps: (a) approximate the intersected points with a NURBS curve [6]; (b) calculate the normal vectors of points in the curve and obtain the offset points according to the offset distance; (c) approximate the offset points with another NURBS curve. The flow of this work is shown in Fig. 4.

Tool path generation of interior-filling. After the generation of the boundary contouring and related tool paths, the tool paths of the internal area of the layer should be generated to complete the fabrication of the sliced layer. There are four steps for this process: (a) orientation determination; (b) offset of boundary contouring; (c) computation of intersected points; (d) tool path generation.

(1) Orientation determination: the orientation of the tool path is the slope of the equidistant lines; the different orientation will result in the different number of line elements which affects the

fabrication efficiency. Since the distance between adjacent tool paths has been determined by the layer thickness (*h*), the number of line elements is decided by the length of a given sliced layer in the direction of normal to the equidistant lines. So choosing the direction with shortest length as the normal direction of the orientation of tool path is an intelligible measure.

(2) Offset of boundary contouring: this step can be accomplished using the same method as mentioned in Section 2.2.1.

(3) Computation of intersected points: these intersected points separate the interior tool paths into two types: lines and tiny arcs.

(4) Tool path generation: According the intersected points, a zigzag tool path can be worked out. However, some problems will appear if we simply use the zigzag tool path as the final path. First, unfilled areas may exist near the turning points because of the inherent shortcoming of direction parallel tool path. And at the same time, excessive filled area exist in the other side of the corner. Second, the abrupt corners between line and tiny arc will result in the sharp alternation of speed of nozzle/print head, which can jeopardize the surface smoothness according to Eq. 4. Hence, the original tool paths need some adjustments in the corners to alleviate the influence of the mentioned issues.

Tool path adjustment. Changing direction abruptly of the tool path is the root of the occurrence of unfilled areas and unstable speed. At each turn, the filled strip of two adjacent tool paths travelling in different directions might not have enough overlapping rate, a scallop appears between them. Besides, as the forming principle of FDM shown in Fig. 2, an excessive filled area will inherently appear. The Fig.5 displays the two phenomena. The areas ① and ② are excessive filled areas and the area ③ is the unfilled area because of the insufficient overlapping rate between the adjacent tool paths.

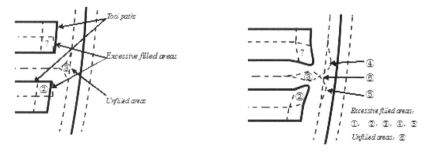

Fig. 5 Fabrication deficiency using original too paths   Fig. 6 Adjustment of the original tool paths

To improve the surface accuracy of sliced layer, the original tool paths at the turn need some modifications to avoid the sharp angle. One feasible approach is shown is Fig. 5. The aim of the modification is increasing the overlapping rate of area ③ and decreasing the overlapping rate of area ① and ②. At these turns, the tool path is extended toward outside so that the scallop can be alleviated to a permissible value. In Fig. 6, the dimension of the "deficiency" areas has been decreased at the cost of the growing number of excessive filled areas. The scatter of the "deficiency" areas can avoid some individuals with great error, and the stack in the rim of strip can compensate the shortage of forming materials such as ④and ⑤. Hence, the surface smoothness can be improved accordingly. Meanwhile, the adjustment of the tool path eliminates the sharp angles at turns, which is beneficial to the surface quality by transiting turns smoothly.

Speed requirements in the tool paths. In the slicing process, the thickness of each sliced layer (h) should keep uniform for the further fabrication. According to Eq. 4 the speed of each layer (v) should keep constant to meet the uniform thickness. And the uniform speed can guarantee the same overlapping rate between adjacent paths based on Eq. 3.

However, the acceleration/deceleration process is inevitable during fabrication process. Sharp corners, which can cause speed alternation, have been avoided through the tool paths adjustment. Nevertheless, the starting/termination of the fabrication of each layer needs an acceleration/deceleration process. This problem can't be solved by improving the tool paths because of the complication and uncertainty unless some techniques are employed in nozzle/print head to control the extruding speed of forming material.

Implementation

The proposed tool path generation approach is implemented with C++ on a PC. After an adaptive slicing process, a sliced layer with the thickness of h (determined by the forming material, FDM machine, specified error tolerance, etc) is shown in Fig. 7. The fabrication part is a flower model, and the boundary contour is obtained by approximating a polyhedral model. The tool path generation of boundary contouring and the tool path generation of interior-filling are accomplished in the following steps, which are displayed as Fig. 7 (b) and (c). And Fig. 7 (d) is the detail of a corner in the final tool paths after adjustment.

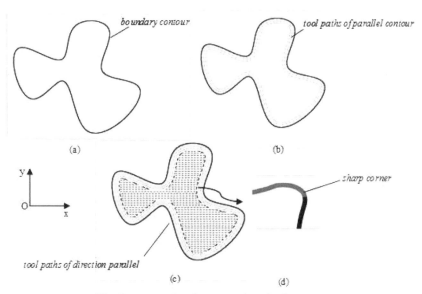

Fig. 7 case study of proposed method

Conclusion

This paper presents an adaptive tool path generation for FDM technologies with a hybrid tool path strategy, and an adjustment approach on the original tool paths to improve the fabrication quality of each sliced layer. The proposed method considers both the surface accuracy and fabrication efficiency. The hybrid tool path generation consists of four steps: adaptive slicing from a 3D model, tool path generation of boundary contouring, tool path generation of interior-filling and tool path adjustment.

The geometric accuracy of original CAD models is maintained through the adaptive slicing and the tool paths which are parallel with the contours. Besides, the tool path adjustment and the speed requirements can guarantee the smoothness of each sliced layer. It's an adaptive process to strike a balance between fabrication quality and building time by choosing the speed of nozzle/print head, the number of the contour parallel tool paths, orientation of direction parallel tool paths, etc. based on the accuracy requirement and the geometrical information of the model.

Future study will be undertaken in the aspect of adaptive tool path orientation of each sliced layer to further improve the geometrical quality. The inherent deficiency in the smoothness of sliced layers can be compensated mutually to guarantee the final surface quality.

Acknowledgements

This paper is sponsored by the Group of scientific and technical innovation of Zhejiang Province (2009R50008), Zhejiang Provincial Natural Science Foundation of China (No. Y1100281, LY12E05018).

Reference

[1]   J.P. Kruth, Material increase manufacturing by rapid prototyping technologies, Ann CIRP. 40 (1991) 603 – 617.

[2]   E. Sabourin, S.A. Houser, J.H. Bohn, Accurate exterior/fast interior layered manufacturing, Rapid Prototyping J. 3 (1997) 44–52.

[3]   G.Q. Jin, W.D. Li, C.F. Tsai, L. Wang, Adaptive tool-path generation of rapid prototyping for complex product models, J. Manuf. Syst. 30 (2001) 154–164.

[4] W.Y. Ma, W.C. But, P.R. He, NURBS-based adaptive slicing for efficient rapid prototyping, Comput. Aided Design. 36 (2004) 1309–1325.

[5]   Z.W. Zhao, Z.W. Luc, Adaptive direct slicing of the solid model for rapid prototyping, Int. J. Prod. Res. 38 (2000) 69-83.

[6]   Y.J. Zhang, M.R. Yu, Computing offsets of point clouds using direct point offsets for tool-path generation, J. Eng. Manuf. 226 (2012) 52-65.

Advanced Materials Research Vol. 819 (2013) pp 13-19
© (2013) Trans Tech Publications, Switzerland
doi:10.4028/www.scientific.net/AMR.819.13

An Insight into the Analytical Models of Granular Particle Damping

D. Q. Wang[1,a], C. J. Wu[1] and R. C. Yang[1]

[1]School of Mechanical Engineering, Xi'an Jiaotong University, Xi'an, 710049, P.R.China

[a]445wdq@163.com(Corresponding author)

Keywords: Granular Particle Damping; Vibration Analysis; Multiphase Flow Theory (MPF); DEM Simulation; Energy Dissipation.

Abstract: Granular particle damping technique is a means for achieving high structural damping by the use of metal particles filled into an enclosure which is attached to the structure in a region of high vibration levels. The particle dampers are now preferred over traditional dampers due to the stability, robustness, cost effectiveness and the lower noise level than the impact damper. Such a promising technique has been used successfully in many fields over the past 20 years. In this paper, a state-of-art review on the development of modeling for particle damping is presented. The fundamentals and individual features of three main mathematical models of the granular particle damping are briefly summarized, i.e. the lumped mass model, the Discrete Element Method (DEM) and the approach based on the multiphase flow (MPF) theory of gas-particle. It is worth noting that an improved analytical model of the particle damping based on MPF theory is also introduced. The co-simulation of the COMSOL Multiphysics live link for MATLAB is conducted using this improved model. It can be shown that this model makes the complicated modeling problem more simply and offers the possibility to analyze the more complex particle-damping vibrating system.

Introduction

Particle damping is a technique of providing damping with granular particles placed in the enclosure attached to or embedded in the holes drilled in the vibrating structure[1, 2]. The particle dampers are derived from the older impact dampers designed to operate by the use of a single body inside an enclosure[3, 4]. This emerging technology can replace the widely used viscous and viscoelastic dampers in particular applications where extreme temperatures (either low or high) are involved. Such a technology is primarily used in the aerospace industry[5-7]. The particle dampers are now preferred over impact dampers due to the lower noise level when they vibrate[8].

The underlying principle of granular damping is the dissipation of vibratory energy through inelastic collisions and frictional interactions among the granules and between the granules and the enclosure wall[9]. It is promising in many applications especially under harsh conditions, however, the mechanism for granular damping is very complicated and highly nonlinear[10]. With the rapid development of the particle damping technique, the desired requirement of developing a mathematical model for general description of the granular damping has long been a challenging issue for researchers in this field. Although at this stage, the comprehensive constitutive relationships for modeling have not been fully established, several models which have been proved useful and are available for predicting the particle damping behavior. In the proceeding sections, three main models, i.e. the lumped mass model, the DEM simulation and the approach based on MPF theory, are subsequently addressed.

The lumped mass model

The intension of this model is that the particle bed (i.e. the aggregation of granules) is assumed to be a mass, which moves unidirectionally in a frictionless enclosure and collides completely inelastically with enclosure ceiling/floor, i.e. zero coefficient of restitution and sustained contact of the particle bed on the structure. The analysis is conducted for the periodic motions of the damper system with the colliding motion of particle bed against the floor or ceiling of enclosure. There may be frictional interaction between particles and enclosure as well as the particles themselves. The granular motion will be described by two kinds of equations, one is corresponding to the case where the granules are in contact with the enclosure floor or ceiling (relative rest motion) and the damper mass moves together with the vibrating body after impact on the end stops, and another equation is consistent with the case of no contact with the enclosure floor or ceiling (separate motion) while in the separation state, the damper moves freely within a clearance[11].

Many researchers have analyzed the damping characteristics of the particle damper by this method. The vertical vibrating flexible structure subjected to a harmonic excitation under gravity was conducted by Inoue et al[11], and the experimental tests verify the analysis results. For multiple degree-of-freedom (MDOF) system (the particle bed is also assumed to be a lumped mass), such an approach is also applicable to investigate the damping performance of multiple particle dampers. Numerical results show that when they are applied simultaneously to multi-body vibrating structure, good agreement was observed between the numerical results and experiments with particle dampers attached to different locations of a three degree-of-freedom system[12]. For a cantilevered beam with the damping enclosure attached to its free end, it is convenient to reduce the continuous beam to an equivalent SDOF system. Similarly, all the particles move as a lumped mass, i.e. the relative motion between the particles is neglected, the effect of acceleration amplitude and clearance inside the enclosure were studied by Friend et al.[13], the damping was found to be highly non-linear, i.e. amplitude dependent. An experimental study of the dynamic behavior of a passive damping system is investigated by Trigui et al.[14], the influence of some system parameters, such as clearance and intensity of excitation, on the specific damping capacity was experimentally examined.

For the lumped mass model, an encouraging result is that the model is simplicity and can capture the essential physics of particle damper. However, the restitution and frictional coefficient of the particles, the diameter of particle, and the inter-particle collisions making contribution to the energy dissipation will not be considered. Actually, aforementioned parameters affect the properties of particle damping evidently. Some researchers[15, 16] provide information of particle motions within the container hole and help explain their associated damping characteristics. A significant damping effect is achieved, not only due to a large number of collisions between particles and the container walls, but also the vibration energy dissipation through impact and friction. As a result, these factors involving inter-collisions and inter-friction should be taken into consideration in order to simulate the true operation condition for particle dampers.

The Discrete Element Method

The Discrete Element Method (DEM) which belongs to the particle dynamics simulation, was first developed by Cundall and Strack 30 years ago[17]. It is a technique similar to that used for modeling molecular dynamics, where the motion of individual particle is computed and particle-particle contact and particle-wall contact can be detected. Interaction forces between the individual particles and the cavity walls are calculated based on Newton's Second Law. Much of the pioneering work using the DEM simulation to simulate the behavior of granular materials was performed by Cundall and Strack[18, 19]. The procedure is an explicit process with small-step iterative computation cycles such that during one time step, the disturbances cannot propagate from any particle further than its immediate neighbors. As a result, at a single time step, the resultant forces on any particle are determined exclusively by its interaction with the particles with which it is in contact. This feature makes it possible to follow the nonlinear interaction of a large number of particles without excessive memory or the need for an iterative procedure.

The DEM simulation has been extensively developed over the years to evaluate the dissipative properties of granular materials. Bai et al[20] used the DEM simulation investigated and compared the damping mechanisms of a piston-type thrust damper and a box type oscillation damper. Similar analytical studies have been performed by Fowler[9, 21] to investigate the effects of mass, viscoelasticity of the particles, orientation with respect to gravity, and excitation amplitude. The damping performance of a multiunit particle damper in a horizontally vibrating system based on DEM was presented by Saeki[22]. This method makes it possible to consider particle parameters such as the particle size, number of particles and the friction between individual particles. And the validity of this numerical method is examined by a comparison with the experimental results. It is shown that the damping performance chiefly depends on the relationship between the mass ratio and the number of cavities. Mao et al.[23] utilized this technique for the modeling and characterization of particle damping, and the experimental results verified the simulation results. It is noted that the particle damping can achieve a very high value of specific damping capacity. Simulations of coefficient of restitution and interface friction to predict energy dissipation were also made by Wong et al.[24] by means of this method. A variety of different particles were used to acquire approximate values of the coefficient of friction and the coefficient of restitution. This work concludes that the dominant mechanism of energy dissipation of particle damping is dependent on the approximation of the material parameters, in the dampers studied.

The DEM simulation can capture the complex interactions of the dissipation mechanisms in a particle damper. It has been incorporated into a comprehensive particle damper design methodology which allows particle damping to be implemented without extensive trial-and-error testing. Owing to these advantages, this method is very convenient to investigate the effects of system parameters on the characteristics of particle damping. However, the DEM simulation suffers from complicated dynamic model and highly time-consuming computation, which makes it difficult to perform parametric analysis when the number of granules is large, and to apply for the vibration analysis of the complicated continuity systems with particle damping.

The approach based on the multiphase flow theory of gas-particle

The granular particles enclosed in a cavity of a vibrating structure can be considered as a multiphase flow of gas-particle with low Reynolds number where the particle concentration is high i.e. the flow is dense (Fan and Zhu[25]). When the gas-solid flow in a multiphase system is dominated by the inter-particle collisions, the stresses and other dynamic properties of the solid phase can be postulated to be analogous to those of gas molecules. Thus, the kinetic theory of gases is adopted in the modeling of dense gas-solid flows[25]. In this theory, it is assumed that only considering the transport of mass, momentum, energy of the particles should be taken into account and velocities of the molecules are assumed to obey the Maxwell-Boltzmann distribution, where the interstitial gas plays little role in the momentum transport of solids. As a result, an analogy between particle-particle interactions and molecular interactions is to be directly applicable. In spite of the energy dissipation due to inelastic collisions in the model, the elastic collision condition theory is indirectly quoted, when the particle concentration is high, the shear motion of particles leads to inter-particle collisions.

Based on above theory, the transfer of momentum between particles can be described in terms of a pseudo-shear stress and the viscosity of particle-particle interactions. For inelastic particles and a simple shear flow such as a laminar flow, the effective viscosity due to inter-particle collisions can be derived from the kinetic theory of dense multiphase flow.

Recently, some researchers have performed limited studies to mathematically evaluate the dissipative properties of granular materials using this approach. Our previous work (Wu et al.[26]) originally introduce the MPF theory to evaluate the characteristics of granular particle damping. It is convenient to investigate the performance of particle damping in terms of the effective viscosity. The numerical and experimental studies show that the particles are helpful to add damping for attenuating the vibration responses of the host structures. Fang and Tang[27] further utilize this model to carry out detailed studies under various forced excitation levels, packing ratios and enclosure dimensions, and the damping effect due to different energy dissipation mechanisms is quantitatively analyzed.

This novel method develops a new perspective to predict the vibration responses of particle-damping system. It offers the possibility for studying the vibration characteristics of more complicated continuity systems with particle damping as a result of greatly reduced modeling complexity and computational cost.

It should be noted that in our original model[26], where the energy dissipation due to inter-granular collisions is quantified as an effective viscosity, and the frictional effects (i.e. the particles-particles and particles-cavity wall) are only represented by the Coulomb friction damping. However, its accuracy of prediction has yet to be improved as a result of a simply friction model being considered. The objective of the next section is to introduce an improved analytical model for particle damping using the MPF theory base on our previous work[26].

An improved analytical modeling based on MPF theory

In view of above limitations of our previous work[26], an improved analytical model based on the MPF theory is also developed in our proceeding work,[28, 29], where the expression of equivalent viscous damping for inter-particle frictions is introduced instead of the one of Coulomb friction damping based on the Hertz contact theory in Ref. [26].

Two typical examples, i.e. the free vibration of a cantilever particle-damping beam (equivalent SDOF system) and the harmonic forced vibration of a SDOF system with particle damping are used to verify this improved model. The comparisons show that the predictions of the improved model match well with the experimental results in Ref. [26] or DEM simulations in Ref. [27]than that of the original model, for appropriate mass packing ratios and excitation levels.

In addition, this improved model combined with FEM is also applied for predicting the forced response of a typical continuum structure, i.e. the cantilever particle-damping beam (see Ref. [29]). Numerical results show that the particle damping is a kind of nonlinear damping, which is highly depended on the vibration amplitude. When the particle damper is exerted on different positions of the beam, an obvious change of the damping properties is presented. The ideal damping effect can be achieved by putting the particle damper on the position of the structure with high-level vibration[29].

Considered the time-consumption for self-programming of FEM and the necessary for extending such an improved model into the vibroacoustic characteristics analysis of more complicated continuity structures with particle damping, the Multiphysics software COMSOL is also introduced in our on-going work. For the sake of brevity, here a similar particle-damping beam as Ref. [26] is also considered as an attempt. Figs.1-2 show that the comparisons of the velocity responses of the beam between the COMSOL simulation and the analytical results [28]. It is worth noting that the well agreement between both results indicates that the improved model combined with COMSOL software is quite qualified for the vibroacoustic prediction of more complicated continuity structures with particle damping, as a result of its strong computation capability for handling multiphysics couplings of COMSOL software. Certainly, future work still need to make this improved model a valuable tool.

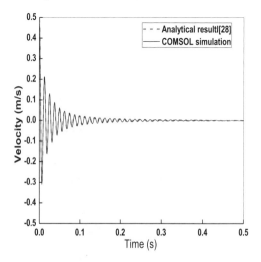
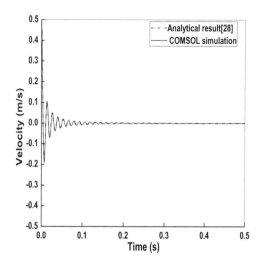

Fig.1. Velocity response of the beam with particles when $\alpha_{mp} = 50\%$

Fig.2. Velocity response of the beam with particles when $\alpha_{mp} = 75\%$.

Concluding Remarks

In this paper, three kinds of mathematical models to predict the vibration responses of the granular damping have been reviewed and an improved analytical model using the MPF theory is stated to describe the particle damping characteristics based on our previous work. It is encouraging to note that the co-simulation of COMSOL Multiphysics with this improved model will broaden our horizons in the design and application for particle damper. In addition, it can be lay a theoretical foundation for the vibroacoustic response prediction for complicated particle-damping composite structures.

Acknowledgments

The work described in this paper was supported by NSFC(Natural Science Foundation of China) ~ Project No.51075316 and Program for Changjiang Scholars and Innovative Research Team in University(PCSIRT).

References

[1]   J. R. Fricke, Lodengraf damping: An advanced vibration damping technology, Journal of Sound and Vibration. 34 (2000) 22-27.

[2]   T. Chen, et al., Dissipation mechanisms of non-obstructive particle damping using discrete element method, Proceedings of SPIE International Symposium on Smart Structures and Materials, 2001,Newport, California, 5-8 March.

[3]   P. Lieber and D. P. Jensen, An acceleration damper: development, design and some applications, Transactions of the American Society of Mechanical Engineers. 67 (1945) 523-530.

[4]   M. R. Duncan, C. R. Wassgren, and C. M. Krousgrill, The damping performance of a single particle impact damper, Journal of Sound and Vibration. 286 (2005) 123-144.

[5]   R. Ehrgott, H. V. Panossian, and G. Davis, Modeling techniques for evaluating the effectiveness of particle damping in turbomachinery, International Journal of Heat and Fluid Flow. 28 (2009) 161-177.

[6]   S. S. Simonian, V. S. Camelo, and J. D. Sienkiewicz, Disturbance suppression using particle dampers, 49th AIAA/ASME/ASCE/AHS/ASC Structures, Structural Dynamics, and Materials Conference, 2008,Schaumburg, Illinois, 7-10 April.

[7]   H. V. Panossian, Non-obstructive particle damping experience and capabilities, Proceedings of IMAC-XX : a conference on structural dynamics 2002,Los Angeles, California 4-7 February

[8]   M. Sánchez, G. Rosenthal, and L. A. Pugnaloni, Universal response of optimal granular damping devices, Journal of Sound and Vibration. 331 (2012) 4389-4394.

[9]   B. L. Fowler, E. M. Flint, and S. E. Olson, Effectiveness and predictability of particle damping, SPIE's 7th Annual International Symposium on Smart Structures and Materials, 2000,Melbourne, Australia, 13-15 December.

[10]   E. M. Flint, Experimental measurements of particle damping effectiveness under centrifugal loads, Proceedings of the 4th National Turbine Engine High Cycle Fatigue Conference, 1999, Monterey, California, 9-11 February

[11]   M. Inoue, et al., Effectiveness of the Particle Damper with Granular Materials, 3rd International Conference on Integrity, Reliability and Failure, 2009,Porto, Portugal 20-24 July

[12]   I. Yokomichi, T. Yoshito, and S. N. Yun, Particle Damping with Granular Materials for Multi-body System, ICSV 15 International Congress on Sound and Vibration, 2008,Daejeon, Korea, 6-10 July.

[13]   R. D. Friend and V. K. Kinra, Particle impact damping, Journal of Sound and Vibration. 233 (2000) 93-118.

[14]   M. Trigui, et al., An experimental study of a multi-particle impact damper, Proceedings of the Institution of Mechanical Engineers, Part C: Journal of Mechanical Engineering Science. 223 (2009) 2029-2038.

[15]   C. Saluena, T. Poschel, and S. E. Esipov, Dissipative properties of vibrated granular materials, Physical Review E. 59 (1998) 4422-4425.

[16]   Z. Lu, X. Lu, and S. F. Masri, Studies of the performance of particle dampers under dynamic loads, Journal of Sound and Vibration. 329 (2010) 5415-5433.

[17]   P. A. Cundall and O. D. L. Strack, A discrete numerical model for granular assemblies, Geotechnique. 29 (1979) 47-65.

[18]   V. P. Legeza, Dynamics of vibroprotective systems with roller dampers of low-frequency vibrations, Strength of Materials. 36 (2004) 185-194.

[19]   P. A. Cundall and O. D. L. Strack, The development of constitutive laws for soil using the distinct element method, Numerical Methods in Geomechanics. 1 (1979) 289-317.

[20]   X. M. Bai, et al., Investigation of particle damping mechanism via particle dynamics simulations, Granular Matter. 11 (2009) 417-429.

[21]   B. L. Fowler, E. M. Flint, and S. E. Olson, Design methodology for particle damping, SPIE's 8th Annual International Symposium on Smart Structures and Materials, 2001,Newport, California, 5-8 March.

[22]   M. Saeki, Analytical study of multi-particle damping, Journal of Sound and Vibration. 281 (2005) 1133-1144.

[23]   K. Mao and M. Y. Wang, Simulation and Characterization of Particle Damping in Transient Vibrations, Journal of Vibration and Acoustics. 126 (2004) 202-211.

[24]   C. X. Wong, M. C. Daniel, and J. A. Rongong, Energy dissipation prediction of particle dampers, Journal of Sound and Vibration. 319 (2009) 91-118.

[25]   L. S. Fan and C. Zhu, Principles of gas-solid flows, Cambridge University Press, Cambridge, 1998.

[26]   C. J. Wu, W. H. Liao, and M. Y. Wang, Modeling of granular particle damping using multiphase flow theory of gas-particle, Journal of Vibration and Acoustics. 126 (2004) 196-201.

[27]   X. Fang and J. Tang, Granular damping in forced vibration: qualitative and quantitative analyses, Journal of Vibration and Acoustics. 128 (2006) 489-500.

[28]   C. J. Wu, R. C. Yang, and D. Q. Wang, An improved of granular particle damping using multiphase flow theory of gas-particle, 20th International Congress on Sound & Vibration, 2013,Bangkok, Thailand, 7-11 July.

[29]   C. J. Wu, Yang, R. C., and Wang, D. Q, Prediction on vibration response of a cantilever particle-damping beam based on two-phase flow theory of gas-particle, Chinese Journal of Mechanical Engineering. 49 (2013) 53-61.

Advanced Materials Research Vol. 819 (2013) pp 20-23
© (2013) Trans Tech Publications, Switzerland
doi:10.4028/www.scientific.net/AMR.819.20

Analysis on the Innovation and Application of Materials in Green Design

Liu Chang [1, a], Zhao Fan [1, b]

[1]Tianjin University of commerce, Tianjin 300134, China

[a]cheese_liu@163.com, [a]zf0079976@126.com

Key words: green design; green materials; dematerialization

Abstract: With the rapid pace of urbanization, more and more people begin to pay attention to environment protection. Green design has long been an area of concern. The foundation of design is the material. The paper firstly emphasizes the benefits of the green design and then explores the development and application of new materials. From the cases listed in the paper, it is not difficult to find out the social progress driving force and the new energy generation promoted by the design. Besides these, this paper also presents the methods of green design, the switch from materialization to dematerialization, the design theory drawn from the essence of traditional green materials, the popularity of the advanced forming technology of green materials and smart materials and the significance of the visual, tactile experience and users' psychology in design. At last, this paper generalizes the basic function and great value of the material's role in green design.

Introduction

The world's urbanization is moving onward driven by the rapid development of economy and industrialization. After the First and Second Industrial Revolution and the Third Scientific-Technological Revolution, technology should not only be used to improve people's material life, but also to create a better living environment for human beings.

Nowadays, many companies only design the products to meet the consumers' demands, ignoring environmental effects such as huge natural resource consumption, destroyed environment and increasingly serious environmental pollution which are threatening the people's life on Earth. Green design, coming from the people's reflection on their environmental and ecological damages caused by modern science and technology, is intended to develop more environmentally benign products and processes by comprehensive utilization of resources and environment protection. It emphasizes on the harmonious coexistence of man and nature, adapts to the theme of sustainable development and reflects the return of the designers' ethical and social responsibility. Now it is getting more and more attention by the features of low energy consumption, low pollution, and low emission.

The application and development of new material

Social progress driven by new material. At the World Expo in London in 1851, Crystal Palace was the super-large architecture which was mainly built by steel and glass. In 1867, at Expo in Paris, the gardener Monnières brought boats and pots made of reinforced concrete which now still plays an important role in the world's civil engineering.

At Shanghai World Expo in 2010, Italy used a kind of new diverse material—the transparent concrete. It has different functions on different occasions. It can sense the temperature and humidity inside and outside the pavilion regardless of the day or night which is very suitable for exhibition buildings. The invention creates a new way to save electricity and has a positive impact on the effective energy conservation.

The invention of new material can change people's original understanding and their way of life hence to reduce unnecessary energy consumption. For example, new material can be reused so as to reduce the cost on transportation and manufacture; it can self-regulate temperature according to different weather conditions, which helps to decrease the coal consumption in winter heating and summer cooling, reduce non-renewable energy consumption and carbon dioxide emission; it can maximize the use of natural lighting by controlling the transparency of itself and create different lighting conditions, which saves electricity for lighting in an enclosed environment.

The sixth of China's national seven strategies says that new material industry should vigorously develop new functional materials, advanced structural materials and composite materials, carry out researches of common basic materials and promote the industrialization of their production, and establish a sound identification and statistics system which can guide the restructure in material industry. This strategy has fully illustrated that material innovation plays a vital role in the development and construction of the national economy.

New energy generation fostered by new material .Future energy pattern is likely to be a mixture of various technologies, adopting flexible technology in line with different regional conditions. The experts of Intergovernmental Panel on Climate Change (IPCC) said that at the end of this century, the proportion of wind, solar, hydro and other clean renewable energy sources in the world's energy is still less than 12%, while the proportion of coal may still be up to 50%. Now, more and more scientists begin to explore in the field of biotechnology. For example, some microorganisms can produce hydrogen by eating carbon and similarly plant photosynthesis is a kind of a transformation from adverse carbon into favorable oxygen. Therefore, the key is efficient conversion of energy.

At Shanghai World Expo in 2010, Switzerland Pavilion features a 17- meter-high interactive intelligent curtain dotted with the red balls made from biological resin which was extracted from soybean fiber; each ball was embedded with a solar battery and could work with energy around the pavilion, such as the solar energy and the camera flash light. These 11,000 chloroplast solar cells could independently produce and store energy and its use in the form of LED lights. When the Expos was over, the facade would be biodegraded. The design concept was intended to show the connections between internal and external surroundings of the pavilion and visitors could experience the fun of energy creation and consumption by participating in the processes.

The methods of energy generation and utilization in urban construction can be humanized rather than industrialized with improvement of citizen's environmental consciousness. Environment protection should become a part of the daily life such as collecting daily exercise energy for lighting, transforming sound or heat energy to drive the bus.

New material can relieve the oil crisis caused by dramatic increase of vehicles, reduce carbon dioxide emission coming from waste gas of cars and decrease GDP loss resulted in traffic congestion.

New material innovation: from materialization to dematerialization. The theory that man is an integral part of nature has a long history in China's ancient civilization, considering that not only in nature but also in the whole world there is no identification of being useful or not. While exposed to human beings of different minds and technical levels, substances will be divided into material and waste. Researches should be conducted on increasing added-value of new material and expanding the definition of it so as to change people's traditional acknowledgment of material and find more innovation and application space.

Traditional green materials. Green design involves a particular framework for considering environmental issues and pollution prevention in the design phase, aiming at environmental protection as the design goal and starting point, minimizing the adverse impact on the environment and maximizing economic and environmental benefits.

The core of green design is the "6R" principles: research, reserve, reduce, recycle, reuse and regeneration. It concerns not only low energy consumption and low emission but also the recycling and reuse of products and components.

In research, the first step is to find out right material for the product which indicates that green material is the basis for green design. Green material is environmentally responsible with high energy efficiency and little environment pollution. It has the features as follows: firstly, it is innocuous without pollution, radiation and noise which prevent damages to health impact and environment from production stage to the whole life-cycle. Secondly, it has excellent processing performance which helps to increase material efficiency and reduce labor costs and energy consumption in the production. Thirdly, it is renewable and recyclable, conductive to reduce energy consumption and the permanent waste formation.

The common green materials include wood, bamboo, ceramics, nano and inorganic polymer geo-polymer materials. They are all easy to recycle and degrade with good environmental compatibility.

Advanced forming technology of green material and smart materials. Now the popular green advanced forming technology and equipment include lightweight mechanical equipment, digitized casting, 3D printing, and composite materials and so on.

The most important advantage of 3D printing technology is that, on one hand, it can generate the product's any part of any shape from the computer graphics data without machine work or mold which shortens the product development cycle and simplify the process of manufacturing so as to reduce costs and material waste. On the other hand, it can create the shapes which are impossible for traditional technology with low costs on complex molds.

Human beings' pursuit of material is endless. In 2013 Massachusetts Institute of Technology launched a 4D printer which shows the automatic deformation capacity of composite materials. The materials were "printed" by 3D printer with a plastic stick and a layer of absorbent smart material. The automatic deformation capacity can make the parts of the products self-assembled in the underground pipelines where workers have difficulty in coming into. In addition, this technology may also be applied to furniture, bicycles, cars, even the buildings.

The intelligence of green material is not only reflected on its forming capacity but also on its own nature. From the view of parallelism features of the design process, the intelligent equipments can save resources to shorten the design cycle in the early stage and the continuity of the intelligent characteristics can reduce labor costs and extra energy consumption in the later stage which creates a new definition of new material.

Visual, tactile and experimental characteristics of material in design. What science and technology can control are objectivity and rationality while human beings are subjective and emotional, which is difficult to manipulate. The characteristics considered in the process of design only show its low energy consumption. However, the 80 percent of information people got comes from their eyes. So the first visual impression of the material is so important that people can have more deep feelings from the visual experience when they use the product.

The contacts between people and the world are superficial. For people's eyes, material is just a symbol because they cannot get any other information without touching it. Some appearances of the material are natural and some are specially treated.

People's understanding of material mainly comes from their sight and touch. Usually, visual experience is always existed but tactile sensation usually occurs only in the touch. When they cannot do that, they can get the information by the previous experience.

The unit size of materials can influence the tactile sensation because people use hands especially fingertips to touch. So when the size is beyond our touch limitation, the tactile experience will disappear. There are also some materials whose sizes are appropriate, but what people perceived is its sub-layer such as the webbed plaited material. If the unit size of the uneven surface is too small, even though the material needs visual experience, tactile sensation should be used to check the material texture.

People can feel the material without touch but with sight which can evoke the tactile sensations. If a person only feels the material with his eyes and the visual effects are as real as possible, the information he gets will be the same no matter it is natural or processed. At the moment, what he cares are the feelings the material brings to him that sometimes come from previous experience or the curiosity aroused by the new different substance.

The pursuit of new materials has been lasting from ancient China. In the Qianlong Era of the Qing Dynasty, Chinese people had already got the porcelains of animal shapes as well as imitation bronze, lacquer, boxes of marble, wood silk look and so on.

So the author holds the view that the emphasis should be put on the researches which works on finding out materials' special properties and improving energy efficiency. Designers need to know the features of products and characteristics of its material. Also, they still have to figure out that the feelings the product gives to customers are on the visual level or tactile. For example, texture wallpaper can show the texture of wood without wood, which helps to reduce logging and waste of material. At the same time, the customers also get the similar feeling.

Dematerialized new material. With the development of science and technology, people can feel the product not only by hands but also on electronic devices where the features may be influenced by sight, touch and psychological factors. What on the electronic devices is an image and the feelings only come from sight and previous experience, having nothing to do with touch. That is the difference.

It is common to see the imitation of material on electronic equipment's. For instance, one of Apple's IOS Human Interface Guidelines says that a great user interface (UI) follows human interface design principles that are based on the way people think and work, not on the capabilities of the device. A UI that is unattractive, convoluted, or illogical can make even a great app seem like a chore to use. But a beautiful, intuitive, compelling UI enhances an app's functionality and inspires a positive emotional attachment in users.

Now there has been a great development of human-machine interface which the information exchange between human and computer needs a real physical interface no more. All orders and feedback can be simultaneously captured by the sound and movement of the camera. The virtual human-computer interface brings the design to a new era when the utilization and innovation of new material has been raised from materialization to dematerialization which has no waste of resources and environmental pollution. In the future, maybe we can abandon the mouse and keyboard so as to reduce electronic wastes.

Conclusion

Materials are the foundation of green design. While designing, people need to make sure the features, shaping methods and manufacturing processes and comply with the principles for the use of green materials. In addition, with the development of science and technology, people should expand new materials' added-values and definitions and focus on the switch of materials from materialization to dematerialization, which shortens manufacturing design cycles and reduces the additional energy consumption. All these should be on the basis of consumers' demands for materials for instance the new materials should provide perfect visual, tactile and , more deeply, psychological experience.

References

[1] Kenya HARA, Dialogue in Design, Jinan: Shandong People's Publishing House, 2006.11

[2] Kenya HARA, Masayo AVE, Dialogue in Design, Jinan: Shandong People's Publishing House, 2009.12

[3] Peng,Chen.Sorry,Out of Oil.The Outlook Magazine [M],2010,102:50-81

[4] Id commune, Today's decoration in design, http://www.hi-id.com/?p=2741

[5] Robert Clay,Beautiful Thing: An Introduction to Design, Jinan: Shandong pictorial publishing house,2010.6

Advanced Materials Research Vol. 819 (2013) pp 24-28
© (2013) Trans Tech Publications, Switzerland
doi:10.4028/www.scientific.net/AMR.819.24

Dynamic Characteristics Analysis of Gantry Machining Center Structure

Tang Yanyun[1,a], Weng Zeyu[1,b], Fang Lika[1,c], Gao Xiang[1], Hu Jiande[1],
Tang Xuezhe[1] and Zhen Yanqing[1]

[1] College of Mechanical Engineering, Zhejiang University of Technology, Hangzhou 310014, China

[a]tangyanyun1989@126.com, [b]wengzy8888@163.com, [c]475409756@qq.com

Keywords: Gantry machining center, Dynamic performance, Relative excitation,Weak link analysis

Abstract. The dynamic performance of modern machine tools is more and more highly emphasized. The research purpose of this paper is to research dynamic performance of gantry machining center through modal experiment. In this paper, the theory of relative excitation test on the machine tool is expounded, and relative excitation test is carried on gantry machining center and then the modal parameters of this machine tool are obtained through experimental modal analysis. Analyze the weak link of this gantry machining center, show that the stiffness of Z direction of Z to feed component of this machine tool is insufficient and the stiffness of X, Y direction of ram is insufficient.

Introduction

CNC machine tool is the important equipment in modern machinery manufacturing industry, it is related to the strategic position and the comprehensive national strength level of a country. Along with the development of modern machine tool to high speed, high precision direction, the requirement of machine dynamic performance become more and more high, so the research on the machine dynamic performance of structure is received widespread attention [1,2]. The analysis method of modal test is a kind of important method to research on the machine dynamic performance, the machine dynamic performance obtained from analysis on modal test is accurate and reliable [3,4,5].

In the mechanical structure, the machine tool is a kind of mechanical system with related double input, the working way is to produce relative motion with each other between cutting tool and workpiece in order to achieve all the processing requirements. In the working process, the incentive condition of machine tool structure is not single point incentive and it also do not meet linear independence requirement of multipoint incentive, this way of excitation is a special kind of relative excitation [6]. Test on the machine tool in the stationary state, vibration generator is installed between cutting tool and workpiece, simulate the cutting force by alternating force produced by the vibration generator ,at the same time the cutting force is acting on the cutting tool and workpiece, therefore this way is closer to the actual working condition than the single point excitation.

Modal Test Theory of Relative Excitation

Assuming the machine tool is a linear viscosity damping system with freedom of n degrees, the relationship between the input of system namely excitation force and the output of system namely displacement response can be expressed as follows:

$$\{X\} = [H]\{F\} \qquad (1)$$

Where, $\{x\}$ refers to the displacement vector $\{n \times 1\}$, $\{F\}$ refers to the force vector $\{n \times 1\}$, $[H]$ refers to the transfer function matrix $\{n \times n\}$。

When it is single point excitation and supposing excitation force is applied on point d, the equation of the response is recorded as:

$$\{X\} = \{H_d\}F_d \tag{2}$$

Where, $\{H_d\}$ is the first d column transfer function of the transfer function matrix $[H]$ of system, basing on the theory of modal analysis, it is known that modal parameters can be identified from the transfer function $H_{id}(s)$ by least square curve fitting method. The best fitting analytical equation is following:

$$H_{id}(S) = \sum_{K=1}^{m} \left[\frac{R_{Kid}}{2j(S - P_K)} - \frac{R^*_{Kid}}{2j(S - P^*_K)} \right] \tag{3}$$

In this equation $P_k = -\sigma_k + j\omega_k$, P_k is the pole point, σ_k is the damping, ω_k is the natural frequency of system, R_k is the complex residue, R^*_k is the conjugate value of R_k, m is the modal number of the system identified.

If the vibration generator is installed between simulation cutting bar and working table, excitation force acts on simulation cutting bar and react on working table, we make action point letter of d, reaction point letter of e, then can be denoted,

$$\{F\} = \{0,0,F_d,0,\cdots,F_e,0,\cdots 0\}^T \tag{4}$$

Equation (1) can be transferred to the following equation:

$$\{x\} = \{H_d\}F_d + \{H_e\}F_e \tag{5}$$

Because the frequency spectrum of excitation force is same with the frequency spectrum of reaction force, namely, $F_d = F_e$ thus equation (4) can be expressed as

$$\{x\} = (\{H_d\} + \{H_e\})F_d = \{\overline{H}_d\}F_d \tag{6}$$

Equation(6) shows that structure transfer function obtained by method of relative excitation can be written as the form of the sum of two parts:

$$\overline{H}_{id}(S) = H_{id}(S) + H_{ie}(S) = \sum_{K=1}^{m} \left[\frac{\overline{R}_{Kid}}{2j(S - P_K)} - \frac{\overline{R}^*_{Kid}}{2j(S - P^*_K)} \right] \tag{7}$$

Where $\overline{R}_{Kid} = R_{Kid} + R_{Kie}$ is the relative residue, if introduce the relationship between residue and modal vector $\{\Phi\}$, $R_{K_{ij}} = \phi_{Ki}\phi_{Kj}$, the equation (7) will be described by modal parameter :

$$\overline{H}_{id}(S) = \sum_{K=1}^{m} \left[\frac{\phi_{Ki}(\phi_{Kd} + \phi_{Ke})}{2j(S - P_K)} + \frac{\phi_{Ki}^{*}(\phi_{Kd}^{*} + \phi_{Ke}^{*})}{2j(S - P_K^{*})} \right] \tag{8}$$

When it is relative excitation, the relationship between residue and modal vector $\{\overline{\phi}\}$ is given by

$$\overline{R}_{Kid} = \overline{\phi}_{Ki} \cdot \overline{\phi}_{Kd} = \phi_{Ki}(\phi_{Kd} + \phi_{Ke}).$$ Thus the residue at excitation point is denoted as

$$\overline{R}_{Kdd} = \overline{\phi}_{Kd} \cdot \overline{\phi}_{Kd} = \phi_{Kd}(\phi_{Kd} + \phi_{Ke}). \tag{9}$$

According to the following method, modal vector matrix $\{\overline{\phi}\}$ can be calculated.

$$\overline{\phi}_d = \sqrt{R_{dd}} = \sqrt{\phi_d(\phi_d + \phi_e)} \ , \quad \overline{\phi}_i = \frac{\overline{R}_{id}}{\overline{\phi}_e} = \phi_i \sqrt{1 + \frac{\phi_c}{\phi_d}} = \alpha \cdot \phi_i \ , \quad \alpha = \sqrt{1 + \phi_e / \phi_d} \ ,$$ where α is the

weighting coefficient. Using the above relational equations, the transfer function of relative excitation also can be represented as

$$\overline{H}_{id}(S) = \sum_{K=1}^{m} \left[\frac{\alpha_K^2 \cdot R_{Kid}}{2j(S - P_K)} - \frac{\alpha_K^{*2} \cdot R_{Kid}^{*}}{2j(S - P_K^{*})} \right] \tag{10}$$

Comparing equation(3) with equation(10), it is known that the natural frequency ω_k and the

damping σ_k respectively identified by single point excitation and relative excitation are the same,

but modal vector are not identical, their relationship can be seen in: $[\overline{\phi}] = [\beta][\phi]$.

Relative Excitation Modal Test on Gantry Machining Center

Test Scheme. Gantry machining center is mainly composed of bed, column, beam, ram, sliding saddle, spindle and working table, etc. During relative excitation, the vibration generator is installed between the working table and simulation cutting tool, drive the vibration generator through the steady sine sweep signal by power amplifier, equal and opposite vibration force occurs between working table and simulation cutting tool. Measure the size of the vibration force by force sensors and obtain the response of each point of machine tool by three to acceleration sensor. After getting transfer function between the vibration point and each point through dynamic signal analyzer, go on having modal parameter identification by importing modal analysis software. There are 779 points layed out on the gantry machining center, the nodes of the grid in Fig. 1 is the measuring points.

Modal Analysis. Through the modal test of relative excitation, we can get displacement transfer function of the number of 779×3. In Fig. 2, displacement transfer function curve of excitation points in the simulation cutting tool is showed. It can be seen from the curve that the displacement amplitude of low frequency part is relatively much bigger than the high frequency part.

Identify modal parameter by importing the transfer function data into modal analysis software, then we can get modal parameters such as natural frequency,damping ratio and vibration mode, etc . Fig. 3, Fig. 4 and Fig. 5 is the vibration mode for first three orders respectively .

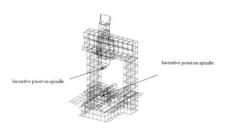
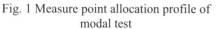
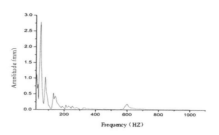

Fig. 1 Measure point allocation profile of modal test

Fig. 2 Transfer function curve of excitation point

In the first order modal, the natural frequency is 46.6 Hz, the damping ratio is 0.26 and large vibration exists in Z direction of Z to feed unit of this machine tool ;

In the second order modal, the natural frequency is 89.2 Hz, the damping ratio is 6.90 and large vibration also exists in Z direction of Z to feed unit of this machine tool and large bending vibration happens in the Y direction of ram parts of this machine tool ;

In the third order modal, the natural frequency is 125.5 Hz, the damping ratio is 2.23 and large bending vibration happens in the X direction of ram parts of this machine tool.

Fig. 3 Main vibration mode of the first step

Fig.4 Main vibration mode of the second step

Fig. 5 Main vibration mode of the third step

The Weak Link Analysis of Gantry Machining Center Structure. From the result of modal analysis of whole gantry machining center, the displacement of the first three order modal vibration is relatively bigger, which is the weak mode that influences the dynamic performance of this gantry machining center.

Stiffness K_z in Z direction of Z to feed component of machine tool is mainly composed in series of axial stiffness K_s of ball screw pair, axial contact stiffness K_j of nut screw, axial stiffness K_α of bearings and axial stiffness K_c of bearing bolster.

$$\frac{1}{K_z} = \frac{1}{K_s} + \frac{1}{K_j} + \frac{1}{K_\alpha} + \frac{1}{K_c} \tag{11}$$

So stiffness K_z in Z direction of Z to feed component of machine tool is smaller than all stiffness of its components. The main factor that impacts the stiffness K_z in Z direction depends on the smallest one of K_s, K_j, K_α and K_c, but we can be sure that the insufficiency of stiffness K_z in Z direction of Z to feed component of machine tool is the reason generating modal vibration.

In the Y direction of ram parts big bending vibration exists, which is caused by the insufficiency of the bending rigidity in the Y direction of ram parts; In the X direction of ram parts big bending vibration exists, which is caused by the insufficiency of the bending rigidity in the Y direction of ram parts. Ram is a kind of non-closed frame structure, both in the front of the ram and in the two sides of ram, there are many open mouths, which are the main reasons of bending stiffness insufficiency in Y direction and X direction of ram parts.

Conclusions

In this paper, the relative vibration theory is discussed, and a modal test of relative excitation is carried on the gantry machining center. Natural frequency, damping ratio, vibration modal and other parameters of machine tool are obtained through the modal parameter identification, the first three modes are analyzed, then get the weak links of gantry machining center. The weak links affect the machining performance of the machine tool are following: large vibration displacement exists in Z direction due to the insufficiency of stiffness K_z in Z direction of Z to feed component of machine tool; bending rigidity of ram parts in X, Y direction is insufficient, which leads to large bending deformation in X, Y direction of ram parts.

Acknowledgment

The project is supported by the Open Fund of Advanced Manufacturing Technology and Equipment of Important Disciplines of Zhejiang Province (No.20100705).

Reference

[1] Fang Li: Dynamic Analysis and Optimization Design of High-precise Numerical Controlled Machine Tool[D]. Nanjing: Southeast University, 2008.3.

[2] Huasheng Yan: Research and Application on Key Technologies of Dynamic Design and Optimization on Machine Tool[D]. Xiamen:Xiamen University, 2008.6.

[3] Zufeng Liang: Modal Test Analysis of Spindle System of TH6350 Machining Center[D]. Kunming: Kunming University of Science and Technology, 2003.

[4] Jinxing Zheng, Mingjun Zhang and Qingxin Meng: Experimental Modal Analysis and its Application in Structure Dynamic Characters of CNC Machine Tools[J]. Measurement and Control Technology, 2007, 26 (12) : 28 - 31.

[5] I.Zaghbani, Songmene k: Estimation of Machine Tool Dynamic Parameters During Machine Operation Through Operational Modal Analysis[J]. International Journal of Machine Tools &Manufacture. 2009(49):947-957.

[6] Baoyang Lin: Application of Relative Excitation to Modal Analysis of Machine Tools. Journal of Chongqing University [J], 1989 (12) : 14-19.

Advanced Materials Research Vol. 819 (2013) pp 29-32
© *(2013) Trans Tech Publications, Switzerland*
doi:10.4028/www.scientific.net/AMR.819.29

Establishment of Cutting Model for Three-axis Surface Machining Based on SolidCAM

Fuxun Lin[1,a], Qichen Wang[1,b], Kaifa Wu[1,c], Ruoyu Liang[1,d], Xian Wang[1,e]

[1]Key Laboratory of Mechanism Theory and Equipment Design of Ministry of Education, Tianjin University, Tianjin 300072, China

[a]lfx3107@163.com, [b]wangq5@miamioh.edu, [c]sdwkf.good@163.com, [d]lryasa@yahoo.cn, [e]wangxian320@163.com

Keywords: SolidCAM, Three-axis machining, Straight line type, Cutting model.

Abstract. For different three-axis machining characteristics, SolidCAM provides many machining strategies including linear, correction, spiral, imitation cutting and so on. This paper chooses three-axis revolving curved surface parts as the research object and reference three-axis stereo processing combining wrap-around roughing and linear processing. Finally, the cutting model of this machining strategy based on parameters has been established to provide a reference for the determination of cutting parameters.

Introduction

SolidCAM is a CAM software developed by SolidCAM company in Israel, seamlessly connected with SolidWorks. All machining operations can be defined, calculated and verified, without leaving the SolidWorks window. Its own 2.5 axis, 3 axis and multi-axis machining modules are sufficient to meet the processing requirements for various features. We can combine different processing strategies flexibly to simplify the settings of the related working step and improve the machining efficiency. According to the widely used three-axis machining, zhu puts forward a method to generate the tool path in three-axis CNC roughing based on filled-end mill, combining vertex bias algorithm and phased bias geometry. This method can generate wholly interference-free roughing tool paths for STL data models and keep each layer tool path in a same plane[1]. Liu proposed a tool path generation algorithm in three-axis CNC roughing based on three-dimensional scattered measurement data points directly. This method can avoid the complex process of constructing surface by measuring points and generating the tool paths by the surface[2]. While the above algorithms are mainly suitable for three-axis stratified method and can't effectively improve the surface quality. This paper adopts the linear processing scheme of three axis linkage and establishs cutting models based on the cutting parameters. This processing scheme provides a reference for the processing path of relevant characteristics and basis for the optimization of cutting parameters.

The establishment of three-axis roughing cutting model

The stratified cutting mode used in three-axis roughing can cut most of the rough material quickly. This method is often used as the first step in three-axis and multi-axis machining[3]. According to the part model shown in Fig. 1, we choose the lower half of rotary body portion as the research object and build the coordinate system in the center of rotation of the bottom surface of the model. Set the intersection of the three arrows in Fig. 1 as the origin of coordinate. The direction of red, green, blue arrows are respectively the positive direction of the X axis, Y axis and Z axis.

The bus of the rotating body part of the model in Y/Z coordinate plane and the relative dimension is shown in Fig. 2. The equation of the arc in Fig. 2 is $(y-60)^2 + (z-60)^2 = 50^2, y \subset [20,45]$. Rotate this arc 180 degrees around Z axis can get a rotating body. According to this principle, the volume formula of the this model is

$$V_0 = \int_{12.5}^{30} \frac{\pi}{2} y^2 dz + 12.5 \times \frac{\pi \times 45^2}{2} \tag{1}$$

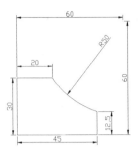

Fig. 1 Part Model   Fig. 2 Bus

According to this principle, it is easy to know the equation of the rotating surface is

$$\left(\sqrt{x^2+y^2}-60\right)^2+(z-60)^2=50^2, y\subset[-45,-20]\cup[20,45], x\subset[20,45], z\subset[12.5,30] \tag{2}$$

The angular coordinates of this equation is expressed as

$$\begin{cases} x=\rho\cos\alpha \\ y=\rho\sin\alpha \quad ,\rho\subset[20,45],\alpha\subset\left[-\dfrac{\pi}{2},\dfrac{\pi}{2}\right],\beta\subset\left[\dfrac{\pi}{10},\dfrac{3\pi}{10}\right] \\ z=60-60\cos\beta \end{cases} \tag{3}$$

Obviously the geometry is formed rotating around the Z axis by the part graphics and the equation of this rotation surface. The three-axis rough machining scheme of the figure can be surrounded. We choose flat bottom vertical cutter whose diameter is D as the cutting tool to improve the efficiency. The parameters of the tool are show in Fig. 3. The blank shape is a cylinder body with the same bottom as the part, as shown in Fig. 4. In order to illustrate the processing effect, the boundary of the model is set for cutting range and increasing a compensation value. The compensation value is considered for the residual during cylindrical cutter cutting in the boundary. In case of resection completely, we can set the diameter of the tool as the compensation value. To illustrate the surround-processing strategy, set the amount of feed is $\delta=5$, the offset of the surface is 0 and the diameter of the tool is $D=2\times R=6$. The effect of rough machining is shown in Fig .5 and the yellow part in this figure is the step with the same height in surround-processing strategy.

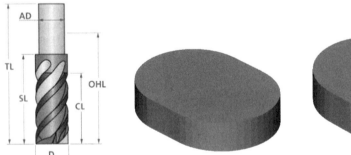
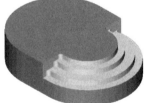

Fig. 3 Cutter Model   Fig. 4 Blank Model   Fig. 5 Rough Machining

We can know that surround-processing is feeding down continuously from the top of the workpiece taking feed δ as the unit from the shape after machining in Fig .5. In Fig .2, the whole cutting depth of the part is 30-12.5=17.5mm. Set the cutting allowance is ε, then the first cutting depth is $\delta-\varepsilon$ from the top of the workpiece. The other cutting depth is δ and the number of cutting layer is $\left[\dfrac{17.5}{\delta}\right]$. The sigh $\left[\dfrac{17.5}{\delta}\right]$ indicates the maximum integer which not greater than $\dfrac{17.5}{\delta}$. From the above, we can know the cutting depth of the ith layer is $\delta i-\varepsilon, 1\le i\le m$. The fan-shaped area of the *ith* cutting layer is $S_i = \dfrac{\pi}{2}\left\{45^2 - \left[60 - \sqrt{50^2 - (\delta i - \varepsilon + 30)^2}\right]^2\right\}$. Then the model volume

removed in roughing stage is $V_1 = S_1 \times (\delta - \varepsilon) + \sum\limits_{i=2}^{m} S_i \times \delta, 1 \le i \le m$. Generally speaking, the

requirement of surface roughness in roughing is $Ra = 1.6 \sim 3.2$, so we can calculate the least cutting volume $\min V_1$ under the premise of $\delta i - \varepsilon \le Ra$. The cutting parameters calculated in this condition can be the reference data for processing .

The establishment of three-axis semifinishing cutting model

The model of three-axis semifinishing is linear. The principle is that there are many equidistant linear distributed on the top surface of cutting area. The distance between adjacent two straight lines is path interval γ, shown in Fig. 8. Consider the value of the tool radius, project all equidistant lines on the surface of the part. Then we can get the tool cutting paths shown in Fig. 6. The yellow lines are the tool paths and the red lines are the path of fast feed and rapid exit. This method can remove the scrap on the bottom of the rotating body to the utmost extent. This way can effectively reduce the roughness of the bottom contrast imitation and correction and be suitable for the rough machining stage of multi-axis machining. The scrap at the top of the rotating body can be removed by improving the overlap rate in multi-axis imitation. The effect of semifinishing is shown in Fig. 7.

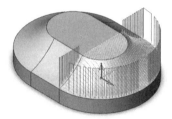  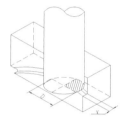

Fig. 6 Tool Path in Semifinishing   Fig. 7 Semifinishing   Fig. 8 Cutting Model of the Tool

Assume the cutting allowance of roughing can always keep semifinishing cutting blank in the process of cutting. From Fig. 6 we can see tool path are distributed on the cutting plane parallel to the X/Z coordinate plane. The curves can be divided into two groups: in one group the top boundary contour is no less than the tool radius in cutting plane and the other is less. It is easy to known than the

tool path are $j = \begin{cases} \left[\dfrac{90}{\gamma}\right] + 1, the \ \ remainder \ \ of \ \ \dfrac{90}{\gamma} = 0 \\ \left[\dfrac{90}{\gamma}\right] + 2, the \ \ remainder \ \ of \ \ \dfrac{90}{\gamma} \ne 0 \end{cases}$ according to path interval value γ. Regulate

the positive direction of y axis is the starting point of the cutting plane. The equation of the *kth* cutting plane is $y_k = 45 - \gamma(k-1), k \subset [1, j]$. Combined with Eq. 2, when the distance is bigger than the

tool radius the equation of tool path is $R + 60 - \sqrt{50^2 - (z - 60)^2} = \sqrt{x^2 + y_k^2}$. Assuming $\sqrt{x^2 + y_k^2} = t$

[4], the equation above can be rewritten as $\begin{cases} z = 60 - \sqrt{50^2 - (R + 60 - t)^2} \\ x = \sqrt{t^2 - y_k^2} \end{cases}, t = \begin{cases} [r + R, 45], y_k < r + R \\ [y_k, 45], y_k \ge r + R \end{cases}$.

In the above formula, r is the radius of the top boss. In this paper, $r = 20$.

As we know, the cutting process of the tool can be considered as 1/4 arc cutting edge scanning around the tool paths. The actual cutting width is equal to the path interval value γ. By the trigonometric formulas, the area of not cutting is

$$s = \frac{\arccos\left(\frac{R-\gamma}{R}\right)}{360} \pi R^2 - \frac{1}{2}(R-\gamma)\left(\sqrt{R^2 - (R-\gamma)^2}\right) \tag{4}$$

As the circular tool exists an empty cutting part, so the actual cutting scrap section is not rectangular but there is a transition fillet in each path interval γ. The volume of the transition fillet in y_k section is $V_2 = \int_{t_0}^{t_1} sdz$. The t_0 and t_1 are respectively the value range of t. Then the volume of the blank cut in semifinishing is

$$V_3 = \int_{t_3}^{t_4} \lambda(30 - z)dx - V_1 - V_2 \tag{5}$$

Search the requirement of surface roughness in semifinishing is $Ra = 0.8 \sim 1.6$ refer to roughing. Then the cutting parameters which meet the requirement of roughness and $\min V_3$ can be chosen as the reference data.

Conclusions

The cutting model of three-axis roughing and semifinishing built in this paper can be applied to the three-axis final machining and multi-axis roughing. According to the requirement of surface quality, we can determine the tool diameter D, each feed δ, cutting allowance ε, path interval γ and so on by this cutting model. This model can also be established to consider the cutting time. At last, combine the scrap model with time model to limit the cutting parameters. Choose achieving the maximum amount of cutting in the shortest time as the goal to optimize the cutting parameters

Acknowledgements

This work is financially supported by the CNC generation of mechanical product innovation demonstration project of Tianjin (2013BAF06B00), The recovery method of weak information and study of algorithm hardened design based on generalized parameter adjustment stochastic resonance (20100032110006) and High-end CNC machining chatter online monitoring and optimization control technology research (12JCQNJC02500). Please communicate with the corresponding author Fuxun Lin, if there are any questions in this paper.

References

[1] Hu Zhu, Lifang Zhang, Rough Machining Tool Path Generation for STL Data Model. Journal of North University of China. (Natural Science Edition). Vol. 2 (2010), p. 100-p. 103.

[2] Siwei Liu, Liyan Zhang, Shenglan Liu, and Ji Du, Automatic Tool Path Generation of Three-axis NC Rough Machining Based on Scattered Measured Data. Machine Building &Automation. Vol. 34 (2005), p. 6-p. 10.

[3] Weidong Xu, Junying Shen, and Gang Ji, SolidCAM iMachining 3D solutions. Modern Manufacturing. Vol. 11(2012), p. 67-p. 69.

[4] Sida Pan, Application of Equation Driven Curve in SolidWorks. Digital Design. Vol. 4(2010), p. 52-p. 54.

Advanced Materials Research Vol. 819 (2013) pp 33-37
© *(2013) Trans Tech Publications, Switzerland*
doi:10.4028/www.scientific.net/AMR.819.33

Machined NURBS Surface Description Using on-Machine Probing Data

Lai Jintao[1,a], Xinhua Yao[1,b], Wenli Yu[2,c], FU Jianzhong[1,d]

[1]Department of Mechanical Engineering, Zhejiang University, Hangzhou, China;

[2]College of Information Engineering, QuZhou College of Technology, QuZhou, China

[a]pocaomao88@gmail.com, [b]yaoxinhuazju@gmail.com, [c]yujimmy@163.com, [d]fjz@zju.edu.cn

Keywords: NURBS surface machining; on-machine probing; surface description

Abstract: The error distribution of a machined NURBS surface is complex to describe since the machined surface isn't that easy to obtain. A new method for machined NURBS surface description based on inspection database of on-machine probing is presented. The main idea is to approximate the machined surface using the inspection data and the basis functions of the nominal NURBS surface and to get the error distribution of the surface. A virtual part with a NURBS surface is presented and its error caused by the interaction of the machine-tool-workpiece is obtained using the FEA method at the inspection points. A machined NURBS surface is achieved and the error distribution of the surface is expressed. This method provides a new way in error compensation and it will improve the machining accuracy of a NURBS surface.

Introduction

Nowadays, NURBS surface is widely used in the design process of industrial parts. The machining accuracy of the machined surface is affected by a combination of the error sources such as the thermal deflection, the geometrical deviation and the deflection of the machine-tool-workpiece system. The evaluation of surface machining error maintains a tough thing for it's impossible to detect the machining error everywhere in the whole surface area. To get a fast description of the machined surface is necessary to evaluate the quality of the surface.

With the developments of the on-machine measurement (OMM) technology, machining error can be effectively detected through the process-intermittent gauging by on-machine probing. This provides the database for the error compensation in CNC machining. Many researches have been carried out on the compensation of machining errors using the OMM data. M.W. Cho[1] presented an integrated machining error compensation method using OMM data to effectively reduce the machining error in the end-milling process. J.P. Choi[2] utilized an OMM with a touch probe to reduce the machining error of a three-axis machine tool. R. Guiassa[3] worked for a predictive compliance based model to compensate the machining error considering the interaction of machine-tool-workpiece system. B.C. Jiang[4] presented a computer aided feature-based statistical concept to calculate sufficient measurement points in on-machine measuring with coordinate measuring machine (CMM). Hua Qiu[5] proposed a measurement technique on machining centers for 3D free-form contours. W.T. Lei[6] suggested a new way for error compensation by approximating the error compensation function with the basis functions of the setting NURBS path. Most work has done for the error compensation in machining of a contour line, the error inspection and compensation in 3d surface machining is worthy discussing. In this paper, a novel method to obtain the machined surface of a NURBS surface is presented and the machined error distribution can be expressed with the machined surface description and the nominal surface designed. The number of inspection points is reduced compare to the strategy of inspecting the whole surface uniformly[5], and the efficiency of inspecting process is increased.

A new method based on inspection database of on-machine probing is presented to describe machined NURBS surface. The rest of this paper is organized as follows. First, a general NURBS surface is presented and its reconstruction method is introduced. Secondly, a virtual part based on the given surface is created and its machining error caused by action of cutting force is obtained using the FEA method at the inspection points. Then, the machined surface is presented and a conclusion follows.

NURBS surface

NURBS surface provides a uniform for the standard surface and arbitrary surface, and it's widely used in the CAD/CAM and graphics area.

A $p \times q$ NURBS surface can be expressed by Eq.1,

$$S(u,v) = \frac{\displaystyle\sum_{i=0}^{m}\sum_{j=0}^{n} w_{i,j} d_{i,j} N_{i,p}(u) N_{j,q}(v)}{\displaystyle\sum_{i=0}^{m}\sum_{j=0}^{n} w_{i,j} N_{i,p}(u) N_{j,q}(v)} \tag{1}$$

where $d_{i,j}, i=0,1,\cdots,m; j=0,1,\cdots,n$ are the control points; $w_{i,j}$ is the weight of the control points; $N_{i,p}(u)$ and $N_{j,q}(v)$ are the b-spline basis functions in u and v directions.

Table.1 The knot vectors of the given surface

u	0	0	0	0	0.5	1.0	1.5	2.0	2.0	2.0	2.0
v	0	0	0	0	0.5	1.0	1.5	2.0	2.0	2.0	2.0

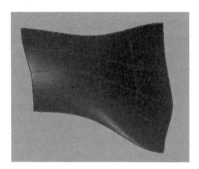 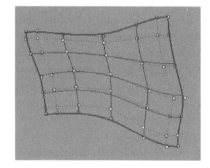

Fig.1 NUBRS surface(left); Control points of the surface(right)

Fig.1 (left) shows a 4×4 NURBS surface with 49 control points. The knot vectors of the surface are shown in Table.1. The expression of the NURBS surface indicates that the surface can be easily modified by adjusting the control points and the corresponding weight. The adjustment of a control point and its weight just modifies the local area of the surface near the control point.

Machining error simulation

The deflection of the machine-tool-workpiece system under the action of cutting forces leads to machining error in parts, particularly in parts of low-rigidity. In this case, the errors caused by the action of the cutting force are considered to be the primary errors due to the low-rigidity of the part.

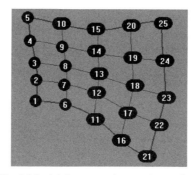

Fig.2 Machining error inspection points

A virtual part is created based on the given surface as shown in Fig.3. The dimension at the origin is 8mm and the node 24 is thinnest point with a dimension of 3mm. The machining error caused by the deflection of the machine-tool-workpiece system is proposed to be the main error source. The machining errors at the inspection points (shown in Fig.2) are obtained by the FEA method.

The material of the part is the aluminum (AL2014) and the cutting force in z direction (F) is set to be 100N. Fig.3 shows the loaded force at the inspection point and the mesh generation of the part in Ansys. The machining errors at the inspection points are simulated (Table.2).

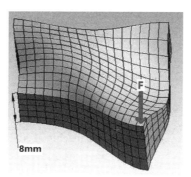

Fig.3 The loaded force at the inspection point and the mesh generation of the part

Table.2 The machining errors

node	1	2	3	4	5	6	7	8	9
error [mm]	0.0311	0.0308	0.0359	0.0282	0.0207	0.0439	0.0494	0.0519	0.0412
node	10	11	12	13	14	15	16	17	18
error [mm]	0.0303	0.0617	0.0807	0.0824	0.0624	0.0523	0.1477	0.1492	0.1671
node	19	20	21	22	23	24	25		
error [mm]	0.1239	0.1019	0.2448	0.2279	0.2735	0.3091	0.1627		

Machined surface description

The deflection of the machine-tool-workpiece leads to machining error in parts by leaving uncut material on the machined surface. The machined surface (S_m) can be approximated by adjusting the parameters of the nominal surface (S_n) according to the database measured by on-machine probing (the simulated data is used here). The flow chart to get the machined surface is shown in Fig.4. The point at the nominal surface is $P_m(x_n, y_n, z_n)$, the measured point is (x_m, y_m, z_m), the control point is (x_c, y_c, z_c) and the calculated point is (x_p, y_p, z_p). The machining error distribution of the surface can be expressed as:

$$E = S_m - S_n \tag{2}$$

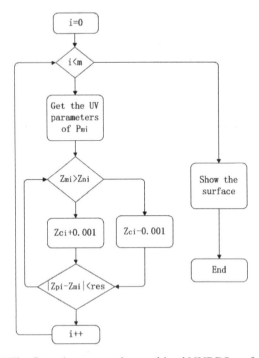

Fig.4 The flow chart to get the machined NURBS surface

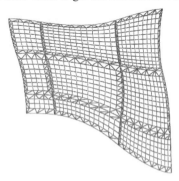

Fig.5 The approximated machined surface

The approximated surface of S_m is shown using the OpenGL library (shown in Fig.5). The compensated surface for finish machining can also be obtained in the similar way and it will helpful to improve the machining accuracy of a NURBS surface.

Conclusion

The error compensation in machining parts with NURBS surfaces is more complicated than it in machining parts with regular graphics. A new method for machined surface description in NURBS surface machining is discussed in the paper. The machined surface is obtained by adjusting the parameters of the nominal surface according to the machining error simulated in Ansys. In fact, with the help of on-machine measurement system, the machining error can be obtained directly in machining process, which will be more accurate and more credible. The machined NURBS surface will be more accurately described. And the machining accuracy will be greatly improved while the error expressed in the surface is compensated in the finish machining.

Acknowledgements

This work was financially supported by the Program for Zhejiang Leading Team of S&T Innovation (No.2009R50008),National Nature Science Foundation of China (No.51175461)，and Zhejiang Provincial Natural Science Foundation of China (No. Y1100281).

References

[1] M.W. Cho, G.H. Kim, T.I. Seo, Y.C. Hong, H.H. Cheng, Integrated machining error compensation method using OMM data and modified PNN algorithm, International Journal of Machine Tools and Manufacture, 46 (2006) 1417-1427.

[2] J.P. Choi, B.K. Min, S.J. Lee, Reduction of machining errors of a three-axis machine tool by on-machine measurement and error compensation system, Journal of Materials Processing Technology, 155 (2004) 2056-2064.

[3] R. Guiassa, J.R.R. Mayer, Predictive compliance based model for compensation in multi-pass milling by on-machine probing, CIRP Annals - Manufacturing Technology, 60 (2011) 391-394.

[4] B.C. Jiang, S.D. Chiu, Form tolerance-based measurement points determination with CMM, Journal of Intelligent Manufacturing, 13 (2002) 101-108.

[5] Hua Qiu, Hironobu Nisitani, Akio Kubo, Yong Yue, Autonomous form measurement on machining centers for free-form surfaces, International Journal of Machine Tools & Manufacture, 44 (2004) 961-969.

[6] W.T. Lei, M.P. Sung, NURBS-based fast geometric error compensation for CNC machine tools, International Journal of Machine Tools & Manufacture, 48 (2008) 307-319.

Advanced Materials Research Vol. 819 (2013) pp 38-42
© *(2013) Trans Tech Publications, Switzerland*
doi:10.4028/www.scientific.net/AMR.819.38

Methods of Gear Damage Assessment Based on Modal Parameter Identification

Jinbao Ma [1, a], Jianyu Zhang [1] and Xinbo Liu [1]

[1]Key Laboratory of Advanced Manufacturing Technology, Beijing University of Technology, Beijing 100124, CHINA

[a]email, shadow8273@163.com

Key words: Modal parameter identification, Correlation analysis, $COMAC$, Fault diagnosis

Abstract. With the evolution and degradation of mechanical fault, changes of the structural inherent characteristics will directly affect the overall response of system. Spur gear, which worked as the research object, is to be explored on the changes of modal parameters under different damage state. Optimum driving-point mobility and modal parameter identification is achieved by comprehensive utilization of experimental modal analysis and finite element analysis. MAC is used to determine the experiment results is whether accurate or not. Then comparing with the differences of modal parameters, the preliminary judgment of gear damage can be made. According to the experimental data of different gears, the $COMAC$ is taken to complete the correlation analysis and to judge the degree of the damage. The results shows that $COMAC$ provide an effective basis for the identification of vibration mechanism and vibration characteristic of fault gear.

Introduction

Modal analysis is a modern method to study the dynamic characteristics, and belongs to system identification methods in the field of engineering vibration. One of the commonly used methods is to combine numerical modal analysis and experimental modal analysis to solve the engineering structure vibration problems. Modal parameter is inherent characteristic of the elastic structure. If the main modal characteristics are identified in a sensitive frequency range, the actual vibration response of the structure of this band can be predicted under the external or internal vibration[1]. Taking the spur gear as an example, the actual vibration response characteristic is distinguished, and the influence of damage on the modal shapes is also considered.

Pre -experimental Analysis of Gear

Section Headings. The section headings are in boldface capital and lowercase letters. Second level headings are typed as part of the succeeding paragraph (like the subsection heading of this paragraph).

Pre-experimental analysis can determine whether the pre-selected excitation points are enough to pick up sufficient modes or not. The basic idea is to minimize the number of excitation points under the condition that each mode can be identified clearly. As a result, not only the experimental workload can be reduced, but also more satisfactory results, such as ideal orientation of excitation point, can be obtained. Pre-experimental analysis can provide a reference for the modal test and improve SNR, so that the test accuracy can be ensured[2].

Finite Element Model for Modal Simulation. The gear under investigation is a spur gear, whose module is 5 and teeth number is 30. The finite element model of gear is established through ABAQUS platform, in which the gear for modal simulation is in free condition. Therefore, the first 6 order modes representing rigid body motion are ignored.

Thereafter, the modal frequencies and mode shapes can be obtained by FEM simulation, which can provide a reference for sampling frequency determination and the set-up of excitation points and measuring point. The simulation results are thus shown in Table.1.

Pre–experiment. The position of the optimum driving-point should far away from each modal nodes at the same time when multi-modes are presented. Normalized modal shape is made out by preliminary experiments, which can help to determine the position of the optimum driving-point[3].

Table 1 Results of FEM simulation

Order	1	2	3	4	5
Frequency(Hz)	2554.4	4885.3	6158.3	7698.4	8624.7
Damping ratio	0	0	0	0	0

The method is as follows: a, To find normalized modal shape through preliminary experiments. b, The first few order modes being valuable is weighted according to the importance of each mode. For example, if the importance of the rth order mode is m times than others, then the rth row of modal shapes is divided by $m.c$, The size of matrix element reflects the distance from modal node, i.e. smaller number means closer distance. The number, as minimal as possible, is singled out and lined in a row. The position of the optimum driving-point corresponds with the biggest element in this row. In test, the position of the optimum driving-point is point 31, which can be used to obtain more mode characteristics.

Modal Testing and Analysis

Modal Testing. The suitable layout of measuring point and excitation points is determined by simulation and preliminary experiments. The result is shown in Fig. 1.

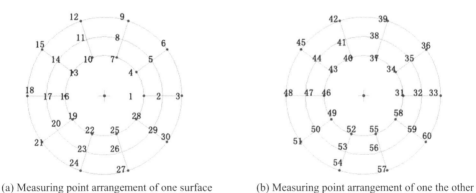

(a) Measuring point arrangement of one surface   (b) Measuring point arrangement of one the other surface

Fig. 1 Measuring point arrangement on gear surface

The chosen excitation method is MISO, and sponge pad support is used to simulate free state. The main modes occur on the normal direction of gear, so that the discussed mode in this paper is only normal. The above mentioned optimum driving-point (point 31) is used to collect acceleration signal on a healthy gear and another tooth broken gear.

Extraction of Modal Parameters. LSCF method[4-6]is used in this paper to extract the modal parameters, such as natural frequencies and modes, in which symmetric modes are merged in identification process. The comparison between healthy and broken gear is shown in Table.2 and Table.3.

Obviously, there are differences between healthy and broken gear on natural frequency and damping radio. It shows that the modal parameters are changed before and after the failure. So that the changes of natural frequency and damping radio can be as a signal to determine whether there is a failure. However, it is not able to distinguish certain damage accurately.

Table 2 Comparison on natural frequency of healthy and damage gear

Order	1	2	3	4	5
Healthy	2563.445	4811.97	6178.761	7684.55	8598.666
Broken	2585.25	4858.084	6206.963	7551.385	8635.839

Table 3 Comparison on damping ratio of healthy and damage gear

Order	1	2	3	4	5
Healthy	0.0885	0.3175	0.5824	0.3548	0.6485
Broken	0.1957	0.287	0.0987	1.3489	1.0781

To assess the orthogonality of experimental modal shapes, the correlation analysis based on Modal Assurance Criterion (MAC)[7] is utilized. MAC value represents the similarity between different modes, whose calculation method is shown in Eq.(1).

$$MAC = \frac{\left|\{\psi^*\}_r^T\{\psi\}_s\right|^2}{\left(\{\psi^*\}_r^T\{\psi\}_r\right)\left(\{\psi^*\}_s^T\{\psi\}_s\right)} \tag{1}$$

Where,

$\{\psi_r\}$ and $\{\psi_r\}$——Two experimental modal vectors.

If the vectors are same, then $MAC \approx 1$. On the contrary, $MAC \approx 0$ means the modal vectors are different.

Fig. 2 and Fig.3 show the verification results of modal shapes.

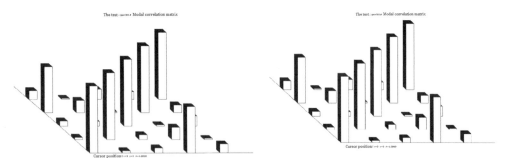

Fig. 2 MAC verification for healthy gear   Fig. 3 MAC verification for tooth broken gear

It is indicated in Fig.2 that the first mode is similar with the fourth-order mode for healthy gear. The reason for that is the fourth-order mode contains mainly the radial deformation, which is just similar to the first order mode in the axial vibration. The MAC values of other modes are acceptable. Low correlation indicates that different order modal shape is independent. Modes shapes are illustrated in Fig.4. In which, Fig.4 shows the 1st modal displacement for healthy and broken gear, and Fig.5 represents that of 2st modal displacement. Comparing Fig.4 and Fig.5 (the numbers represent separately the corresponding measuring point, it is easy to find that the modal displacement has changed before and after fault, especially on the position of nearly broken tooth (point16 and 18). It can be concluded that the location of vibration type node has changed when damage appears. This feature can certainly be used to predict the damage location.

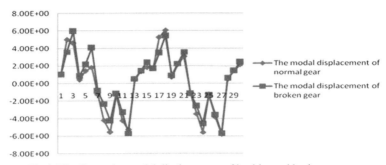

Fig.4 The first order modal displacement of healthy and broken gear

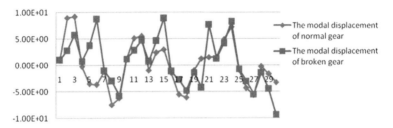

Fig.5 The third order modal displacement of healthy and broken gear

Correlation Analysis of same point in different conditions

Lieven and Ewins proposed an improved *MAC* algorithm, i.e. Coordinate Modal Assurance Criterion(COMAC)[8-10],whose calculation can be completed by Eq.(2).

$$COMAC(k) = \frac{\left(\sum_{r=1}^{n} |\psi_{u_r}(k) \cdot \psi_{d_r}(k)|\right)^2}{\sum_{r=1}^{n} \psi_{u_r}^2(k) \sum_{r=1}^{n} \psi_{d_r}^2(k)} \tag{2}$$

Where, $\psi_{u_r}(k)$ and $\psi_{d_r}(k)$ are vectors of point K in different modes of healthy and faulty conditions.

In *COMAC* value calculation, just some typical measuring points are selected to measure the modal value. When the fault doesn't exist, thus *COMAC* $= 1$,which indicates that displacement mode before and after failure is completely correlative. On the other hand, if damaged, then the mode shape must change, which will result *COMAC* $\neq 1$. Obviously, the *COMAC* value is more reliable than modal parameters in the identification of local fault. Furthermore, this value can even be used to localize the damage, which will be illustrated in future research.

Low *COMAC* value represents large difference between healthy and faulty gear. It can be seen in Fig.5 that the *COMAC* values of point 16 and 18 are smallest than others. Actually, these two points are nearer to the broken location than other points. The result shows that *COMAC* is effective to identify damage location.

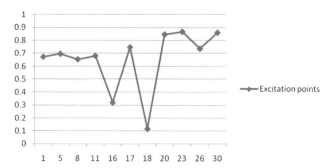

Fig. 6 *COMAC* value of some excitation point between healthy and broken gear

Conclusions

The dynamic characteristic of the structure is directly related to structural parameters, and structural damage will inevitably cause corresponding change of the dynamic characteristics.

Modal parameters are extracted from normal and fault states, then compared the differences of the natural frequency and damping ratio, preliminary judgment that whether failure occurs. Then, according to changes of position of modal displacement, the fault is to distinguished occurred. Finally, *COMAC* value is to show degree of fault. The results show that it is effective to identified the position and degree of fault. However, the method is subject to further consideration to locate the damage position accuracy.

References

[1] Thonmson. W.T. Theory of Vibration with Application, Prentice-hall, Inc, London 1981.

[2] Ewins.D.J: Modal Testing. Theory and Practice, Research Studies Press LTD 1984.

[3] Xiqing Zhang, Changle Xiang, Huilui and Fu Zhong Chen. Modal testing study on a complicated housing structure based on pre-test analysis, Journal of vibration and shock.

[4] Ward Heylen, Stefan Lammens and Paul Sas, Modal Analysis Theory and Testing (Kathoieke Universiteit Leuven, faculty of Engineering, Department of Mechanical Engineering, Division of Production Engineering, Mcchine Design and Automation,1998.

[5] Hui He. Applications of the PllyMax Method inAerocrafts Modal Parameter Estimation Aero Weaponry (2010).

[6] Xiaoping Xie, Xu Han, Changde Wu and Fei Lei. Experimental Modal Analysis for a Car Body-in-white Based on PolyMAX Method, Automotive Engineering, 31 (2009) 440-446.

[7] Debao Li, Qiuhai Lu. The Experimental Modal Analysis and the Application, Science Press, China 2001.

[8] Yingpei Gu, Changdeng and Fusheng Wu: Modal Analysis and Damage Diagnosis of Structure ,Southeast University, China, 2008.

[9] Wenha Shen, Senwen Zhang, Changyou Li, Yuhua Cao and Yaoqiang Kang, Research of structural damage detecting parameters based on experimental mode shapes, Journal of Jinan University(Natural Science), 29 (2008) 268-271.

[10]Tao Lin. Research on COMAC-like Indicator for Bridge Structure's Damage Detection, The World of Building Materials, 32 (2011).

Advanced Materials Research Vol. 819 (2013) pp 43-47
© *(2013) Trans Tech Publications, Switzerland*
doi:10.4028/www.scientific.net/AMR.819.43

Milling Machine Spindle Dynamic Analysis

Yu Li[1,a], Qichen Wang[1,b], Yinming Ge[1,c]

[1]Key Laboratory of Mechanism Theory and Equipment Design of Ministry of Education, Tianjin University, Tianjin300072, China.

[a]guangxuanly@163.com, [b]wangq5@miamioh.edu, [c]geyinming@126.com

Keywords: FE model, static stiffness trials, mode frequency, cutting trials

Abstract. This paper uses the FE (finite element) and experimental hybrid modeling method to establish the dynamic model of the spindle system. Calculate and analyze how static stiffness, span, and cantilever length affects mode frequency and mode shapes. Analyze the effect of the flywheel on dynamic performance based on harmonic respond. Moreover, the effect of the flywheel is well testified by experiment.

Introduction

Mechanical equipment manufacturing industry has been an important material basis for national economic development. It's also an important manifestation of the country's comprehensive national strength. The size of China's mechanical equipment manufacturing industry has already been in the forefront of the world, but still showed a tendency of 'big but not strong'.

With the development of the spiral bevel gear cutting machines from low speed wet cutting to dry cutting without cooling fluid, there are more high requirements on the performance of the spindle, especially dynamic performances. Hence, it is very important to establish the spindle system dynamic model, and to analyze the dynamic characteristics.

Theory

Generally, complex continuous elastic structure is divided into finite element. Assume discrete structure with N degrees of freedom, mass matrix M, damping matrix C, stiffness matrix K, incentive matrix F (t), and the dynamic equation like Eq. 1 is set up.

$$M\ddot{u}(t) + C\dot{u}(t) + Kx(t) = F(t) \tag{1}$$

Assume linear free vibration without damping, the solution of equation is as follows:

$$u = a\sin(\omega t + \varphi) \tag{2}$$

Then it's easy to get Eq. 3. Mode frequency ω_i (i=1,2,…,N and $\omega_i \leq \omega_{i+1}$) and mode shapes $A^{(i)}$ can be received by solving it.

$$\left(K - \omega^2 M\right)A = 0 \tag{3}$$

Static stiffness trials

Loading force and the relative displacement of the various components are measured as shown in fig. 1. The spring axial stiffness is solved by springs in series and least square. The average axial static stiffness is 6.0517e7 N/m. Here take the axial stiffness 6.1e7 N/m. Because of bearing NSK32040 with contact angle pi / 9, the radial support stiffness should be slightly higher, assumed 1.5e8 N/m.

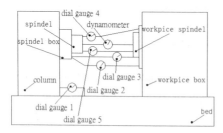
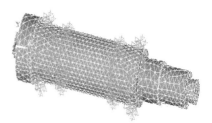

Fig. 1 Theory of static stiffness test   Fig. 2 The FE model of spindle

Modal analysis based on ANSYS

To reduce the amount of calculation and improve work efficiency, the spindle system should be simplified. Therefore, ignore some process structure, such as the assembly shoulder, screw holes and so on; fillets are replaced by right angles; reducing the shorter bosses by combining without changing mass center and mass; radial and axial stiffness of bearings is constant, and ignore their angular stiffness and mass; each bearing is simplified 12 radial spring dampers and 12 axial spring dampers, and ignore bearing quality.

Density 7800 kg/ (m*m*m), elastic modulus 2.1e11Pa, Poisson ratio 0.3, solid92 and combine14 is applied. So each radial spring damper stiffness is 2.01 e7 N/m, each axial is 2.542 e6 N/m. Here damping coefficient of combine14 is set 0.5. One node in combine14 is coupled with the node in the spindle, the other one is fixed. The final element model is fig. 2. The first 6 natural frequencies and mode shapes are solved as shown in Table 1. Because milling cuter plate with 28 cutters have speed 180r/min-260r/min, namely the excitation frequency 84Hz - 121Hz. So the second mode has an important effect on the dynamic performances.

Table 1: Mode frequency and mode shapes

NO	Modes[Hz]	Mode shapes
1	0	Torsion around y axis
2	91.549	translation along x axis
3	173.20	swing around the front support in the xoy plane
4	173.28	swing around the front support in the xoz plane
5	184.97	swing around the rear support in the xoz plane
6	185.13	wing around the rear support in the xoy plane

The cantilevered length L sequentially take 0.114,0.119,0.129,0.144,0.159 and 0.169m, with others unchanging.; span length sequentially take 0.210, 0.260, 0.310, 0.360, 0.410, 0.460, 0.480 and 0.490m, with others unchanging; Traditionally, the front stiffness is usually higher than the rear support, so change the rear supporting radial stiffness k with the front stiffness 2.01e7N/m. The relative results are like fig. 3.

The following conclusions are easily drawn. In the given range of L, mode frequency corresponding to swinging around the front support increases, and that corresponding to swinging around the front support decreases; in the given range of S, mode frequency corresponding to swinging around the front support increases , and that corresponding to swinging around the front support first decreases, then increases; with k increasing, mode frequency corresponding to swinging around the rear front is unchanged; radial stiffness is uncoupled to axial stiffness.

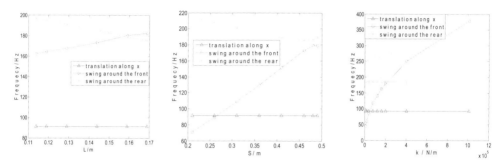

Fig. 3 The relation between mode frequency and design parameter

Harmonic analysis and cutting trials

One FE model including cutter, and the other one including cutter and flywheel are both established. Harmonic response in 84Hz, 98Hz, 107.3HZ, and 121.3Hz is respectively solved based on ANSYS. Extract the response at two key points like Fig. 4. The magnitude of harmonic response in the Y direction with flywheel is higher, and the magnitude in the Z direction without flywheel is higher.

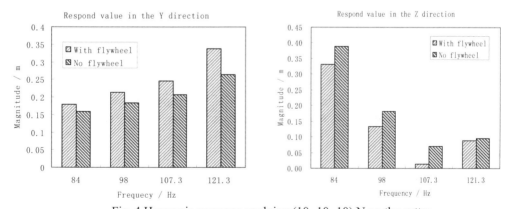

Fig. 4 Harmonic response applying (10,-10,-10) N on the cutter

Two groups of cutting trials of spiral bevel gear are implemented. One is with flywheel, the other is not. The vibration acceleration signals are collected on the spindle box near the front bearing as fig. 5 by the instrument like fig. 6. Here displacement is obtained by the integral two times. The RMS value of displacement as vibration evaluation is solved in the different spindle speed as fig. 7. We can easily draw the same conclusion with the RMS value.

Conclusions

The paper successfully establishes the dynamic model of the spindle system. AS the span length or the cantilevered length increases, frequency varies in the opposite trend between corresponding to swinging around the front support and corresponding to swinging around the front support. Cutting experiment validates the varying trend of harmonic response value with flywheel and without flywheel based on ANSYS. Moreover, reliability of the hybrid modeling is also testified.

Fig. 5 Sensor placement

Fig. 6 Portable Dynamic Signal Analyzer

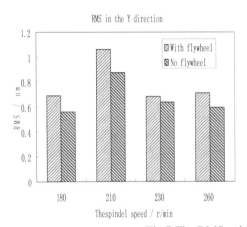

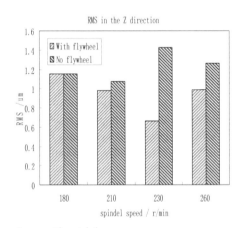

Fig.7 The RMS value based on cutting trials

Acknowledgements

This work was financially supported by the Science and Technology Supporting Key Project of Tianjin (12ZCZDGX01600), and the Science Planning Project of Fujian, China (2012H1008).

References

[1] Beizhi Li, Shang Liu, Jianguo Yang, and Xi Ru: Machinery Design & Manufacture, vol.2 (2011), p.163. In Chinese.

[2] Teng Hu, Jian Fu, and Yulai Zheng: Journal of Xihua University, vol.30 (2011), p.28. In Chinese.

[3] Qiming Zheng, Gang Zhou, and Qinghai Lao: Journal of Shaanxi University of Science & Technology, vol.23 (2005) , p.109. In Chinese.

[4] Zhiyong Yang, Hua Zhou, and Honglian Wang: Equipment Manufactring Technology, vol.5(2010), p.6. In Chinese.

[5] HagiuGD, GafitanuMD: Wear, vol.211(1997), p.22.

[6] Marc Simnofske , Annika Raatz , Ju¨rgen Hesselbach: Prod. Eng. Res. Devel. 3 (2009):461–468.

[7] Yongliang Chen, Peihua Gu: Chinese Journal of Mechanical Eengineering V01.21, No. 3, (2008)p.7.

[8] Jia   Man,   Lianhong   Zhang,   Yongliang   Chen   :   China Mechanical Engineering V01(2010)51–54

[9] Ghorbanali Moslemipour, T. S. Lee: J Intell Manuf23 (2012):1849–1860.

Advanced Materials Research Vol. 819 (2013) pp 48-54
© (2013) Trans Tech Publications, Switzerland
doi:10.4028/www.scientific.net/AMR.819.48

Modal Analysis and Numerical Solution in Cable Drilling System

C.G. Bu[a], J.W. Li and B. Long

China University of Geosciences Beijing 100083 China

[a]bucg@cugb.edu.cn

Keywords: cable drill; four-bar linkage; buffering spring; modal frequency

Abstract. Cable percussive drill is an important drilling machine in the construction of large diameter pile foundation, and complicated dynamic system. However, the design of cable percussive drill is mainly relied on empirical methods or simplified formulas, the dynamic characteristics of coupling system has not been fully understood. Based on the special conditions, the two degrees of freedom dynamic system coupling with buffering mechanism and bit is analyzed, and the vibration differential equation is built. Finally, the modal frequency of system is obtained by numerical method, and the correlation between the modal frequency of system and the structure parameters is built. The result develops the design theory of cable percussive drill and provides valuable preferences for the optimal design.

Introduction

Traditional cable drill belonging to no circular single cable percussive drilling process, there are obvious weaknesses to broke rock repeatedly. China University of Geosciences (Beijing) has designed the double cable synchronous hoist innovatively, and the synchronous hoist is applied successfully to the traditional percussive drills. The cable percussive drill with reverse circulation is endowed with new life [1,2]. It has an advantage of simple operation, low cost construction, and continuous discharge debris. So it is applied to construct pile in the soil, sand, especially the gravel layer, bedrock, and other complex formations. In China, the percussive reverse circulation drill is still widely used in the water drilling, geological exploring and mining, infrastructure of railway, road and bridge [3,4].

However, the research focuses mainly on the impact mechanism of cable percussive drills, and traditional research methods [5] make the impact mechanism be simplified the slider-crank mechanism in order to analyze impact motion. Yu [6] only considered buffer spring with four-bar linkages, did not consider the inertia force and wire rope deformation influencing on dynamics system. So far there is not full understanding such as the relationship between the modal frequency of the system and component parameters dynamic characteristics.

By analyzing the two degrees of freedom dynamic system coupling with buffering mechanism and bit, the vibration differential equation is built, the modal frequency of system is obtained by numerical method, and the correlation between the modal frequency of system and the structure parameters is built based on GCD-1500 cable drill.

The Dynamic System Model Coupling with Buffering Mechanism and Bit

When percussive reverse circulation drills are working, the motion relationship of the bit is consisted of the carrier motion of four-bar linkage and the relative motion of spring buffer. While the four-bar linkage is in a different location, the spring buffer and rope stiffness should be taken into account. We can research the local multi-degree freedom buffering mechanism including spring buffer, wire rope and drill bit, and analyze the relative motion of drill bit synthetic motion in the spring-mass system. Through the establishing of a simplified dynamics system model and solving it, the modal frequency of the relative motion of the local multi-degree of freedom system will be obtained.

The Simplifying of the System. The four-bar mechanism working in a certain position, the pressure wheel and the buffer mechanism at this moment along the spring shaft direction are in the balance.

The percussive mechanism of GCD-1500 cable drill is shown in Fig.1, and is simplified dynamic model coupling with buffering mechanism and bit, as shown in Fig.2. The angle between the vertical direction and the rope, which is determined by the pressuring wheel and top wheel, is β at an arbitrary position. The angle between any position of the impact beam and the horizontal position is ϕ. The plasmids m_1 and m_2 move along the reciprocating linear motion of spring-axis direction and the vertical direction respectively. As long as the two independent coordinates x_1 and x_2 can be applied to position m_1 and m_2 in any instantaneous position, the dynamic system coupling with buffering mechanism and bit has two degrees of freedom.

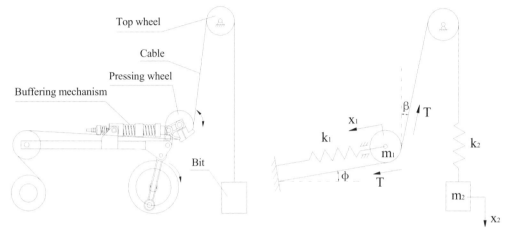

Fig. 1   Percussive mechanism of GCD-1500   Fig. 2   Dynamic model coupling with buffering
               cable drill   mechanism and bit

Basic Assumptions. In order to facilitate the analysis of the dynamic model of simplified discrete system with two degrees of freedom, the three assumptions are given as follows:

(1) The damping effect of each contact between the cable and wheels is not considered.

(2) Assume that variation of the cable cut point position between the top wheel and pressuring wheel is omitted.

(3) Because the derrick is high relatively, while the plasmid m_1 is fro beeline relative motion along the axial direction of the spring, the variation of declination angle β of cable between the pressure wheel and the top wheel is ignored.

System Modeling. The selection of the generalized coordinates and the direction in dynamical systems is shown as Fig.2, taking in the positive direction of the acceleration and force with the positive direction of coordinates. In any time of the vibration, the displacement of m_1 and m_2 is determined to the generalized displacement, x_1 and x_2, respectively. According to Newton's second law, the vibration differential equation of the system is

$$\begin{cases} m_1\ddot{x}_1 + k_1 x_1 = T - T\sin(\beta + \phi) \\ m_2\ddot{x}_2 + k_2(x_2 - x_1) = 0 \end{cases} \tag{1}$$

The rope tension T is

$$T = m_2 g - m_2\ddot{x}_2 \tag{2}$$

Where: m_1—the equivalent mass of pressure wheels; m_2—the equivalent mass of cable and bit; k_1—the equivalent stiffness of spring buffer; k_2—the equivalent vertical stiffness of rope in the drilling hole;

Decree $C = 1 - \sin(\beta + \phi)$, the result of Eq. 2 substituting in Eq. 1 is

$$\begin{cases} m_1\ddot{x}_1 + m_2 C\ddot{x}_2 + k_1 x_1 = m_2 gC \\ m_2\ddot{x}_2 - k_2 x_1 + k_2 x_2 = 0 \end{cases} \tag{3}$$

The form of system equations written in matrix is

$$\begin{pmatrix} m_1 & m_2 C \\ 0 & m_2 \end{pmatrix}\begin{pmatrix} \ddot{x}_1 \\ \ddot{x}_2 \end{pmatrix} + \begin{pmatrix} k_1 & 0 \\ -k_2 & k_2 \end{pmatrix}\begin{pmatrix} x_1 \\ x_2 \end{pmatrix} = \begin{pmatrix} m_2 gC \\ 0 \end{pmatrix} \tag{4}$$

The above expressed as the standard multi - degree of freedom system dynamics equations is:

$$M\ddot{X} + KX = F \tag{5}$$

Where: M—System mass matrix; K—System stiffness matrix; F—Column matrix of force Applied to the system by the external excitation.

The mass matrix and the stiffness matrix are determined by the nature of the system itself. In other words, it is only related to the system mass, stiffness, has nothing to do with external excitation and the initial conditions.

The displacement matrix and acceleration matrix of the system respectively are:

$$X = \{x_1 \;\; x_2\}^T, \ddot{X} = \{\ddot{x}_1 \;\; \ddot{x}_2\}^T \tag{6}$$

The system dynamics equations involved above is non-homogeneous linear equations with constant coefficients.

The Modal Frequency of System. Considering the free vibration damping system of two degrees of freedom, the kinetic equation is

$$M\ddot{X} + KX = 0 \tag{7}$$

Assuming the form of the dynamics differential equations above is:

$$X_i = A^{(i)}\sin(\omega_{2i}t + \varphi_i) \quad i = 1, 2 \tag{8}$$

Where: ω_{2i}^2, φ_i—the i modes modal frequency and phase angle; X_i—the i modes displacement array, $X_i = \{x_1 \;\; x_2\}_i^T$; $A^{(i)}$—the i modes maximum displacement or amplitude vector, $A^{(i)} = \{A_1^{(i)} \;\; A_2^{(i)}\}^T$.

Formula 8 substituting into formula 7, taking into account $\sin(\omega_{2i}t + \varphi_i)$ does not equal to zero, and obtain the following algebraic equations:

$$\left(K - \omega_{2i}^2 M\right)A^{(i)} = 0 \tag{9}$$

Due to $A^{(i)} \neq 0$, according to the theory of the existence of solutions of linear algebraic homogeneous equations, the equations above solvability conditions is the determinant of the coefficient matrix is zero [7].

$$\left|K - \omega_{2i}^2 M\right| = 0 \tag{10}$$

Formula (10) is the characteristic equation of the system of kinetic equations. Substitute the system mass matrix and the stiffness matrix into Eq. 10, the result is:

$$\begin{vmatrix} k_1 - \omega_{2i}^2 m_1 & -\omega_{2i}^2 m_2 C \\ -k_2 & k_2 - \omega_{2i}^2 m_2 \end{vmatrix} = 0 \tag{11}$$

Simplify and expand available by the characteristic Eq. 11, the result is:

$$m_1 m_2 \omega_{2i}^4 - D\omega_{2i}^2 + k_1 k_2 = 0 \quad i = 1,2 \tag{12}$$

Where: $D = m_2 k_1 + Cm_2 k_2 + m_1 k_2$

Hence,

$$\omega_{2i}^2 = \frac{D \mp \sqrt{D^2 - 4m_1 m_2 k_1 k_2}}{2m_1 m_2} \tag{13}$$

The characteristic solutions of ω_{21}^2 and ω_{22}^2 can be solved by Eq. 13, modal circular frequency ω_{21} and ω_{22} can be got after the open square.

The relationship of frequency and circular frequency is

$$f = \frac{\omega}{2\pi} \tag{14}$$

The Calculation of Each Component Parameter

By the analysis above, there are requirements to solve the modal frequencies of the two degrees of freedom spring-mass system coupling with the buffering mechanism and bit. Firstly the parameter of each component is needed to calculate.

(1) The equivalent mass m_1 of pressure wheel

The equivalent mass of pressure wheel is converted by the translational mass which is consisted with the mass of top wheel, pressure wheel, rocker arm, inner buffer spring, outer buffer spring, outer return spring, spring shaft and coil.

The top wheel rotates around mass centre axis, the pressure wheel not only rotate but also translate, the rocker arm rotates around the intersection point of impact beam, the mass of three components convert into the mass of pressure wheels according to the same principle as kinetic energy is not change. The inner buffer spring, outer buffer spring and the outer return spring belong to elastic element, according to the mechanical vibration [8], one-third of its mass equivalently converts to the pressure wheel, the mass of the spring shaft and coil is directly transformed to the pressure wheel. Ultimately the equivalent mass of pressure wheel is obtained by analysis and calculation $m_1 = 1219.25\,\text{kg}$.

(2) The equivalent stiffness k1 of spring buffer

Considering the preload of spring, the buffering mechanism simplifies the buffering mechanism for the system of spring and mass as shown in Fig 3.

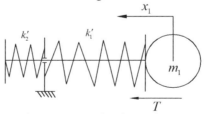

Fig. 3 System of spring and mass

The buffer mechanism is composed of an inner, outer buffer spring and the return spring; system equivalent stiffness was calculated according to multiple series and parallel connection above the spring. The formula of equivalent stiffness of the spring series and parallel systems are obtained respectively:

$$k_c = \frac{k_{11} k_{12}}{k_{11} + k_{12}} \tag{15}$$

$$k_b = k_{11} + k_{12} \tag{16}$$

The equivalent stiffness k_1' of buffer spring groups on both sides are calculated by equivalent spring series and parallel, the stiffness of the inner buffer spring and the outer buffer spring is 389.09×10^3 N/m and 854.06×10^3 N/m respectively, according to the Eq. 15 and Eq. 16, it is available to get the equivalent stiffness of the buffer spring group on both sides.

The equivalent stiffness of the return spring on both sides is calculated by the spring series equivalently, the outer return spring stiffness is 655.98×10^3 N/m. According to the Eq. 16, the equivalent stiffness of both sides of the return spring group is obtained.

$$k_2' = 1311.96 \times 10^3 \, \text{N/m}$$

Finally the equivalent stiffness of spring buffer is obtained.

$$k_1 = k_1' + k_2' = 2555.11 \times 10^3 \, \text{N/m}$$

(3) The equivalent mass m_2 of bit

The equivalent mass of bit is equal to the mass of bit itself, which weighs 3500 kg, minus the buoyancy of the bit in mud, and then additional the equivalent mass of suspending rope.

When the bit in the drilling mud moves up and down, it has been subject to the buoyancy which is keeping vertically upward, the formula of bit buoyancy is:

$$F_b = \rho v g \tag{17}$$

ρ—the mud density, v—the bit volume, g—Acceleration of gravity.

Thus buoyancy, which is 448.72kg, can be equivalent to bit. Cable belongs to elastic element, bit equivalent mass will be add one-third mass of the suspending cable. The length of the cable is the total of the two part cable outside the drill hole, which is 20m, and inside the drill hole. Its reference weight is 235.90kg/100m. The equivalent mass m_2 of bit is calculated in different depth, as shown in Table 1.

(4) The equivalent stiffness k_2 of cable

The longitudinal equivalent stiffness of single cable, the length is L, is calculated according to the following formula.

$$K = \frac{EA}{L} \tag{18}$$

Where: E—the cable elastic modulus; A—Total Cross sectional area of the cable.

Because of two cables suspending synchronously the same bit, the formula 16 of equivalent stiffness of the two-cable-paralleled spring system can obtained the equivalent stiffness k_2 of cable in different deep hole, as shown in Table 1.

When the impact beam of four-bar linkage are in the top, lower dead center, which is in the horizontal position, and the middle position, which is between the top and lower dead center, respectively, the numerical value of β, ϕ and C is shown in Table 2.

Table 1 Equivalent mass of bit and equivalent stiffness of cable in different hole depth

Drilling hole depth[m]	Cable length [m]	Equivalent stiffness of cable[N/m]	Equivalent mass of bit[kg]
10	30	3346.0×10^3	3098.46
30	50	2007.6×10^3	3129.92
50	70	1434.0×10^3	3161.37
70	90	1115.3×10^3	3192.82
90	110	912.5×10^3	3224.27

Table 2 Parameters β, ϕ, C of four-bar linkage in different locations

Parameters	Lower dead center	Middle point	Top dead center
β [degree]	12.85	14.56	16.27
ϕ [degree]	0	7.38	14.75
C	0.7776	0.6264	0.4846

Numerical Solution of Modal Frequency in System

According to the above analysis of two degree of freedom dynamic system coupling the buffer mechanism with bit and the method of calculating the modal frequency, plugging m_1, m_2, k_1, k_2 and C into formula 13, and formula 14, at the same time, we can obtain the two order modal frequency (f_1 and f_2) of the spring-mass system which have two local degrees of freedom when four-bar linkage system is in the different locations.

Table 3 Modal frequency of four-bar linkage system in different locations & different hole depth

Drilling Depth [m]	First-order modal frequency[Hz]			Second-order modal frequency[Hz]		
	LDC	MP	TDC	LDC	MP	TDC
10	3.44	3.62	3.81	11.07	10.53	9.98
30	3.05	3.18	3.31	9.61	9.23	8.86
50	2.76	2.85	2.94	8.95	8.67	8.39
70	2.53	2.60	2.67	8.58	8.35	8.13
90	2.34	2.39	2.45	8.34	8.15	7.96

Where: LDC—Lower dead center; MP—Middle point; TDC—Top dead center.

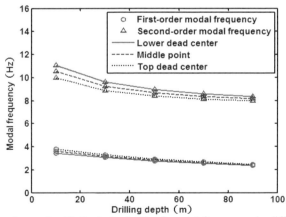

Fig. 4 Curves changed with the hole depth of modal frequency in different locations

Table 3 and Fig.4 show that both first-order and second-order modal frequency slowly become smaller as the drill hole depth increases. The range of first-order modal frequency is narrower, while the range of second-order modal frequency is broader. No matter how the hole depth changes, the equivalent stiffness of buffering mechanism and the equivalent mass of pressure wheel will not change, the equivalent mass of bit will change a little, but the equivalent stiffness of suspending cable will change obviously. So the relation of modal frequency of spring-mass system and parameters of each component can be obtained.

The first-order modal frequency is mainly determined by spring buffer and bit, while the second-order modal frequency is mainly affected by the cable and the bit.

While the drilling depth is shallower, the position of the rocker arm has a smaller effect on the system modal frequency. While the drilling hole becomes deeper, the position of the rocker arm almost has no effect on the system modal frequency.

Summary

Establishing the vibration differential equation of two degrees of freedom dynamic system which couple with the spring buffer and bit, and using numerical method to solve the modal frequency of the system, the modal frequency of the system will change with the drill hole depth while the rocker arm of the four-bar linkage system is in different locations. The relation between modal frequency of multi-degrees of freedom system and parameters of each component is obtained; the first-order modal frequency is mainly determined by spring buffer and bit, while the second-order modal frequency is mainly affected by two suspending cables and bit. The modal analysis has practical significance on the improving performance of the cable impact drilling system. In the process of optimization design, the driving round speed should avoid the natural circular frequencies in order to prevent the resonance and prolong the life of the buffering mechanism.

Acknowledgment

The authors would like to acknowledge the National Natural Science Foundation of China for their financial support for this project under Grand No. 51275493.

References

[1] Z.S. Yu, Brief analysis of the Development of Steel Cable Type Impact Drill Rig, Journal of Mechanical engineering. 3 (2002) 30-34.

[2] C.G. Bu, Design of Synchronous Hoist in Percussive Reverse Circulation Drill, Journal of Mining machinery. 9(2011) 9-11.

[3] D.C. Hu, Primary Discussion on the Reverse Circulation Percussive Drilling Technology. Journal of West exploration engineering. 3(2000) 96-97.

[4] Z.Y. Li, Status Quo and Trend of Percussive Reverse Circulation Pile Hole Drill Rigs in China, Journal of Exploration engineering. 1(2001) 60-61.

[5] H.M. Yang, Drilling equipment, Geological publishing house, Beijing 1988.

[6] P. Yu, The Development of CHF-20 Percussive-Rotary Reverse-circulation Drill- research and simulation analysis of working mechanism, Jilin university, Changchun 2005.

[7] B.C. Wen, Mechanical vibration, Metallurgical industry press, Beijing 1999.

[8] F.S. Tse, I.E. Morse, R.T. Hinkle, Mechanical Vibration Theory and Applications, Allyn and Bacon Inc. 1978

Advanced Materials Research Vol. 819 (2013) pp 55-58
© (2013) Trans Tech Publications, Switzerland
doi:10.4028/www.scientific.net/AMR.819.55

Modeling Cutter Engagement Region for Triangular Mesh

Gan Wenfeng[1, a], Shen Hongyao[1, b], Lin Zhiwei[1, c], Chen Zhiyu[1, d],

Fu Jianzhong[1, e]

[1]Department of Mechanical Engineering, Zhejiang University, Hangzhou, 310027, P. R. China

[a]gwf421@zju.edu.cn, [b]shenhongyao@zju.edu.cn (corresponding author), [c]zjtzhylin1986@163.com, [d]21125021@zju.edu.cn, [e]fjz@zju.edu.cn

Keywords: List the keywords covered in your paper. These keywords will also be used by the publisher to produce a keyword index.

Abstract. Cutter Engagement Region (CER) is the part of tool that contacts with stock, where a series of energy exchange and material transportation take place. A full understanding of its shape and area is indispensible for cutting force calibration, chatter analysis, and tool wear prevention. It can provide a criterion for efficient tool-path planning as well. In literature, however, there is no available modeling of CER. We hereby propose a modeling method for ball-end tool with design surface represented in triangular mesh. First, design surface is offset to form a stock surface. Next, compute the intersection between every triangular facet and the tool. Then, connect all the intersecting arcs to form the boundary of CER. Finally, compute the area of CER. Experimental validation for the proposed method is still needed.

Introduction

All Cutter Engagement Region (CER), the region where the tool contacts with stock material, has received considerable attention in recent years in light of its importance to cutting force calibration, chatter analysis, and energy transfer analysis. It can also provide a criterion for tool path planning. It is meaningful to find out its shape and area.

P. Leeand Y. Altintas [1] first proposed the concept of Differential Cutter Edge. Cutter edge is derived by angle parameter into a series of small edge elements. Then consider the tangent, radial, and axial force on each element. Finally, integrate all forces along the whole edge curve. It was an innovative model except the fact that only the edges were considered to be in contact with stock, which is not the case in actual cutting process. Although the above weakness makes Differential Cutter Edge an incomplete model, I. Lazoglu [2] inherited the idea and developed it by meshing the work piece into an assembly of boxes. Lamikiz [3] considered the intersecting arcs between the vertical profile of the tool and the work piece . On each vertical section around the axis, different arcs are found. The combination of these arcs forms the CER. The advantage of this model is that the whole tool surface is taken into consideration. However, discretization of cutter into a series of vertical section weighs heavily on calculation. Also, it is difficult to calculate the area of CER with such method. Li [4] proposed the concept of Virtual Cutter Edge, VCE. Assume the tool surface as a torus, the intersecting curve between work piece surface and that torus is defined as VCE. VCE is considered as the boundary of CER. When the former is calculated, the latter can be easily found. Nevertheless, the problem of CER area is left untouched in authors' modeling.

As far as our review shows, no effective modeling method for CER and its area is available in literature. We hereby propose a modeling method for ball-end tool in five-axis milling. In this paper, design surface is represented in triangular mesh for its flexibility and popularity in tool path planning, cutting analysis, and CAM fields. OFF and STL are the most commonly used formats for tessellated meshes.

Problem Description and Theoretical Analysis

The main goal of this paper can be stated as follows: to find the area of CER on cutter contact points CCP with a given tool S_{tool}. According to APT (Automatically Programmed Tools) model, a cutter can be represented as a revolved surface, as denoted in Fig. 1-1. To find the boundary of CER means to find out the intersect between cutter surface S_{tool} and stock surface S_{stock}. Since stock is modeled by triangular mesh, one can assert that boundary ∂S_{CER} must be a sequence of arcs lying on S_{tool}. After ∂S_{CER} is determined, the shape and area of S_{CER} can be computed accordingly.

To accomplish this goal, a 4-step method is proposed. The flowchart of the whole process is shown in Fig. 2-1. First, an OFF file of deign surface together with cutter parameters and tool posture are input as data for upcoming procedures. Moreover, a stock surface must be constructed such that its intersection with cutter surface can be easily calculated. Then, every tri-facet on stock surface is intersected with tool surface, namely a sphere with radius R. Intersection process includes a 3-phase judgment procedure. After that, a set of unorganized arcs is produced. They should be sorted and connected into a piecewise smooth curve to form the boundary of CER. In the final step, surface integral is performed to calculate the area of CER.

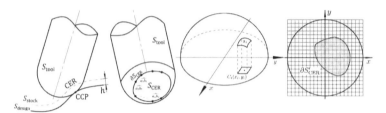

Fig. 1. (1) Definition of CER. (2) Boundary of CER in piecewise arcs. (3) Projection of CER facet to planar cell. (4) Summing up of CER area.

Stock Surface Construction

Stock surface S_{stock} is constructed by offsetting design surface S_{design} by h, denoting the cutting depth. Here we use a simple offset method such that every vertex of triangular mesh is offset along its normal direction. More sophisticated algorithms can also employed to acquire smoother S_{stock} and to achieve higher offset accuracy. However, it is not the main goal of our method, so we will leave that issue to more pertinent papers.

Intersection between Tri-facets and Tool Surface

In order to compute the area of CER, the boundary of it must be determine before-hand. The boundary of CER ∂S_{CER} is a piecewise smooth curveconsist of intersecting arcs, as shown in Fig. 1-2. We will discuss in detail how to find out the intersecting arcs.

Step1: Distance between Ball Center and Tri-facet Plane

First, distance d from ball center O to tri-facet plane Σ is calculated through a vector method

$$d = \overrightarrow{OV_1} \cdot \boldsymbol{n}, \tag{1}$$

in which $\boldsymbol{n} = \overrightarrow{V_1V_2} \times \overrightarrow{V_2V_3}$. Then we judge whether $d < R$. If so, go to the next step; else start a new intersection judging routine with another tri-facet.

Step2: Minkowski Sum to Determine whether Intersection Occurred

Intersection problem is now a two dimensional problem. A *Minkowski Sum* method is employed. In plane Σ, tri-facet becomes a triangle $\triangle V_1V_2V_3$, while Ω is projected as a circle O^* with radius r. Offset $\triangle V_1V_2V_3$ by r to construct a region $\mathbb{M}$. This is called Minkowski Sum. Next, determine whether circle center O^* lies inside $\mathbb{M}$. If so, $\triangle V_1V_2V_3$ intersects with circle O^*.

Step3: Intersecting Arc Sections between Triangle and Circle

When a facet fulfilled the two conditions above, it is ready to preform intersecting process. First, for each edge in $\triangle V_1V_2V_3$, find the intersect points. There are three conditions for the number of points: no point, one point, and two different points. Without loss of generality, take V_1V_2 for instance, the above relationshipcan be expressed in Eq. 2

$$\text{if } \Delta = b^2 - 4ac \begin{cases} > 0 & \text{two different points} \\ = 0 & \text{one point} \\ < 0 & \text{no point} \end{cases}, \tag{2}$$

in which $a = \overrightarrow{V_1V_2}^2 = |\overrightarrow{V_1V_2}|^2$, $b = 2 \cdot (\overrightarrow{O^*V_1} \cdot \overrightarrow{V_1V_2})$, $c = \overrightarrow{O^*V_1}^2 - r^2 = |\overrightarrow{O^*V_1}|^2 - r^2$.

For $\Delta \geqslant 0$, Eq. 3 calculates intersecting point(s)

$$\vec{P} = \vec{V_1} + \lambda \cdot \overrightarrow{V_1V_2}, \tag{3}$$

where $\lambda = (-b \pm \sqrt{\Delta})/(2a)$.

Intersecting points on edge V_2V_3 and V_3V_1 can be calculated in similar manner.

When intersecting points $\{P_i | i = 1, 2, ...\}$ on circle O^* are all found, they should be sorted around center O^* in counter-clockwise to find out arcs. First take out P_1 from $\{P_i | i = 1, 2, ...\}$ and construct vector $\overrightarrow{O^*P_1}$. Then, sort the resting points P_i according to the angle between every $\overrightarrow{O^*P_i}$ and $\overrightarrow{O^*P_1}$. Now that intersecting points are sorted, a whole circle O^* is divided into several arcs. Next is to find out which arcs are inside triangle while others not. Assume Q_i as the mid-point of $\overset{\frown}{P_iP_{i+1}}$. Check if Q_i is always on the left side of directed edge $\overrightarrow{V_1V_2}$, $\overrightarrow{V_2V_3}$, and $\overrightarrow{V_3V_1}$. If yes, that it means arc $\overset{\frown}{P_iP_{i+1}}$ is inside triangle $\Delta V_1V_2V_3$, i.e. an intersecting arc. Mark it as $\overrightarrow{a_1b_1}$. If not, it is not an intersecting arc. Find out all the intersecting arcs of triangular mesh S_{stock} likewise.

Connection of Arcs

The set of intersecting arcs is denoted as $\{\overset{\frown}{a_ib_i} \mid i = 1, 2, 3...\}$. However, they are now scattered randomly in space. In order to find the boundary ∂S_{CER}, they should be sorted and connected.

The commonly used *Dictionary Sorting* is employed here. All arcs are sorted by a 6-tuple $[x_{a_i}, y_{a_i}, z_{a_i}, x_{b_i}, y_{b_i}, z_{b_i}]$, where $[x_{a_i}, y_{a_i}, z_{a_i}]$ are the coordinates of their starting points a_i, and $[x_{b_i}, y_{b_i}, z_{b_i}]$ are those of ending points b_i. After sorted, arcs can be connected by just comparing starting points and ending points. Such algorithm is not time efficient, with the complex of $O(n^2)$, but it is useful in our application.

Area Calculation

With ∂S_{CER} found, the area calculation of CER becomes easier. For ball-end cutter, first construct a mapping from half-sphere to the plane that cross the ball center, as shown in Fig. 1-3. Establish a local coordinate system in which origin is placed on the circle center. The calculation of the area of CER is now converted to a 2-dimensional integration problem, shown in Eq. 1-4.

$$S_{\text{CER}} = \iint\limits_{S_{\text{CER}}} ds = \iint\limits_{S'_{\text{CER}}} \sqrt{\frac{R^2 + 3(x^2 + y^2)}{R^2 - x^2 - y^2}} dxdy. \tag{4}$$

However, due to the fact that ∂S_{CER} is a piecewise smooth curve, its image $\partial S'_{\text{CER}}$ on the plane is also piecewise smooth. One intuitive thought is to discretize the whole plane into $n \times m$ rectangle cells. The number n and m are chosen according to calculation accuracy. This means the length and height of every cell respectively subject to $dx = 2R/n$ and $dy = 2R/m$. As a result, area of CER can be found by summing up the area of occupied cells s_i, i.e.

$$S_{\text{CER}} = \sum_{\text{cells}} s_i, \text{ with } s_i = \frac{4R^2}{nm} \frac{R}{\sqrt{R^2 - x_i^2 - y_i^2}}. \tag{5}$$

Computer Simulation and Result Analysis

The proposed CER modeling process is implemented in C++ language and run on a PC with Pentium(R) Dual-core CPU E6300 @ 2.80GHz 2.80GHz and 2.00GB RAM.

An OFF model *Nefertiti* consist of 4565 vertices and 8992 tri-facets is input as S_{design} , as Fig. 2-2 shows. Offsetting it by $h = 1\text{mm}$ to construct S_{stock} . Ball end tool with $R = 2\text{mm}$ is placed on it. Three arbitrary positions and postures are selected for area calculation, shown in Fig.2-3.

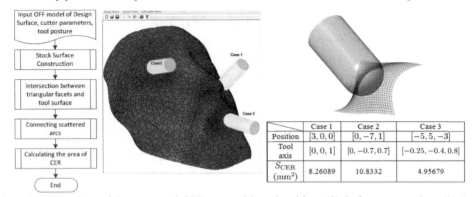

	Case 1	Case 2	Case 3
Position	$[3,0,0]$	$[0,-7,1]$	$[-5,5,-3]$
Tool axis	$[0,0,1]$	$[0,-0.7,0.7]$	$[-0.25,-0.4,0.8]$
S_{CER} (mm^2)	8.26089	10.8332	4.95679

Fig.2. (1) Flowchart of the proposed CER recognition algorithm. (2) Software snapshot. (3) Area calculation results.

Conclusion and Further Discussion

A complete Cutter Engagement Region modeling algorithm is proposed in this paper. Compared to other models, not only the cutter edge but the whole tool tip part is taken into consideration, which makes our model a suitable one for force calibration, chatter analysis, and etc. These, however, does not necessarily mean the proposed method is flawless. A few pitfalls are listed here for later development.

First, in roughing, stock surface may not be the offset of design surface, but the remaining of last cut. Theproposed algorithmis still valid as long as stock surface can be represented by triangular mesh. Second, our method is not time efficient enough, with a calculation time of over 82 seconds for every case. This is certainly a burden for triangular meshes with more vertices and facets. All in all, experimental validation for the proposed method is still needed.

Acknowledgments

This work was financially supported by the Program for Zhejiang Leading Team of S&T Innovation (No.2009R50008), National Nature Science Foundation of China (No. 51105335).

References

[1] P. Lee, Y. Altintas, Prediction of ball-end milling forces from orthogonal cutting data, International Journal of Machine Tools & Manufacture, 36 (1996) 1059-1072.

[2] I. Lazoglu, Sculpture surface machining: a generalized model of ball-end milling force system, International Journal of Machine Tools & Manufacture, 43 (2003) 453-462.

[3] A. Lamikiz, L.N.L. de Lacalle, J.A. Sanchez, M.A. Salgado, Cutting force estimation in sculptured surface milling, International Journal of Machine Tools & Manufacture, 44 (2004) 1511-1526.

[4] Z. Li, W. Chen, A global cutter positioning method for multi-axis machining of sculptured surfaces, International Journal of Machine Tools & Manufacture, 46 (2006) 1428-1434.

Advanced Materials Research Vol. 819 (2013) pp 59-64
© *(2013) Trans Tech Publications, Switzerland*
doi:10.4028/www.scientific.net/AMR.819.59

Performance Evaluation and Prediction of Escalator Structure using FEM-based Analysis

Jiancai Zhao[1, a], Xi Chen[2, b], Zeyu Zhao[3, c]

[1]School of Mechanical Engineering, Shanghai Jiaotong University, Shanghai 200240, China

[2]School of Electronics and Information Engineering, Sichuan University, Chengdu 610064, China

[3]College of Engineering and Applied Science, University of Cincinnati, Cincinnati OH 45221, USA

[a]zhaojc@sjtu.edu.cn, [b]594127785@qq.com, [c]zeyuzhao1990@gmail.com

Keywords: Escalator truss, Performance evaluation, Stiffness, Strength, Finite element method

Abstract. The paper applied Finite Element Method (FEM) to study performance of escalator structure. A case study is given to demonstrate the method of the FEM. To validate the results of FEM, some experiments were conducted. The results show that all the test samples meet the requirements of the truss, and the experimental results verify the numerical calculation results.

Introduction

FEM is a numerical technique for finding approximate solution of partial differential equation. It is widely applied in the automotive, biomechanical, aeronautical, manufacturing and other industries for identifying the critical regions, finding the root causes of failure, optimizing of structures and materials, and reducing manufacturing costs. More and more escalator manufactures realize that FEM is a useful and effective method in the product design and development.

Many FEA (Finite Element Analysis) groups have been founded to aid new product design and analyze the causes of the service failure of the escalator. Bangash [1] introduced the design analysis of the elements of escalators using the finite element technique in Section IV of the book "Lifts, elevators, escalators and moving walkways/ travelators". The design of elements forming escalators is useful for engineers, technologists, specialists in lifts, escalators or moving walkways. Luh and Lin [2] developed an algorithm to find optimal truss structures. A two-stage approach was adopted in this method to achieve the minimum weight under stress, deflection and kinematic stability constraints. Azid et al [3] proposed the use of conventional GA in layout optimization. This approach was applied to Michell's truss problems and was extended to include dual stress-displacements constraints. Wong et al [4] developed an optimization approach for design management by adapting the Balanced Scorecard methodology. This method was a versatile aid to the existing practice of prioritizing client requirements. Qiao and Zhu [5] studied the issues of uniform strength design and torsion strength design for structure optimization. These studies resulted in the conclusion that chord could be used with profile of different cross-section moment of inertia. Zhang et al [6] analyzed the static strength and dynamic characteristics of the steel structure of an escalator by means of the finite element method. This method could be used in the early design of a product to ensure product design quality. Fang and Zhu [7] developed light-weight methods to optimize the metal structure. This method obtained satisfactory weight reduction while maintaining the necessary stiffness and strength of the metal structure. Table 1 showed maximal deflection and stress when steadily decrease the cross-sectional area of the lower chord.

Wu [8] introduced a method to calculate the high-strength bolted connection of the escalator truss sticky segments. Through force analysis of the high-strength bolts, a verification model was developed and the steps and methods of verification were summarized.

Azid et al [9] determined that the optimal topology of a Michell's truss was only applicable up to a particular distance ratio between the loading point to the line joining the supports and the span of the supports. Once this critical ratio was exceeded, the optimum topology of the Michell's truss changed. It was observed that it was possible to demarcate the region of two different types of optimum topologies by a linear relation. This showed that the height of the loading point and span of the supporting joints had significant contributions in determining the optimal topology of the structure.

Other research concentrated on the escalator system, such as the chain drive, anti-entrapping device, and so on. Liu et al [10] developed a kinematics and dynamics simulation model in RecurDyn based on chain transmission kinematics theory. This model is intended as a guide for product development and improvement. Zhu and Hong [11] discussed the safety rules for escalators. Six kinds of anti-entrapping devices were presented and it was noted that some design parameters for these devices must be quantified. Hu et al [12] developed two methods to improve the life of the main round through analyzing the force factor effect on the escalator driver and its stress distribution.

Table 1 Maximal deflection and stress when steadily decrease the lower chord cross-sectional area

Lower chord shape [mm]	Maximal stress[MPa]	Maximal deflection [mm]	Ratio of decreased weight [%]
L125×75×10	93. 8	8. 47	0
L110×70×10	93. 8	8. 67	1. 78
L110×70×8	93. 7	9. 03	4. 9
L100×63×8	93. 7	9. 21	6. 22
L100×63×7	93. 7	9. 43	7. 64
L90×56×7	101	9. 64	8. 84

The previous researches, however, seldom study the critical regions on the truss and conduct corresponding experiment to validate the results of FEM, due to its complex structure and large size.

The research presented in this paper aims to overcome the above weakness. The subject of this study is the truss of a heavy duty escalator. Escalators, which are mainly applied at the public transportation segment e.g., airports and railway stations, must be designed to meet the requirements of the EN115 international design safety codes. The truss is the most important support in a heavy duty escalator. Its strength is the best determining factor of the safeness of the design of an escalator. Improvement in the quality of the truss design can improve the safety of the escalator. The FEM is applied in this paper to predict the critical properties of the escalator, namely, stiffness and strength of a truss, through numerical calculations, find the most critical parts of the truss, and to evaluate truss performance.

A case study is provided to demonstrate the procedure of the proposed FEM, which carried out by MSC Patran/Nastran software. Through the numerical analysis, the performance of the truss is evaluated. To validate the results of FEM, some experiments were conducted.

Methods
Truss Profile. The truss is constructed of welded frameworks, which made mainly of angle profiles such as L-profiles and U-profiles. There is a soffit plate and many transversal beams, which put together form horizontal frames and make the whole escalator a stiff space truss. This three-dimensional stiffness prevents the upper chord from buckling. Fig. 2 shows the profile of the truss.

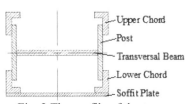

Fig. 2 The profile of the truss

Finite Element Model of Truss. A finite element model includes meshing, material properties, element properties and boundary conditions.

Meshing. Meshing is the process of creating elements from curves and surfaces. In this process, lines will be meshed into BAR2 elements and surfaces will be set in Quad elements by IsoMesh method.

Material Properties. Since trusses are made of steel, for this model, truss material will be considered to be Q235B for most of parts or components, and Q235C for cold-rolled parts. When higher strength truss materials are employed, Q345B and Q345C are used. Therefore, the material parameters of the truss in the calculation can be set as:
Young's modulus=206000MPa; Poisson's ratio=0.3; Density=7.85E-09 ton/mm³.

Boundary Conditions

Displacement Constraint. According to the actual installation on-site, the constraints are as shown in Table 2.

Table 2 Constraints of truss

Coordinate	Left single side		Right single side	
	UP	LP	UP	LP
X	0	F	0	F
Y	0	0	0	0
Z	0	0	F	F

Legend: UP-the point in the upper landing, LP-the point in the lower landing, F-the degree of freedom in this direction is free.

Loads. In order to ensure analysis accuracy, the following loads will be applied in the model according to DIN 18800: Truss gravity (denoted as L1), distributed forces (L2), passenger load (L3), static load (L4), dynamic loads (L5).

Loads on FEA Model

Load Assumptions. The loads are applied to the FEM model according to the real function. Some safety factors listed in Table 3 are given in line with the requirement of DIN 18800.

Table 3 Load assumptions

No	Symbol	Factor	Description	Type of loads
1	L1	1.35	Gravity	Fixed value
2	L2	1.35	Distributed forces	Variable
3	L3	1.50	Passenger load	Variable
4	L4	1.35	Static loads	Fixed value
5	L5	1.50	Dynamic loads	Fixed value

Load Cases. There are several verification criteria to be considered due to the different load cases. Load cases must be considered according to the requirements of Table 4.

Table 4 Load cases

No	Load case	Combination	Comment
1	LC1	L3	Deflection verification by EN115
2	LC2	1.35(L1+L2+L4) +1.5(L3+L5)	Criteria for design and research

Results and discussion

The truss shall be designed for rigidity and strength to carry the passenger capacity load and machinery components. By using FEA, we can predict many properties of the truss, for example, the stiffness prediction denoted by the displacement and the strength prediction denoted by the stress.

Beam Strength Prediction. The location and value of the maximum stress of the beam are shown in Table 5. As seen in Table 5, the maximum Von Mises is 232MPa, which is less than 235MPa, the yield limit of Q235. Hence the entire beam structure is safe and reliable under load case LC1 and LC2. But under load case LC2, the tension stress is near the safety limit of 235MPa. This should be noted and observed.

Soffit Plate Strength Prediction. The soffit plate is 3mm thick sheet steel plate. The strength of the soffit plate can be predicted from the FEA results. The location, value and distribution of the maximum stress of surfaces are shown in Table 6. As seen in Table 6, the entire soffit plate structure is safe and reliable at either load.

Table 5 Beam stress

LA	Tension stress [MPa]			Max. shear stress [MPa]		
	C	A	R	C	A	R
LC1	72	235	0.41	36	135	0.27
LC2	232	235	0.99	116	135	0.86

Note: C=calculated value, A=allowable value, R=stress ratio. The following forms have the same meanings.

Table 6 Surface stress

LA	Tension stress [MPa]			Max. shear stress [MPa]		
	C	A	R	C	A	R
LC1	62	235	0.26	30	135	0.24
LC2	217	235	0.92	100	135	0.86

Truss Stiffness Prediction. The location, value and distribution of the maximum displacement of the truss are shown in Table 7. As seen in Table 7, the maximum displacement occurred in the middle of the truss. This is in line with the actual real-world situation.

Table 7 Truss displacement

LA	Location	Max. displacement in X [mm]	Max. displacement in Y [mm]	Max. displacement in Z [mm]
LC1	Incline	4.1	-14.2	-1.0

According to escalator standard EN115, the stiffness of the escalator truss can be analyzed under load case LC1, that is, the maximum deflection is less than 1/1000 of the distance between two supports.

Under load case LC1, the maximum displacement is 14.2mm. The span of the escalator truss is 16860 mm. So, 14.2mm< 16860mm/1000 =16.9mm. The truss can satisfy the stiffness requirement according to EN115.

Identification of the Key Parts of the Truss and Their Internal Force Prediction. From the FEA results and actual real-world conditions, there are two critical regions in the truss. One is the supporting angle steel, which is the part that connects the escalator with the environment and bears the entire weight of the escalator. The supporting reaction force of the supporting angle steel is at its maximum. Another is the splice at which different inclined sections are connected for the convenience of the transport. The displacement here is at its maximum. This should be focused on for preventive maintenance as this structure is very weak.

The maximum internal force of the supporting angle steel is 138kN. Therefore, the safe load limits are set as 138kN in the escalator truss structure experiment.

Escalator truss structure experiment

Because the whole escalator truss structure is too large to do an experiment to verify the results, the key parts of the truss will be considered instead. In order to verify the accuracy of the FEA model, the experiment is carried out: strength tests of the supporting angle steel.

Strength Test of Supporting Angle Steel. The aim of this test is to verify the strength of supporting angle steel; to determine if it can satisfy the design requirements and stand up under its maximum carrying capacity. The test assembly for the supporting angle steel is shown in Fig. 3. In the test device, one force sensor (denoted as FS-1), five displacement sensors (DS-n, n=1,2,…5), one displacement data acquisition instrument and one load reading instrument are used.

The results of this experiment are as follows:

When F1 is no more than 138kN, the load and displacement curves of sample No.1, 2, or 3 do not make an abrupt change. This shows that the strength of the supporting angle steel is adequate at this load.

When F1 is greater than 138kN, sample No.1 makes an abrupt change at about 300kN as it experiences plastic strain. Sample No.2 and 3 do not make an obvious abrupt change.

The data shows that the strength of the supporting angle limits at slightly higher than 300kN. The safety factor of the supporting angle is calculated to be 2.17 through dividing 300kN by 138kN.

The Results Comparison between the Numerical Calculation and the Experiment. From the numerical calculations, the maximum force that the supporting angle can withstand is 138kN- this is equivalent to an applied load of 138kN on an equivalent structure. The corresponding relationship between displacement sensors and the truss numerical model is shown in Fig.4.

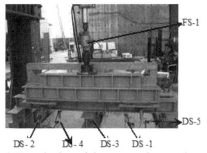

Fig.3 Test device of the supporting angle steel

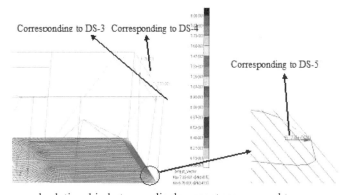

Fig. 4 Correspond relationship between displacement sensors and truss numerical model

As seen in Fig. 4, with numerical calculation, the results of displacement corresponding to DS-3, 4, and 5 in the same direction are 0.733mm, 0.675mm, and 6.74mm respectively. The comparison between the strength test of the supporting angle and numerical calculations can be referred to Table 8, where the maximum error is 8.9% between the experimental results and numerical calculations.

Table 8 Comparison between the experimental and the numerical calculations

	DS-3[mm]	DS-4[mm]	DS-5[mm]
Sample No.1	0.59	0.57	/
Sample No.2	0.75	/	6.48
Sample No.3	0.79	0.67	8.11
Average	0.71	0.62	7.3
Numerical calculations	0.733	0.675	6.74
Error percent	3.2%	8.9%	7.7%

Conclusions

This paper has conducted FEA on the truss of a heavy duty escalator. Based on the FEA results, the performance of the escalator truss is evaluated as follows:

The maximum Von Mises of the beam under load case LC2 is 232MPa, which is near the yield limit of Q235. This should be noted and observed.

The maximum stress of the soffit plate under load case LC2 is 217MPa, which is less than the yield limit of Q235. So the entire soffit plate structure is safe.

The maximum displacement of the truss is 14.2mm under load case LC1, which is less than 1/1000 of the span of the escalator truss. The truss can satisfy the stiffness requirements required by EN115.

The supporting angle steel and the splice are relatively critical regions of the truss. The maximum internal force of the supporting angle steel is 138kN and that of the splice is 247kN.

The numerical calculation has been validated through experiment. The result of the experiment shows that all test samples meet the requirements for use in the truss. The safety factor of the supporting angle is 2.17. The maximum error between the experimental results and numerical calculations is 8.9%, which is within the acceptable range according to industrial experience.

References

[1] M.Y.H. Bangash, T. Bangash, Lifts, elevators, escalators and moving walkways/ travelators, Taylor & Francis, London; New York, c2007.

[2] G.C. Luh, C.Y. Lin, Optimal design of truss structures using ant algorithm, Struct. and Multidiscip. O., 36, 4 (2008) 365-379.

[3] I.A. Azid, A.S.K. Kwan, K.N. Seetharamu, A GA-based technique for layout optimization of truss with stress and displacement constraints. Int. J. Numer. Methods Eng., 53, 7(2001)1641-1674.

[4] F.W.H. Wong, P.T.I Lam, E.H.W. Chan, Optimising design objectives using the Balanced Scorecard approach, Design Studies, 30, 4(2009)369-392.

[5] Y.H. Qiao, C.M. Zhu, Points to need attention in design of escalator metal structure, Hoisting and Conveying Machinery, 7(2003)18-21.

[6] Q. Zhang, D. B. Zhu, Y. He, Finite element analysis of metal structure of escalators, Hoisting and Conveying Machinery, 2(2008)73-76.

[7] X.M. Fang & C.M. Zhu, Research for the light-weight methods for the metal structure of escalator. J. Mech. Strength, 30, 3(2008)433-436.

[8] Q.S Wu, The checking of the bolted connection of the truss sticky segments of the escalator. China Elevator, 13, 4(2002)20-24.

[9] I.A. Azid, A.S.K. Kwan, K.N. Seetharamu, The effect of radius/height ratio on truss optimization. Comput. Struct., 82, 11-12(2004) 857-861.

[10] Y. Liu, J.H. Li, & Q. Zhang, Escalator chain drives simulation analysis. Hoisting and Conveying Machinery, 5(2008)70-73.

[11] C.M. Zhu & Z.Y. Hong, Safety problem with both sides of escalator steps. Hoisting and Conveying Machinery, 6(2000)28-30.

[12] Z.F. Hu, L.Y. Yan & G.M. Liu, The force factor analysis and stress calculation of the escalator's drive. Acta Agriculture Universitatis Jiangxiensis, 26, 8(2004)647-649.

Advanced Materials Research Vol. 819 (2013) pp 65-70
© (2013) Trans Tech Publications, Switzerland
doi:10.4028/www.scientific.net/AMR.819.65

Precision straightening method of thin-walled seamless steel pipes

Li Lianjin[1, a] and Yang Jia[1,b]

[1]School of Mechanical Engineering, Tianjin University of Commerce, China

[a]Lilianjin@tjcu.edu.cn, [b]984932975@qq.com

Keywords: Oil casing; Elastic-plastic theory; Straightening; Flattening rate

Abstract. With the highest product performance, quality and reliability requirements, oil casing is the most demanding service conditions in the oil and gas fields. However, when solving with thin-walled seamless steel pipes by the existing straightening theory, the cross section distortion phenomenon shall not be considered, and when the proportion of diameter to wall thickness increases, an ideal straightening effect can't be obtained. With the elastic-plastic bending straightening theory, this paper analyzes straightening process of thin-walled seamless steel pipes, studies the relationship of the flattening rate, deflection and straightening force in thin-walled pipes straightening, deduces the best reduction calculation method in the practical six roller skew rolling straightening machine, and in the meantime, results of several steel pipes flattening rate are compared.

Introduction

The oil casing pipe straightening generally uses the method of skew roller straightening[1]. Mechanics of materials methods is commonly used to calculate straightening flattening rate. Sometimes bending straightening and flattening correction synthesis pressure is too large, leading to the contact stress between the roll and the casing beyond the yield limit of the pipes body, then generating the spiral indentation and surface depressions in the surface of thin-wall pipe. The poor stability of the cross-sectional structure of the thin-walled pipes can deepen the defect, seriously affecting the dimensional accuracy and surface quality of the sleeve. To this end, scholars have studied the theory and practice of straightening reduction calculation. For example, Zhu Meizhen et al studied the deflection model of straightening using elastic-plastic bending theory, and proposed the calculation of flattening rate to the six inclined roller pipes straightening machine[2]; Li Lianjin researched the relationship between straightening flattening rate and error correcting, to solve the optimal selection of straightening flattening rate[3]. But due to the cross-section distortion of the thin-walled pipes involving steel pipe material nonlinearity and contact nonlinearity, theoretical research and flattening rate calculation of straightening process are full of difficulties, and the surface defects of thin-walled pipes caused by cross section distortion are not considered in the literature. And therefore, the existing straightening method is needed to be refined, feasible precision straightening procedures to be formulate in order to avoid thin-walled cross section distortion phenomenon appeared in the process of oil casing pipe straightening.

Theoretical analysis

After casing's entering into straightening rolls, bending and flattening deformation occur in the roll under pressure. With ante displacement of the casing's screw, the rolls length of contact is fully straightened and rectified within the scope of the local bending and elliptic. Casing in the force situation of six roller straightening machine is shown in Fig.1 Diagram, P_1, P_2 and P_3 are respectively different roll straightening force, M for the straightening torque.

In straightening process, the casing is straightened by the method of combining the bending straightening with correction of rolled flat ellipse, including bending straightening for making the steel pipe longitudinal evenly bending by repeated straightened, and crushing the steel pipes repeatedly flattening deformation in order to correct and improve the straightness and roundness.

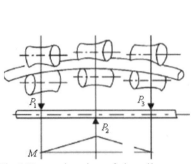

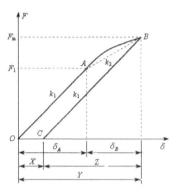

Fig.1 Force situation of six roller
straightening machine

Fig.2 The theoretical model of
straightening process

The calculation of the optimum straightening flattening rate. In the process of straightening casing, three pairs of straightening rolls unit constitutes a three point bending deformation, and the elastic plastic sleeve straightening process model is shown by the load deflection curve in Fig.2. Straightening process is divided into three stages: OA section for elastic loading stage, A is the points of maximum elastic deformation of the material; AB section for the elastic-plastic deformation stage, unloading after loading to point B; BC for elastic rebound phase, after unloading steel bushing elastic rebound to point C. According to Hooke's law, the slope of elastic rebound and elastic loading phase are the same, with k_1 represented. As shown in Fig.2, the best condition of straightening casing is that the initial bending amount X is equal to the length of OC segment, corresponding to the straightening flattening rate as Y.

Shown in Fig.2, the geometric relationships can be derived for the formula of the optimum straightening flattening rate.

$$Y = \frac{k_1}{k_1 - k_2} X + \frac{F_1}{k_1} \tag{1}$$

Where F_1—the Maximum elastic deformation of the load;

k_1, k_2—respectively for the slope of OA, AB segment.

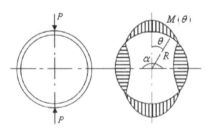

Fig.3 Force situation of flattened oval correction

Best corrected calculation of the flattening rate. The stress of the casing corrected is shown in Fig.3. Under the effect of clamping force produced by the pair of rolls, the casing occurs elliptic deformation partly, driving the greater correction of deformation of the tubular body in the direction of the major axis of the ellipse. As seen in Fig.2, in radians α range M (bending moments) can be expressed as:

$$M(\theta) = P \cdot R \left[\frac{1}{2} - \frac{\cos\left(\alpha/2 - \theta\right)}{\sin\left(\alpha/2\right)} \right] \tag{2}$$

Where P—The clamping force acting on the cross section of unit width steel ring;
$\quad\quad\alpha$—The distribution of the angle of the axis of symmetry;
$\quad\quad\theta$—Radians α range at any angle.

The calculation of flattening rate

Applying the elastoplastic mechanical methods and mathematical means, the relationship can be determined theoretically between the bending moments, the force and the deflection in the process of straightening, and the best flattening rate can be obtained. However, when straightening and correction jointly act on the casing surface, the role of resultant force sometimes exceeds the yield limit result in appearing the deboss on the casing surface, which can easily occur in the maximum force. Therefore, in a traditional way, straightening and corrective flattening rate shall be respectively calculated by elastoplastic mechanical method, the contact stress of rolls and thin-walled pipes shall be verified, and the existing straightening theory shall be perfected in order to ensure the dimensional accuracy and surface quality of the thin-wall pipes.

The times of straightening. Provided the distance between roll and casing was L, and when casing rotating, forward lead was $S = \pi d \cdot \tan\alpha_d$ (α_d was the the angle between the corrected centerline and roll axis). When casing rotated one week, bending deformation occurred twice, and the times of bending straightening were $i_1 = 2L/S$; When the flattening deformation of casing cross section occurred four times, the straightening times were $i = 4L/S$.

The calculation method of flattening rate. (1) Bending deformation curvature

The process of casing straightening is equivalent to simply supported beam withstand the concentrated force, and the bending deformation curvature y″ meets (3):

$$y'' = \frac{M(x)}{E \cdot I} \tag{3}$$

Where y—the deflection at x, mm; $M(x)$—the bending moment at x, N·m; E—Elastic modulus, MPa; I—Material cross-sectional moment of inertia, mm4.

(2) Elastic limit deformation curvature

In order to analyze the different flattening rate impact of straightening quality, it's necessary to determine the curvature of the initial deformation of the plastically deformed. When straightening the casing, the elastic limit state of casing is the starting point of the plastic deformation. The casing is still in the elastic state at the start of the plastic deformation, and according to the principle of functionally equivalent elastic deformation, limit deformation curvature can be solved, and it is:

$$C_e = \frac{2\sigma_S}{E \cdot J \cdot d} \tag{4}$$

Where d—the outer diameter of the casing, mm; σ_S—the elastic limit of the casing, MPa.

The deflection of elastic limit δ is:

$$\delta = \frac{\sigma_S \cdot l^2}{6E \cdot d} \tag{5}$$

Where l—the bearing distance of roll.

The rolling force of straightening is:

$$P = \frac{48J \cdot \delta \cdot \sigma_E}{l^3} \tag{6}$$

Where J—Moment of inertia in thin-walled pipes, mm⁴; σ_E—Yield limit of casing material, MPa.

(3) Anti-bending rate

The flattening rate of casing straightening is obtained by adjusting the relative locations of the feed and flattening roller pair in the middle of straightening machine, and associated with the straightening sleeve is anti-bending rate. According to the curved state of the sleeve in the straightening machine, we can determine the relationship between the flattening rate and curvature C_W for:

$$C_W = 12y/l^2 \tag{7}$$

Where y—Flattening rate of straightening roll.

(4) Residual curvature

When straightening sleeve, deformation residual curvature ΔC_1 calculated by:

$$\Delta C_1 = C_0 \pm (C_W - C_e) \tag{8}$$

Where C_0—Sleeve original curvature; C_e—Variable quantity of the deformation curvature.

By the formula (8) can be obtained straightness error Δ as follows:

$$\Delta = \Delta C_1 \cdot l^2 / 12 \tag{9}$$

The calculation method about correction flattening rate. (1) Corrective torque and the elastic flattening force

The original curvature of the sleeve cross-section is $2/d_0$, and corrected sleeve will occur elastoplastic deformation on the effect of straightening rolls. The relationship of the corrective torque and the curvature variation is:

$$M = EJ \cdot \left(\frac{1}{d} - \frac{1}{d_0} \right) = EJ \cdot \Delta C \tag{10}$$

Where ΔC—The curvature variation, 1/mm; d—The (actual) outer diameter of the casing, mm; d_0—The nominal outer diameter of the casing, mm.

According to the internal force balance of casing bending deformation, the maximum elastic flattening force can be derived by means of elastoplastic mechanical methods.

$$F = 1.08bt^2 \cdot \sigma_E / d \tag{11}$$

Where F—The maximum elastic flattening force; b—Contact width of casing and straightening roll, mm; t—The casing wall's thickness, mm; d—The casing's outer diameter, mm.

(2) Elastic limit flattening rate

According to the principle of functionally equivalent elastic deformation, elastic limit flattening rate δ_e of standard casing can be solved, and it is:

$$\delta_e = 0.2325 \cdot \frac{\sigma_S \cdot (d_0 - t)^2}{E \cdot t} \tag{12}$$

(3) The original curvature-variation of casing

Provided casing before correction is elliptical, the diameter difference of the long and short axis is $\triangle d$, and the original curvature respectively is:

$$\Delta C_{01} = \left[\left(1 - \frac{\Delta d}{2d_0} \right) \middle/ \left(1 + \frac{\Delta d}{2d_0} \right)^2 - 1 \right] \cdot \frac{2}{d_0} \qquad \Delta C_{02} = \left[\left(1 + \frac{\Delta d}{2d_0} \right) \middle/ \left(1 - \frac{\Delta d}{2d_0} \right)^2 - 1 \right] \cdot \frac{2}{d_0} \tag{13}$$

(4) The variation of Anti-bending rate

Provided casing is substantially oval in the correction process, and the ratio of the quantity is approximately 0.92 between the increase amount of the horizontal direction diameter with the vertical direction of the flattening rate. Therefore, if the vertical direction flattening rate in correction is δ_c the increase amount of the horizontal direction diameter is $0.92\delta_c$. Thus, the minor axis and long axis respectively are ($d_{max} - \delta_c$) and ($d_{min} + 0.92\delta_c$), and when it is corrected for relative standard pipe of the anti-bending rate, the variation of initial bending rate are respectively:

$$\Delta C_{w1} = \frac{2}{d_{w2}} - \frac{2}{d_0} = \frac{2 \cdot (d_{min} + 0.92\delta_c)}{(d_{max} - \delta_c)^2} - \frac{2}{d_0} \qquad \Delta C_{w2} = \frac{2}{d_{w1}} - \frac{2}{d_0} = \frac{2 \cdot (d_{max} - \delta_c)}{(d_{min} + 0.92\delta_c)^2} - \frac{2}{d_0} \tag{14}$$

(5) The curvature-variation of total deformation

The curvature-variation of total deformation are the algebraic sum between the original curvature and the variation of anti-bending rate (calculation needs to be the difference between both positive and negative direction).

$$\Delta C_{21} = \Delta C_{01} \pm \Delta C_{W1} \qquad \Delta C_{22} = \Delta C_{02} \pm \Delta C_{W2} \tag{15}$$

(6) The variation of residual curvature

The variations of residual curvature are:

$$\Delta C_{C1} = \Delta C_{21} - \Delta C_y \qquad \Delta C_{C2} = \Delta C_{22} - \Delta C_y \tag{16}$$

Where ΔC_y—The curvature-variation of elastic recovery, 1/mm.

(7) The variation of residual ovality

Put the formula (16) into the formula (13), and the relationship corrected between residual curvature and ellipse axis diameter difference, is:

$$\Delta C_{C1} = \Delta C_{01} = \left[\left(1 - \frac{\Delta d_{01}}{2d_0} \right) \Big/ \left(1 + \frac{\Delta d_{01}}{2d_0} \right) - 1 \right] \cdot \frac{2}{d_0} \tag{17}$$

According to the formula (17) after simplification, the residual ellipse axis diameter differences of casing corrected are:

$$\Delta d_{01} = -\frac{d_0^2 \cdot \Delta C_C}{3 + d_0 \cdot \Delta C_C} \qquad \Delta d_{02} = \frac{d_0^2 \cdot \Delta C_C}{3 + d_0 \cdot \Delta C_C} \tag{18}$$

The minimum pressure of casing surface depression. When rolls contact thin-walled pipes, the contact stress p0 can be calculated by the Hertz theory.

$$P_0 = \left(\frac{24P \cdot E^2}{\pi^3 \cdot d^2} \right)^{1/3} \tag{19}$$

Where: P_0—The contact load between rolls and thin-walled pipes, MPa.

In the process of straightening roll, the maximum shear stress in the contact stress field occurs at the bottom of casing surface in the intermediate roll. By Tresca criterion, the yield limit is 1.6 times than the material yield stress. The minimum pressure of casing surface depression is:

$$P_0 = \frac{\pi^3 \cdot d^2}{24E^2} \cdot (\sigma_E)^3 \tag{20}$$

In the straightening process of thin-walled pipes, resultant force between straightening force by bending-straightening flattening rate calculated and corrected force by flattened oval corrected flattening rate is lower than the minimum allowable pressure in the Hertz stress, which can avoid the sunken surface in the thin-walled pipes.

Example of flattening rate determined

The current thin-walled pipes straightening flattening rate procedures are by reference to the method given by the DEMAG company (formula (21)). Flattening reduction procedures are drawn up on the basis of experiments. For the different specifications of the pipes, the maximum allowable flattening-reduction rate are shown in Table 1, while Flattening reduction is calculated by formula (22):

$$K = 0.36 \frac{\sigma_S \cdot l^2}{E \cdot d} \tag{21}$$

$$\delta = \varepsilon \cdot d \tag{22}$$

Table 1 The current allowed maximum flattening reduction rate %

diameter / mm	wall thickness / mm					
	<6	6~8	8~10	10~12	12~15	16~20
≤299	3.5	3.0	2.5	2.0	1.6	1.0
>299~426	-	2.5	2.0	1.7	1.3	-
>426	-	1.9	1.5	1.2	-	-

The basic parameters of straightening machine: the distance between the straightening roll l=1200 mm, roller length L=660 mm, roll diameter D=480 mm, rolling diameter 114~273 mm, rolling temperature 500 °C, rolling speed 10~72 m/min. In this paper, we study two kinds of different specifications sleeve of rolling process, the material of steel pipes is 29CrMo44V, which mechanical performance are: σ_S=36.92 MPa, σ_E=39.79 MPa, E=21920 MPa. The size of steel pipes are respectively: d_1=219.08 mm, t_1=6.71 mm; d_2=273.05 mm, t_2=10.16 mm.

Using the DEMAG company deflection method, experience flattening reduction calculation, the method of straightening flattening rate and corrected flattening rate which computer optimized, the calculated methods considering revised straightening flattening rate by the effects of hertzian contact stress and corrected flattening rate, flattening rate can be calculated in two kinds of different specifications casing and the results are shown in Table 2.

Table 2 The flattening rate of different specifications casing by different methods mm

pipes diameter	DEMAG straightening	experience correction	calculated straightening	calculated correction	corrected straightening	revised straightening
273.05	12.79	6.83	10.87	5.21	10.20	4.89
219.08	15.94	6.57	12.16	5.13	11.31	4.77

Using straightening (or correct) method proposed in this paper, products' quality testing after more than a year, the results show that straightness error of thin-walled pipes are less than 0.8mm/m, the maximum ovality less than 0.53%, and meet the requirements of use and national standard, and no surface depression.

Conclusion

(1) By applying the elastic-plastic theory, it analyzes the relationship between flattening rate, deflection and straightening force when thin-walled casing is straighten, and deduces the calculation of the optimum straightening flattening rate. Seen from the calculation method, the flattening rate of casing is closely related with its dimensions and mechanical performance.

(2) By the elastic-plastic theory, it makes a quantitative calculation of straightening and corrective flattening rate of thin-walled casing during the process of straightening, obtains the best flattening rate under the condition of guaranteeing resultant force under pressure less than Hertz contact pressure, which provides theory basis for improving the straightening quality of thin-walled casing and formulating operation rules.

(3) By adopting procedure-making methods recommended by DEMAG Company, optimized and revised by computer, it calculates several rolling reductions of thin-walled casing, and by testing instruments, it also puts forward the calculation method of corrective flattening rate that could improve straightening quality, which is directly used in production.

References

[1] Aiwen Cao, Guoliang Xiong. Development of Pressure Straightening Technology[J]. China Metalforming Equipment & Manufacturing Technology, 2007, (1):9-12.

[2] Meizhen Zhu, Yuhau Huang, Chunxiang Wang. Study of pressing flat down quantity of six tilt rollers straightener[J]. Journal of Baotou University of Iron And Steel Technology, 1998, 17(1): 33-35.

[3] Lianjin Li. Study of reduction rate for cross-roll seamless steel tube straightener[J]. Forging & Stamping Technology, 2008, 33(5):82-84.

Advanced Materials Research Vol. 819 (2013) pp 71-75
© (2013) Trans Tech Publications, Switzerland
doi:10.4028/www.scientific.net/AMR.819.71

Research on the Dynamic Characteristics of NC Boring Machine Spindle System Based on Finite Element Analysis

Jiao Feng[1,a]   Sun Guangming[1,b]   Liu Jianhui[1,c]

[1]School of Mechanical and Power Engineering,Henan Polytechnic University,Jiaozuo,454000,China

[a]jiaofeng@hpu.edu.cn,[b]sunguangming001@163.com,[c]jhliu@hpu.edu.cn

Keywords: Spindle, Dynamic characteristics, Bearing dynamic stiffness, FEM

Abstract. In this paper, elastic hydrodynamic lubrication theory was introduced into bearing dynamic stiffness calculation process to take into account the impact of the oil film on the rolling stiffness. Dynamic characteristics of NC boring machine spindle system dedicated to the manufacturing of bearing retainer were studied based on finite element analysis software. The rationality of the structural design of spindle was verified in the paper. The research provided a theoretical basis for the improvement of machine performance and dynamic design of NC boring machine spindle.

Introduction

Spindle is the important parts of a machine tool. Vibration will occur for the affection of both static force and the interference of cutting force when it working. Once resonance was created, machining precision and service life will be affected. So the design of spindle should improve its anti-vibration characteristics as far as possible. The lower order frequencies and mode shapes are the evaluation indexes of the vibration characteristics, so it is necessary to analyze the dynamic characteristics [1].

Bearing stiffness is the basis of analysis of spindle vibration performance. The factors involved in stiffness calculation of rolling bearing are very complex. The present study mostly adopts the method of building approximation model. The contact between the rolling bearing roller and raceway is regarded as a pure Hertz contact in the static state and then the static stiffness of bearing was obtained. However, in order to improve the service life of rolling bearings, lubricating oil in practical work mostly exists, forming the elastic hydrodynamic lubrication state. So the analysis based on pure Hertz contact theory will not be accurate enough. Comprehensively considering between Hertz theory and elastic hydrodynamic lubrication theory, one of the references has put forward a more reasonable calculating method for the radial stiffness of bearing [2]. In this paper, the dynamic stiffness of support bearing in consideration of the impact of the oil film on the rolling bearing was obtained. And the modal and harmonic response of the spindle was analyzed based on finite element analysis.

Dynamic stiffness of bearing

In view of the present problems existing in the research of bearing stiffness, a more reasonable bearing stiffness calculation formula was derived through combining the Hertz theory and elastic hydrodynamic pressure lubrication theory, shown as Eq. 1 [2].

$$K = \frac{1}{\dfrac{1}{K_f} + \dfrac{1}{K_c}} = \frac{1}{0.13 C F_r^{-1.13} + B \ln F_r + A + B} \tag{1}$$

where

$$A = \frac{8.16}{E'Zl}[\frac{1}{\pi}\ln 0.39E'Zl(R_1+r)+1.15-0.5\ln\frac{8.16rR_2}{E'(R_2-r)Zl}]$$

$$B = -6.68\frac{1}{E'Zl}$$

$$C = 0.28\alpha^{0.54}(\eta_0 n_i)^{0.7}r^{0.43}(R_1+r)^{0.7}[(1-\gamma)^{1.13}(1+\gamma)^{0.7}+(1+\gamma)^{1.13}(1-\gamma)^{0.7}]E'^{-0.03}Z^{0.13}l^{0.13}$$

$$\gamma = \frac{r}{R_1+r}$$

The symbols in the above formulas are explained as follows:

K_f --Oil film stiffness of Rolling bearing, $K_f = (0.13CF_r^{-1.13})^{-1}$; K_c --Contact stiffness in the stationary state of rolling bearing, $K_c = \frac{1}{A+B+B\ln F_r}$; r --Radius of roller; R_1, R_2 --Radius of cone and cup respectively; Z--Number of roller; E' --Effective elastic mould, $E' = \frac{E}{1-v^2}$; F_r --Radial force; l --Effective contact length; α --Viscous pressure coefficient of lubricating oil; η_0 --Dynamic viscosity of lubricating oil under atmospheric pressure.

Establishment of the finite element model of the bearing spindle system

The spindle system is composed of the spindle box, spindle, front support bearing, double-directional thrust bearing and back support bearing, as shown in Fig. 1, where the front support bearing and the back support bearing are cylindrical roller bearings. In addition, the double-directional thrust bearing is used in parallel with the front support bearing to bear the big axial force in boring.

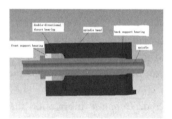

Fig.1 Schematic diagram of the spindle system

The bearing is simplified as an elastic support, simulated with combin14 spring–damper elements during FEA. Four springs are uniformly distributed in the circumferential direction of the each cylindrical roller bearing [3]. Each spring is simulated with one spring–damping element while double-directional thrust bearing is simulated by axial spring-damper elements. After ignoring some less important factors, the mechanical model of spring-damper elements of spindle support are established and shown in Fig 2.

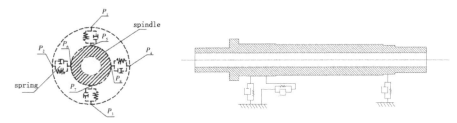

Fig.2 Mechanical model of spring-damper elements of spindle support

The material of the spindle is 45 steel whose density is 7800 kg/m^3, elastic modulus is 2.06E11 Pa, poisson's ratio is 0.3, and solid 95 element is used in the analysis. In the process of applying constraints, node P_1, P_2, P_3 and P_4 should be consolidated fully to limit the axial degree of freedom of P_5, P_6, P_7 and P_8. The finite element model of spindle is shown in Fig 3.

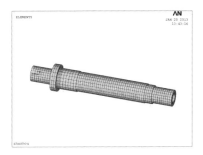

Fig.3 Finite element model of spindle

According to the Eq. 1, the radial stiffness of front support bearing $K_1 \approx 0.95 \times 10^9$ N/m, the radial stiffness of back support bearing $K_2 \approx 0.96 \times 10^9$ N/m, and the shear stiffness of double-directional thrust bearing $K_1 \approx 0.83 \times 10^9$ N/m.

Modal analysis

Block Lanczos method was adopted in modal analysis of the spindle. The first six orders natural frequency and the corresponding mode shapes were obtained, as listed in Table. 1 and shown in Fig 4.

Table.1 The natural frequency of previous sixth-order and vibration mode described

Modal order	Natural frequency /Hz	Description of vibration mode
1	0	Rotating around X axis
2	597.76	Central part bending in YZ plane
3	805.65	Both ends swing in YZ plane
4	1084.9	Central part bending + both ends swing
5	1104.0	Twisting around X axis +Central part bending
6	1478.2	Front end wobbling

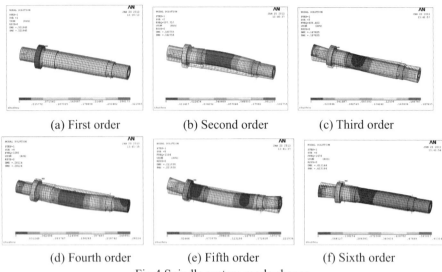

(a) First order   (b) Second order   (c) Third order

(d) Fourth order   (e) Fifth order   (f) Sixth order

Fig.4 Spindle system mode shapes

Analysis of results: It can be seen from the Fig.4 that first-order frequency is equal to 0 Hz, which means that the spindle can rotating around the central axis as a whole, and axial movement and the radial bending do not happen. The fourth order frequency is close to the fifth, which can be explained that the structure of spindle is symmetrical.

Relationship between spindle speed and frequency is expressed as $n=60f$ [5]. So the natural frequency of the spindle can be converted to the critical speed and the results are shown in Table. 2.

Table. 2 First six orders critical speed

Order	1	2	3	4	5	6
Frequency /Hz	0	597.76	805.65	1084.9	1104.0	1478.2
Speed /(rpm)	0	35865.6	48339	65094	66240	88692

Analysis of results: When the spindle working, the speed range is 800-4000 *r/min*, spindle speed can be seen outside the speed of each order in the Table.2, avoiding the resonance frequency effectively. The analysis result illustrates that structural design was reasonable and the machining accuracy can be guaranteed effectively.

Harmonic response analysis

The spindle system will be in periodic vibration state under the impact of cutting forces when it works [6], so it is necessary to make a harmonic response analysis of the spindle system. In consideration of the processing characteristics of this boring machine tool for bearing retainer, the key point of the harmonic response analysis should be a certain node on the front face of cutting tool. Based on modal analysis results, parameters are set as frequency range 0-1500 Hz, spindle cutting force 1951 N, torque 36.83 N·m. The loads are applied to the front spindle surface. Then the response displacement curve of the key node in X, Y and Z directions will be obtained as shown in Fig5.

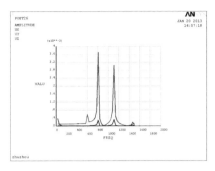

Fig.5 Amplitude-frequency characteristic curve of the pitch point

In Fig 5, the black curve is amplitude-frequency characteristic curve in direction UX, while the red curve is in UY direction, and the blue one is in UX direction. Through the amplitude-frequency characteristic curves, it can be seen that there are two peaks in both directions of UX and UY within the frequency range 0-1500Hz, and the peak frequencis are corresponding to the third order natural frequency 805.65Hz and fourth-order natural frequency 1084.9 Hz. It also can be seen that the displacement in UZ direction may be negligible. Spindle's maximum speed is much lower than the third and fourth-order natural frequency, so there will be no resonance when it works.

Summary

Dynamic characteristics of NC boring machine spindle system dedicated to the manufacturing of bearing retainer were studied based on finite element analysis. In order to improve the analysis accuracy, elastic hydrodynamic lubrication theory is introduced to bearing dynamic stiffness calculation process to take into account the impact of the oil film on the rolling stiffness. Modal analysis result shows that the resonance will not occur during the spindle speed range; harmonic response analysis result proves that the design of spindle structure is reasonable and the machining accuracy can be guaranteed effectively. The research results provide a theoretical basis for the optimization design of spindle system.

Acknowledgments

The paper is sponsored by Program for Science & Technology Innovation Talents in Universities of Henan Province (No. 2010HASTIT031) and Henan Provincial Key Discipline.

References

[1]Wen Huaixing, CNC milling machine design. Chemical Industry Press, Beijing, 2006.

[2]Wu Hao, AN Qi, Calculation on Stiffness of Cylindrical Roller Bearing with EHL, J. Bearing.(1) 2008, 1-4.

[3]Zhang Yaoman, LIU Chunshi, Study on the Finite Element Modeling Method of Spindle Assemble of High Speed NC Machine Tool, J. Design and Research.(9)2008, 76-80.

[4]LI Yan，LI Huiqin， Dynamic analysis to the spindle of plane miller for carbon block based on FEA, J. Machinery Design & Manufacture. (1)2011, 64-66.

[5]Hu Shijun, Han Jian, Dynamic Characteristic Analysis of Crankshaft of C14125 Connecting Rod Based on Ansys, J. Coal Mine Machinery.(2)2012, 95-96.

[6]Tang Hengling, Machine tool dynamics,China Machine Press,BeiJing, 1983.

Advanced Materials Research Vol. 819 (2013) pp 76-80
© *(2013) Trans Tech Publications, Switzerland*
doi:10.4028/www.scientific.net/AMR.819.76

Research on Thermal Error Compensation Instrument Based on Thermal Modal Analysis for NC Machine Tools

Bo YANG[1], Yi WANG[1], Wenli YU[2], XinhuaYAO[1],Jianzhong FU[1,a]

[1]Department of Mechanical Engineering, Zhejiang University, Hangzhou 310027, China
[2]College of Information Engineering, QuZhou College of Technology, QuZhou 324000, China
[a]fjz@zju.edu.cn

Keywords: thermal modal analysis, PLC, temperature transmitter, thermal error compensation

Abstract: Great efforts have been made to improve the accuracy of NC machine tools, within which thermal error compensation is one of the most efficient ways. A new thermal error compensation instrument which is based on thermal modal analysis for NC machine tools is introduced in this paper. OMRON'sCJ2M-CPU11 is used as microcontroller, and SAILING TECHNOLOGY's STA-A08 temperature measuring modules as temperature transmitter. Through hardware and software design, high precision and stability can be achieved. By measuring several key points' temperature and making use of a thermal error compensation theory, real-time thermal error compensation can be output to the machine tool, thus thermal error can be reduced.

Introduction

Errors due to the increment in the temperature of the machine tool elements cause relative displacement between the workpiece and the tool during the machining process and finally have an influence on the accuracy of the work being produced, known as the thermal errors [1]. The thermal deformation error accounts about 40%~70% of the total error in machine tools [2].

Error compensation technology is a key technology to improve the machining precision. In order to effectively implement the error compensation, an appropriate error modal is indispensable. Lots of methods have been developed, including finite element method [3], time series model [4, 5], and etc. Since it is hard to obtain the thermal error between the tool tip and the workpiece directly, we often use some empirical models which relate the thermal errors to the temperature field of the machine have been pre-established and then the thermal errors can be estimated on line using the empirical models by monitoring the machine temperature field [6].

This paper develops a digital temperature measurement and compensating system based on PLC. The PLC controller gets the key points' temperature values of the machine tool through Pt. thermal resistance. The PLC will calculate the thermal error and sends the corresponding thermal error compensation values to CNC machine tools.

Thermal Modal Analysis

As long as the heat source remains unchanged, the temperature field and the thermal displacement of a machine tool won't change. The discrete heat conduction equation of a solid system is

$$[c]\{\frac{dT}{dt}\}+[h]\{T\}=\{p\} \tag{1}$$

Where
   $[h]$ is the thermal impedance matrix,
   $[c]$ is the heat capacity matrix,
   $\{p\}$ is the heat load matrix.

Eq.1 is a differential equation of $\{T\}$.The basic purpose of thermal modal analysis is to obtain a temperature curve fitting by solving Eq.1. In order to get the curve, we should get a Temperature-time curve previously, where our system will come in handy [7].

Technical Requirements of Thermal error compensation System
According to the mechanism of thermal error compensation, the thermal error compensation System basicity has three tasks. At first, it measures the temperature of key points on the machine tool. Secondly, it calculates the thermal error with the compensation algorithm. Finally, the error data are imported to the machine tool's control system. The system shown in Fig.1 represents the system's basic compensation mechanism.

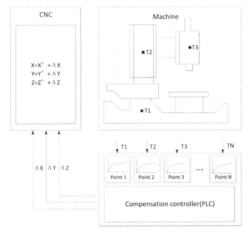

Fig.1Thermal error compensation mechanism(3-Axis machine tool as an example. T: temperatures on key points; ΔX: compensation value of axis X derived from embedded compensation algorithm)
Measurement of the temperature. The temperature transmitter we choose has 8 data input ports. Each of them is connected to a PT100 temperature sensor. The transmitter can output the temperature values with ASCII data. The temperature data will be converted into floating format for subsequent processing, during which the thermal error compensation mechanism will be applied and compensation values will be output into the machine tool.
Thermal error compensation mechanism. The temperature values obtained by temperature sensors which are arranged on the key points of CNC machine tools are processed by the temperature transmitter and then sent to the PLC through RS485 interface.

The thermal errors and the squareness errors are scalar thermal error components and they have been modeled as [6]

$$\delta = a_0 + \sum_{i=1}^{n} b_i \Delta T_i + \sum_{i=1}^{n}\sum_{j=1}^{n} c_{ij}\Delta T_i \Delta T_j + ... \qquad (2)$$

Where
  δ is the thermal error,
  a_0, b_i, c_{ij} are the coefficients of the regression model,
  ΔT_i, ΔT_j: temperature rising at some key points on the machine.
The quadratic term in the formulation, meaning the $\Delta T_i \Delta T_j$ term is relatively negligible comparing to the terms before it, so we can conveniently omit it, and the formulation will become

$$\delta = a_0 + \sum_{i=1}^{n} b_i \Delta T_i \tag{3}$$

Instrument Design and Implement

Hardware design. The hardware design is shown in Fig.2. To guarantee that the system has a reliable performance, we choose OMRON's CJ2M-CPU11 as microcontroller, two temperature measurement modules as temperature transmitters and a touch screen is introduced for human-computer exchange.

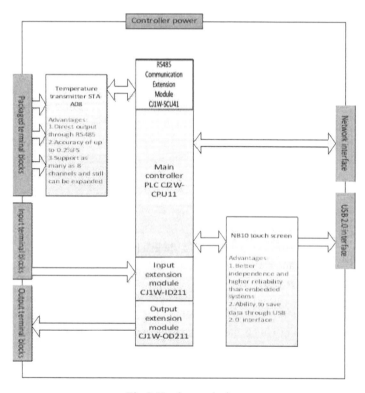

Fig.2 Hardware design

Software design. The idea mentioned above can be summarized as following steps:

1) Measuring the temperatures.
2) Data communicating between the PLC and the temperature transmitter.
3) Data processing and format converting.
4) Calculating compensation values.
5) Outputting the compensation values.

According to the steps above, the PLC program should be divided into corresponding sections [8]. The relationship between each section is shown in Fig. 3.

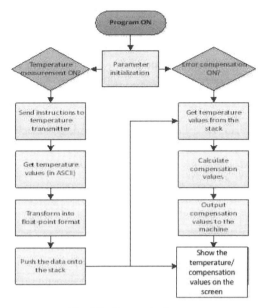

Fig.3 Software design

Instrument implement. The thermal error compensation instrument has an interface as Fig.4. During the experiment process, we find the system performances very well, and the results of experiments show that the system has high precision and stability. Fig.5 shows the setup for the experiment.

During the experiment, we found the operation parameters of the system are as below:

Temperature range: 0~100°C

Compensation range: ±200μm

Resolution: 0.1°C

Sampling period: ≥1s

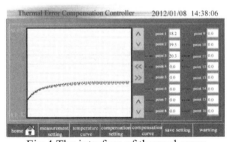

Fig.4 The interface of thermal error
compensation system

Fig.5 The setup for thermal error compensation
experiment

Conclusions

In order to improve the precision of NC machine tools, a new thermal error compensation instrument based on thermal modal analysis for NC machine tools is introduced in this paper. This system has the advantage of high dependability and quick response. By measuring several key points' temperature and making use of a thermal error compensation theory, real-time thermal error compensation can be output to the machine tool, thus thermal error can be reduced.

Acknowledgements

This work was financially supported by the Program for Zhejiang Leading Team of S&T Innovation (No.2009R50008), National Nature Science Foundation of China (No. 51105336), Zhejiang Provincial Natural Science Foundation of China (No.Y1100281) and Educational Commission of Zhejiang Province of China (No.Y200909224).

Reference

[1] R. Ramesh, M.A. Mannan, A.N. Poo, Error compensation in machine tools—a review: part II: thermal errors, Int. J. Mach. Tools Manuf. 40 (2000) 1257–1284.

[2] M. Weck, P. McKeown, R. Bonse, Reduction and compensation of thermal error in machine tools, Annals of CIRP. 44 (1995) 589–598.

[3] B. Denkena, C. Schmidt, M. Krüger, Experimental investigation and modeling of thermal and mechanical influences on shape deviations in machining structural parts, Int. J. Mach. Tools Manuf. 50 (2010) 1015–1021.

[4] Y.X. Li, H.C. Tong, Application of time series analysis to thermal error modeling on NC machine tools, Journal of Sichuan University 37 (2006) 74–78.

[5] S.M. Pandit, S.M. Wu, Time series and system analysis with applications, Wiley, New York Trans ASME. 115 (1983) 472–479.

[6] J.S. Chen, J.X. Yuan, Real-time Compensation for Time-variant Volumetric Errors on a Machining Center. J EngInd. 115 (1993) 472-479.

[7] Z.C. Chen, Research on Thermal modal theory and Thermal State Precision Control Strategy for Machine Tools, Thesis, Zhejiang University (1989).

[8] H.Y. Shen, System of Numeral Temperature Testing Based on ARM and DS18B20, Mechanical & Electrical Engineering Magazine. 22 (2005) 50-53.

Advanced Materials Research Vol. 819 (2013) pp 81-85
© (2013) Trans Tech Publications, Switzerland
doi:10.4028/www.scientific.net/AMR.819.81

Simulation and Experimental Study on Rear Frame Strength of Winch Lashing Car

Yue Jingtao[1,a], Pu Hui[1], Tao Xianghe[2]

[1]Military Transportation Institute, Tianjin 300161, China;.

[2]PostgraduateTraining Brigade, Military Transportation University,Tianjin 300161, China

[a]20082008@vip.sina.com

Keywords: winch lashing car; rear frame; finite element simulation; strength test

Abstract: Heavy general transport vehicles are recommended as lashing points during rescue operations of winch. Taking SX2190 heavy transport vehicles frame as the experimental object, this paper introduces contents and methods of the test, and gets evaluation result of the rear frame through finite element simulation and experimental validation of the real frame-towing hook system for winch lashing car.

Introduction

Heavy general transport vehicles are recommended as lashing points during rescue operations of winch[1]. The winch system should be linked to the rear frame traction hook of the lashing car. The working winch has to provide pulling force of 20t-level up to the highest, so the strength of lashing car's rear frame under such heavy load will affect the winch rescue work directly. This thesis takes SX2190 transport truck as experimental subject, to simulate and compute and test its rear frame strength. The 1:1 model of the tested SX2190 truck rear frame is showed as Fig 1.

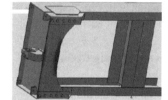

Fig.1 Physical map of rear frame   Fig.2 Solidworks model of rear frame

The frame modeling and strength simulation

The building of frame finite element model. Sketch the SX2190 rear frame, and use Solidworks software build solid model according to the measurement as figure 2. Under the premise of ensuring the accuracy, appropriately simplify the frame's geometric structure, and lead the solid model into ANSYS. Because the frame belongs to composite slab structure, analyze it with shell unit (shell 63), and divide it into unit grid of 40mm. The material of frame is 16Mn low alloy steel, and it's mechanics parameters is as following: Poisson's ratio, $\mu = 0.3$; Modulus of Elasticity, $E = 2.1 \times 10^5 MPa$; Density, $\rho = 7800 kg / m^3$; yield strength, $\sigma_s = 350 MPa$. The frame takes static stress analysis in pulling, so X,Y,Z direction full displacement restraint should exert on the front of the stringer.

The simulation analysis of the rear frame in static force. The analysis is carried according to the load and constraint conditions in the experiment. For the field condition of the experiment, punch holes of $6 \times \phi 20$ on the left and right stringers, and exert full restraint on the inner wall of the 12 holes. Exert concentrated load of 15t, 20t, 25t, 30t on the traction hook of the frame, and carry the finite simulation computing, we can get the cloud picture of the frame stress and deformation distribution in all kinds of load. The stressed frame and the displacement cloud picture of the frame in 30t overload tension is showed as figure3,4,5.

Fig.3 Frame stress distribution of 30t(external)   Fig.4 Frame stress distribution of 30t(internal)

Fig.5 Frame deformation maps of 30t

With ANSYS to simulate, we can preliminarily learn the distribution of the stress deformation condition on the surface of the frame. After the analysis of rear frame stress distribution picture as figure 4,5 and deformation picture as figure 6, we will learn that when frame is loaded with winch tension, the frame relatively forced position is the hooking holes of the frame rear-beam assembly A and inside-outside flat position of rear beam, and the frame deformation of corresponding positions is more severe. In continuous overload tension of 30t which is the biggest, the frame doesn't has obvious plastic deformation or destruction, while the stress of some parts is relatively high as 478.39MPa, and the maximum distortion energy guidelines stress reaches the yield limit.

The experiment of frame strength[2]

Static test of SX2190 heavy general transport vehicles' rear frame could be compared with the simulation computing to improve the calculation accuracy. With its data, we can improve and optimize frame structure [3].

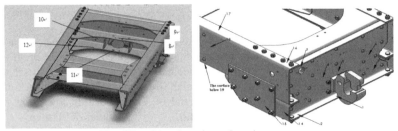

Fig 6 Patching location of strain rosette

Reference standard and equipment. This frame strength test is strictly abides by GBT 6792-2009 "Measure method of stress and deformation for bus skeleton" and GBT 12534-90 "General rules of road test method". The test site is indoor in normal atmospheric temperature. The instruments and appliances involved contain resistance strain gauge, resistance strain rosette, shield cable, steel rule, calipers and stopwatch.

The determination of the location of the strain gages. According to the simulation analysis of finite element software ANSYS, we can preliminarily fix the strain distribution. On the basis of it, pick out the position in relatively high stress and stick foil gauge. The specific position as showed as figure 6. The test result is showed as table1.According to the actual test conditions, both of the last two stretches reach the maximum tonnage 30t.

Table1 Stress values of important test points

	test point1			test point4			test point5		
rally value	Simulation value	test value1	test value2	Simulation value	test value1	test value2	Simulation value	test value1	test value2
15t/equ	120.4	121	132	85.97	72	86	71.514	55	80
20t/equ	177.59	173	193	112.09	101	118	98.102	81	113
30t/equ	268.2	335	353	168.14	164	193	139.6	159	195

	test point8			test point11			test point12		
rally value	Simulation value	test value1	test value2	Simulation value	test value1	test value2	Simulation value	test value1	test value2
15t/equ	176.63	-165	-155	183.28	-201	-186	135.95	-148	-152.5
20t/equ	197.29	-220	-207	223.05	-272	-249	145.95	-179	-173.2
30t/equ	342	-332	-316	379.98	-407	-394	196.68	-255	-233

The frame forces are mainly concentrated around the loading position, and the focus analysis measurement points (1,4,5,8,11,12) are shown in Table 1(measured value 1-the first frame after 25t pulling force stretch; measured value 2-the second frame which not pre-pulls).

The stress under maximum distortion energy criterion. The stress value measured by experiment is frame surface stress, but Equivalent stress can better reflect the frame real force stress. The important measuring point's Equivalent stress are counted in Table 2.

Table 2 The key points stress values under fourth strength theory (Mpa)

	test point1	test point4	test point5	test point8	test point11	test point12
rally value	Simulation value	Simulation value	Simulation value	Simulation value	Simulation value	Simulation value
15t/equ	120.4	85.97	71.514	176.63	183.28	135.95
20t/equ	177.59	112.09	98.102	197.29	223.05	145.95
30t/equ	268.2	168.14	139.6	342	379.98	196.68

Table3 Error of important test points (Mpa)

Rally value	test point1			test point4			test point5		
	Simulation value	error1 %	error2 %	Simulation value	error1 %	error2 %	simulation value	error1 %	error2 %
15t	123.43	-2.0	6.5	83.9	-16.5	2.44	87.5	-59.1	-9.38
20t	143.8	16.88	25.5	126.92	-25.66	-7.56	114	-40.74	-0.885
30t	266.5	20.45	24.5	179.19	-9.26	7.16	177.17	-11.43	9.14

rally value	test point8			test point11			test point12		
	simulation value	error1 %	error2 %	simulation value	error1 %	error2 %	simulation value	error1 %	error2 %
15t	-183.6	-11.27	-18.45	-184.02	8.45	1.06	-131.2	11.35	14.6
20t	-210.1	4.5	-1.498	-228.92	15.838	8.06	-142.3	20.5	10.1
30t	-360.5	-8.58	-14.08	-380.51	24.95	22.97	-194.7	23.65	11.8

Processing and Analysis of Experimental Data

Analysis of experimental data. By the simulation and test of ANSYS the distribution of the stress field is basically the same. In the ANSYS simulation, the established model size, the material properties , the loading position and constraints are basically the same as the actual test conditions, which the simulation having a true reference and reliability. Simulation and measured values in the following analysis both are the frame surface stress. Because strain flower ranges larger and value of simulation analysis takes a small value unit, the values of the measuring points are taken many times to calculate average values in order to make the simulation results more reliable. The measured value is the larger value σ_1 of the principal stress determined in accordance with the experimentally measured strain.

According to the plane strain analysis theory, if a point's linear strain to any of three directions is known, its main strain and main direction can be calculated which its main stress can be calculated.

Error analysis. For the measured point of small stressed values, the loaded pulling forces rope is thick among physics experiment, even slightly different direction of the pulling force can lead to a big difference of the two measured values and simulation values. Therefore, only the error range of the larger forced measured points has a valuable reference. The following Table 3 concludes the error range of important measuring points of large stress value. wherein error 1 is the error of the measurement value1 and error 2 is the error of the measurement value2. Positive value represents that the measured value is greater than the simulation value, whereas negative value indicates the measured value is less than the simulation value.

As can be seen from the above analysis that most of the measured values and simulation values are more consistent, but part of the measuring points have a large difference and the reasons are as follows:

Frame model in finite element analysis is made some simplification, such as bolts and rivets all made to simplify;

The constraints between Simulation and actual test have differences; The direction of the force will vary as the load changes in the planar direction and direction of the force is always in the horizontal direction toward the rope straightened in the test, so the direction of the force of the simulation and testing has a slight difference;

Pulling force loads as concentrated loads which are perpendicular to the loading plane in simulation and frame deforms when loaded force is;

The thickness of 502 glue has some influence to test accuracy when stickers strain rosette;

The use of dial indicators and recorders will generate a magnetic field which has an impact on strain gauge resistance which in turn affects the accuracy of some testing points.

Above the testing points, the first testing frame and the second testing frame has a different original state (the first testing frame has loaded 25t pulling force but the second testing frame does not have any pre-pulling) and strain gauge position has small difference in the process of experiment which will have different pulling force (pulling force accuracy is 0.1t) and direction.

Conclusion

As the ANSYS simulation analysis and field trial testing shows that the frame - towing hook system can be able to withstand 20t sustained tensile force which reached the designing strength. In a maximum 30t continuous pulling force, frame has no significant plastic deformation or damage, but part of the region of stress value is a little big which the stress of maximum distortion energy criterion in the simulation reaches the yield limit. The simulation and experimental test achieves the desired objectives which can provide a theoretical and experimental analysis of reference for other vehicles of the same type of simulation.

References

[1] Chang Shuchun,Li Lixing,Wang Jin. The Self-help and Mutual Aiding Methods When Military Wheeled Vehicles Silt Trap[J]. Automobile Application, 2010(5):39-40.

[2]Chen Xianming.The Stress Analysis of Semi-trailer Frame [J]. SPECIAL PURPOSE VEHICLE, 1994(4):17-19.

[3]Zhao Qian,Hao Ziyong.Mechanical Loading Calculation and Measured Analysis of Automobile Frame[J].Tianjin Auto, 1999(4):22-25.

[4] Zhan Junyong, Huang Jianmin .ANSYS Parametric Modeling by Solidworks Importing [J]. MW Metal Cutting, 2010(4):71-72.

Advanced Materials Research Vol. 819 (2013) pp 86-90
© (2013) Trans Tech Publications, Switzerland
doi:10.4028/www.scientific.net/AMR.819.86

Study on the curve reconstructing in the process of blade repairing

Tao Wang[1,a], Yiliu Liu[2,b], Jie Tang[1,c], Hao Wang[3,d] and Liwen Wang[1,e]

[1]Aeronautical Automation College, Civil Aviation University of China, Tianjin, China, 300300

[2]Department of Production and Quality Engineering, Norwegian University of Science and Technology, Trondheim, Norway, 7050

[3]CETO, Civil Aviation University of China, Tianjin, China, 300300

[a]wangtaotdme@163.com, (corresponding author), [b]leoyiliu@yahoo.com.cn, [c]tangjie@cauc.edu.cn ,[d]wanghao@cauc.edu.cn, [e]lwwang@cauc.edu.cn

Keywords: blade; aero-engine; cross section curve

Abstract: the aero-blades are often easy to be damaged because of the severe working conditions, and the invading of out objects. Repairing is preferred to replacing with a new one due to the very high expense. The mathematical model of cross section curve is very important to the process of repairing the blade and to the analysis for the working performance of repaired blade. The method of remodeling the curve of cross section is studied including scanning the broken blade, denoising the data points, analyzing the points of cross section and dividing them into four assembles, and fitting them with adaptive mathematical equations. Lastly, the future research plans are given.

Introduction

Life spans of blades in turbines always play dominant roles in determining the minimum lifecycle costs of modern aero-engines. The aero-blades are often easy to be damaged because of the severe working condition with high pressure and temperature and the invading of out objects such as birds, sand, ice, tephra and so on. Some broken blades collected by the authors are illustrated in Fig.1, including the turbine blades and compressor blades. Although breaks of blades may have critical influence on the flying safety of airplanes, replacements of broken blades with new ones are often unwelcome due to the very high expense, and thus it is often a natural choice for operators to make repairs on those broken components.

According to the survey by Zhang et al. [1], repairs and enhancements of hot components, such as blades, in aero-turbines have been studied since 1970's. In 2003, AROSATEC (Automated Repair and Overhaul System for Aero Turbine Engine Components) project was initiated by EU to improve existing repairs with self-adaption processing technology, and develop a data management system so to connect all maintenance procedures [2]. More recently, an integrated automatic blade repair solution was proposed in reference [3] and [4], where the geometry method is used to reconstruct the profiles of blade tips. In reference [5] and [6], a robotic grinding and polishing system for automatic repair of blades is developed, and an optimized profile fitting method based on the template is introduced. Wang Tao et al [7] have done the researches on obtaining the boundary points of blades with the method of varied step lengths according to curvature, plotting the envelope curves with a third power B spline and establishing the 3-D digital model of blade. In the reference [8], Wang Tao et al have studied a CNC system with controlling multi-axis simultaneously, which can be used to manufacturing or repairing the broken blades.

It is not hard to conclude that most of current researches on blade repair models adapt geometry to reconstruct broken blades. Such a repair cannot establish the parameterized mathematic model of the curve of un-broken cross section of the blade, and thus provides little input to post-repair studies on working performance of the repaired blade. Thus, how to reconstruct the mathematical model of the curve of un-broken cross section of the blade will be proposed in this paper. The paper is organized as following: Data points of experiment blade is collected firstly, and then the noise in the data points is eliminated， lastly the mathematical models of un-broken cross section of blade.

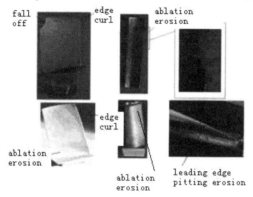

Fig.1 Some broken turbines and compressor blades

Data point collection

An experiment blade in an aero-engine is scanned with the laser tracer.　As shown in Fig 2(a), the upper left corner of the blade has been broken. Fig 2(b) illustrates the original point cloud, which includes amount of noise points. Fig 2(c) is a partially enlarged view of Fig 2(b), showing the data point cloud in the scan for the blade in more details. It is obvious based on Fig 2(b) and Fig 2(c) that the blade is broken in the upper left corner.

a. Experiment blade　　b. Blade point cloud　　c. Partially enlarged view

Fig.2 Experiment blade and its point cloud

Elimination of Noise

Invalid noise points are unavoidable in actual points collected in measurements. These noise points should be eliminated before following operations. The blade point cloud obtained by the laser tracer can be divided into ordered point cloud and unordered point cloud. As a result, denoising should be conducted for these two parts respectively.

Denoising and smoothing of blade point cloud. A space grid method of is used to remove noise in the unordered point cloud, setting a threshold of data points and removing those less than the threshold. Detailed procedures are as following:

Load the file of the blade point cloud, save all 3-coordinate points into an array, and reveal the maximum and minimum values of the coordinate X, Y and Z. Then, a cuboid including all points is formed in parallel with the coordinate. It is further divided into small cube grids according the density of data points. Identify the grid for each point, and record their reference numbers in the linear table corresponding to the grid. Fig. 3(a) illustrates the effect of denoising unordered points

Denoising and smoothing of ordered point cloud. For the ordered point cloud, the smoothing filtering method is widely used for eliminating noise. In this method, there are three algorithms: average filtering, median filtering and Gaussian filtering. Since Gaussian filtering algorithm uses Gaussian distribution for weights in the specified range, with smaller average effects, and therefore it can keep the original shape better while filtering, it is adapted in this study to deal with the ordered point cloud. Fig. 3(b) illustrates the effect of denoising ordered points

a. Denoising unordered points.   b. Denoising ordered points.

Fig.3 The effect drawing of the blade point cloud after denoising.

Curve mathematic remodeling of the cross section of blade

As shown in Fig. 5, the point cloud of a cross section of the blade includes four parts: LE, TE, PS and SS point cloud. LE and TE are circular curves, while PS and SS are polynomial curves. AS a result, the curvatures of LE and TE are constants. On the other hand, curvatures of PS and SS are variables that are much smaller than those of LE and TE. A parabolic equation: $y=ax^2+bx+c$, can be applied to calculate the average curvature. In Fig. 6, curvatures of different parts are compared. It can be found that curvatures on LE and TE are obviously larger than those on PS and SS. In fact, there are some differences in curvatures at LE and TE, due to errors in the measurement and some uncertain factors such as wear in the use.

The fitting equation of the circular curves on LE and TE is

$x^2+y^2+ax+by+c=0$

Number the point clouds on LE as #1, #2, ... For example, one circle can be fitted with #1, #4, #7, the second one is with #2, #5, #8, and the third one is with #3, #6, #9. Following these procedures, all pointed on LE can be assessed, and many circles can be obtained, in which the minimum one is regarded as the base circle of LE. Similarly, the base circle of TE can be found. The point where the base circle is located is the base point set. Extend the base point set forward and backward, and set the deviation of the new point from the fitting circle and two thresholds of fitting errors of the curve. Then, calculate the fitting errors of points *a* and *b* given the requirement of deviation. The point with smaller fitting error will be added in the point set. If only one point satisfies the requirement of deviation, it is included in the point set. When neither of the two points can satisfy the requirement, the fitting process ends, and the two circles obtained are those where LE and TE are located. Their circular curves can be determined by two boundary points in each point set.

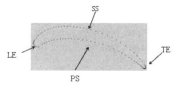

Fig.5 Point cloud of a cross section.   Fig.6 Curvature analyses of curves.

For PS and SS, a 5-order polynomial equation is applied

$$y=a_0+a_1x+a_2x^2+a_3x^3+a_4x^4+a_5x^5$$

Its first order derivative is

$$y`=a_1+2a_2x+3a_3x^2+4a_4x^3+5a_5x^4$$

It is necessary to select 6 known coordinates so to determine the parameters of the polynomial. In addition, since SS (or PS) is tangent with TE and LE, there are two constraints of tangency. Only 4 data points are enough, plus 2 constraints, to determine the 6 parameters. The selection of data points is similar with that for LE and TE, as a result, all points on SS (or PS) can be assessed, and the group of parameters with the smallest fitting errors is the final result. Fig. 7 illustrates the curve of the blade in a cross section with the method mentioned above, where the points are those used in the final fitting curve.

Fig.7 The curve of the blade in a cross section.

Conclusion

The methods of eliminating the noise point are researched after the blade is scanned through laser tracer. Then the data points of a cross section of un-broken part of experiment blade are divided into four gathers: LE, TE, SS and PS, by analyzing the curvature. Lastly, the circular curves equation and the 5-order polynomial equation are adopted to fitting the four gathers respectively. In the future, it will be researched that the mathematical model of the cross section curve of broken part of blade is reconstructing and how the broken part of blade is repaired.

Acknowledgments

This work was supported in part by the Research Starting Funds of Civil Aviation University of China (Grant No. 09QD05S), the Basic Science-research Funds of National University (Grant No.ZXH2011C005) and the National Important Special Funds for Science and Technology (Grant No.2013ZX04001071).

Reference

[1] Zhang S, Hou JB, Li XH. A review on repairs of hot-end parts in high thrust aero-engines. Proceedings of 2007 Summit of Key Techniques of Large Airplanes and Seminar of Chinese Aviation Academy

[2] Information on http://www.arosatec.com/front_content.html

[3] Jian Gao, Xin Chen, Oguzhan Yilmaaz, Nabil Gindy. An integrated adaptive repair solution for complex aerospace components through geometry reconstruction. International Journal of Advanced Manufacture Technology , 2008(36):1170-1179

[4] Oguzhan Yilmaz, Nabil Gindy, Jian Gao. A repair and overhaul methodology for aeroengine components. Robotics and Computer-Integrated Manufacturing, 2009, 8 Advanced Manufacture Technology , 2008(36):1170-1179

[5] H. Huang, Z. M. Gong, X. Q. Chen and L. Zhou, SMART Robotic System for 3D Profile Turbine Vane Airfoil Repair. International Journal of Advanced Manufacture Technology, 2003(21):275-283.

[6] H. Huang, Z.M. Gong, X.Q. Chen, L. Zhou, Robotic grinding and polishing for turbine-vane overhaul. Journal of Materials Processing Technology 2002(127): 140-145

[7] Wang Tao, Liu Yiliu, Wang Liwen, Wang Hao, Tang Jie. Digitally Reverse Modeling for the Repair of Blades in Aero-engines. Applied Mechanics and Materials, 2012,141:258-263

[8] Tao Wang, Liwen Wang, Qingjian Liu. A three-ply reconfigurable CNC system based on FPGA and field-bus. Int J Adv Manuf Technol, 2011,57(5-8):671-682

Advanced Materials Research Vol. 819 (2013) pp 91-94
© (2013) Trans Tech Publications, Switzerland
doi:10.4028/www.scientific.net/AMR.819.91

The Application of Biomimetic Materials on Industry Design Research

Li Xiaodong[1, a]

[1]Tianjin University of Commerce, Tianjin 300134, China

[a]lxd_8111@163.com

Key words: biomimetic materials; industry design; application

Abstract. For thousands of years, people have been in nature to make use of our natural materials, among them some are harder than iron and steel, some are superior than the best optical device, what let human more amazing is that the nature's reasonable application of these materials. Through the study and research of nature, humans create the biomimetic material to meet the demand of industrial design; they were widely used in the field of industrial design according to the different characteristics. Application of these new materials opened up a brand-new road for the development of industrial design.

Introduction

Nature is full of good designs as examples, for the industrial design, they are inexhaustible, inexhaustible "treasure-house of design". The good design choices of the nature, some structure is exquisite, some material is reasonable, function is complete, and embodies the nature of adaptive law on everything. In the early days of humanity, we rely on the natural material to build our home, make clothing, and for various purposes. In the past few years, scientists have started with a new perspective of the nature; they found that through the study of nature, we can find warmer, more solid, more fashion, and more environmentally friendly materials to change our lives. Learning from nature brings the development of industrial design into a new world full of surprises.

Biomimetic materials and industrial design

Biomimetic materials refer to materials that imitate various biological characteristics or features to develop. Usually those that modeled on the operation of the living system and the structure law of biological materials to design and manufacture of artificial materials called biomimetic materials. There is a close relationship between biomimetic materials and industrial design [1]. Through the form ,Industrial design express the appearance of the product characteristics, material as a carrier of the ideology, make products to express more abundant and vivid, no industrial product can exist alone without material. Material as partner of industrial designers, bring inspiration for the designer, the same product use different material will produce different effect, and give people different feelings too. Material is full of spirituality; it can give people endless imagination. So the material in the industrial design is integrated into the design process of thoughts and feelings, making industrial design product become the medium of communication with the customers [2]. Therefore, with the deep understanding and research of human to nature, and constantly learn from the natural survival. Biomimetic materials have become resources for scientists and industry designers to develop and utilize. New biomimetic material application will also open a new path of innovation for the development of industrial design.

Features biomimetic material application in industrial design

All creatures on earth in nature, after millions of years' evolution, through the selection, they have a variety of functions to adapt to the natural environment, and this evolutionary degree is almost perfect [3]. Human modified materials by imitation and innovation of biological function, through the combination of science and nature, create more new materials for human. Scientists have found that some aspects of the function of animal and plant, far more beyond the existing of human science and technology. All kinds of creatures survive in nature run their special functions to thrive in the rough and complex environment; this is a carefully selected result of nature. By imitating their special function in the nature, a bridge between biology and industrial design is set up, it also provide effective design solutions for industrial design.

Beetle live in the Namib Desert, covered in tiny bumps, they can also be in here and have a lot of breeding in the hot desert almost doesn't rain. Their survival is the only water every morning mist rose from the sea. So the first thing to do of these beetles every day is climb to the top of the sand dunes, intercept the water mist. The tiny bump, help them to condense water vapor on the body. Raised tip can capture the water in the desert; the water will be caught along the channel between bump textures to drop into the mouth of the beetle. The survival strategy in the desert, become new ideas for a company to design for refugee camp. These tents designed condense moisture from the air in the morning, so that people in disaster area can drink, even in regions with ground water supplies shortage. Learning law of nature survival way offers help for human's frequent water crisis, as shown in figure 1.

Fig.1 The beetle in Namib Desert

When it comes to the highest efficiency to through underwater, there is no doubt that it is shark; shark's skin makes it seems can get twice the result with half the effort. We will make shark's skin surface zoom in hundreds, you will find that the central part has a ridged bumps, the protruding parts locked a layer of water on the surface of the skin, reduces the resistance when it swims. Nowadays, swimsuit manufacturers adhered something like shark leather puffs on the surface of the swimsuit and analyzes the swimmer's form. Like sharks, setting ridged projection in the most effective location, this design can reduce nearly 4% of the water resistance. At the 2004 Athens Olympics, Australian swimmer, wear this swimsuit for the first time and far ahead in the race, as shown in figure 2.

Fig.2 Shark and bionic sharkskin swimsuit

Visual features of biomimetic material in the application of industrial design

With the excessive use of natural resources by human, some plants and animals were used as goods, making some plants and animals endangered, destroying the ecological balance. So people use artificial biomimetic materials to replace natural materials to meet the demand of people's pursuit of natural material for industrial products. The biomimetic material is changing role gradually in the development of industrial design [4]. For example, the best-selling industrial products such as the leather bags, belts, and handbags made of crocodile, snake skin in the international market, because of the high economic benefits, making these creatures killed. Basing on the protection of nature, according to the skin characteristics and simple sense of these plants and animals, using some new techniques of biomimetic material, industrial designers make the product resembles natural plant and animal skins. This biomimetic material that is able to mass produce, not only meets the demand of consumers to a certain extent, but also protects the natural life at the same time.

As another example, blue flash butterfly, living in the tropical rain forest, shining with light blue wings, like the lighthouse in the tropical rain forest, you can see it in 500 meters outside, this is the result of its evolution, in order to attract mates. Surprisingly, however, there is no existence of the pigment on the wings of a butterfly. Ways for them to make the color like soap bubbles, this flickering color is caused by the emitting light inside and outside of the bubbles surface, and will change as the change of the viewing angle. But the color of the blue flash butterfly is stronger than the color of the bubble; it is the result of its every piece of microscopic structure on the wings' scales. These tiny scales are covered with ridges on the surface and each ridge contains many corneous layers, the stratum cuticle of the spacing is the wavelength of the blue light, so when light irradiates on the corneous layer, only the blue light is reflected. The same spacing between cuticles means continuous blue light, produce interference phenomena. This not only strengthens the light, and to strengthen the color. Japanese manufacturers have applied this principle to the fabric. According to the blue flash butterfly wing scales' corneous layer structure arrangement, they produce a material called "Moore Buddha fiber" made of nylon and polyester fiber. The material glaring interference color that never fades. Now, the cosmetics company has also developed lipstick with the same effect. The revelation that blue flash butterfly brings to humankind, continues to be applied to the industrial design field, as shown in figure 3.

Fig.3 Blue flash butterfly and Moore Buddha fiber materials

Intelligent characteristics of biomimetic materials in the application of industrial design

Nature provides the biomimetic material inexhaustible precious prototypes, new materials imitated by biological characteristics, can even to perception, memory, and performs some behavior of the intelligent materials, and have gradually been applied to the industrial design. That's for pine cones imitation. When pine cones mature, a gradual shift in color from green, purple, yellow, light brown or dark brown. Shortly after the most mature of pine cones, their scales will open, seeds fall off quickly. This phenomenon is discovered by designers, in this research, because the cones scales are constituted by hard fiber material of resin, but from the top, the fiber layers are arranged in a different direction, when the scales after drying, the fibers of the bottom is stronger than the contraction of the top, it split pine cones, and release the seeds. By understanding the working principle, architect finds clothing fiber which has the same performance. When the fiber gets wet because of sweat, part of the fiber will open so that plays the effect of ventilation. This material is applied in industrial design flexibly, as shown in figure 4.

Fig.4 Pine cones and intelligent biomimetic materials

The application of biomimetic material structure characteristics in industrial design

Biomimetic structure materials are mainly to study the problems of how to apply the material that is made according to the internal structures of the principle between organisms and nature in actual design reasonably. Such as the structure of the cellular saves material most, and the capacity is very big, the robustness is rather strong. People adopt all kinds of materials to simulate and manufacturing of the honeycomb sandwich plate structure, light weight, great strength, do not readily conduct heat and sound, the best material to make the spacecraft, the space shuttle, satellites and so on. Like, ceramic has good qualities such as corrosion-resistant, wear-resisting, and high temperature resistant, but its brittleness limits its application in many fields. Scientists have been studying how to improve its toughness, through the study of the shell in nature, and imitate the structure, humans seem to have found the answer. Abalone shell is one of the hardest wear-resisting materials in nature. Scientists compared the abalone shell to brick wall, ultra-thin layers of calcium carbonate as bricks, organic protein layer seems to cement that combine calcium carbonate layers together. Abalone shell strength is not worse than advanced ceramics, but it is not as fragile as ceramics [5]. Through imitating the multilayered structure of abalone shell, scientists produce strong ceramic. These ceramic is not only used in pots among our life, but also has been widely used in industrial field, as shown in figure 5.

Fig.5 Ceramic tool

Conclusions

The biomimetic materials, as correspondence of the human production activities and nature, are gradually promote the industrial design to a new peak. With the in-depth study of human to nature, breaking the conventional mode of thinking, we will open a door to the natural world, and find a more excellent way for the biomimetic material application in industrial design.

References

[1] Fang Yan, Sun Gang, Cong Qian, Ren Lun-quan.Advances in Researches on Biomimetic Materials, J. Transactions of the Chinese society for agricultural machinery,37(2006) 163-166.

[2] Lang Li-juan，Jiang Wen, Innovative Production Design Based on M aterial Characteristics, Journal of Guangdong University of Technology (Social Sciences Edition), 12(2012) 77-78.

[3] Xu Chi-ye, Discussion on bionic design in industrial design, J. The leading journal for enterprise, (2011) 263-264.

[4] Gao Rui-tao, Cao Yu-hua,Zhang Xiao-kai, Analysis on the Role Changes of Material in Industrial Design, J. Packagng engneering,27(2006) 315-317.

[5] Kun ling, from the shape to the spirit likeness, a new generation of biomimetic materials, J.Knowledge is power, 10(2002) 10-11.

Advanced Materials Research Vol. 819 (2013) pp 95-99
© (2013) Trans Tech Publications, Switzerland
doi:10.4028/www.scientific.net/AMR.819.95

The Curved Surface Fitting and Optimization of Scattered Points' Data Based on the Given Surface Tolerance and Fairing

Zhiqiang Zhang[1,a], Wenjin Wang[1,b], Zhang Jing[1,c], Zhao Jian[1,d], Liying Sun[1,e]

[1]School of Control and Mechanical Engineering, Tianjin Institute of Urban Construction, Tianjin, 300384, China

[a]zhiqiangzhang1@163.com, [b]wangwenjin71@yahoo.com.cn, [c]emmyzj@163.com,

[d]zhaojian_tju@yahoo.com.cn, [e]sun_liying@163.com

Keywords: Curved Surface; Fitting; Evaluating function; Optimization; Genetic Algorithm

Abstract: The curved surface simple expression which met a given tolerance and fairing can be obtained under the conduction that the surface tolerance and fairing had been predetermined by the fitting optimization of the scattered points of the origin curved surface. The genetic factor and evaluation function are designed for the optimized expression of surface is acquired. The validity of the solution is verified by comparing the computed error between fitting optimization surface and the original surface through the numerical simulation.

Introduction

The problem of the curved surface fitting and optimization can be described as that the curved spline surface $s(u,v)$ which is expressed as compact as possible (for example: the control points are as few as possible for the use of expressing the curved spline surface) by the given scattered data points, the vector parameter, the *Tol* (tolerance of curved surface) and the *Eor* (fairing of curved surface)[1,2]. The NC code length can be greatly reduced and the machining speed and machining accuracy can be improved greatly during the in the process of the NC machining process for succinct expression of the complex curved surface [3,4].

For the reason that the problem of the curved surface fitting optimization by scan points **is** solved relatively mature and the problem of the scattered points is not solved perfectly. In this paper, how to solve the problem of the curved surface fitting optimization by the scattered points is researched implemented.

Genetic Algorithm and the expression of Nurbs curved surface

Genetic Algorithm is a randomized adaptive search algorithm. The anticipant termination conditions are met by the elimination of incompetent individuals through the generations of the competition. Genetic Algorithm is widely used in areas such as function optimization, optimization of neural network, etc. The global optimal solution can be approximated with probability 1 by Genetic Algorithm if the condition of the hereditary generations is not limited in theory. Genetic Algorithm is used for solving the unconstrained optimization problem originally and it can be used for solving the problem with multi-objective and multi-constrained conditions。

The Nurbs curved surface can be expressed as:

$$s = s(u,v) = \frac{\sum_{i=0}^{m}\sum_{j=0}^{n} w_{i,j} d_{i,j} N_{i,k}(u) N_{j,l}(v)}{\sum_{i=0}^{m}\sum_{j=0}^{n} w_{i,j} N_{i,k}(u) N_{j,l}(v)} \tag{1}$$

Where $w_{i,j}$ are weights($i=0,1, ..., m$; $j=0,1, ..., n$), $d_{i,j}$ are control points($i=0,1, ..., m$; $j=0,1, ..., n$), $N_{i,k}(u)$ and $N_{j,l}(v)$ are standard B spline functions, k and l are order of the spline, u and v are node vector.

The curved surface fitting optimization Algorithm

The selection of genetic gene and the management of gene code. Nurbs curve or Nurbs surface is completely determined by Eq. 1 if the control points, weights and vector function are determined. As the node vector can be calculated if the control points and weights are identified with the relatively mature way such as Riesenfeld and Hartley Judd algorithm, so the control points and weights are regarded as the genetic genes and the Riesenfeld and Hartley Judd algorithm is used to calculate the vector value of Internal node.

There are two methods mainly for genes encoding: binary code and real value code. The real value code is always used for the multivariable optimization problem because the higher probability of crossover and mutation probability can be used to improve the optimization in this code. So the real values of parameters codes are used for the curved surface fitting optimization problem.

The Constraints in the process of calculation. The main constraints in the process of the calculation are:

1)The value of each weight is between 0 and 1, i.e, $0 \le w_{i,j} \le 1$;

2)The weights of four angle points of the curved surface are stipulated specially : $w_{0,0}, w_{0,n}, w_{m,0}, w_{m,n} > 0$;

3) The tolerance of fitting optimization surface is less than a given value(*Tol*);

4) The fairing of fitting optimization surface is less than a given value(*Eor*);

5) The number of Control points of two parameters direction is higher than the degree, and it is increased or decreased according to the integral change.

The calculation of evaluating function and Fitness function. To ensure the fitting optimization surface expression is concise enough, and the tolerance between the fitting optimization surface and the origin curved surface within the given value(*Tol*), and the fairing of fitting optimization surface is not more than the given value(*Eor*) , Therefore the tolerance, fairing and the number of control points should be all included in the evaluation function. The evaluation function can be expressed as:

$$F = D + \alpha E_r + \gamma J \tag{2}$$

Where D stands for the tolerance between the fitting optimization surface and the origin curved surface, E_r is the fairing of fitting optimization surface, J is the number of control points, α is the coefficient of E_r ($0 < \alpha < 1$)and γ is the coefficient of J ($0 < \gamma < 1$).

To ensure that the minimum value of F can be calculated, the fitness function can be expressed as:

$$FF = 1/(1 + F) \tag{3}$$

The calculation of node vector. The node vector values can be calculated by the Hartley – Judd algorithm if the number of control points and its coordinates are selected, the node vector values can be expressed as:

$$u_i - u_{i-1} = \frac{\sum_{j=i-k}^{i-1} |d_{i,j} - d_{i,j-1}|}{\sum_{s=k+1}^{n+1} \sum_{j=s-k}^{s-1} |d_{i,j} - d_{i,j-1}|} \tag{4}$$

The calculation of tolerance. Tolerance (D) is the sum of all the distance between each scattered data point P_i and the corresponding point that the distance is shortest from the scattered data point to the fitting optimization surface. And D can be expressed as:

$$D = \sum_{i=1}^{N}\sum_{k=1}^{N}(P_i - s(u_k,v_k))^2 \qquad i=1,2\cdots N; k=1,2\cdots N \tag{5}$$

The calculation of fairing. According to the thin plate elastic deformation equation in mechanics of elasticity, the energy function of the fitting optimization surface can be expressed as:

$$E_r = \iint \left[\begin{pmatrix} \alpha_{11}s_u^2 + 2\alpha_{12}s_u s_v + \alpha_{22}s_u^2 + \\ \beta_{11}s_{uu}^2 + 2\beta_{12}s_{uv}^2 + \beta_{22}s_{vv}^2 \end{pmatrix} - 2sf(u,v) \right] dudv \tag{6}$$

Where s_u is the first-order partial derivatives along the u direction of the fitting optimization curved surface $s(u,v)$, s_v is the first-order partial derivatives along the v direction of the fitting optimization curved surface $s(u,v)$, s_{uu} is the second-order partial derivatives along the u direction of the fitting optimization curved surface $s(u,v)$, s_{vv} is the second-order partial derivatives along the v direction of the fitting optimization curved surface $s(u,v)$, s_{uv} is the mixed partial derivative of the fitting optimization curved surface $s(u,v)$, α_{ij} and β_{ij} (i, j =0,1) is a a given constant which depends on the material characteristics, $f(u,v)$ stands for the force on the body in the elastic deformation equation(the external load). Effect of these parameters on the surface can be found in the literature.

Combined with the characteristics of the fitting optimization curved surface problem, $A(u,v)$ and $W(u,v)$ can be expressed as:

$$\begin{cases} A(u,v) = \sum\limits_{i=1}^{m}\sum\limits_{j=1}^{n} w_{i,j}d_{i,j}N_{i,k}(u)N_{j,l}(v) \\ W(u,v) = \sum\limits_{i=1}^{m}\sum\limits_{j=1}^{n} w_{i,j}N_{i,k}(u)N_{j,l}(v) \end{cases} \tag{7}$$

Where s_u, s_v, s_{uu}, s_{vv} and s_{uv} can be expressed as:

$$\begin{cases} s_u = \dfrac{A_u - W_u s}{W} \\[2mm] s_v = \dfrac{A_v - W_v s}{W} \\[2mm] s_{uu} = \dfrac{A_{uu} - 2W_u s_u - W_{uu}s}{W} \\[2mm] s_{vv} = \dfrac{A_{vv} - 2W_v s_v - W_{vv}s}{W} \\[2mm] s_{uv} = \dfrac{A_{uv} - W_u s_v - W_v s_u - W_{uv}s}{W} \end{cases} \tag{8}$$

Where A_u is the first-order partial derivatives along the u direction of the function $A(u,v)$, A_{uu} is the second-order partial derivatives along the u direction of the function $A(u,v)$, W_u is the first-order partial derivatives along the u direction of the function $W(u,v)$, W_{uu} is the second-order partial derivatives along the u direction of the function $W(u,v)$, A_{uv} is the mixed partial derivative of the function $A(u,v)$, W_{uv} is the mixed partial derivative of the function $W(u,v)$. The fairing can be calculated by substituted Eq.8 into Eq.6.

The main steps of calculation

(1) The scattered data points $P_i(i=1,2\cdots N)$ and other initial parameters (k , l , Tol , Eor , m , n) are input, where k is the spline degree of the vector u direction, l is the spline degree of the vector v direction, Tol , is the tolerance, Eor is the fairing, m is the initial number of control points along the vector u direction, n is the initial number of control points along the vector v direction.

(2) The fitting optimization curved surface is generated by Genetic Algorithm.

1) The parameters (the hereditary generations, population, crossover probability, mutation probability) are determined.

2) The population is initialized. The weight of each control point is generated randomly and encoded, thereby the genetic individual is generated, and the whole population can be initialized by this method.

3) The evaluation function and fitness function can be calculated by the Eq. 2 and Eq. 3.

4) The fitness of each individual in the population is calculated, and the optimal individual is the largest individual in this generation, so the optimal individual of this generation is selected. The new global optimal individual can be selected by comparing the fitness of the optimal individual with the fitness of the old global optimal individual.

5)The global optimal individual and hereditary generations are judged, if the optimal individual fitness is 1, then results is output and the program is ended. If the hereditary generations reaches the value which was given at the beginning of the program, then the program is gone to step (4), otherwise the program is gone to step 7).

6) The genetic, crossover, mutation of the chromosomes are operated.

7) The new population is generated, and the number of genetic generations plus 1, then the program is gone to step 3).

(3) The tolerance (D) between the fitting optimization surface and the origin curved surface is calculated. If D is greater than the given tolerance Tol , the program is gone to the next step, otherwise, the program is gone to step (5).

(4)the fairing(E_r) of the fitting optimization surface is calculated. If E_r is greater than the given fair Eor , ting, the program is gone to the next step, Otherwise, the program is ended and the result is output..

(5) the number of control points number is increased, $n=n+1$,and the program is gone to step (2), if D still does not meet the conditions of tolerance or E_r still does not meet the conditions of fairing conditions, $m=m+1$, the program is gone to step (2).

Simulation example

The result of the fitting optimization surface is shown in Figure 1. Where "* " stands for the given scattered points, the number of scattered points is 269. The initial parameters are: $m=3$, $n=3$, $k=3$, $l=3$, $Tol=0.8$, $Eor=0.85$, the crossover rate is 0.9, the mutation rate is 0.13, and the largest genetic generations is 80.

The results of the fitting optimization surface are $m=7$, $n=9$, $D=0.7576$, $E_r=0.7932$.so the fitting surface consists of control point 63.

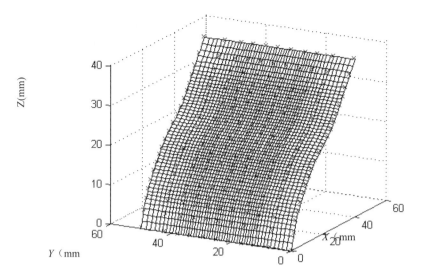

Fig.1 The fitting optimization surface based on scatted points

Conclusions

(1) The fitting optimization curved surface can be drawn by Genetic Algorithm under the conduction that the population size is set proper, the genetic factor and the evaluation function are selected reasonably. The simulation indicates that the constrained conductions are met fully by the fitting optimization curved surface.

(2) The number of the control points for the fitting optimization curved surface increases dramatically if the *Tol* or *Eor* is not given reasonably in the simulation process.

Acknowledgement

1. This work was financially supported by the Technology Development Foundation of Tianjin Education Committee (Grant No. 20120410, Project name: Research on high speed and high precision of complex surface toolpath planning error based on feedback compensation).

2. This work was financially supported by the Natural Science Fund of Tianjin (Grant No. 11JCYBJC 06200, Project name: The Theory and Method of Nondestructive Examination for Mechanical Features of Pipeline Weld by Laser Shock Processing)

3. This work was financially supported by The National Natural Science Foundation of China for Young Scientists (Grant No. 51105271, Project name: The Research on The Method of Getting Manufacturing Quality Features for Complex Curved Surfaces Based on Spatial Data Mining)

References

[1] Galleily J E, An Overview of Genetic Algorithms, J . Kybemetics, 1992 ,21(6):26~30.

[2] Zhang X, Jiang X, Scott P.J, A new free-form surface fitting method for precision coordinate metrology, J . Wear, 2009, 266(5-6) :543-547.

[3] Grimm Cindy, Ju Tao, Phan Ly, Adaptive smooth surface fitting with manifolds, J .Visual Computer, 2009 ,25,(5-7): 589-597.

[4] Park, Hyungjun, B-spline surface fitting based on adaptive knot placement using dominant columns, J. CAD Computer Aided Design,2011, 43(3): 258-264.

Advanced Materials Research Vol. 819 (2013) pp 100-104
© (2013) Trans Tech Publications, Switzerland
doi:10.4028/www.scientific.net/AMR.819.100

The Derivation and Simulation of Curved Tooth Face Gear Tooth Theoretical Contract-Point Trace Line Equations

Peng Xueyu[1,a] , Li Qing[1,b] and Wang TaiYong[1,c]

[1]Tianjin Key Laboratory of Equipment Design and Manufacturing Technology, Tianjin University, Tianjin 300072，China

[a]yuxuepeng1358@163.com,[b]qingat2007@yahoo.com.cn,[c] tywang@189.cn

Keyword: The curved tooth face gear, Contract-point trace line, Simulation

Abstract. The face gear tooth surface theoretical equation, based on the mesh of curved tooth face gear and involute worm, was deduced by means of differential geometry, meshing theory and so on. According to the conditions of the gear meshing, studying the ideal contract-point trace line theoretical equation under the conditions of no machining errors, installation errors and so on. By solving the equations and simulating in SOLIDWORKS, finally the tooth contact situation of face gear and cylindrical worm in the meshing process was got.

Introduction

The face gear drive is a kind of gear meshing drive with cylindrical gear and bevel gear[1]. The curved tooth face gear drive is that the curved tooth face gear meshes with the cylindrical worm. This transmission pairs has more widespread application prospect in higher reliable request equipment of national defense and other aspects.

Compared with the common spur or helical face gear, the curved tooth face gear transmission has advantages of higher contact ratio, larger carrying capacity, better stationarity and so on.

Face gear tooth surface equations

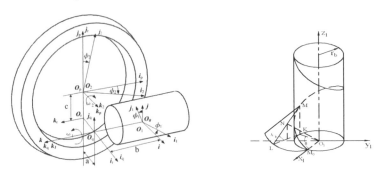

Fig.1 Face gear drive coordinate system Fig.2 Formation of flank I involute helicoid

In the gear meshing analysis process, using the following four coordinate systems(Fig.1): two fixed coordinate systems of the initial position of the worm and the face gear: $S_0 - O_0, i, j, k$ and $S_p - O_p, i_p, j_p, k_p$, two moving coordinate systems of the worm and the face gear: $S_1 - O_1, i_1, j_1, k_1$ and $S_2 - O_2, i_2, j_2, k_2$. The origins of coordinates O_p and O_2 coincide, as well as O_0 and O_1. The axes k_p and k_2 coincide with the rotational axis of the face gear, and axes k and k_1 coincide with the rotational axis of the worm. $S_q - O_q, i_q, j_q, k_q$ and $S_r - O_r, i_r, j_r, k_r$ are two auxiliary coordinate systems.

Shown in Fig.2, it is the coordinate system of laevo rotatory ZI-type cylinder worm. For any point N on the tooth surface, it satisfies the following equations [2]:

$$\overline{O_1N} = \overline{O_1K} + \overline{KM} + \overline{MN} \tag{1}$$

$$\begin{cases} \overline{O_1K} = r_b\left(\cos\theta \vec{i_1} - \sin\theta \vec{j_1}\right), \overline{KM} = p\theta \vec{k_1} \\ \overline{MN} = u\cos\lambda_b\left(\sin\theta \vec{i_1} + \cos\theta \vec{j_1}\right) - u\sin\lambda_b \vec{k_1} \end{cases} \tag{2}$$

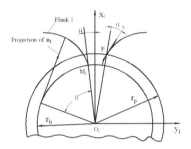

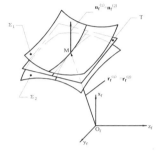

Fig. 3 End section of involute worm   Fig. 4 Point contact mesh of conjugate tooth surfaces

In the Eq. (2), r_b is the cylindrical worm base circle radius, λ_b is the lead angle of the helix, $p=r_b*\tan\lambda_b$ is the helical parameter, and variables u and θ are the parameters of the tooth surface. Shown in Fig.3, in the end section of $z_1=0$, axis x_1 is the symmetry axis. It can be drawn from the Fig.3 that $\mu=\omega_t/2r_p - inv\alpha_t$. Where, ω_t is the alveolar width on the pitch cylinder as well as within the end section and α_t is the profile angle of the end section: $\alpha_t = \arccos\left(r_b/r_p\right)$. From involute trigonometric relationship, $inv\alpha_t = \tan\alpha_t - \alpha_t$. By calculating, the final expression of laevo rotatory cylindrical worm flank I:

$$r_1(u,\theta) = \begin{bmatrix} x_1 & y_1 & z_1 & 1 \end{bmatrix}^T = \begin{bmatrix} r_b\cos(\theta+\mu)+u\cos\lambda_b\sin(\theta+\mu) \\ -r_b\sin(\theta+\mu)+u\cos\lambda_b\cos(\theta+\mu) \\ -u\sin\lambda_b+\theta r_b\tan\lambda_b \\ 1 \end{bmatrix} \tag{3}$$

The unit normal vector equation of flank I in the coordinate system S_1:

$$\vec{n_1} = \frac{\left(\partial\vec{r_1}/\partial u\right)\times\left(\partial\vec{r_1}/\partial\theta\right)}{\left|\left(\partial\vec{r_1}/\partial u\right)\times\left(\partial\vec{r_1}/\partial\theta\right)\right|} = \begin{bmatrix} -\sin\lambda_b\sin(\theta+\mu) \\ -\sin\lambda_b\cos(\theta+\mu) \\ -\cos\lambda_b \end{bmatrix} \tag{4}$$

The tooth surface equations of the other flank can be obtained by similar calculation. The tooth surface equations of face gear can be obtained as follows:

$$\begin{cases} \vec{r}_2(u,\theta,\phi_1) = M_{21}\vec{r}_1(u,\theta) \\ f(u,\theta,\phi_1) = 0 \end{cases} \tag{5}$$

Where, M_{21} is a coordinate transformation matrix, which expresses the transformation from coordinate system S_1 to S_2. $f(u,\theta,\phi_1)=0$ is the meshing equation. That solving $f(u,\theta,\phi_1)=0$ is based on $\vec{n}_1 \cdot \vec{v}_1^{(12)} = 0$. Where, $\vec{n}_1$ is the normal vector of involute cylindrical worm tooth surface, $\vec{v}_1^{(12)}$ is the relative velocity vector at the gear contact points.

$$M_{21} = M_{2p}M_{pr}M_{rq}M_{q0}M_{01}$$
$$= \begin{bmatrix} -\sin\phi_1\sin\phi_2 & -\cos\phi_1\sin\phi_2 & -\cos\phi_2 & b\cos\phi_2 + c\sin\phi_2 \\ \sin\phi_1\cos\phi_2 & \cos\phi_1\cos\phi_2 & -\sin\phi_2 & b\sin\phi_2 - c\cos\phi_2 \\ \cos\phi_1 & -\sin\phi_1 & 0 & a \\ 0 & 0 & 0 & 1 \end{bmatrix} \tag{6}$$

Simultaneous Eq. (3), Eq. (4), Eq. (5) and Eq. (6) to get the tooth surface equations of the curved tooth face gear:

$$\vec{r}_2(u,\theta,\phi_1,\phi_2) = \begin{bmatrix} \left(\pm r_b \cdot \sin A \mp u \cdot \cos \lambda_b \cdot \cos A + c\right)\sin\phi_2 - \left(\mp u \cdot \sin \lambda_b \pm r_b \cdot \theta \cdot \tan \lambda_b - b\right)\cos\phi_2 \\ -\left(\mp u \cdot \sin \lambda_b \pm r_b \cdot \theta \cdot \tan \lambda_b - b\right)\sin\phi_2 + \left(\mp r_b \cdot \sin A \pm u \cdot \cos \lambda_b \cdot \cos A - c\right)\cos\phi_2 \\ r_b \cdot \cos A + u \cdot \cos \lambda_b \cdot \sin A + a \\ 1 \end{bmatrix} \tag{7}$$

Where, $A = \theta + \mu \mp \phi_1$.

Mesh of face gear

Tooth contact analysis is intended to limit contract-point trace line in local and reasonable gear engagement region by changing some parameters, so that the gear drive can be more stable and accurate, and has less noise. For non-orthogonal face gear transmission, if the number of teeth of cutting tool and pinion is the same, it can realize line contact when it meshes with the pinion. Refer to the point contact mesh in bevel gear transmission and in order to make the non-orthogonal face gear the point contact transmission, the number of teeth of pinion is usually 1 to 3 less than the cutting tool[3].

The tooth contact point trace line can be calculated by tooth contact analysis, and all the coordinate systems in calculation should be in the same fixed machine tool coordinate system. Fig. 4 shows the point contact of two conjugate tooth surfaces. The radius vector and normal vector of the two tooth surfaces are equal at the meshing point M. ϕ_1 and ϕ_2 represent the angle of the pinion and gear in the process of engagement. According to the theory of gear meshing, the conjugate tooth surfaces ,which are point contact, should satisfy the following equations at meshing point:

$$\vec{r}_f^{(1)}(u,\theta,\phi_1) - \vec{r}_f^{(2)}(u,\theta,\phi_1,\phi_2,\phi_2) = 0 \tag{9}$$

$$\vec{n}_f^{(1)}\left(u,\theta,\varphi_1\right)-\vec{n}_f^{(2)}\left(u,\theta,\phi_1,\phi_2,\varphi_2\right)=0 \tag{10}$$

Eq. (10) can be expressed as two scalar equations because of $\left|\vec{n}_f^{(1)}\right|=\left|\vec{n}_f^{(2)}\right|=1$, therefore the equations consisting of Eq. (9), Eq. (10) and meshing equation can be represented by the following 6 non-linear equations:

$$f_i\left(u,\theta,\phi_1,\phi_2,\varphi_1,\varphi_2\right)=0,\quad i=1,...,6 \tag{11}$$

There should be a point satisfying the Eq. (11) and the equations also satisfy that Jacobian is not equal to zero since the tooth surfaces of gear and pinion mesh.

$$\frac{D(f_1,f_2,f_3,f_4,f_5,f_6)}{D(u,\theta,\varphi_1,\varphi_2,\phi_1,\phi_2)}=\begin{vmatrix} \dfrac{\partial f_1}{\partial u} & \dfrac{\partial f_1}{\partial \theta} & \dfrac{\partial f_1}{\partial \varphi_1} & \dfrac{\partial f_1}{\partial \varphi_2} & \dfrac{\partial f_1}{\partial \phi_1} & \dfrac{\partial f_1}{\partial \phi_2} \\ \vdots & \vdots & \vdots & \vdots & \vdots & \vdots \\ \dfrac{\partial f_6}{\partial u} & \dfrac{\partial f_6}{\partial \theta} & \dfrac{\partial f_6}{\partial \varphi_1} & \dfrac{\partial f_6}{\partial \varphi_2} & \dfrac{\partial f_6}{\partial \phi_1} & \dfrac{\partial f_6}{\partial \phi_2} \end{vmatrix}\neq 0 \tag{12}$$

Since the mathematical computation of equations is non-linear, according to engineering science computational method, the method of non-linear iteration is used to solve the problems.

Results

As shown in Fig. 1 and Fig. 4, shaft angle error, axial offset error of face gear and so on are not taken into account in the meshing process of the involute worm and non-orthogonal curved tooth face gear, so the transmission pair is engaged with the ideal contact points. Modeling the face gear and involute worm in 3D software SOLIDWORKS and getting the contact points and contact trace of face gear in Fig. 5 and Fig. 6.

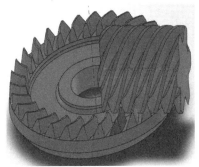 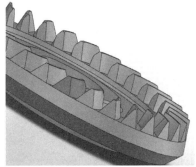

Fig. 5 Assembly drawing of face gear and involute worm   Fig. 6 Contact trace of face gear

Conclusion

Deriving the tooth surface equations of the curved tooth face gear by known involute worm tooth surface equations, building the three-dimensional models of pinion and gear and obtaining the contact trace equations in the absence of errors and contact points on the tooth surface of face gear in the condition of single tooth meshing, that would lay foundations for the later derivation and adjustment of face gear contact trace in practical cases and be beneficial for the further study of curved tooth face gear drive.

Acknowledgements

This work is supported by National Science and Technology Support Program(2013BAF06B00), and Tianjin application foundation and frontier technology research program(12JCQNJC02500 and 13JCZDJC34000).

References

[1] Yanzhong Wang, Wei Xiong and Li Zhang, Tooth Surface Equations and Tooth Contact Analysis of Face Gear, J. MACHINE TOOL & HYDRAULICS. 12 (2007) 7-9.

[2] F.L. Litvin, Gear Geometry and Applied Theory, Shanghai, Shanghai Scientific and Technical Publishing House, 2008.

[3] R.P. Zhu, S.C. Pan and D.P. Cao, Current State and Development of Research on Face Gear Drive, J. Journal of Nanjing University of Aeronautics and Astronautics. 29 (1997) 357-362.

[4]Jingcai Li. Study on Basic Applied Technology in the Process of Digitized Manufacturing for Spiral Bevel Gear[D].Tianjin: Tianjin University, 2008:42-44.

Advanced Materials Research Vol. 819 (2013) pp 105-109
© *(2013) Trans Tech Publications, Switzerland*
doi:10.4028/www.scientific.net/AMR.819.105

The Influence of the Laser Cutting System Performance on Cutting Quality

Ye Chang[1,2,a] ,Ji Jinjun[1,2],Yin HongMei[1,2]

[1]Huaian College of Information Technology, Huai'an 223003, China

[2]Jiangsu Engineering Technology Research and Development Center of Electronic Products Equipment Manufacturing

[a]whxsyhgf@163.com

Key words: laser cutting, system performance, influence, optimal parameter

Abstract: Laser cutting is widely applied in the modern industry. With the development of production, there is a higher demand for the improvement of cutting quality. Hence, investigating the influence of laser cutting parameters on the sheet quality is of important significance in the quality control of the sheet processing. So, this research mainly investigated the influence of the cutting system performance on the cutting quality. First, the experiment was conducted to investigate these system parameters such as mode of light velocity, polarity, nozzle, and airflow etc. Then, by contrasting the experimental data, the optimal system parameter was obtained.

Introduction

Laser cutting has been widely applied in the modern industry. The laser cutting equipments account for more than 70% of the whole laser processing equipments. With the development of production and the application of new processes, there is higher demand for the improvement of cutting quality. Thus, the requirements to the selection of the cutting process standard are also much higher. The laser cutting quality is influenced by many factors such as the laser cutting system performance (simply called as LCSP in the rest sections), laser cutting parameters etc. Among the influencing factors, LCSP must be taken into consideration in the laser purchase and the initial period of the system establishment of laser cutting. And the parameter setting of LCSP has great influence on the cutting quality. But the influence of LCSP on the cutting quality is rarely investigated in our country at present. Therefore, it is especially important to select parameters such as mode structure, polarity, nozzle, and airflow etc.. Thus, we can obtain better cutting effect; and we can also investigate each item in the system parameters of laser cutting comprehensively and systematically [1].

The evaluation indexes of laser cutting quality [2, 3]

The main evaluation indexes of cutting quality.

The laser cutting defects. Overburning: When laser power is too large, or cutting speed is too slow, the melting range of the workpiece is larger than the blowdown range of high pressure draft. As a result, the molten metal cannot be completely blown down by airflow, which causes overburning, as demonstrated in Figure 1.

Adhering slag: The auxiliary gas flow cannot completely blow down the molten or vaporized materials produced in the cutting process. Consequently, the slag adheres to the lower border of the cut surface. This phenomenon is called as adhering slag, as illustrated in Figure 2.

The measurable indexes of cutting quality. Cut surface roughness. Cut surface roughness is an important index reflecting cutting quality and this index is denoted by Rz.

Kerf width. Kerf width mainly depends on the mode structure and the diameter of the focal spot. Besides, cutting parameters also have certain influence on kerf width.

Kerf taper. Occasionally, the cutting parameters are inappropriately selected or there is inadequate auxiliary gas pressure. In this situation, the kerf is likely to present the taper shape with the top wide and the bottom narrow, as described in Figure 3. But for the plate cutting, this problem is easy to be solved.

 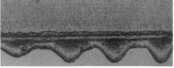 

Fig.1 Overburning   Fig.2 Adhering slag   Fig.3 Kerf taper

The evaluation benchmark of cut surface roughness. For the laser cutting of the sheet with the thickness of over 2 mm, the cut surface roughness is mal-distributed and presents great difference in the direction of the thickness. Besides, the morphology of the cut surface is divided into two distinct parts. The upper surface is even and smooth, with orderly, compact cutting stripes and small roughness height. In contrast, the uneven lower surface shows disorder stripes and large roughness height, as illustrated in Figure 4. This indicates that the roughness height on the cut surface reaches to the maximum value at the place close to the lower border. So, the lower border is the weak link of cut surface quality. Hence, the evaluation of cut surface quality should take the lower surface as the benchmark. Currently, the roughness at 1/3 from the lower border is mainly adopted as the benchmark in our country.

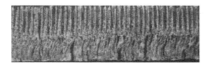 

a) The sheet thickness of the pulse laser cutting   b) The sheet thickness of the pulse laser cutting
   is 2 mm; p=600 W, f=150 HZ, v=15 mm/s.   is 2 mm; p=600 W, f=150 HZ, v=15 mm/s.
Fig.4 The cut surface morphology in different cutting parameters

Main factors influencing the cutting quality

The main factors influencing the laser cutting quality are divided into two types: the influence of LCSP and the influence of laser cutting parameters. This paper mainly investigated the influence of LCSP on the cutting quality.

The influence of the light velocity mode on the cutting quality. Two types of optical modes (TEM00 and TEM01) were adopted respectively to cut the sheets with different thickness. The roughness on the cut surface was presented in the table 1.

Table 1 The data records of cut surface roughness under different optical modes

Material number	Material name	Material thickness mm	Laser mode	Auxiliary gas	Roughness on the cut surface under the optical mode TEM_{00}, $R_Z/\mu m$	Roughness on the cut surface under the optical mode TEM_{01}, $R_Z/\mu m$
001	Low-carbon steel	2	pulse	oxygen	10	20
002	Low-carbon steel	3	pulse	oxygen	12	22
003	Low-carbon steel	4	pulse	oxygen	15	25
004	Low-carbon steel	5	pulse	oxygen	18	34
005	Low-carbon steel	6	pulse	oxygen	26	46

The analysis on the Table 1 shows that to obtain better cutting quality and higher cutting efficiency, the laser with TEM_{00} mode is preferred. At least, the laser mode should not exceed TEM_{01}.

The influence of beam polarization on the cutting quality. Linearly polarized light was adopted to conduct the cutting. The cutting direction varied with the change of the beam polarization direction. Thus, the laser absorptance of the cutting front also presents variation. Accordingly, the laser cutting effect was thereby influenced.

Table 2. The relation between the beam polarization direction and the cutting quality [4]

Material number	Material name	Laser mode	Auxiliary gas	The relation between the polarization direction and the kerf Kerf shape	Kerf evaluation
006	Low-carbon steel	pulse	oxygen	Parallel	The kerf was narrow, with low kerf verticality and roughness. The cutting speed was high.
007	Low-carbon steel	pulse	oxygen	Perpendicular	The kerf was wide, with large surface roughness. The cutting speed declined.。
008	Low-carbon steel	pulse	oxygen	Beveled	The kerf slanted with the angle.
009	Low-carbon steel	pulse	oxygen	Circularly polarized	The kerf was even and smooth.

Table 2 indicates that when the circularly polarized light was adopted to conduct the cutting, the kerf was the most uniform and smooth.

The influence of the nozzle and the airflow on cutting quality.

The influence of the nozzle aperture on cutting quality. Under certain auxiliary gas pressure, there is an optimal aperture range. Although auxiliary gas can eliminate the products of melting within the kerf, the excessively small or large aperture can influence the elimination process. Accordingly, the cutting speed is negatively affected. Laser power of 1500 w was selected in the experiment. Then, the cutting experiment in different aperture size was conducted to the low-carbon steel and the cemented carbide both with the thickness of 2 mm. Afterwards, the cutting speed was contrasted. The experimental data records were presented in Table 3 and Table 4.

Table 3 The experimental records 1 of the relation between the cutting speed and the nozzle aperture

Material number	Material name	Material thickness mm	Laser power w	Auxiliary gas	Gas pressure/MPa	Nozzle aperture/mm	Cutting speed mm/s
010	Low-carbon steel	2	1500	oxygen	0.14	1.0	96
011	Low-carbon steel	2	1500	oxygen	0.14	1.5	**102**
012	Low-carbon steel	2	1500	oxygen	0.14	2.0	96
013	Low-carbon steel	2	1500	oxygen	0.14	2.5	92
014	Low-carbon steel	2	1500	oxygen	0.14	3.0	87

Table 4. The experimental records 2 of the relation between the cutting speed and the nozzle aperture

Material number	Material name	Material thickness mm	Laser power w	Auxiliary gas	Gas pressure/MPa	Nozzle aperture/mm	Cutting speed m/min
015	Cemented carbide	2	1500	oxygen	0.4	1.0	0.9
016	Cemented carbide	2	1500	oxygen	0.4	1.5	**1.5**
017	Cemented carbide	2	1500	oxygen	0.4	2.0	1.3
018	Cemented carbide	2	1500	oxygen	0.4	2.5	1.0

In addition, the aperture can also influence the top kerf width and the heat affected zone (HAZ), as demonstrated in Figure 5. With the increase of the aperture, the kerf becomes broad, while the HAZ becomes narrow.

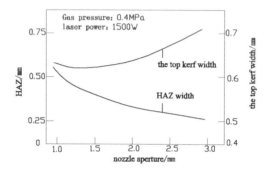

Figure 5. The influence of the nozzle aperture on the top kerf width and the HAZ of the low-carbon steel with the thickness of 2 mm at the limiting speed [5]

Table 3, Table 4, and Figure 5 show that under certain gas pressure, when the aperture is 1.5 mm, the highest cutting speed and better cutting quality can be obtained for the low-carbon steel and the cemented carbide steel both with the thickness of 2 mm.

The influence of the nozzle gas pressure on cutting quality.   Two types of laser powers (1500 w and 1700 w) and two types of sheets with different thickness (2 mm and 3 mm) were selected respectively. Then, the influence of different oxygen pressure on the cutting speed was investigated.

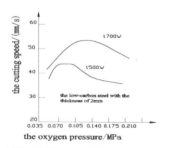

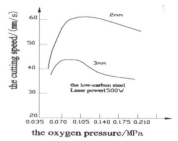

Fig.6 The influence of the oxygen pressure on the cutting speed in different laser powers

Fig.7 The influence of the oxygen pressure on the cutting speed in different sheet thickness [5]

The figures show that increasing the gas pressure can improve the cutting speed. But when the maximum value of the cutting speed is reached, the continuous pressure increase leads to the decrease of the cutting speed. When the laser power of 1500 w is adopted to cut the sheet with the thickness of 2 mm, the highest cutting speed is achieved with oxygen pressure ranging from 0.07 Mpa to 0.10 Mpa.

Conclusions

When 1500 w carbon dioxide laser is used to cut the low-carbon steel with the thickness of 2 mm, the optimal cutting quality is obtained with the following system parameters chosen. Circularly polarized light of the mode TEM_{00} should be selected. And the conical nozzle aperture is set to 1.5 mm, while the oxygen pressure is set to 0.09 Mpa.

The experiment proves that the laser power and LCSP also have great influence on the cutting quality in the laser cutting process. Specifically, mode structure, beam polarization, the nozzle, and the airflow can all exert certain influence on the cutting quality. To ensure the product technicality and obtain good cutting quality, appropriate cutting system parameters need to be selected. Furthermore, it is also necessary to shorten the cutting time as much as possible under the premise of the guaranteed quality. So, we can improve the product economy and the work efficiency, reduce the cost.

With the improvement of the laser cutting process, the laser cutting has been transformed from the test into the industrial productivity. It can also be developed from the general industrial application into the fine processing field.

References

[1] Chen Wuzhu. The laser welding and the cutting quality control. Beijing: China Machine Press, 2010.

[2] Zhang Yongqiang. The coaxial visual inspection and control of the laser cutting quality. Beijing: Dept. of Mechanical Engineering, Tsinghua University, 2006.

[3] Zhang Yongqiang, Wu Yanhua, Chen Wuzhu et al. Online evaluation methods of the laser cutting quality. Chinese Journal of Lasers, 2006, 33 (11): 1581-1584.

[4] Yan He et al. The processing and the application of the high power laser. Tianjin: Tianjin Science & Technology Press, 1992.

[5] Mukherjee K, et al. Lase in metallurgy. Chicago:Metallurgical Society of AIME, 1981.

Advanced Materials Research Vol. 819 (2013) pp 110-114
© (2013) Trans Tech Publications, Switzerland
doi:10.4028/www.scientific.net/AMR.819.110

The research of the epicycloids bevel gear cutting based on the common six-axis machine

Wang Yong[1, 2,a], Wang TaiYong[1,b], Lin FuXun[1,c], Lu ZhiLi [1,d], Wang Dong [1,e]

[1]Tianjin Key Laboratory of Equipment Design and Manufacturing Technology, Tianjin University China

[2]School of Mechanical Engineering, Tianjin University of Commerce, China

[a]wylxl2001@163.com, [b]tywang@189.cn, [c]lfx3107@163.com, [d]812270432@qq.com, [e]845073206@qq.com (wylxl2001@163.com)

Keywords: Epicycloids bevel gear; VERICUT; machining simulation.

Abstract: Based on the traditional cradle-type epicycloids bevel gear machining theory, the mathematical model of the epicycloids spiral bevel gear machining is created. Then, by applying coordinate transformation from the cradle-type machine to a CNC one, a six-axis virtual CNC machine model is established. At the same time, the gear cutting simulation is carried out under VERICUT according the properties of the general six-axis CNC and the relevant configuration parameters. By means of this method, the accuracy of the epicycloids bevel gear cutting system based on a general six-axis CNC can be well verified. Finally, a cutting experiment is conducted.

Introduction

As an efficient, stable, and high-load transmission component, a spiral bevel gear is widely used in ships, aircrafts, helicopters, cars and other equipment, of which the demand is very large. The traditional mechanical spiral bevel gear machine is gradually phased out due to its limitations. The application of the theory and technology of the digital processing by use of CNC spiral bevel gear has become a trend in the current spiral bevel manufacturing industry.

The traditional bevel gear processing theory is based on the cradle-type machine with complex structure which leads to the difficulties in adjusting the machine and cutter. While the adjustment time for bevel gears processing is shortened by using a six-axis CNC machine with its three linear motions and the three rotary motions. In order to the application of the six-axis CNC processing theory, it is necessary to study the parameters adjustment and conversion between traditional mechanical machine and CNC machine. The face hobbing technology has been developed by the Gleason and Klingenberg Company [1-3]. In domestic, the research on face hobbing technology adopting formate method has developed [4-5], but it is insufficient about the generated method for the epicycloids bevel gear cutting.

In this paper, through coordinate transformation, the six-axis motion relations are transformed to the CNC machine. Then the gear processing simulation on VERICUT proves the transformation to be correct. In order to reduce costs and improve the utilization of the machine, the epicycloids bevel gear is manufactured by the common six-axis CNC system embedded the bevel gear module.

The transformation Theory of CNC

In the transformation of machine motion, it should be ensured that the relative position and direction between the workpiece and cutter remain unchanged. According to the adjustment parameters of the cutter and machine on the mechanical gear cutting machine (SKM2), the position vector from the center of cutter to the origin of the workpiece coordinate system is gotten by use of the coordinate transformation, and then convert it to a six-axis CNC machine coordinate system. At the same time, the angle between the cutter axis and the workpiece axis stays unchanged in the coordinate transformation, the adjusting angle A between workpiece rotation axis and the blade tip plane of the cutter is introduced in six-axis CNC machine. Next, the cutter rotation angle W and the workpiece rotation angle C of the CNC machine can be analyzed according to the velocity ratio.

Thus, the movement transformation from a mechanical cradle bevel gear machine to the six-axis CNC machine is achieved.

When the spiral bevel gears are processed by the traditional mechanical cradle bevel gear machine, the workpiece has generating movement with the generating gear which derived from the cutter rotation driven by cradle. The following Fig.1 shows the processing of the gears by use of tilt method in the traditional cradle-type machine.

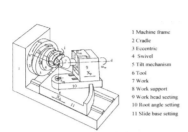

1 Machine frame
2 Cradle
3 Eccentric
4 Swivel
5 Tilt mechanism
6 Tool
7 Work
8 Work support
9 Work head seeting
10 Root angle setting
11 Slide base setting

Fig.1 Kinematical model of the gear generators   Fig.2 Bevel gear machining coordinate system

To transfer the traditional cradle-type spiral bevel gears machine to six-axis CNC machine, the gear manufacturing coordinate system is established as is shown in Fig.2, Coordinate system $\sum m = \{O_m,\ i_m,\ j_m,\ k_m\}$ is the machine coordinate system, in which axis coincides with the cradle axis. Coordinate system $\sum t = \{O_t,\ i_t,\ j_t,\ k_t\}$ is the cutter coordinate system, in which k_t axis coincides with the cutter axis. Coordinate system $\sum s = \{O_s,\ i_s,\ j_s,\ k_s\}$ is the workpiece coordinate system, in which k_s axis coincides with the workpiece axis, the symbols in the figure are as follows:

O_m—Center of machine O_t—center of cutter O_s—Crossing point

$\overline{k_t}$—Vector of cutter axial $\overline{k_s}$—Vector of workpiece axial δ_M—Root angle

q—Cradle angle q_0——Initial cradle angle S_p—Radial displacement

Δq—Cradle angle variation H—Horizonal setting i_{01}—Ratio of roll

V—Vertical setting E—Vertical displacement X_{B1}—Slide base

β—titl angle of cutter

The vector of cutter axis $\overline{k_t}$ can be seen as a vector $\overline{k_m}$ rotates β around the axis i_t .

$$\begin{cases} \overline{i_t} = \cos(q-\eta)\overline{i} - \sin(q-\eta)\overline{j} \\ \overline{k_m} = \overline{k} \\ \overline{k_t} = \overline{k_m}(\overline{i_t} - \beta)^R \end{cases}$$

. (1)

By calculation, the vector of cutter axis $\overline{k_t}$ is:

$$\overline{k_t} = \sin\beta \cdot \sin(q-\eta)\overline{i} + \sin\beta \cdot \cos(q-\eta)\overline{j} + \cos\beta\overline{k} \tag{2}$$

The vector of workpiece axis $\overline{k_s}$ is:

$$\overline{k_s} = \cos \delta_M \overline{i} + \sin \delta_M \overline{j}$$ (3)

Thus, the position vector $\overline{O_sO_t}$ is from the cutter origin O_t to the workpiece origin O_s:

$$\begin{cases} \overline{O_sO_t} = \overline{OO_t} - \overline{OO_s} \\ \overline{OO_s} = H \cdot \cos \delta_M \overline{i} - E\overline{j} + (X_{B1} + H \cdot \sin \delta_M)\overline{k} \\ \overline{OO_t} = S_p \cdot \cos q \overline{i} - S_p \cdot \sin q \overline{j} \\ \overline{O_sO_t} = (S_p \cdot \cos q - H \cdot \cos \delta_M)\overline{i} + (E - S_p \cdot \sin q)\overline{j} - (X_{B1} + H \cdot \sin \delta_M)\overline{k} \end{cases}$$ (4)

Finally, the vector $\overline{T_t}$ of cutter axis in the new coordinate system is

$$\overline{T_t} = \overline{O_sO_t}(\overline{k_s}, \ \Delta A)^R (\overline{j_s}, \ \varphi)^R$$ (5)

In equation 5, the 'R' is the vector rotation formula from reference [1]. The cutter speed is constant in the processing, $C = \omega t = \theta$ can be set. C is the cutter rotation angle. Based on the analysis of above, the rotation angle of workpiece used for indexing is:

$$A_1 = [(q - q_0) + \theta](z_0 / z_i)$$ (6)

The rotation angle of the workpiece for generating movement is:

$$A_2 = i_{01}(q - q_0) + \Delta \alpha$$ (7)

In which, z_0 is the groups of the cutter blades, z_i is the teeth number of gear. If let crossing point of gear as the origin of CNC machine, the expressions of every motion axis are:

$$\begin{cases} X_T = \overline{T_t} \cdot \overline{i} \\ Y_T = \overline{T_t} \cdot \overline{j} \\ Z_T = \overline{T_t} \cdot \overline{k} \\ A_T = A_1 + A_2 \\ B_T = \dfrac{\pi}{2} - \Delta B \end{cases}$$ (8)

According to the above theory, by using adjustable parameters in cradle-type spiral bevel gear machine, the cutter position and orientation relative to the workpiece can be determined. Through the coordinate transformation, these parameters can be converted to a Six-axis CNC system to achieve the relative motions of the cutter and workpiece. Then, the NC code of pinion cutting can be generated through Visual C++ software.

Epicycloids bevel gear cutting simulation and cutting experiment

General six-axis numerical machine, increased by dual rotary tables in the three-axis vertical CNC machine coupled with continuous indexing spindle control, constitutes the six-axis CNC machine. X, Y, and Z are the three linear axes. In dual rotary tables, axis A is the root cone adjustment axis, axis C for the workpiece rotation axis, axis W for the cutter rotation axis, wherein the axis C and axis W as the electronic gear control. The position of axis Z is adjusted to the plane of the blade tip tangent to the root cone plane of the workpiece. As to epicycloids cycloid gear processing, the axis of the workpiece rotates synchronously with the cutter axis, so it is needed for setting the electronic gear function in numerical control system. In the PMAC control, the position of the workpiece axis follows the position of the cutter axis. At the same time, According to the rotation speed of the mechanical machine cradle and the radial displacement of the cutter, the feed rate of X axis and Y axis is gotten. In order to simulate the complicated CNC machine, a six-axis machine model is established by use of Pro/E. Machine drive chain as follows:

Fig.3 Drive chain for the six-axis machine   Fig.4 Cutter head model

The cutter modeling is an important part for epicycloids bevel gear machining simulation. Because there is no required special cutter in VERICUT tool library, it is needed to create EN cutter geometry model. The blade is modeling by calculating the key data points. All blades obtained by the circular array are divided into five groups, the cutter shown in Figure 4.

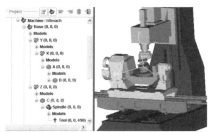

Fig.5 Virtual bevel gear machine on VERICUT   Fig.6 Epicycloids bevel gear machining

After the completion of the above work, it comes to the simulation stage. By setting the relevant parameters in VERICUT, the real-time cutting motions including cutter and workpiece are illustrated on the screen. At the same time, G codes monitoring window can be paged up. During simulations, and the errors, if exist, in the processing are displayed when the simulation ends. Figure 5 shows the bevel gear machining simulation.

According to the CNC conversion method, the epicycloids bevel gear machining module is embedded the CNC system. Figure 6 shows the epicycloids bevel gear cutting.

Conclusions

(1) Through analyzing the epicycloids bevel gear machining principle on the cradle-type machine, these position relations about the cutter, work gear cradle are gotten. Then, through coordinate transformation, theses position relations are transformed to the six-axis CNC machine. Lastly, the gear processing experiment proves the transformation to be correct.

(2) The epicycloids bevel gear machining simulation is conducted by use of VERICUT software. It can verify the accuracy and feasibility of the CNC theory by checking the NC code.

Acknowledgements

This work was financially supported by the National Key Technology Support Program (2013BAF06B00), the Key Technologies R & D Program of Tianjin (12ZCZDGX01600) and Tianjin Application Foundation and Frontier Technology Research Program (13JCZDJC34000).

References

[1] Wang Xiaochun, Wu Lianyin, Li Bin. Study on kinematic transformation from traditional machine tool to Free-Form ones based on spatial kinematics. Chinese Journal of Mechanical Engineering, 2001, 37(4): 93~ 98.

[2] Qi Fan, Kinematical Simulation of Face Hobbing Indexing and Tooth Surface Generation of Spiral Bevel and Hypoid Gears, Gear Technology, 2006(1), p. 30-38.

[3] Yi-Pei Shih, Zhang-Hua Fong, Mathematical Model for a Universal Face Hobbing Hypoid Gear Generator. ASME Journal of Mechanical Design, 129(2007), p. 38-47.

[4] Meilin Li, Xiaolong Shen, Yongxiang Li and Nanlin Yu. The Cutting Experiment of Full CNC Epicycloids Bevel Gear Cutting Machine, 2nd International Conference on Electronic & Mechanical Engineering and Information Technology, EMEIT-2012

[5] Shaowu Nie, Xiaozhong Deng,Jianxin Su. CNC Machine Model of Epicycloid Bevel Gear and Cutting Experiment Based on Formate Method, Advanced Materials Research Vols. 542-543 (2012) p.1157-1162

Advanced Materials Research Vol. 819 (2013) pp 115-119
© *(2013) Trans Tech Publications, Switzerland*
doi:10.4028/www.scientific.net/AMR.819.115

The Vibration Isolation Effect Research of the Floating Raft Isolation System Based on the Adjustable Flexibility of Foundation

Wang Hui[1,a],Weng Zeyu[1,b],Xiang Gan[1,c],Lu Bo[1],Ding Honggang[1],You Hongwu[1]

[1]College of Mechanical Engineering, Zhejiang University of Technology, Hangzhou, 310014, China;

[a]wanghuiwyh@163.com, [b]wengzy8888@163.com, [c]xiangganzgd@163.com

Keywords: floating raft isolation system, vibration isolation effect, flexibility of foundation, ADAMS

Abstract: The floating raft isolation system is widely used in the field of marine engineering for its vibration isolation effect. Along with the application of light thinning structure of the ship, the flexibility of foundation of floating raft isolation system makes the vibration isolation effect vary widely between the practical floating raft isolation system and its theoretical results. In order to research the vibration isolation effect of the floating raft isolation system on different flexibility of foundation, the floating raft isolation system with flexible foundation is designed in this paper, and the adjustable flexibility of foundation is achieved by using elastic beams. With simulation and analysis of test system in ADAMS, the results of the relationship between flexibility and vibration isolation effect are obtained.

Introduction

The flexibility of foundation is increasingly prominent along with the application of light thinning structure in the field of marine engineering, so that the vibration isolation effect of the floating raft isolation system varies greatly with theoretical results, which promote the wider study of non-rigidity of vibration isolation system. The dynamic model of floating raft isolation system which based on flexible foundation is more in line with the engineering practice, and provides effective means of theoretical analysis for a wider range of engineering design and vibration isolation effect evaluation, and safeguard for the implementation of a high level of vibration isolation[1]. It is of great significance to research on the flexibility of foundation of isolation system for the development of the floating raft isolation system.

The purpose of vibration isolation is to reduce the transmission of vibration. For specific vibration isolation design of engineering practice, we are most concerned about how much attenuation or control is obtained about the vibration level of protected object after the vibration isolation. In the vibration isolation design, optimization design of structural parameters of the system is expanded around vibration isolation efficiency usually. Therefore, the determination of evaluation index is the key content of performance evaluation system. Due to the displacement, velocity and acceleration of the vibration isolation system can be measured, and easily to deal with, so it is more convenient to use the vibration level difference as the evaluation index of vibration isolation effect, also can better reflect vibration isolation effect of the floating raft isolation system[2].

Designing the Floating Raft Isolation System with Flexible Foundation

The impedance converter[3] and rectangular plate are adopted to simulate the flexible foundation in existing research, but it is inconvenient to adjust the flexibility, so multiple elastic beams is given in this paper to simulate the flexible foundation, which can continuously adjust the flexibility, and make experimental study conveniently, in order to study the influence of the flexibility of foundation on the vibration characteristics of the floating raft isolation system.

Fig. 1 Diagram of supported elastic beam

The diagram of supported elastic beam under the action of force is shown in Fig. 1. Under the action of force F, the deflection of the other end of beam is[4]:

$$w = \frac{FL^3}{3EI} \tag{1}$$

The equivalent stiffness of elastic beam is:

$$K = \frac{F}{w} = \frac{3EI}{L^3} \tag{2}$$

Where, E is the Modulus of elasticity; $I = \frac{ah^3}{12}$ is the area moment of inertia of elastic beams; L is the distance between two points of force application on elastic beam; h is the thickness of the cross section of elastic beam; a is the length of the cross section of elastic beam. As shown in Eq.2, the stiffness of elastic beams is inversely proportional to L^3, and is proportional to h^3.

Table 1 Specific parameter of the floating raft isolation system

Subsystem	Parameter	Value
Mass 1	Quality [*Kg*]	93
Mass 2	Quality [*Kg*]	93
The vibration isolator Supporting mass 1	Stiffness [*N/mm*]	270
The vibration isolator Supporting mass 2	Stiffness [*N/mm*]	270
intermediate	Quality [*Kg*]	130
The vibration isolator Supporting intermediate	Stiffness [*N/mm*]	792

The specific parameters of floating raft vibration isolation system are shown in Table 1. According to the above parameters in the table, we can build a model for analysis, and establish real system for testing.

The Test Model of Floating Raft Isolation System

According to the design of the floating raft isolation system and virtual prototype technology, a dynamic model is built by using dynamic simulation software ADAMS. In the design of mechanical vibration isolation system, we generally use ADAMS/View module to model the system, but comparing with other CAD software, the ADAMS/View modeling capability is weak, so we always save 3D CAD model as Parasolid (filename extension. "X_T") format, and import the model into

ADAMS/View through the model data exchange interface. After successfully importing model, we can set attribute, constraint the motion of the components, and define the applied load, finally simulate the system[5]. This paper builds Pro/e model, and import graphic file into ADAMS/View with Parasolid format, then set spring and damping elements, and add to kinematic constraints between the mass and the intermediate raft, then complete the establishment of the floating raft isolation system geometric model. The geometric model in ADAMS/View is shown in Fig. 2. In the figure, the top layer is the mass, the second layer is the intermediate raft, and the bottom layer is the base.

Fig. 2 Model of floating raft isolation system with flexible foundation

According to the actual kinematic relationship among parts, the kinematic pairs are simplified to the kinematic constraints in ADAMS[6]. The constraints include the following: the guiding device between the mass and the intermediate raft is fixed on the intermediate; the guiding device between the intermediate raft and the base is fixed on the base; the motion of the mass is simplified to sliding constraints for vertical direction; the motion of the intermediate raft is simplified to sliding constraints for vertical direction. To simulating the flexible foundation of the floating raft isolation system, there are four vibration isolation components under the base four-terminal for supporting.

Simulation and Analysis

ADAMS/Vibration, which are ADAMS software's own vibration analysis module, is used to vibration analysis of floating raft isolation system[7] in this paper. Based on the relationship between the stiffness   and multiple elastic beams' simply supported distance of the floating raft isolation system, when the distance L is taken $74.69mm$, $83.79mm$, $100mm$ or $149.39mm$, the corresponding foundation stiffness is $48000N/mm$, $34000N/mm$, $20000N/mm$, $6000N/mm$. By simulating this system in ADAMS, Fig. 3 and Fig. 4 show the vibration acceleration frequency response of the mass and the vibration acceleration frequency response of the base with different foundation stiffness.

The vibration acceleration frequency response of the mass in Fig. 4 is almost coincident, which illustrates that the foundation flexibility of the floating raft isolation system impact the mass vibration acceleration less. As shown in Fig. 5, the first order natural frequency and the second order natural frequency of different foundation stiffness are nearly constant, and the third order natural frequency exists large change that the frequency go backwards in turn with the decrease of the stiffness, such as the formant value of foundation stiffness being 48000 N/mm move back 92.06 Hz comparing with the stiffness being 6000 N/mm. With the increase of foundation stiffness, vibration acceleration frequency response overall declines, and the acceleration response of the base is almost coincident when frequency is greater than 300 Hz. The results show that, in less than 300

Hz frequency range, the greater the foundation stiffness is, the smaller the vibration of the base passed from the vibration source of this system is, which means the vibration effect of the system is better.

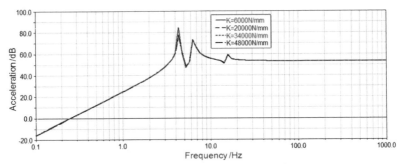

Fig. 3 Acceleration frequency response of the mass

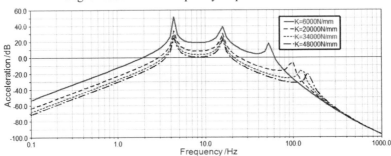

Fig. 4 Acceleration frequency response of the base

According to the definition of vibration level difference, the vibration level difference graph of different flexible foundation of floating raft isolation system is shown in Fig. 5.

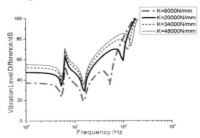

Fig. 5 Influence of the foundation stiffness to the vibration isolation effect

As shown in Fig. 5, the vibration level differences of acceleration of floating raft isolation system are all positive, so the vibration isolation device is effective. Vibration level difference fall to the trough in the first two order resonance, and the vibration isolation effect is the worst at moment, so we prefer to make the natural frequency of the floating raft isolation system be away from the vibration frequency of the mass as far as possible. From Fig. 5, we can also conclude that the vibration effect of the floating raft isolation system increases with the improvement of the stiffness of foundation.

According to the simulation model and other design parameters, the picture of the floating raft isolation system is shown in Fig. 6. There are guide devices between each layer, which is composed of cylindrical guide rod and clamp brackets; flexible foundation is composed of multiple elastic beams; measuring points of the acceleration sensor arrange in the middle of the mass and base. Through test and data analysis, the influence of the foundation stiffness on the vibration isolation effect of the floating raft vibration isolation system is basically identical with the simulation results.

Fig. 6 The floating raft isolation system

Conclusions

(1) This paper simulates the floating raft isolation system by using ADAMS software to avoid the complicated formula derivation, which makes dynamic modeling and analyze vibration characteristic of this system be conveniently.

(2) The flexibility of floating raft isolation system is more obvious in the high frequency vibration, and there is strong dynamic coupling effect between each other, which seriously impacts the vibration isolation effect.

(3) The vibration effect of the floating raft isolation system increases with the improvement of the stiffness of foundation.

Acknowledgment

The project is supported by the Project Fund of Science and Technology Department of Zhejiang Province (applied research on nonprofit technology) (No.2012C21089).

References

[1] K.J. Song, W.B. Zhang and J.C. Niu: Chinese Journal of Mechanical Engineering, Vol.39 (2003) No.9, p23-28.

[2] F.C. Hu, Y. Cai and Q.M. Zhong: Noise and Vibration Control, Vol.27 (2007) No.5, p10-13.

[3] Y.Y. Gu: *Research of the Design Method of the Vibration Isolation System Based on Elastic Base* (Shanghai Jiao Tong University, Shanghai 2009).

[4] H.W. Liu: *Mechanics of Materials* (Higher Education Press, Beijing 2005).

[5] C.Z. Jia, J.H. Yin and W.X. Xue: *MD-DADAMS Virtual Prototype from Entry to the Master* (Mechanical Industry Press, Beijing 2010).

[6] Q. Fu, G.Y. Gao and Y.Q. Zheng: Light Industry Machinery, Vol.29 (2011) No.6, p36-39.

[7] D.M. Chen, C.F. Huai and K.T. Zhang: *Proficient in Virtual Prototype Technology of ADAMS2005/2007* (Chemical Industry Press, Beijing 2010).

Advanced Materials Research Vol. 819 (2013) pp 120-124
© (2013) Trans Tech Publications, Switzerland
doi:10.4028/www.scientific.net/AMR.819.120

Track Smoothness of Moving Axis Considering Kinematical Characteristics of Machine Tool

Cai Yonglin[1,a], Yang Zhimin[1,b], Li Jianyong[1,c] and Huang Chao[1,d]

[1]School of Mechanical, Electronic and Control Engineering,

Beijing Jiaotong University, Beijing, 100044, China

[a] ylcai@bjtu.edu.cn (corresponding author), [b] 11121410@bjtu.edu.cn, [c] jyli@bjtu.edu.cn,
[d]11125667@bjtu.edu.cn

Keywords: track smoothness, kinematical characteristics, NC machining

Abstract. In multi-axis NC (Numerical Control) machining for a complex part, high feed rate may bring shock to moving axes and influence the machining quality of the part. Firstly, the paper works out velocity, acceleration and jerk kinematical characteristics according to the displacement curve of a moving axis. Secondly, find out the acceleration curve segments which need to be smoothed according to the value of jerk, and smooth these acceleration curve segments with B-spline method. Then calculate the smoothed velocity and displacement through numerical integration method to get the new displacement curve of the moving axis. Finally, a test example is given to demonstrate the promising use of the proposed solution. The method presented by this paper can effectively decrease the acceleration shock of machine tool and improve the machining quality of part surface.

Introduction

In high feed rate multi-axis NC machining for a complex part, because of the complex shape of the part, the serious velocity and acceleration shock can always be seen, which can influence the machining quality of the part surface. Especially, the phenomenon is more obvious when the rigidity of machine tool is insufficient [1]. Because part machining needs certain allowance, analyzing the velocity and acceleration performance and smoothing moving axis track of machine tool is an effective way to decrease and avoid the velocity and acceleration shock of machine tool under conditions of permit precision.

The essence of track smoothness is to smooth data. Existing methods for data smoothness include least square method, energy method, global smoothing method and so on [2-4], and the data smoothed by these methods are smooth themselves. Wolfgang [5] smoothes the first and second difference curve, and improved the original points by integration method. In NC machining, when the feed rate is very high, although the data of the tool path are smooth, the surface of part is still rough. The author analyzes the reason is the acceleration shock after eliminate factors. As shown in Fig.1, (a) is the displacement curve of a translational axis, (b) is the velocity curve and (c) is the acceleration curve. In Fig.1 although the displacement curve is smooth, the acceleration curve has mutation.

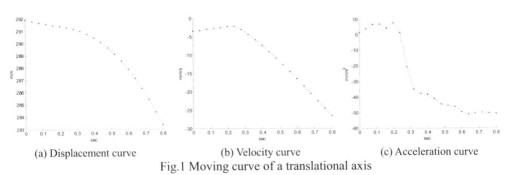

(a) Displacement curve   (b) Velocity curve   (c) Acceleration curve

Fig.1 Moving curve of a translational axis

Analysis of kinematical characteristics of moving axis

Hypothesise the machine tool moves from $(X_{i-1}, Y_{i-1}, Z_{i-1}, A_{i-1}, C_{i-1})$ to $(X_i, Y_i, Z_i, A_i, C_i)$ with the feed rate F after it implements the ith row code. So, the resultant displacement of the three translational axes is expressed as

$$\Delta S_i = \sqrt{(X_i - X_{i-1})^2 + (Y_i - Y_{i-1})^2 + (Z_i - Z_{i-1})^2} . \tag{1}$$

Presently, most existing NC systems regard program feed rate F as the resultant velocity of translational axes when plan feed rate. So, under the condition, the move time between the two adjacent rows of NC codes can be obtained as following

$$t_i = \frac{\Delta S_i}{F} . \tag{2}$$

Velocity of each axis is

$$v_i = \frac{\Delta_i}{t_i}, \tag{3}$$

where, Δ_i is the difference between two adjacent rows of NC codes. To a translational axis, it is displacement difference; to a rotation axis, it is angle difference.
Acceleration of each axis is

$$a_i = \frac{v_i - v_{i-1}}{t_i} . \tag{4}$$

Jerk of each axis is

$$J_i = \frac{a_i - a_{i-1}}{t_i} . \tag{5}$$

Track smoothness of moving axis based on acceleration characteristics

Firstly, smooth the acceleration curve; secondly, make second-order integral for the smoothed curve. In this way, the smooth displacement curve can be obtained.

Smoothness of acceleration curve

Existing least square method and energy method can be used to globally smooth acceleration curve but the smoothed curve may be relatively different from the original curve so that the part distorts after machining. If use these methods to locally smooth curve, some gap may exist in the smoothed curve, which also can bring relatively serious acceleration shock. B-spline curve has the polishing characteristic which means the order is higher the curve is smoother with identical control points. Besides, the end-points of control polygons coincide with that of the curve. With the two characteristics, regard the original acceleration curve as control point curve and regard the calculated B-spline curve as smoothed acceleration curve. The continuity problem of the whole curve is solved at the same time.

(1) Analysing the amount of movement of moving axis

Hypothesise the displacement curve of a moving axis is (t_i, s_i), $i=0, 1, 2, \cdots, n$; where, t_i is time, s_i is displacement. According to Eq. (3), (4) and (5), calculate velocity, acceleration and jerk and obtain the velocity curve (t_i, f_i), the acceleration curve (t_i, a_i) and the jerk curve (t_i, J_i).

(2) Determining curve segments requiring to be smoothed

Determine the critical value of jerk J_c. The critical value is related to the structure and rigidity of machine tool and it can be obtained through experiments. Those points in jerk curve whose values are greater than the critical value need to be smoothed.

If $J_{l~m} > J_c$, the $i=l, l+1, ..., m$ curve segments in acceleration curve (t_i, a_i) need to be smoothed.

(3) Smoothing local acceleration curve

For data (t_i, a_i), $i=l, l+1, ..., m$, note them as $q_j(t_j, a_j)$, where $j=0, 1, ..., m-l$. Let q_j be B-spline control points and let curve order $k=3$, so B-spline curve can be obtained by following formula:

$$p = \sum_{j=0}^{m-l} q_j \cdot N_{j,k}(t) \tag{6}$$

where, $N_{j,k}(t)$ is the basis function of B-spline,

$$
\begin{cases}
N_{j,0}(t) = \begin{cases} 1, t \in [t_j, t_{j+1}] \\ 0, other \end{cases} \\
N_{j,k}(t) = \dfrac{t - t_j}{t_{j+k} - t_i} N_{j,k-1}(t) + \dfrac{t_{j+k+1} - t}{t_{j+k+1} - t_{j+1}} N_{j+1,k-1}(t) \cdot \\
\qquad\qquad let: \dfrac{0}{0} = 0
\end{cases}
$$

Let t_i be variable quantity, use Eq. (6) to calculate points p_i in B-spline curve. So, curve (t_i, p_i) is the smoothed acceleration curve.

Move track calculation of moving axis

After obtaining the smoothed acceleration curve, use integral method to calculate velocity and displacement value.

The smoothed velocity f_{si} can be obtained by formula

$$f_{si} = f_{si-1} + p_i \cdot \Delta t_i, \tag{7}$$

where, $\Delta t_i = t_{i+1} - t_i$, $i=l, l+1, ..., m$.

As shown in Fig.2, for the velocity curve calculated by Eq. (7), when $i=l-1, f_{sl-1}=f_{l-1}$; but when $i=m$, because of accumulated error, $f_{sm} \neq f_m$, that is the smoothed start point coincides with the original, yet the smoothed end point does not coincides with the original, as shown in Fig.2 (a).

(a) Before processing   (b) After processing

Fig.2 Processing end point of velocity curve

Therefore, let the difference between the smoothed end point and the original in velocity curve be $d_f = f_{sm} - f_m$. Distribute d_f to every point f_{si} $(i=l+1, l+2, ..., m-1)$ in smoothed curve and calculate the difference after distribution by $d_f \times w_i$, where w_i is weight. w_i can be obtained through Bernstein basis function. Both start and end point of Bernstein basis function are 0 and the maximum value is located in middle part, so it can satisfy the distribution demand of d_f, as shown in Fig.2 (b).

After calculate velocity curve by Eq. (7), displacement curve can be calculated in the same idea,

$$s_{si} = s_{si-1} + f_{si} \cdot \Delta t_i, \tag{8}$$

where, $\Delta t_i = t_{i+1} - t_i$, $i=l, l+1, \ldots, m$.

Identically, there is difference between smoothed end point and original end point in track curve $d_s = s_{sm} - s_m$. By the similar method, distribute d_s to the whole curve, then the ultima displacement curve is the smoothed track curve of the moving axis.

Calculation example

In high feed rate five-axis NC machining for a mould surface, the movement data of X axis are shown in Table 1.

Table 1 Movement data of X-axis before smoothing

No.	Time[s]	Displacement[mm]	Velocity[mm/s]	Acceleration[mm/s²]	Jerk[mm/s³]
1	0	291.8962			
2	0.04	291.7660	-3.2550		
3	0.08	291.6464	-2.9900	6.6250	
4	0.12	291.5379	-2.7125	6.9375	7.8125
5	0.16	291.4363	-2.5400	4.3125	-65.6250
6	0.20	291.3468	-2.2375	7.5625	81.2500
7	0.24	291.2594	-2.1850	1.3125	-156.2500
8	0.28	291.1391	-3.0075	-20.5625	-546.8750
9	0.32	290.9630	-4.4025	-34.8750	-359.3750
10	0.36	290.7265	-5.9125	-37.7500	-71.8750
11	0.40	290.4285	-7.4500	-38.4375	-17.1875
12	0.44	290.0651	-9.0850	-40.8750	-62.5000
13	0.48	289.6303	-10.8700	-44.6250	-92.1875
14	0.52	289.1230	-12.6825	-45.3125	-17.1875
15	0.56	288.5421	-14.5225	-46.0000	-17.1875
16	0.60	287.8820	-16.5000	-49.4375	-85.9375

According to the experiment of the machine tool, let the critical value of jerk be $J_c = 200$ mm/s³, so the 8th and 9th jerk value are greater than J_c. Let the B-apline order $k=3$, use the above method to smooth acceleration curve (t_i, a_i), where $i=5,7, \ldots, 11$, and obtain velocity and displacement data, shown in Table 2.

Table 2 Movement data of X axis after smoothing

No.	Time [s]	Displacement[mm]	Velocity[mm/s]	Acceleration[mm/s2]	Jerk[mm/s3]
1	0	291.8962			
2	0.04	291.7660	-3.2550		
3	0.08	291.6464	-2.9900	6.6250	
4	0.12	291.5379	-2.7125	6.9375	7.8125
5	0.16	291.4363	-2.5400	4.3125	-65.6250
6	0.20	291.3465	-2.2375	7.5625	81.2500
7	0.24	291.2568	-2.1850	1.3125	-156.2500
8	0.28	291.1357	-2.9250	-18.5000	-495.3125
9	0.32	290.9618	-4.1975	-31.8125	-332.8125
10	0.36	290.7268	-5.6775	-36.9375	-128.1250
11	0.40	290.4285	-7.2125	-38.4375	-37.5000
12	0.44	290.0651	-9.0850	-40.8750	-62.5000
13	0.48	289.6303	-10.8700	-44.6250	-92.1875
14	0.52	289.1230	-12.6825	-45.3125	-17.1875
15	0.56	288.5421	-14.5225	-46.0000	-17.1875
16	0.60	287.8820	-16.5000	-49.4375	-85.9375

As the Table 1 and the Table 2 show, the displacement change appears in the 8th row before and after smoothing, which is 0.0034mm. The adjustment of displacement make the maximum acceleration value change from 20.5625mm/s² to 18.5mm/s², so the acceleration shock obviously decrease.

Figure 3 describes kinematical characteristics before and after smoothing, the points which jerk more than critical value of 200 mm/s3 are smoothed. Although the displacement is almost same, the two curves overlap as shown in fig 3(a), the velocity, acceleration and jerk curve are different, see fig 3 (b), (c) and (d).

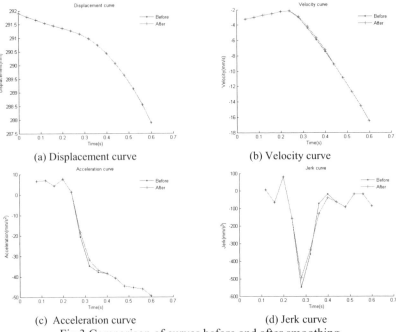

(a) Displacement curve   (b) Velocity curve

(c) Acceleration curve   (d) Jerk curve

Fig.3 Comparison of curves before and after smoothing

Conculsion

Under conditions of permit precision, the smoothness method presented by this paper for moving axes track can effectively decrease the acceleration shock of machine tool and improve the machining quality of part surface through fine adjusting the displacement of translational axes and rotation axes. The method has been used in the programming software of five-axis NC machining for complex module surface developed by the author, and obtain a good effect.

Acknowledgements

This research was supported by the Ministry of Education of the People's Republic of China (grant number 625010351). Their support is greatly appreciated.

References

[1] ZHENG Yan, BI Qingzhen, WANG Yuhan, Method of Feed Rate Planning in Five-axis Machining, Modular Machine Tool & Automatic Manufacturing Technique. 2011(7)1-4.

[2] Zhu Xinxiong, Modeling Technology of Free-Form Curve and Surface, Science Press, Beijing, 2000.

[3] Gerald Farin, Curve and Surface for CAGD: A Practical Guide (fifth edition), Academic Press, San Francisico, 2002.

[4] Jiang Dawei, Li Anping, Shape preserving Least-squares Approximation by B-spline Faired Curve, Chinese Journal of Engineering Mathematics. 17(1)125-128.

[5] Wolfgang Renz, Interactive Smoothing Digitized Point Data, Computer aided design. 14(5) 267-269.

Advanced Materials Research Vol. 819 (2013) pp 125-128
© *(2013) Trans Tech Publications, Switzerland*
doi:10.4028/www.scientific.net/AMR.819.125

Numerical Simulation of Hydro-forming Process of Shaped Tube

Chen Jie[1]

[1]Department of Mechano-electronic Engineering, Suzhou Vocational University, Suzhou 215104, China

[a]chenjie35@vip.sina.com

Key word: Shaped tube; Hydro-forming; Numerical Simulation

Abstract: In the forming process of section tube using hydro forming, distributed regulation of the strain, loading rate and destabilization have a large influence to forming. The forming process is simulated by means of FEM in this paper. Simulation results shows that strain are maximal at the convexity and concave of die. And with pressure increasing, the phenomenon of destabilization would occur on some side of its convexity randomly, the forming process of section tub can be divided into two obvious stages. Besides, critical pressure of two stages is pointed out. The simulation results are compared with the experiment results, and the correctness of the simulation is verified.

Introduction

The Hydro-forming technology is a new metal forming technology[1]. It was proposed and has been conducted in 1980s, and applied to the industrial production at the beginning of the 90's [2~3]. In the past ten years, especially in recent years, the hydro-forming technology is developing rapidly with the development of the computer control technology and the high pressure hydraulic system[1~3]. Now, many kinds of structure can be produced on a large scale by this technology. Compared with the traditional technology, such as stamping and welding formed technology, the hydro-forming product has a light weight and high quality. So it is widely applied to the forming process of the complex shape structure tube with sealant cross-section in the automobile industry, the aviation industry and the oil industry. Its basic principle of forming is that after the standard pipe is pre-bended and pre-formed, the high-pressured fluid is injected, and the fitting tube is formed in the seal mold cavity [4~7].

Finite element method is used in this paper. The forming process of the shaped tube is simulated and analyzed by means of FEM. The simulation results are compared with the experiment results, and the correctness of the simulation is verified.

Simulation Algorithm

By using the Lagrangian describing increment method, the dynamic increment equilibrium equations of a moved object at time tn Based on the principle of virtual work is shown in Eq.1 [6].

$$\int_{\Omega}[\delta\varepsilon_n]^T\sigma_n d\Omega-\int_{\Omega}[\delta u_n]^T[b_n-\rho_n\ddot{u}_n-c_n\dot{u}_n]d\Omega-\int_{\Gamma t}[\delta u_n]^T d\Gamma=0 \tag{1}$$

Where, virtual displacement δu_n corresponds with $\delta\varepsilon_n$, b_n is the body force, t_n is the surface force, δ_n is the stress, ρ_n is the mass density, and c_n is the damping parameters. "."is the time differential item, The region, "Ω", includes the boundary force of tn, Γ_t and the boundary Γ_u of the boundary displace u_n. After the Space discretization, Eq.1 can be expressed with matrix as Eq.2.

$$M\ddot{d}_n+C\dot{d}_n+P_n=f_n \tag{2}$$

Where, M is the lumpy mass matrix, C is the total damping matrix, P_n For interface resistance total node force, f_n is the node force vector combined the function physical force and the surface , $\overline{d}_n$ is the total particle acceleration and $\dot{d}_n$ is the total node speed.

By using the centered difference, its acceleration can be expressed as following:

$$\ddot{d}_n \approx a_n = \{d_{n+1} - 2d_n + d_{n-1}\} / (\Delta t)^2 \tag{3}$$

The time step is Δt, the speed is expressed as Eq.4.

$$\ddot{d}_n \approx v_n = \{d_{n+1} - d_{n-1}\} / (2\Delta_t) \tag{4}$$

$$d_{n+1} = g(d_n, d) \tag{5}$$

Eq.5 indicates that the displacement at time $(t_n + \Delta_t)$ can be expressed explicitly by using the displacement at time t_n and $(t_n - \Delta_t)$. Namely, the equation of motion quality matrix and a damping are diagonalized, the current time displacement solution does not need carry on the iteration, and previous time acceleration solution does not need the simultaneous equation group.

Because explicit solution does not need consider the astringent question, and solve the simultaneous equation group, and form the total rigidity matrix also. Besides, solution is quick, and the storage space taken is few, so it is an effective algorithm to solve these kinds of non-linear question. Based on mentioned above, explicit solution method is used for loading process in this paper [8~9].

Hydro-Forming Example of Shaped Tube

Simulation Model of Shaped Tube. The hydro-forming process of shaped tube is that inside of the tube or outside is sealed by the mold, as a whole, a side without the mold is sealed in the vessel. Then the liquid is injected, under the hydraulic pressure function, the tubing is pasted to the mold, thus shaped tube is formed. This simulation process has involved many non-linear analysis technologies, such as the geometry, the material and the boundary condition and so on.

The common tubing is the axial symmetrical characteristic, and the shape of the mold is considered, for the convenience of the simulation, a quarter-lateral section of the tube is chosen in the finite element simulation. Because the components length is longer, the longitudinal stress quite is smaller in the hydro-forming process. The 20mm length pipe with the thickness of 8.89mm in the longitudinal direction is taken for analysis. The tube material that chosed is 25Mn, which is heat treated. Its property value is listed in Table.1.

Table 1 Material property value

Name	Value
Young's module E/GPa	210
Poisson's ratio γ	0.3
Initial Yield strength σ_s/MPa	270
Density ρ/kg·m^{-1}	7.6×10^3

The tube is meshed by the hexahedron "Full Integration", this element has eight nodes as shown in Fig.1. The mold is regarded as rigid parts and the workpiece as a flexible part (see Fig.2). For convenient analysis, the mesh in the FEM must be uniform.

The coulomb law of friction is used in the simulation process, the friction coefficient is 0.12. The boundary condition after the simplification is obtained as shown in Fig.3: Apply1 constrains the longitudinal move of an end which the tube and the mold contacted, apply2 restrains another end along the crosswise migration, and apply3 is the hydraulic pressure acting on the tube which loaded from 0 to 70MPa in 400s. Fig.4 shows the profile figure of the whole model.

Fig.1 Finite element model of the tube   Fig.2 Finite element model of the mold   Fig.3 Boundary condition

Simulation Results and Discussion. In the forming process of shaped tube, the distortion mainly concentrates in the convexity and concave of the mold. Fig.5 shows the colored strain nephogram at 232s. It can be seen that the strain in the convexity and concave of the mold obviously is bigger than other position of the tube. Thereinto, the strain at the convexity are the biggest, because the process is equal to a constant pressure in the convexity, the bending moment caused by the pressure in the concave relative to the material in the convexity is the biggest. It can be seen from Fig.6, with pressure increasing, the bending moment caused in the convexity is growing bigger and bigger. At 287.2s, the material of a side in the convexity is growing unsteady: The thickness arrives thin drastically, by simulating many times, the rule is discovered that unsteady phenomenon always stochastically appears in any side in the convexity, which can be slow down by mending the amplitude of the pressure.

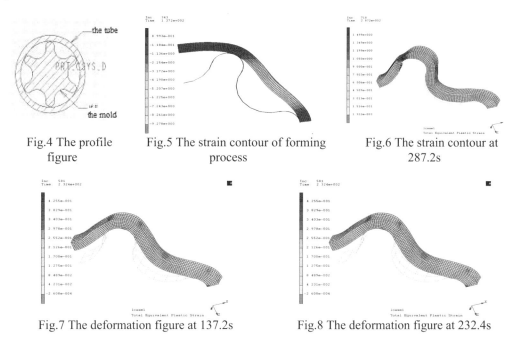

Fig.4 The profile figure   Fig.5 The strain contour of forming process   Fig.6 The strain contour at 287.2s

Fig.7 The deformation figure at 137.2s   Fig.8 The deformation figure at 232.4s

The forming simulation process of shaped tube obviously can be divided two stages as shown in Fig.7 and Fig.8: the initial distortion stage and the continuing distortion stage. The simulation results show that the initial distortion pressure is24 MPa, and the continuing distortion pressure is 40.7 MPa. The deformation strengthening in the distortion process results in the result.

Experimental Validation. In the test, when the pressure is about 64 MPa, unsteady phenomenon appears in any side of the mold convexity. While the simulation pressure is 51.2 MPa, both is close. Similarly the forming process obviously has two distortion stages after the test: The initial distortion pressure is 32 MPa, and the continuing distortion pressure is 52.2 MPa. The forming process agrees well with the simulation process. The error between simulation results and experimental data of The initial distortion pressure is less than 25%, and the continuing distortion pressure is 22%. Considered the simulation process is a simplification process, this result can be accepted.

Conclusion

In this research, relevant methodologies and techniques developed for the hydro-forming of shaped tube have been presented. Research results accumulate a lot of reliable parameters for the technics research of the tube hydro-forming, and provide a basis for the design and forming technics of this kind of components. The application of the CAE technology at the manufacture domain makes the process compact and the production period shortened, thereby, resulting in substantial saving of production costs.

References

[1] T.X.YU, W.Johnson. The Buckling of Annular Plates in Relation to the Deep-drawing press [J]. Int.J.Mech.Sci.1982, 24(3):185-188

[2] Browne D, J.Battikha E. Optimization of Aluminum Sheet Forming using a Flexible die[J]. Journal of Material Processing Technology,1995,55:218-223

[3] Frode Paulsen,Torgeir Welo. Application of numerical simulation in the aluminium-alloy profiles. Journal of Materials Processing Technology, 1996, (58):274-285.

[4] GU Tao; E Daxin; GAO Xiaowei; REN Ying. Finite element simulation and experimental analysis of forming process of pipe-bending. China Metal Forming Equipment & Manufacturing Technology, 2006,1:66-68. (In Chinese)

[5] GU Tao; E Da-xin; REN Ying; GAO Xiao-wei. Finite element simulation and experimental analysis of wall thickness deformation in the process of pipe bending. Die & Mould Industry, 2006, 32(4): 17-20.(In Chinese)

[6] WANG Da-nian. Metal plasticity forming principle [M]. China Machine Press, 1982. (In Chinese)

[7] CHEN Shu-feng, CHEN Zhu-lin. The developing hydro-forming technology of the tube [J]. Construction Machinery. 2001, (8): 35-38. (In Chinese)

[8] H.Guo, D.W.Zuo, S.H.Wang, *et al*. The Application of FEM technology on the deformation analysis of the aero thin-walled frame shape workpiece. Key Engineering Materials, 2006, vol.315-316: 174-179.

[9] Guo Hun, Zuo Dun-wen, Wang Shu-hong, et al. Effect of tool-path on milling accuracy under clamping. Transactions of Nanjing University of Aeronautics & Astronautics, 2005, 22(3): 234-239. (In Chinese)

CHAPTER 2:

Control, Automation and Detection Systems

Advanced Materials Research Vol. 819 (2013) pp 131-135
© *(2013) Trans Tech Publications, Switzerland*
doi:10.4028/www.scientific.net/AMR.819.131

Development of Off-Line Inspection System on Equipment Based on Embedded Linux Technique

Sun Huamin[1, a], Shen Haikuo[1, b], Wang Taiyong[2, c], Nie Meng[1, d]

[1] School of Mechanical, Electronic and Control engineering, Beijing Jiaotong University, Beijing, 100044, PR China

[2] School of Mechanical Engineering, Tianjin University, Tianjin, 300072, PR China

[a]11121407@bjtu.edu.cn, [b]shenhk@bjtu.edu.cn, [c]tywang@189.cn, [d]09116327@bjtu.edu.cn

Keywords: Off-line inspection, Embedded Linux, data acquisition and analysis

Abstract: In order to meet the requirements of the business enterprises for equipment monitoring instruments, an embedded data acquisition and analysis system based on Linux technique has been developed, and its hardware configuration and software structure were described. The system incorporates many functions in a unit, including data acquisition, real-time display, storage, analysis, and communication, etc., and has a high scalability, maintainability and reliability. The experiment results showed that its functions is stable, and can be widely used in the system monitor and diagnosis in various equipment and installations.

Introduction

With the rapid development of modern industry, the structure and the function of the mechanical equipment is more and more complicated and powerful, the degree of automation in production is more and more high, the relationship between each subsystem of the equipment is increasingly close, the loss caused by the equipment failure is extremely huge , so make sure whether mechanical equipment can be safe and reliable motion in the best state, timely discover and predict the occurrence of faults is very necessary[1-2].The traditional off-line monitoring equipment are generally divided into economical type and functional complete type[3]:the price of the economical type is low, but they only have basic signal acquisition function and less signal analysis function, so they are only suitable for regular inspection of equipment; the complete type have the perfect signal analysis function, but most of them are based on PC hardware frame and have large volume and high price, therefore they cannot be easily applied in the field of micro and small[4].In conclusion, the development of the embedded off-line monitoring system with small volume, low cost, perfect signal analysis function and meeting the configurability and open requirements, has become one of the hot and difficult researches in today's advanced manufacturing[5].

Response to the current situation, this research makes full use of the advantages of embedded system, and then develops an embedded equipment condition monitoring system with the modular reconfigurable characteristics. It has built an application development platform for signal acquisition and analysis that uses the YL2410 embedded development board and the S3C2410 embedded microprocessor as the core. The platform has the advantages of the open and scalability, and it is used to provide a reliable basis for the judgment of the running state and the follow-up diagnosis of the equipment. This research starts from building a modular-based embedded data acquisition and analysis system, describes the realization methods and the basic functions of the system in detail, and then elaborates the correctness and feasibility of the embedded data acquisition and analysis system by a monitoring example on the equipment.1 Description of The Pneumatic System

The overall configuration of the embedded data acquisition and analysis system

The monitoring function of this embedded data acquisition and analysis system is mainly achieved through the acquisition and analysis of the vibration signal for the equipment. So different from the general data acquisition and analysis system, it can not only realize the acquisition and the preprocessing of the vibration signal of the equipment, but also realize the waveform display and storage of the acquisition signals. And what's more with a specialized data analysis module which is used to analyze the acquisition data the system can draws the operation state of the equipment. When the users need a further analysis for the acquisition data, the system can also upload the data through the communication interface to a PC, then use the dedicated data analysis software on the PC to diagnose faults [6].

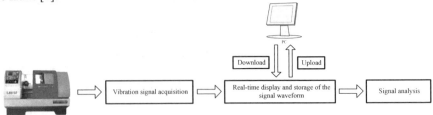

Fig. 1 Function configuration of the embedded data acquisition and analysis system

The hardware configuration of the embedded data acquisition and analysis system

The embedded data acquisition and analysis system uses ARM architecture, its hardware architecture are designed shown in the fig. 2.The system mainly includes signal conditioning module, data acquisition module, and embedded control module. Signal conditioning module mainly completes the analog signal preprocessing.it is mainly used for amplitude amplification, filter or operation for various signals of the voltage, current, etc. That sensor outputs to get the signals that meet users need and the requirements of the signal acquisition card. Data acquisition module is the channel signal into the system, it mainly completes data acquisition, and is responsible for converting the analog signals into digital signals. Embedded control module completes the integration of the whole embedded data acquisition and analysis system and the coordination between each module. And it is the kernel of the whole system. The system chooses embedded motherboard can not only realize the control of the acquisition process and the kernel operation of the data acquisition, but also support many external expansion device, including: liquid crystal display, SD card, keyboard, mouse, RS232 serial port, USB interface, network interface, etc., realize the module control of the LCD display of the data acquisition, the keyboard, the storage and the communication interface.

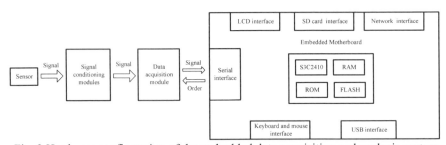

Fig. 2 Hardware configuration of the embedded data acquisition and analysis system

The software configuration of the embedded data acquisition and analysis system

The system software is designed to use Qt / Embedded development tool, with embedded Linux operating system. It takes the idea of object-oriented and modular design, and then make a specific module division according to the different functions of the system. Each module has its specific function, and provides certain services. The system software provides specialized interface functions to realize the connection between the modules, so that each module is not only independent of each other and also keep close contact with each other. The modularity of the system software take the process that realizes system software functions into the process that assembles the function of each module. It makes the system software has a high resistance to external changes, and at the same time also improving the scalability, maintainability and reliability of the system software. Besides, the Qt Designer is a convenient tool to design the GUI visually without writing any source code as to shorten the development cycle [7].

As shown in fig. 3, the system software function modules are mainly composed of four parts, respectively, ①Human-computer interaction module. This part can realize the communication between the users and the system. It is the most direct part for users, including many interfaces like the acquisition parameters setting, the real-time display of the data acquisition and the data analysis results show, etc.;②Data acquisition module. This part can realize the collection and preservation of the vibration and noise signals in the actual operation of the equipment; ③Signal analysis algorithm module. This part can provide Intuitive maps from multiple perspectives, such as time domain and frequency domain. The maps make the users to observe and analyze the collected data more convenient, and make the right judgment for the operation status of the equipment; ④Data management module. According to the users need, one can view the data files stored in the system, and browse, delete the selected files, and at the same time can also transmit data files through the USB port, the serial port, and the network to realize the data interaction with PC.

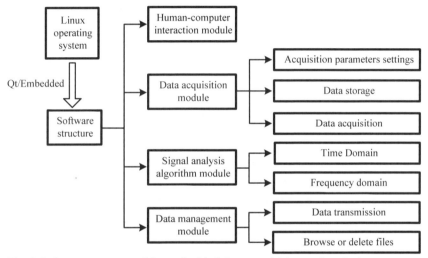

Fig. 3 Software structure of the embedded data acquisition and analysis system

The verification of the embedded data acquisition and analysis system

In order to verify the function of the data acquisition and analysis system, now using the signal generator to generate a sine signal to test the system. The sine signal generated by the generator has the effective value of 1G, the frequency of 200Hz, the screen sampling points of 2048 points, and the analysis frequency of 5000Hz.Then using the data acquisition and analysis system, we can get the original waveform diagram shown in fig. 4,the frequency spectrum diagram shown in fig. 5,and the autocorrelation analysis diagram shown in fig. 6.Fig. 4 is the result show of the time history of the signal, which reflects the waveform of the signal changed over time. From fig. 4, it can be seen that the signal amplitude closes to 1G and basically in line with the original signal. Fig. 5 is the amplitude spectrum analysis of the signal, which reflects the vibration intensity and frequency information of the signal. From fig. 5, it can be seen that the signal frequency is 200Hz, completely in line with the original signal. Fig. 6 is the result show of the autocorrelation analysis of the signal, which reflects the cycle variable of the signal. From fig. 6, it can be seen that the autocorrelation of the signal is still a periodic function, and not happen attenuation without the random noise.

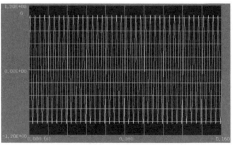

Fig. 4 System structure

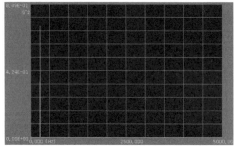

Fig. 5 System structure

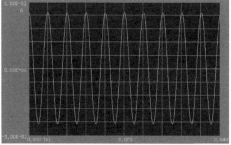

Fig. 6 System structure

By considering the error of the signal generator and the quantization error of acquisition, the experimental results above show that: both the time domain analysis function and the frequency domain analysis function of the data acquisition and analysis system are correct.

Conclusions

1) In order to meet the requirements of the business enterprises for equipment monitoring instruments, this research develops a embedded equipment condition monitoring system with the modular reconfigurable characteristics. It has built an application development platform for condition monitoring that uses the YL2410 embedded development board and the S3C2410 embedded microprocessor as the core. The platform is used to provide a reliable basis for the judgment of the running state and the follow-up diagnosis of the equipment.

2)The embedded data acquisition and analysis system set data acquisition, storage, analysis, communication and other functions in a whole, has many characteristics such as small size, light weight, portability, simple operation ,etc. And the system have validated the function, and obtained good results. In a word, the system can be widely used in the system monitor and diagnosis in various equipment and installations.

References

[1] Huang Weili,Huang Weijian,Wang Fei, et al. Mechanical equipment fault diagnosis technology and its development trend [J], Mining Machinery. Vol.33,No.1,2005,66-68.

[2] Zhong Binglin,Huang Ren.Mechanical Fault Diagnostics[M].BeiJing: Machinery Industry Press, 2002,6-7 .

[3] Zhao Yanju, Wang Taiyong,Xu Yue,et al.Development of portable fault diagnosis instrument based on dual CPU[J].Jounral of Jilin University(Engineering and Technology Edition), Vol.38,No.3,2008,557- 560.

[4] Zu Shuzhi, Wang Taiyong,Deng Xuexin,et al.Protable system for condition monitoring and signal analysis[J].Jounral of Jilin University(Engineering and Technology Edition), Vol.35,No.1,2005, 101-105.

[5] Xu Xiaoli. Progress of monitoring and diagnosis technology of electrical and mechanical equipment [M].BeiJing: China Aerospace Press,2003,11-15

[6]He Huilong,Wang Taiyong,Xu Yonggang,et al.Implementation of fault diagnosis system for mechanical equipment based on internet for plant management[J].Jounral of Jilin University(Engineering and Technology Edition), Vol.36,No.5,2006, 691-695

[7] Jasmin Blanchette, Mark Summerfield. C++ GUI Programming with Qt 4[M]. BeiJing: Electronics Industry Press, 2011,17-19

Advanced Materials Research Vol. 819 (2013) pp 136-139
© (2013) Trans Tech Publications, Switzerland
doi:10.4028/www.scientific.net/AMR.819.136

Design and Implementation of Online Monitoring and Remote Diagnostic System for CNC Machine Tools

Jing luyang[1,a], Wang Taiyong[1,b], Chen Dongxiang[1]and Fang Jingxiang[1]

[1]Key laboratory of mechanism theory and equipment design of ministry of education, Tianjin300072, China

[a]jingluyang@yeah.net, [b]tywang@189.com

Keywords: CNC machine tools, online monitoring, fault diagnosis.

Abstract. With the development of network technology and fault diagnosis technology, monitoring and diagnosis methods for the CNC machine tools had a great change. In this paper, an online monitoring and remote diagnosis system for CNC machine tools was built. The system was consisted of the multi-channel online acquisition system and remote fault diagnosis system. The online acquisition system achieved a real-time monitoring for CNC machine tools. The remote fault diagnosis system provided the management of devices and assistant for experts to analyze data which was uploaded from acquisition system. The system offered real-time state information of CNC machine tools and reduced downtime of machine effectively.

Introduction

CNC machine tools is a complex electromechanical system as a whole set of machines, electric and hydraulic. Due to the complexity of CNC machine tools and diversity of machined parts, it is tedious and difficult to monitor, diagnose and repair faults for CNC machine tools [1]. In order to detect the faults of machine complex system, monitor the processing status, improve the efficiency and reliability of the CNC machining, the design and implementation of monitoring and diagnostic system for CNC machine tools is needed [2].

With the development of network technology and fault diagnosis technology, it is possible to implement the real-time monitoring and remote fault diagnosis system for CNC machine tools. In this paper, a multi-channel online monitoring and remote diagnostic system for CNC machine tools was built. The multi-parameter of CNC machine tools was monitored and faults of CNC machine tools were analyzed and diagnosed by the remote host computer.

System framework

The online monitoring and remote diagnostic system for CNC machine tools was consisted of two parts: the multi-channel online acquisition system, which achieved 32-channel online data acquisition and provided real-time multi-parameter monitoring for CNC machine tools; the remote device management and fault diagnosis system, which provided device management function and remote analysis and diagnosis of faults. The framework of the system was shown in Fig. 1.

Online monitoring system

Multi-threaded and message queues programming design. Multi-threaded and message queues programming was used to solve overflow of the data acquisition card [2]. When sampling started, a buffer storing three message queues was established. At first the data acquisition device obtained a buffer from the ready queue and placed it in the run queue. This buffer stored data according to the specified clock rate. Once the buffer was full, the data acquisition device transferred the buffer from the run queue to the completion queue and thread 1 was started to map full queue data to the user buffer. Thread 2 then took data from the user buffer and completed display, storage and post-processing of the data. The system message queue is shown in Fig. 2.

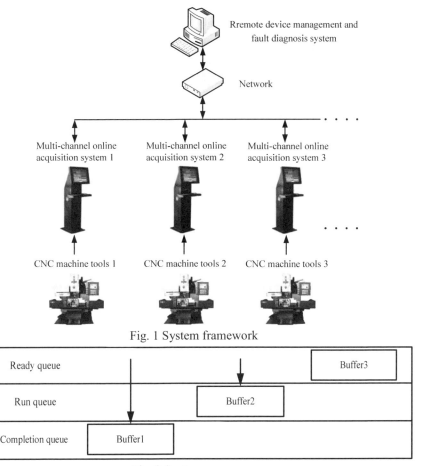

Fig. 1 System framework

Fig. 2 System message queue

Multi-channel online acquisition.A variety of data communication and Multi-channel online monitoring were supported. The sensors picked up data of each machine at work site. Then after signal conditioning circuit data was collected and contiguously stored in card memory. Acquisition screen was real-time displayed in IPC at work site, as shown in Fig. 3. Alarm mode was used. When monitoring value was greater than the alarm value, system would start alarm, providing an easy way to capture sudden failure. Automatic data acquisition, save and remote backup were supported, as shown in Fig. 4. The system was compatible to a variety of sensors, such as acceleration, velocity, pressure, concentration, PH value, the eddy current.

Remote diagnosis system

Remote control module. Remote control module used socket technology to provide the management for uploading or downloading the data of each acquisition terminal. The connection messages of acquisition terminals were listened, which returned system verification information at the same time [3]. The identity of terminals was authenticated.

The Remote control module has the following features:Listening to system verification information of the server and setting the number of concurrent users; Encrypting data packets to ensured the security of the system; Listening the feedback data of acquisition terminals to provided connection status to the acquisition terminal.

   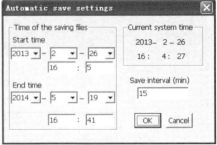

Fig. 3 32-channel online monitoring   Fig. 4 Automatic data save

Device management and fault diagnosis module. Device management and fault diagnosis module used a treeview to provide device management, which was consisted of plant factory, branch factory, workshop equipment, monitoring points and physical quantities. A graphical interface was used and property of each nodes made it easy to check the location information of machine measuring points and classification management of measurement data. At the same time, operatic functions, such as add and delete of measuring points, could be easily operated according to the specific needs. Searching functions for equipment and data were provided, which could count the number of measuring points in the factory, workshop and equipment or queried the alarm data [4].

This module contained the signal pre-processing methods, signal time-domain analysis, signal frequency-domain analysis, signal time-frequency analysis and feature extraction methods. Feature extraction and fault diagnosis methods included parameter alarm, narrowband alarm, characteristic frequency calculation, feature information extraction, neural networks, state prediction and fault diagnosis report [5]. The running interface of signal analysis was shown as Fig. 5. Experts used these functions to analyze the equipment, guiding the site personnel's the commissioning and maintenance.

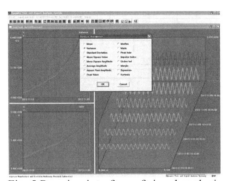

Fig. 5 Running interface of signal analysis

Site test

Cooperated with a company, the system testing on site was achieved. Fig. 6 showed the execution process of the online monitoring and remote diagnostic system for CNC machine tools. According to the tasks, acquisition of online monitoring system was implemented. Then the data was uploaded to the remote device management and fault diagnosis system through network. The professionals used remote device management and fault diagnosis system for data analysis and processing, and generated the corresponding fault diagnosis report which could be auto-transmitted to the work site.

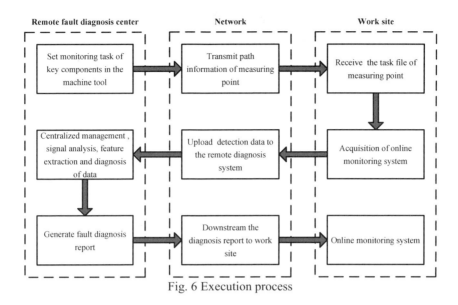

Fig. 6 Execution process

Conclusion

In this paper, sensor technology, computer network technology, fault diagnosis technology and modern equipment management theory were used to build an online monitoring and remote diagnosis system. The online monitoring for CNC machine tools was provided. Device management and remote diagnosis were achieved. The system provided real-time state information of CNC machine tools and assisted experts to diagnose faults, which reduced downtime of machine effectively and improved the level of maintenance.

Acknowledgment

This work was supported by National Science & Technology Pillar Program (2013BAF06B00), Doctoral Fund of Ministry of Education of China (20100032110006), Research Program of Application Foundation and Advanced Technology of Tianjin (12JCQNJC02500).

References

[1] Wang Taiyong, Jiang Yongxiang, Liu Lu, et al, Dynamic monitoring-control and intelligent diagnosis techniques for of complex manufacturing systems, Aeronautical Manufacturing Technology. 13 (2010) 27-29.

[2] Lin Jinzhou, Jiang Dayong, Geng Bo, et al, Research on the remote monitoring and fault diagnosis of CNC system based on network, 2012 2th International Conference on Functional Manufacturing and Mechanical Dynamics, China: Hangzhou. 141 (2012) 465-470.

[3] Shen Aiqun, Nie Zhonghua, Xing Yan, Research on the remote online monitoring and diagnostics system of CNC machine tool, Journal of Manufacturing Automation. 25 (2003) 37-39.

[4] Ma Yutao, Ye Wenhua, Design of data Collecting System of NC Machine Based on ARM and Linux. Machinery Manufacturing and Automation. 4 (2008) 110-112.

[5] He Huilong, Wang Taiyong, Xu Yonggang, Implementation of fault diagnosis system for mechanical equipment based on internet for plant management, Journal of Jinlin University. 5 (2006) 691-695.

Advanced Materials Research Vol. 819 (2013) pp 140-143
© *(2013) Trans Tech Publications, Switzerland*
doi:10.4028/www.scientific.net/AMR.819.140

The Research on Modular Adaptable Design Platform for Non-standard Waste Detection Equipment

Bing Cheng[1,a], Taiyong Wang[1,b], and Xinhua Xiao[1,c]

[1]Key Laboratory of Mechanism Theory and Equipment Design of Ministry of Education, School of Mechanical Engineering, Tianjin University, Tianjin 300072, China

[a]chengbing198805@163.com (corresponding author), [b]tywang@189.cn, [c]xxinhua@126.com

Keywords: adaptable design platform, base type, module division, national specialized standard.

Abstract. The design of non-standard waste detection equipment costs much manpower and material resources. The company need reduce unnecessary duplication of work to respond quickly to market needs. The equipment can be designed on the modular adaptable design platform. The function analysis of waste detection equipment brings the base type of product. Modules are the basic units in product design on the platform. A new product can be quickly designed based on the base type by reconfiguring, upgrading general modules and replacing special modules on the platform.

Introduction

Increasingly fierce market competition, the traditional design and production methods couldn't adapt to the rapid changes in the market demand. Increasing product diversity and personalization becomes an important means to meet customer needs and capture the market [1]. A large number of non-standard equipments appear in the mechanical products with the product personalizing. The traditional design methods can't afford to design the non-standard equipment quickly. So we need increase product external diversity, meanwhile decrease internal diversity [1]. The changes in product demands also contribute to the changes of production mode. The production of personalized products need be organized with small batches or even a single piece. The reduction in the production batch means the increase of production costs to a certain extent. The traditional scale production can not adapt to changes in market demand. Therefore, it is an agent need to solve the problem of designing new products to meet the market in the shortest time and at low costs.

Research on modular technology for waste detection equipment

The vehicle and internal combustion engine industry are developing rapidly in recent years in China. Product quality and safety testing have become increasingly important. Air tightness testing is an important means to ensure the quality and safety [2]. In the automotive industry, cylinder block, valve and gearbox need strict air tightness testing by the waste detection equipment before leaving the factory. It is a great need to design and produce customized waste detection equipment, which can be achieved on the modular adaptable design platform. Its modular technology is mainly just as follows.

Plan product spectrum series

Functional decomposition of the waste detection equipment is based on the function of structure and takes the device under test as the core. The different base types are denoted by the letter A, B, C…….
Different variants of the same basic type are denoted by two English letters. The former is the basic type category, and the latter is the version number of different variants. For example, the different variants of base type A can be denoted by AA, AB, AC……. The frequently used base type codes are shown in Table 1.

Table 1 Codes of base types and variants

the DUT name	Waste detection method	Transmission Type	Mechanical structure	Characteristic Description	Code
JL clutch	KC	1Y, 2C, 12R	Fixed station, Single online		A
JL gearbox	KC	1Y, 2H	Fixed station, Single online	Horizontal mixed compression	AA
air conditioner compressor	KZ	1Z, 2Q, 6Q	Fixed station, Multi-line	gas - liquid booster	B
XH crankcase	KS	1Y, 2Q,5Y,7Q,8Q,	Passing type, Suspension type		C
JL crankcase	KC	3Y,(2Q,2Y,2H),7 Q,8Q			D
LX battery	KC	1Q, 2Q,13R, 9D	Rotary table, 4 station	Rotary table	E

For the different base types of waste detection equipments, design of the new product costs much time. We need divide the whole machine into different modules according to functional decomposition based on the generalized modular design [3] [4]. The new product can be designed on the platform with the least time by reconfiguring, upgrading general modules and replacing special modules. And the design of a new product prototype could be started by module adjustment instead of starting from scratch. The modules and their codes are shown in Table 2.

Table 2 Module division and module codes of the waste detection equipment

Code	Module name	Code	Module name
00	worktable standard library	52	lower fixture
01	worktable	60	waste detection device standard library
10	protective devices standard library	61	air medium waste detection device
11	protective devices	62	helium medium waste detection device
		63	water medium waste detection device
20	main frame standard library	64	oil medium waste detection device
21	main frame		
		70	feed device standard library
30	pressure system standard library		
31	pressure system	80	electrical system standard library
		81	electrical system
40	hydraulic system standard library		
41	hydraulic system	90	attachments and other standard library
		91	packing box and brand
50	fixture standard library	92	supplied accessories
51	upper fixture		

Build the module library

After the functional decomposition and module division, build the module library of the waste detection as general modules and special modules. For the general modules, the module library can be built from a mother module. Extract the driving parameters of the mother module by the program as shown in Fig 1, which is done by secondary development based on SolidWorks with Visual Basic 2008. Pick the drive parameter from the three-dimension graph of the waste detection equipment. Then click the button 'Add a module variable'. A new name of the drive parameter can be given by the button 'Give a new parameter name' if it is needed. The specification code, parameter connector and code segment connector are also can be given by the program and checked if they are unique in the SQL Server database. After the checking, we can get the part instance code. Save all the information into the database.

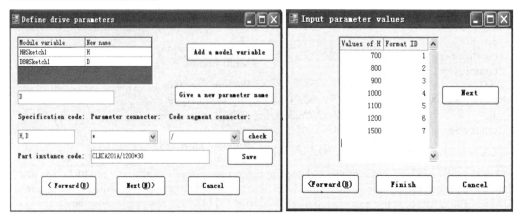

Fig. 1 Define drive parameters   Fig. 2 Input parameter values

After extracting the driving parameters of the mother module, assignment all the parameter values to each parameter by the program shown in Fig 2. All the values of each driving parameter can be inputted into SQL SERVER while checking in the module. Then a generalized module library is created in the platform.

Then the new product can be generated by reconfiguring, upgrading general modules and replacing special modules.

Table 3 Canonical format of the data change records

serial number	1	2	3	4
type	update	add	update	delete
Table name	GB_TYPE_HOLES	GB_DATA_Fastener_B OLT_TTB_LD	GB_DATA_Fastener_ BOLT_TTB_LD	GB_TYPE_N UTS
Type of ID	key		key	key
ID	3		132	17
Field Name	HoleDescriptin Format	key	SIZE	Title
Field values	Tap Drill for %size TapDrills	132	M6	

National Specialized Standard on Toolbox

After building the general module library, the standard parts used on the platform for designing need be built as the special module library. Toolbox is SolidWorks's own library of standard parts. If we take SolidWorks software system as the software platform for product design, Toolbox may be the best choice for standard parts. However, the Toolbox differs from China's national standards and the standards and habits of Chinese enterprises. So it is necessary to do national specialized standard on Toolbox. It is needed to modify the Access database that stores the data of the Toolbox. The method is given as follows.

Normalized treatment the change records and set data change records in the canonical format, as shown in Table 3.

Use the following statement to update the existing data.

Dim cmd2 As New OleDb.OleDbCommand("UPDATE [" & d3 & "] set [" & d6 & "] = '" & d7 _ & "' where [" & d4 & "] = '" & d5 & "'")

If some new data need be added to the database, add a data row using the following statement to the Access first of all.

Dim cmd1 As New OleDb.OleDbCommand("insert into [" & d3 & "] ([" & [d6] & "]) values ('" _ & d7 & "')")

For other data in the new data row, use the data updating statement.Use the following statement to delete data.

Dim cmd3 As New OleDb.OleDbCommand(" UPDATE [" & d3 & "] set [" & d6 & "] = _ null where [" & d4 & "] = " & d5)

d1, d2, d3, d3, d4, d5, d6 and d7 represent the seven data in Table 3 from left to right in one row.

Summary

Non-standard equipments can be quickly designed by reconfiguring, upgrading general modules and replacing special modules on the modular adaptable design platform. This paper gives an example of waste detection equipment to plan product spectrum series and divide the machine into modules. It gives the realization methods of extracting the driving parameters and saving the parameter definition domain of the mother module. It also gives the program of national specialized standard on toolbox.

Acknowledgements

This work was financially supported by National Key Technology R&D Program of the Ministry of Science and Technology of China (2013BAF06B00), the Research Fund for the Doctoral Program of Higher Education of China (20100032110006), Fujian Science and Technology Projects of China (2012H1008), Tianjin Application Foundation and Frontier Technology Research Program (12JCQNJC02500) and Tianjin Municipal Science and Technology Commission of China (12ZCZDGX01600).

References

[1] Cheng Qiang, Liu Zhifeng, Cai Ligang, et al. Planning method for product platform based on axiomatic design. Computer Integrated Manufacturing Systems. 16(2010) 1587-1596.

[2] Zhengde Zhu. The exploration and practice of calibration on the hermetically sealed sample in leak detection. Modular Machine Tool & Automatic Manufacturing Technique. 12(2000) 3-6.

[3] Xu Yanshen，Xu Qianli, Chen Yongliang, et al. Research on generalized modular design method with flexible modules of product structures. Chinese Journal of Mechanical Engineering. 17(2004) 447-480.

[4] Weiguo Gao,Yanshen Xu,Yongliang Chen,Qing Zhang. The principle and method of generalized modular design. Chinese Journal of Mechanical Engineering (CJME). 43(2007) 48-54.

Advanced Materials Research Vol. 819 (2013) pp 144-149
© *(2013) Trans Tech Publications, Switzerland*
doi:10.4028/www.scientific.net/AMR.819.144

Centralized Monitoring Method for Isomeric Heat Treatment Equipments

Xiao Xinhua[1,2,a], Yang Xiaofeng[3,b], Yu Hongbin[1,2,c], Peng Junqiang[1,2,d]

[1]School of Mechanical Engineering, Tianjin Polytechnic University, Tianjin, China

[2]Tianjin Key Laboratory of Advanced Mechatronics Equipment Technology, Tianjin, China

[3]Product Development Department, Tianjin Hongda Textile Machinery CO., LTD, Tianjin, China

[a]xxinhua@126.com,[b]yxf101@126.com,[c]hbyu@yahoo.cn,[d]junqiangpeng@tom.com

Keywords: centralized temperature monitoring; ilsomeric equipment; framework; communicate protocol matrix; dynamic protocol adapter; knowledge reuse

Abstract: Heat treatment equipment are usually harmful to people, they need to be monitored remotely. As the fact that different types of equipment are used in factories, which leads to the problem that monitoring equipment is difficult. Therefore, research on remote centralized temperature monitoring system is necessary. The concept of isomeric equipment is proposed. The framework of monitoring system is established, which is constituted of temperature control instruments, monitoring server and clients. The functions of monitoring system are described, such as setting temperature, receiving temperature, showing the chart of temperature, accumulating and reusing heat treatment experience etc. The key technologies of monitoring system including dynamic protocol adapter and knowledge accumulation and reuse are also explained. Finally, the system is applied in a factory. The engineer can monitor almost all of the heat treatment equipment only by use the system in the office.

Introduction

Temperature is the most important factor impacting on produce quality during the metal heat treatment process. With the development of enterprise, more and more heat treatment equipment are put into production. Monitoring so many different kinds of equipment besides them is difficult and not economical, the reasons as follow:

Monitoring equipment manually cannot record the temperature data instantly; therefore we cannot review the data history and analyze the temperature change rule.

In order to improve the quality during the heat treatment process, inert gases are put into the equipment. These gases usually are harm to the health of monitoring people.

Monitoring and setting on work field needs many people, which increase the cost of production. Incorrect data are set for lack of protect mechanism now and then.

In the heat treatment, we need to adjust temperature accurately according to the process. Monitoring and setting on work field cannot control the temperature based on model. Even there is a similar treatment process, we cannot reuse the temperature setting experience.

The temperature data is not send to the manage department of enterprise and the produce information is not send to the equipment. There is a fault age between the equipment and managers.

Many scholars have research on centralized monitoring system of variant equipment [1~3]. Some manufacturers provide centralized monitoring system for their own equipment. Because of this reason that variant equipment are used even in one enterprise, the centralized monitoring system for one type of equipment is not applied widely. In this paper, the centralized temperature monitoring system of variant heat treatment equipment are researched and applied in a factory, which works well in practice.

Concept of Isomeric Equipment

Temperature control instrument (TCN) is wildly used in industry for its powerful function and good stability. Therefore, automatic heat treatment equipment are almost based on TCN. The control system of heat treatment equipment is shown in Fig.1. According to heat process, proper temperature value is set through the TCN. The furnace temperature obtained by sensors is sent to the TCN and compared with the setting temperature. The TCN will send signal to control the furnace current flow according to the compared result. If the furnace temperature is higher than set value, the TCN will send signal to decrease the furnace current flow in order to lower temperature. Otherwise, if the furnace temperature is lower than set value, the TCN will send signal to increase the furnace current flow in order to raise temperature. Hence, the furnace temperature is always consistent with the setting value.

In order to communicate with other devices, TCN supports serial ports. However, different type of TCN has different communication protocol. Which leads to monitor different types of heat treatment equipment is difficult. The concept of isomeric equipment is that the control system of heat treatment equipment is based on TCN, and the TCN support serial ports with different communication protocol.

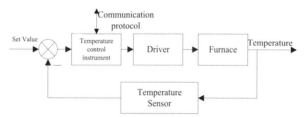

Fig.1 Control system of heat treatment equipment

Principle of Centralized Monitoring

Concept of Dynamic Protocol Adapter. To realize centralized monitoring, we should comprehend the communication protocol between the masters and the slaves. In the monitoring network, the master side means personal computer (PC), the slave side means the TCN. Although the content of communication protocol differs from the others, their communication formats are similar. Take SR90 Series TCN for example [4~5], its communication format is shown as Fig.2.The communication format comprises the basic format portion I, the text portion and the basic format portion II.

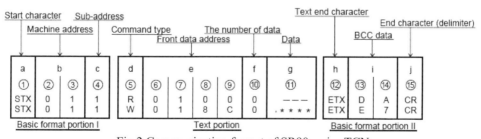

Fig.2 Communication format of SR90 series TCN

For the sake of communicating between the masters and the slaves with different protocols, the dynamic protocol adapter (DPA) is put forward. Its working principle describes as follow:

Machine addresses [6], communication protocols and their relations are stored in database (DB). Therefore, we can add, delete or modify protocol related with special machine address.

The rules of temperature variety are also stored in DB. Combined with communication formats, communication commands are built and sent to TCN, and the furnace temperature are set.

By sending response command to TCN, we can read the furnace temperature and save them in DB.

The engineers can retrieve the temperature history of heat treatment procedure so that find the factors related to produce quality and make better produce planning.

The Structure of DPA. Based on the DPA working principle, the DPA system was designed. Its structure was shown in Fig.3.The TCN address, number, type, Manufacture Company, etc. are stored in TCN information DB. The TCN's communication protocol such as length, format is stored in protocol DB. The temperature data will be set to TCNs and detected from TCNs are stored in temperature DB. The scan frequency, alarm scope and user privilege are stored in parameter DB.

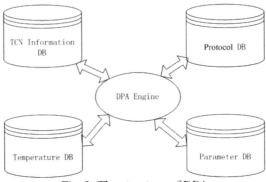

Fig. 3 The structure of DPA

The DPA engine will build communication command according TCN address and it's protocol and send command to each TCN using special frequency. The received data will be stored in temperature DB.

Framework of the Centralized Monitoring System. Based on the dynamic protocol adapter method, the framework of the centralized monitoring system of isomeric equipment is brought forward, shown in Fig.4

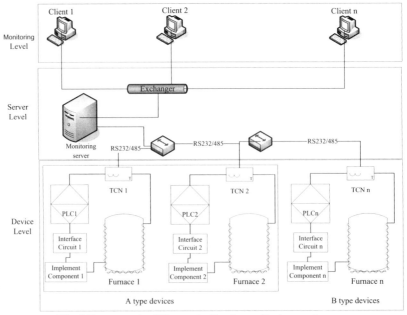

Fig.4 Framework of the centralized monitoring system

The framework comprises three levels.

The first level is device level, which is composed of different types of heat treatment equipments. All of the equipments are connected via RS485, which is supported by TCN. However, there is only RS232 interface on general PC, so RS485/RS232 converters are required to send signal to PC.

The second level is monitoring sever level, which connects devices level and clients. The server level not only in charge of receiving temperature data from devices and saving data to DB, but also getting data from clients and sending them to devices.

The third level is monitoring level, which is composed of PC. Engineers use clients to set data or get data from devices so as to monitoring working condition.

Realization

Functions of Centralized Monitoring System. The function of centralized monitoring system is mainly consists of three Modules. The first module is responsible for maintaining equipment addresses and communication protocols. The second module is responsible for monitoring equipment, including obtaining temperature data from devices and setting the needed temperature value to devices according to the heat treatment process. The third module is responsible for processing the temperature data and displaying data history in chart.

Development of Centralized Monitoring System. Based on Microsoft SQL Server 2000 and Microsoft Visual Basic 6.0, The centralized monitoring system was developed. The procedures of program are listed as follow:

Get DB connection string, open the connection between the monitor client and the server.

Set communication port number, bit rate and open it.

Get TCN address and protocols from the DB, build the monitoring command and send them to the TCNs. If the command type is reading, the TCNs send temperature data to the server and the data will be stored in DB. If the command type is writing, the server will compute the control data and build the command context according the process of heat treatment stored in DB.

Set alarm range for each TCN. When the temperature of furnace is over the limit, the monitor server will build the command for decreasing temperature and send to the TCN. Therefore, the temperature of furnace will fluctuate around the set value.

CASE

We developed a centralized monitoring system of isomeric equipments for a heat treatment enterprise. Fig.5 shows the temperature monitoring UI. The big group box means a kind of furnace. A small group box means a furnace of the specific type. Sixteen furnaces of six types can be monitored at the same time. We can see the real temperature value of the equipments and set data value to them. If the deviation between the real value and setting value excess the accuracy range, the background color of corresponding furnace will be red, if the furnace is not running, the back color will be gray, otherwise it will be blue.

Fig.6. shows real temperature curve during heat treatment process. From this curve, we can analyze the relationship between furnace temperatures and product quality. Therefore enable us adjust the set value correctly in order to improve quality. At the same time, the data history of heat treatment process is an important report for finding and solving problems.

Summary

The paper put forward the concept of isomeric equipment and explains it in detail. Using dynamic protocol adapter technique, isomeric equipment are connected based on RS485/RS232 network. The framework comprising device level, server level and client level is brought up. According the framework, a centralized monitoring system is developed and applied in practice. By using this system, different kinds of heat treatment equipment are monitored easily and data history in produce are saved. Therefore, human resources and work intensity are reduced, furnace temperature control is optimized, and the heat treatment experience and knowledge can be reused conveniently.

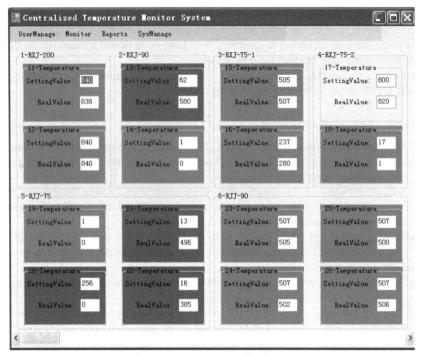

Fig.5 Temperature monitoring UI

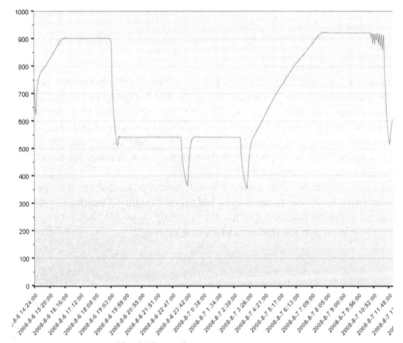

Fig.6 The real temperature curve

Acknowledgement

Tianjin SME Technology Innovation Fund, China (No. 10ZXCXGX09200).

References

[1] SHIMADEN CO., LTD,SR90 Series Digital Controller Communication interface instruction manual, Japan,2001.

[2] ZHANG Hui-xian,ZHAO Yan-feng,Serial Communication Between FP93 Temperature Controller and Computer, Microcomputer Information, 24（2008）130~131， 159.

[3] Liang Xiu-xia,Liu yue,Jin Luxiao,etc, On course parameter on-line monitoring system of process control based on RS485 bus,2009 4th IEEE Conference on Industrial Electronics and Applications,Xi'an, China, 25-27 May 2009

[4] Fu ZY, Pan P, Zeng M, Data acquisition and supervisory system of DSP based on VB International Conference on Informational Technology and Environmental System Science,Henan Polytechn Univ, Jiaozuo, PEOPLES R CHINA, MAY 15-17, 2008

[5] Song Dongdong, Ren Zhenhui,Gu Yanxia,etc.,Design and implement of an intelligent coal mine monitoring system 8th International Conference on Electronic Measurement & Instruments. ICEMI 200,.Xi'an, China, 16-18 August 2007

[6] Li Donaming, Liu Yongfu, A study on intelligent greenhouse temperature measurement system based on RS485BUS International Conference on Agriculture Engineering,Baoding, CHINA, OCT 20-22, 2007

Advanced Materials Research Vol. 819 (2013) pp 150-154
© *(2013) Trans Tech Publications, Switzerland*
doi:10.4028/www.scientific.net/AMR.819.150

Dynamics Analysis of ADCP Carrier and its Mooring System

Zhenjiang Yu[1, a], Yuan Dai[1, b], Peng Huang[1, c], Xiaopeng Zhang[1, d], Zongyu Chang[1, e]

[1]Engineering College, Ocean University of China, Qingdao 266100, China

[a]yuzhenjiang521@126.com, [b]daiyuan8@163.com, [c]huangpeng0532@126.com, [d]pengxiaozhang123@126.com, [e]zongyuchang@ouc.edu.cn (corresponding author)

Keywords: ADCP Carrier, Drag Coefficients, Numerical Simulation, Multi-body Dynamics Analysis

Abstract: Acoustic Doppler Current Profiler (ADCP) is one effective instrument to measure the sea current and is widely utilized in oceanographic research and ocean engineering. Some structure or frame is utilized to carry or hold ADCP in mooring system. This paper studies the hydrodynamics of different kinds of ADCP carriers and the motion of mooring system with ADCP carriers. By applying Fluent software external flow field of ADCP carrier is analyzed. Meanwhile, the velocity distribution and pressure distribution are obtained. Drag coefficients of ADCP carrier are used to analyze the mechanics and profile of mooring system with ADCP. Designing the ADCP carrier and mooring system is very useful which can provide theoretic basis for current data analysis.

Introduction

Acoustic Doppler Current Profiler (ADCP) has developed rapidly in recent decades across the world and was widely utilized in marine exploration, marine environment and so on. The vertical profile of ocean current can be got by ADCP. Generally, the properties of carriers such as drag were measured by experiments. However, the properties could be affected by huge cost and cycle period during experiments. Conditions of flume would be also affected by many factors such as scale effect and interference [1].

When a rolling or pitching movement of ADCP happened, there would be tremendous influence on the measurement [2]. The movement about the ADCP carrier was studied in the paper. Using Fluent software, flow field, drag size and flow field distribution about three different kinds of carriers were obtained [3]. Moreover, pressure distribution and drag coefficient CD of the three kinds of carriers could also be obtained [4, 5]. The whole system was simplified to 12 discrete nodes, and attitude about mooring system was analyzed by Matlab.

Anchored structure with ADCP

The existing ADCP (75 kHz) provided the foundation for study. Its dimensions was shown in Fig.1, Maximum length for 1014mm, Maximum width for 550mm. Sizes of each carrier were based on the existing ADCP (75 kHz). Mooring system which can carry ADCP was shown in Fig. 2.

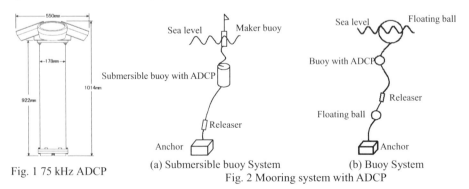

Fig. 1 75 kHz ADCP

(a) Submersible buoy System   (b) Buoy System

Fig. 2 Mooring system with ADCP

(a) Spherical carrier

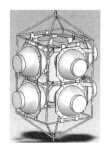
(b) Carrier with eight floating ball
Fig.3 ADCP carriers

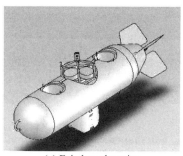
(c) Fairshaped carrier

Three kinds of ADCP carriers were shown in Fig.3. Fairshaped carrier can provide more buoyancy than other carriers but it was also heavier than other carriers. And spherical carriers were commonly used. The transport and install of the carrier with eight floating ball was very easy.

Hydrodynamic dynamics analysis of ADCP carrier

The hydrodynamic dynamics of ADCP was simulated by Fluent with uncoupled implicit algorithm to solve three-dimensional stationary flow. Each Reynolds number was more than 4000, so the turbulence model was chosen. According to the working environment of ADCP, we defined the fluid material physical properties as density $\rho = 1000 kg/m^3$, viscosity coefficient $\mu = 1.518 \times 10^{-3} kg/m \cdot s$. No slipping boundary condition was utilized for wall boundary. The normal velocity was $V_n = 0$ [6].

Spherical carrier. The overall characteristics of the length L was 1000 mm. Maximum width B was 1000 mm. Maximum height H was 1100 m. Fluid area which was rectangular domain was 15L×10B×10H. The grid result was shown in Fig. 4.Pressure distribution was shown in Fig. 5.

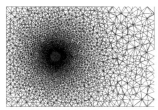
Fig. 4 Grid of carrier

Fig. 5 Pressure distribution

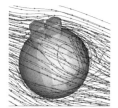
Fig. 6 Streamlines

There were two vortexes behind the floating balls near separation point, as shown in the Fig.6. So pressure between forward and backward became larger. The results of pressure, viscous drag and total drag were 166.78N, 8.64N and 175.42N, respectively. So pressure played an important role in practice.

Carrier with eight floating ball. The overall characteristic of the length L was 1140 mm. Maximum width B was 1140 mm. Maximum height H was 1300 m. Fluid area which was rectangular domain was 15L×10B×10H. The grid result was shown in Fig. 7. Pressure distribution was shown in Fig. 8.

There were two vortexes behind the floating balls near the separation point, as shown in the Fig.9. So pressure between forward and backward became larger. The results of pressure, viscous drag and total drag were 392.25N, 12.20N and 404.45N, respectively.

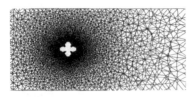  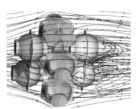

Fig. 7 Grid of carrier   Fig. 8 Pressure distribution   Fig. 9 Streamlines

Fairshaped carrier. The overall characteristic of the length L was 2700 mm. Maximum width B was 640 mm. Maximum height H was 1010 m. Fluid area which was rectangular domain was 15L×10B×10H. The grid result was shown in Fig. 10. Pressure distribution was shown in Fig. 11.

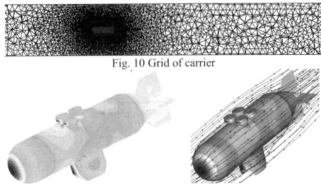

Fig. 10 Grid of carrier

Fig. 11 Pressure distribution   Fig. 12 Streamlines

There were no vortexes behind the floating balls as shown in the Fig. 12. So pressure drag between forward and backward were the lowest. The results of pressure drag, viscous drag and total drag were 59.87N, 16.21N and 76.09N, respectively.

Analysis of drag parameters.

Table 1 Drag parameters of models

Type of carrier	Drag coefficient (N)	Pressure drag (N)	Viscous drag (N)	Total drag (N)
Spherical carrier	0.447	166.78	8.64	175.42
Carrier with eight floating balls	0.7	392.25	12.20	404.45
Fairshaped carrier	0.397	59.87	16.21	76.08

As is shown in Table 1, the drag coefficient and total drag of fairshaped carrier was the lowest. However, drag coefficient and total drag of the carrier with eight floating was the largest. It was clearly that distance between vortex and carrier was very close. So pressure between forward and backward became larger. There was no vortex near the fairshaped carrier's tail, so pressure became smaller. The viscous drag of fairshaped carriers was the largest, because their surface area was the largest. Instead, viscous drag of spherical carriers was the smallest, because their surface area was the smallest. The drag of fairshaped carrier was reduced by 56.62%, compared with spherical carrier.

Multi-body dynamics analysis of mooring system

The attitude of the mooring system was analyzed. The whole system is simplified to 12 discrete nodes using lumped-mass method [7], as shown in Fig.13. Among these nodes, node 1, 2, 4, and 10 was anchor, release, frame of specimen and carrier, respectively. Node 3, 5, 6, 7, 8, 9, 11, and 12 was floating made of glass. Physical parameters of mooring system were shown in Table 2.

Node i	1	2	3	4	5	6
Mass of node m_i (Kg)	1000	75	45	700	26.5	30
Buoyancy B_i (N)	250	202	1140	1952	106	752
Immersed area A_x (m^2)	0.1	0.15	0.375	0.8	0.05	0.25
Immersed area A_y (m^2)	0.5	0.05	0.125	0.2	0.01	0.125
The length of rope L (m)	0	6	3	4	7	6
Elastic coefficient (N/m)	48000	48000	48000	48000	48000	48000
Node i	7	8	9	10	11	12
Mass of node m_i (Kg)	60	75	75		15	15
Buoyancy B_i (N)	1518	1898	1896		372	372
Immersed area A_x (m^2)	0.5	0.625	0.625		0.125	0.125
Immersed area A_y (m^2)	0.125	0.125	0.125		0.125	0.125
The length of rope L (m)	5	6	7	9	5	5
Elastic coefficient (N/m)	48000	48000	48000	48000	48000	48000

Table 2 Physical parameters of system

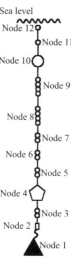

Fig. 13 Mooring system

Dynamic equations
Dynamic equation of model can be obtained in Ref. [7]:

$$(m_i + m_{ai})\frac{d\vec{U}_i}{dt} = m_{vi}\frac{d\vec{U}_{wi}}{dt} + \vec{B}_i + \vec{W}_i + \vec{T}_i + \vec{F}_{Di} + \vec{f}_{ci} \tag{1}$$

Where, m_i is the mass; $\vec{B}_i$ is the buoyancy; m_{ai} is the added mass ($m_{ai} = C_m\rho_w V_i$); m_{vi} is the virtual mass ($m_{vi} = (1 + C_m)\rho_w V_i$); C_m is the coefficient of the added mass; ρ_w is the density of sea water; $\vec{U}_{wi}$ is the velocity of water at node i; $\vec{U}_i$ is the velocity of mass point; $\vec{W}_i$ is the weight; $\vec{T}_i$ is the tension force; $\vec{F}_{Di}$ is the drag force; $\vec{f}_{ci}$ is the interaction forces between components and bottom; Dynamic equations of nodes can be described in matrix form:

$$\begin{pmatrix} m_i + m_{ai} & 0 \\ 0 & m_i + m_{ai} \end{pmatrix}\begin{pmatrix} \ddot{x}_i \\ \ddot{y}_i \end{pmatrix} = \begin{pmatrix} m_{vi}\dot{U}_{wix} + T_{ix} + F_{Dix} + f_{cix} \\ m_{vi}\dot{U}_{wiy} + T_{iy} + F_{Diy} - m_i g + \rho_w V_i(y_i)g + f_{ciy} \end{pmatrix} \tag{2}$$

Analysis of mooring system with ADCP. The attitude of mooring system has been shown in Fig.14 and the offset of carriers has been shown in Table.3. The movement of the fairshaped carrier was least; instead, the movement of the carrier with eight floating was largest. Because buoyancy provided by fairshaped carrier was much more than spherical carrier and carrier with eight floating. Moreover, the immersed area A_x of fairshaped carrier was least. When carriers got the same immersed area A_x, fairshaped carrier could provide larger buoyancy, so the fairshaped carrier would have better adaptability. Meanwhile, the ADCP could get a smaller movement in work.

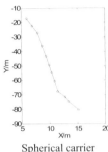
Spherical carrier

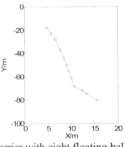
Carrier with eight floating balls

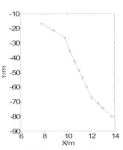
Fairshaped carrier

Fig.14. Attitude of mooring system

Table 3 Coordinate parameters of carriers

Type of carrier	Coordinate of anchor(m)		Coordinate of carrier(m)		Offset(m)
Spherical carrier	X 15.14	Y -80.38	X 7.816	Y -27.27	7.324
Carrier with eight floating balls	X 15.58	Y -80.41	X 6.654	Y -27.87	7.726
Fairshaped carrier	X 13.76	Y -80.28	X 9.785	Y -26.28	3.975

Conclusions

(1) Drag of fairshaped carrier is 0.43 times lower than drag of spherical carrier. Instead, drag of carrier with eight floating is 2.3 times higher than drag of spherical carrier.

(2) Using fairshaped carrier could effectively reduce the drag in water.

(3) The carrier with great buoyancy and small immersed area could enhance the stability.

Acknowledgments

The authors appreciated the support of NSFC (No. 51175484) and Science Foundation of Shandong province (No. ZR2010EM052). The authors are also grateful for the help from Key Lab of Ocean Engineering of Shandong province.

References

[1] Z.Z. Han, J. Wang, X.P. Lan, Engineering examples and applications of fluid simulation calculation, Beijing Institute of Technology Press, China, 2004.

[2] X.F. Huang, X.C. Zhou, B.C. Yuan, The control of the parameters of shipboard with ADCP. Ship Science and Technology, 19(5) (2007) 67-69.

[3] P. Wang, B.W. Song, X.L. Liu, D.H. Yu, J. Jiang, Multidisciplinary Design Optimization and Simulation Validation for Shape Design of Torpedo, Journal of System Simulation, 20(7) (2008) 1915-1918.

[4] J. Hurley, B.D. Young, C.D. Williams, Reducing Drag and Oscillation of Spheres Utilized for Buoyancy in Oceanographic Moorings, Oceanic Technology, 25 (2008) 1823-1833.

[5] L. Zhang, K. Sun, Q.J. Luo. Hydrodynamic design of diversion covers for a tidal-stream hydro turbine, Journal of Harbin Engineering University, 28(7) (2007) 734-737.

[6] L. Sun, G.J. Liu, M. Wang, B. He, Numerical Simulation of the External Flow Field and Geometry Parameters Optimization of AUV Dome on Bow, Computer Simulation, 28(5) (2011) 188-192.

[7] Z.Y. Chang, Y.G. Tang, H.J. Li, J.M. Yang, L. Wang, Analysis for the Deployment of Single--Point Mooring Buoy System Based on Multi-Body Dynamics Method, China Ocean Engineering, 26(3) (2012) 495-506.

Advanced Materials Research Vol. 819 (2013) pp 155-159
© (2013) Trans Tech Publications, Switzerland
doi:10.4028/www.scientific.net/AMR.819.155

Improved Local Mean Decomposition and Its Application to Fault Diagnosis of Train Bearing

Wang Peng[1,2, a], Ma Huaixiang[1,b]

[1] School of Mechanical Engineering, Shijiazhuang Tiedao University, Shijiazhuang, 050043, PR China

[2] Key Laboratory of Traffic Safety and Control in Hebei, Shijiazhuang, 050043, PR China

[a]pengwang@live.cn, [b]mhxwp@126.com (corresponding author)

Keywords: LMD, improvement, kurtosis criterion, fault diagnosis

Abstract. Fault diagnosis of train bearing is an important method to ensure the security of railway. The key to the fault diagnosis is the method of vibration signal demodulation. The local mean decomposition (LMD) is a self-adapted signal processing method which has a good performance in nonlinear nonstationary signal demodulation. The improved LMD method based on kurtosis criterion can prevent errors in the process of calculating the product functions. With the verification of simulation and wheel set experiment, the improvement method has been certified usefully in practical application.

Introduction

The operational safety of bearing is a significant subject of the intensive investigation and research with the increasing demand for the security of train. Vibration signal monitoring as a method of predictive programs have proven to be highly effective. As the major effort in the fault diagnosis of train bearing, vibration signal demodulation seems to be especially important. Some of those demodulation methods have been shown successfully under certain conditions. However, the vibration signal of bearing is a nonlinear, nonstationary signal, those methods may fail to implement signal processing. To be more serious, inappropriate method may lead to inaccurate false alarms or even neglect which have a severe effect to the safety of train.

So far, many techniques have been developed in the vibration signal demodulation, include the resonance demodulation [1], Hilbert transform [2], energy operator demodulation [3] and so on. But there are also some shortages when it comes to multi-component modulation processing. Wavelet transform has a better performance [4], but the selection of wavelet basis is an aporia in application and a critical factor of the result. Empirical Mode Decomposition (EMD) is a self-adaptive signal processing method which can decompose a complicated signal into a series of intrinsic mode functions (IMFs) [5]. By performing Hilbert transform to each IMF, the corresponding instantaneous amplitude and instantaneous frequency can be calculated. Since EMD method has superiority in engineering application with some excellent established cases [6, 7], there are also some problems to be solved such as end effect, the physical interpretations of IMFs, and so on [8, 9].

Local mean decomposition (LMD) was put forward by Jonathan S. Smith in 2005 [10], it can decompose a multi-component signal into a set of product functions (PFs) self-adaptively. The PFs are calculated with local mean and local magnitude of extrema. LMD is ostensibly similar to EMD, but there are effective improvements in end effect. With LMD, the instantaneous frequency can be calculated with PF directly, so the error comes from Hilbert transform can be avoided, negative frequency can also be eliminated [11]. Recently, more and more applications based on LMD with good performance emerged endlessly [12-14].

The paper is organized as follows: A brief introduction to the LMD algorithm is provided in Section 2. The improved LMD algorithm based on kurtosis criterion and simulation signal analysis are introduced in Section 3. The analysis results from train bearing fault vibration signals are given in Section 4. And finally, we offer the conclusions in Section 5.

The introduction of LMD algorithm

The nature of LMD is to demodulating a multi-component signal into a set of PFs with local mean and local magnitude of extrema. For a signal x(t), it can be decomposed as the follow steps:

(1)Determine all the maximum and minimum points as the extrema n_i, calculating the local means m_i and the local magnitudes a_i:

$$m_i = \frac{n_i + n_{i+1}}{2} \tag{1}$$

$$a_i = \frac{|n_i - n_{i+1}|}{2} \tag{2}$$

(2)Plot the m_i, a_i points and link with straight lines respectively and calculating the local mean signal $m_{11}(t)$ and the envelope estimate signal $a_{11}(t)$ using moving averaging.

(3)The frequency modulated signal $s_{11}(t)$ can be separated from x(t) with $m_{11}(t)$ and $a_{11}(t)$:

$$s_{11}(t) = \frac{x(t) - m_{11}(t)}{a_{11}(t)} \tag{3}$$

(4)According to step (1) and step (2), the local magnitude signal of $s_{11}(t)$ can be calculated as $a_{12}(t)$. If $a_{12}(t)=1$, $s_{11}(t)$ is a purely frequency modulated signal. Otherwise $s_{11}(t)$ is regarded as the original signal and the above procedure needs to be repeated n times until $a_{1(q+1)}(t)=1$. But in practice it's hard to be achieved. So a variation δ can be determined in advance, when $1-\delta \leq a_{1(q+1)}(t) \leq 1+\delta$, $s_{1q}(t)$ can be maintained as a purely frequency modulated signal.

(5)The envelope signal of the first product function can be derived by multiplying together the envelope estimate signals:

$$a_1(t) = \prod_{q=1}^{n} a_{1q}(t) \tag{4}$$

(6) The first product function PF_1 can be obtained by multiplying envelope signal $a_1(t)$ and the purely frequency modulated signal $s_{1q}(t)$:

$$PF_1(t) = a_1(t)s_{1q}(t) \tag{5}$$

The instantaneous amplitude of $PF_1(t)$ is the envelope signal $a_1(t)$, the instantaneous frequency can be obtained from $s_{1q}(t)$:

$$\phi_1(t) = arccos(s_{1q}(t)) \tag{6}$$

$$\omega_1(t) = \frac{d\varphi_1(t)}{dt} \tag{7}$$

(7)Subtract PF_1 from the original signal x(t) resulting in a new function $u_1(t)$ which become a new original signal, repeat the procedure above for k times until the $u_p(t)$ is a constant or contains no more oscillations to obtain all the PFs and a monotonic function $u_p(t)$.

$$\begin{cases} u_1(t) = x(t) - PF_1(t) \\ u_2(t) = u_1(t) - PF_2(t) \\ \vdots \\ u_p(t) = u_{p-1}(t) - PF_p(t) \end{cases} \tag{8}$$

Improvement of LMD and simulation signal analysis

According to the algorithm introduced in section 2, acquisition of the purely frequency modulated signal $s_{1n}(t)$ is a critical step. Obviously, decrease the value of δ can increase the accuracy of LMD, but on the other hand, this may increase the end effect. End effect is a great obstacle in LMD, even a great deal of method have been put forward, this effect can not be avoided completely by now. When the qualification of purely frequency modulated signal be strict, more looping times make the end effect more significant. To prevent this phenomenon, another criterion which can stop the loop before endpoint contaminate the signal is necessary.

Kurtosis is any measure of the peakedness of the probability distribution of a real-valued random variable. For a signal X length of N, it's mean value described as μ,σ is the standard deviation, the discretized formula of kurtosis (K) is:

$$K = \frac{1}{N}\sum_{i=1}^{N}(\frac{X_i - \mu}{\sigma})^4 \tag{9}$$

The envelope estimate signal $a_{pq}(t)$ is a smoothed signal, so the kurtosis K_{pq} is decreasing following looping process theoretically. When a increase of K_{pq} appears after n times of loop, the progress should be stopped. PF_p can be obtained as follow:

$$PF_p(t) = s_{pn}(t)\prod_{q=1}^{n} a_{1q}(t) \tag{10}$$

Consider a multi-component signal x(t)

$$x(t) = (1 + 0.5\cos 10\pi t)\cdot\sin(200\pi t + 2\cos 10\pi t) + \sin 2\pi t \cdot \sin 30\pi t$$

Set $\delta = 10^{-4}$, dispose end effect preliminary with similar extrema extension[15], focus on the calculation of PF_2. Decompose x(t) with traditional method, after 3 loops, K_{2q} are described as Table 1. Stop the process at loop 10 manually, the result of PF2 is shown in Fig.1. As a contrast, the process based on the kurtosis criterion which got a result of PFs shown in Fig.2. The result of improved method is more effectively.

Table 1 K_{2q} in the process of PF_2 calculation

	q=1	q=2	q=3	q=4
K_{2q}	-1.491	-1.997	-2.038	11.069

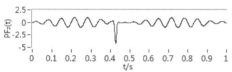

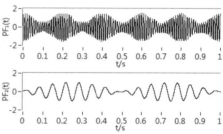

Fig.1 PF$_2$ decomposed by traditional method

Fig.2 PFs decomposed by improved method

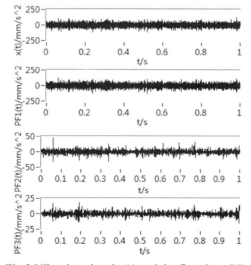

Fig.3 Vibration signal x(t) and the first three PFs

Application to fault diagnosis of train bearing

With the experimental facility of wheel set, a series of vibration acceleration signal of train bearing was acquired. The rotation speed is 469 r/min, the fault character frequency of the bearing with a breakdown in outer ring is 67.327 Hz. Decompose the signal with the improved LMD, the original signal and the result was showed in Fig.3. According to the amplitude spectrum of PF1 which is shown in Fig.4, the main frequency matches to the fault character frequency of the bearing.

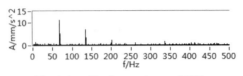

Fig.4 Amplitude spectrum of PF1

Conclusions

According to the analysis of simulation signal and the bearing vibration acceleration signal acquired from experimental facility, the improvement of LMD has been proven to be effective and appropriate for engineering application. End effect has been reduced dramatically, level of automation has been upgraded also. Improvement the orthogonality and the operating efficiency of the LMD program can be research programs in the future.

Acknowledgement

This research is supported by National Key Basic Research Program of China(973 Program)(2012CB723301), National Natural Science Foundation of China (11227201, 11202141, 11172182), Key Program of the Ministry of Railways of China (2011J013-A), Natural Science Foundation of Hebei Province(A2013210013) and Educational Commission of Hebei Province, China (Z2011228, zh2011215).

References

[1] Wenyi W. Early Detection of Gear Tooth Cracking Using The Resonance Demodulation Technique[J]. Mechanical Systems and Signal Processing. 2001, 15(5): 887-903.

[2] Yi Q, Shuren Q, Yongfang M. Research on iterated Hilbert transform and its application in mechanical fault diagnosis[J]. Mechanical Systems and Signal Processing. 2008, 22(8): 1967-1980.

[3] Junsheng C, Dejie Y, Yu Y. The application of energy operator demodulation approach based on EMD in machinery fault diagnosis[J]. Mechanical Systems and Signal Processing. 2007, 21(2): 668-677.

[4] Qiu H, Lee J, Lin J, et al. Wavelet filter-based weak signature detection method and its application on rolling element bearing prognostics[J]. Journal of Sound and Vibration. 2006, 289(4-5): 1066-1090.

[5] Huang N E, Shen Z, Long S R, et al. The empirical mode decomposition and the Hilbert spectrum for nonlinear and non-stationary time series analysis[J]. 1998, 454(1971): 903-995.

[6] Huang N E, Shen Z, Long S R. A new view of nonlinear water waves: the Hilbert spectrum[J]. Annu. Rev. Fluid Mech. 1999(31): 417-457.

[7] Peng Z K, Tse P W, Chu F L. A comparison study of improved Hilbert–Huang transform and wavelet transform: Application to fault diagnosis for rolling bearing[J]. 2005, 19(5): 974-988.

[8] Huang N E, Wu M C, Long S R, et al. A confidence limit for the empirical mode decomposition and Hilbert spectral analysis[J]. Proc. R. Soc. Lond. 2003, A(459): 2317-2345.

[9] Dätig M, Schlurmann T. Performance and limitations of the Hilbert–Huang transformation (HHT) with an application to irregular water waves[J]. Ocean Engineering. 2004, 31(14–15): 1783-1834.

[10] Smith J S. The local mean decomposition and its application to EEG perception data[J]. Journal of the Royal Society Interface. 2005, 2(5): 443-454.

[11] Wang Y X, He Z J, Zi Y Y. A Comparative Study on the Local Mean Decomposition and Empirical Mode Decomposition and Their Applications to Rotating Machinery Health Diagnosis[J]. Journal of Vibration and Acoustics-Transactions of The ASME. 2010, 132(021010): 1-10.

[12] Liu W Y, Zhang W H, Han J G, et al. A new wind turbine fault diagnosis method based on the local mean decomposition[J]. Renewable Energy. 2012, 48: 411-415.

[13] Cheng J S, Zhang K, Yang Y. An order tracking technique for the gear fault diagnosis using local mean decomposition method[J]. Mechanism and Machine Theory. 2012, 55: 67-76.

[14] Wang Y X, He Z J, Xiang J W, et al. Application of local mean decomposition to the surveillance and diagnostics of low-speed helical gearbox[J]. Mechanism and Machine Theory. 2012, 47: 62-73.

[15] Lu S, Xiaojun Z, Zhigang Z, et al. Boundary-extensionmethod in hilbert-huang transform[J]. Journal ofvibration and shock. 2009(08): 168-171.

Advanced Materials Research Vol. 819 (2013) pp 160-164
© *(2013) Trans Tech Publications, Switzerland*
doi:10.4028/www.scientific.net/AMR.819.160

Online monitoring recognition theory based on the time series of chatter

Jiang Yongxiang[1,a], Du Bing[2,b], Zhang Pan[3,c], Deng Sanpeng[1,d], Qi Yuming[1,e]

[1] Tianjin University of Technology and Education,School of Mechanical Engineering , Tianjin, China, 300222;

[2]China Oilfield Services Limited, Tianjin, China, 300451

[3] Tianjin Key Laboratory of Equipment Design and Manufacturing Technology, Tianjin University, Tianjin, China, 300072

[a]jyx1212@qq.com,[b]99359385@qq.com, [c]zhangpan2003@126.com, [d]37003739@qq.com,[e]284075043@qq.com

Keywords: cutting chatter, on-line monitoring, K-S entropy, coarse-grained entropy

Abstract: On-line monitoring recognition for machining chatter is one of the key technologies in manufacturing. Based on the nonlinear chaotic control theory, the vibration signal discrete time series for on-line monitoring indicator is studed. As in chatter the chaotic dynamics process attractor dimension is reduced, the Kolmogorov‑Sinai entropy(K-S) index is extracted to reflected the regularity of workpiece chatter, then the k-S entropy is simplified by coarse‑grained entropy rate(CER), which can easily evaluated as chatter online monitoring threshold value. The milling test shows that the CER have a sharp decline when chatter occurre, and can quickly and accurately forecast chatter.

Introduction

On-line monitoring recognition of machining chatter is one of the key technologies of precision manufacturing. The chatter often occurs quickly, therefore, the forecast threshold should be established to detect chatter and suppress it in embryonic period. Researchers put forward a variety of methods to determine and forecast chatter through online monitoring machining signals such as vibration [1], cutting force [2], noise [3], workpiece stress strain [4] and machining quality [5]. However, current research results have poor timeliness or poor accuracy, if use in remote monitoring, the chatter detecting require more diagnosis time. Therefore, new method on chatter prediction with timeliness as well as accuracy has an important engineering application value.

Based on the nonlinear chaotic control theory, the vibration signal discrete time series for on-line monitoring indicator is studed. As in chatter the chaotic dynamics process attractor dimension is reduced, the K-S entropy index is extracted to reflected the regularity of workpiece chatter, then the k-S entropy is simplified by coarse‑grained entropy rate(CER), which can easily evaluated as chatter online monitoring threshold value. The milling test shows that the CER have a sharp decline when chatter occurre, and can quickly and accurately forecast chatter.

Theoretical modeling of stability forecast based on coarse-grained entropy rate

K-S entropy can be used to define the uncertainty of the system, and is an important characteristic to identify chaotic series. As a measurement of information, entropy K_{N+1} and K_N represent the information at time $N+1$ and N. The time series of the vibration acceleration can be shown as $\{a(t)\} = \{a_1, a_2, \cdots a_N\}$ ($t = 1, 2, \cdots, N$). (K_{N+1}-K_N) denotes the required system information at a_{N+1}, which means that (K_{N+1}-K_N) denotes the loss of information from time $n\tau$ to $(n+1)\tau$. K-S entropy can be defined as the average loss of the information rate, that is

$$K = \lim_{\tau \to \infty} \lim_{\varepsilon \to \infty} \lim_{N \to \infty} \frac{1}{N\tau} \sum_{t=1}^{N} (K_{i+1} - K_i) = -\lim_{\tau \to \infty} \lim_{\varepsilon \to \infty} \lim_{N \to \infty} \frac{1}{N\tau} \sum_{t=1}^{N} P_i \ln P_i \tag{1}$$

Where, P_i is the probability of $a(\tau i)$ falls on the i-th unit a_i at time τ.
Entropy K_q is defined with unknown differential equations of the system as

$$K_q = -\lim_{\varepsilon \to \infty} \lim_{n \to \infty} \left(\frac{1}{q-1} \frac{1}{n\tau} \log \sum_{i=1}^{n} P_i^q \right) \tag{2}$$

Usually, the second-order K_2 is taken as an approximation of the K-S entropy

$$K_2 = -\lim_{\varepsilon \to \infty} \lim_{n \to \infty} \left(\frac{1}{n\tau} \ln \sum_{i=1}^{n} P_i^2 \right) \tag{3}$$

Given the observed time series, phase space is reconstructed and correlation integral is obtained in the m-dimensional phase space

$$C_m(r) = \frac{1}{N^2} \sum_{i,j=1,i\neq j}^{N-m+1} \theta(r - d_{i,j}) \tag{4}$$

Given small r, the function $C_m(r)$ can be approximated according to formula (5)

$$\ln C_m(r) = \ln C + D(m) \ln r \tag{5}$$

$D(m)$ unchange with the increase of embedding dimension in phase space, its attractor correlation dimension D can be defined as

$$D(m) = \lim_{m \to \infty} D(m) \tag{6}$$

From function(4), (5) and (6), then

$$C_m(r) = r^D \exp(-m\tau K_2) \ \genfrac{(}{)}{0pt}{}{m \to \infty}{r \to 0} \tag{7}$$

Take the logarithm of the formula (7), then

$$\ln C_m(r) \sim D \ln r - m\tau K_2 \qquad \ln C_{m+1}(r) \sim D \ln r - (m+1)\tau K_2 \tag{8}$$

K_2 is get from formula (8),

$$K_2(m,r) = \frac{1}{\tau} \ln \frac{C_m(r)}{C_{m+1}(r)} \tag{9}$$

Thus, for a given time series, in its scale-free region, for $m = 2, 3 \ldots$, $C_m(r)$ is obtained according to the formula (8) and K-S entropy can be calculated by the formula (9). When the entropy calculated by formula (9) is no longer changed with m, $K_2(m_0, r)$ is the K-S entropy of the system. m_0 is the minimum of m for $K_2(m_0, r)$ reaching saturation, its reciprocal $1 / K_2$ is the maximum predictable time scale, also is the maximum prediction time of time series. With prediction away from this moment, the predicted credits will be gradually reduced, and will seriously reduce when exceeded the maximum prediction time. In ordered system, $K_2=0$; In random system, $K_2=\infty$; $0 < K2 < \infty$ means chaotic system, the chaotic will more serious when K_2 grows.

Considering the m-dimensional random variables X_1, ..., X_m, the probability distribution of variable X_i is defined as $p(x_i)=\mathrm{Prob}\{X_i=x_i\}$. while the joint probability distribution of m variables X_1, ..., X_m is $p(x_1, ..., x_m)= \mathrm{Prob}(X_1=x_1, ..., X_m=x_m)$, then the related information entropy is defined as

$$H(X_1, \cdots X_m) = -\sum_{X_1 \in \psi_1} \cdots \sum_{X_m \in \psi_m} p(x_1, \cdots x_m) \log p(x_1, \cdots x_m) \tag{10}$$

In random process, entropy rate of $\{X_i\}$ is defined as

$$h = \lim_{m \to \infty} \frac{1}{m} H(X_1, \cdots X_m) \tag{11}$$

The h describes the size of the information obtained by measuring variables X_i.

The concept of interaction information is used to describe average amount of information X_m, and by boundary redundancy R', mutual variable X_1, ..., X_{m-1} are described as

$$R'(X_1, \cdots X_{m-1}; X_m) = \sum_{X_1 \in \psi_1} \cdots \sum_{X_m \in \psi_m} p(x_1, \cdots x_m) \log \frac{p(x_1, \cdots x_m)}{p(x_1, \cdots x_{m-1}) p(x_m)} \tag{12}$$

When m-dim dynamical systems describing by variable X_i, h is the K-S entropy.

m is usually unpredictable for a complex dynamic system. If a representative signal $\{y(t)\}$ is available, variable X_i is y (t) of build by time hystéresis value $y(t)$ as $X_i=y(t+(i-1)\tau)$, where τ is time lag. For fixed signal, marginal redundancy is an m-dimensional function, delay independent of time t is τ, then

$$R'(m, \tau) = R'[y(t), y(t+\tau), \cdots y(t+(m-1)\tau)] \tag{13}$$

Entropy rate h can be close to the marginal redundancy by expanding the range of m and a suitable choice of τ

$$h \approx \frac{R'(m,0) - R'(m,\tau)}{\tau} \tag{14}$$

The approximation accuracy of h depends on the signal to noise ratio (SNR) of the measurement data, when SNR reduced, h is unreliable. Although entropy h cannot be accurately determined, marginal redundancy can be represented by CER

$$CER(m) = \frac{R'(m, \tau_0) - \|R'(m)\|}{\|R'(m)\|} \tag{15}$$

where, $\|R'(m)\|$ is marginal redundancy $\left\|R'(m)\right\| = \dfrac{\displaystyle\sum_{\tau=\tau_0}^{\tau_{max}} R'(m,\tau)}{\tau_{max} - \tau_0}$ (16)

In equ. (16), τ_0 usually set to 0, so that by selecting τ_{max} can make $R'(m, \tau>\tau_{max})\approx0$.
CER is a relative value between [0, τ_{max}], so the interval of stability prediction value τ_{max} often choose [0, 1].

Experiment of CER forecast theoretical in chatter

The CER is calculated by three direction of vibration acceleration a_x, a_y, a_z while n=0-8000 r/min, a_p=10mm, a_e=12 mm，f_z=0.2 mm/tooth and is shown in Fig. 1. In the spindle speed of 2500-2900, 3500-4500, 5000-6000 and 6500-7000r/min, CER values have a sharp decline in the valleys, this phenomenon consistent with the waveform changes shown in Fig. 2. It indicates that sharp decrease of CER can be used to predict chatter. In addition, the value of CER also shows the intensity of chatter, such as chatter in n=2500 r/min is much smaller than that of n=3500 r / min.

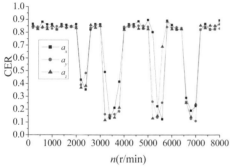
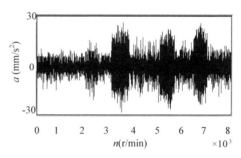

Fig. 1 CER value under a_p=10mm, a_e=2mm,   Fig. 2 a_p=10mm, a_e=2mm, f_z=0.2mm/tooth
f_z=0.2mm/tooth variation with n acceleration waveform with n

Accuracy assessment of the CER in stability forecast

It can be seen from formula (15) that τ_{max} and m are the main factors affecting CER. The phase space reconstruction theory indicates that reasonable values of τ_{max} and m affect the calculation accuracy. This section focuses on analyzing the influence of τ_{max} and N on CER, and the accuracy of forecast is under assessment by choosing different values of τ_{max} and N.
Fig. 3 shows the CER according to the vibration acceleration time series when the τ_{max} =10, 50 and 100, the sampling points N = 65535, which is the same as Fig. 1. The calculation result shows that in different τ_{max}, CER reduce in non-chatter and chatter junction. But when τ_{max}=10, sharp reduce is not obvious, this indicate that short interval value lead to instability of predictive accuracy.
The CER in sampling points N=65535, N=4096 ,N=1024 and τ_{max} = 50 is shown in Fig. 4. The analysis shows that CER value maintains the same characteristics around chatter. Using few sampling points make CER value appears floating change, but didn't appear false prediction. This indicates that under condition of accurate chatter prediction, selecting an appropriate N can reduce the amount of data calculation and improve prediction speed.

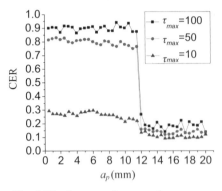

Fig. 3 The impact of τ_{max} on the prediction accuracy of CER

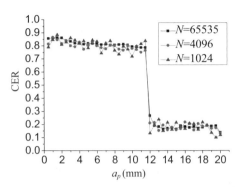

Fig. 4 The impact of sampling points on the prediction accuracy of CER

Conclusion

Based on time series of machining chatter, on-line monitoring to establish discrete dynamic system is studied, and the coarse-grained entropy rate (CER) is extracted as chatter decision threshold value. Stability forecast experiment showed that the characteristics of rapid decrease CER value can be used as a sign of chatter. The analysis of the CER with different τ and N showed that reduce N can qulikly predict chatter accurately.

Acknowledgements

This study is support by Tianjin application foundation and frontier technology research program (12JCQNJC02500).

References

[1] Govekar E, Gradisek J, Kalveram M et al. On Stability and Dynamics of Milling at Small Radial Immersion [J]. CIRP Annals-Manufacturing Technology, 2005, 54(1): 357-362

[2] Kim S J, Lee H U, Cho D W. Prediction of Chatter in Nc Machining Based On a Dynamic Cutting Force Model for Ball End Milling [J]. International Journal of Machine Tools and Manufacture, 2007, 47(12-13): 1827-1838

[3] Liu A M, Peng C, Liu J Z et al. Detection of chatter and prediction of stable cutting zones in high-speed milling [J]. Chinese journal of mechanical engineering, 2007, 43(1): 164-169

[4] Rahman M, Ito Y. A Method to Determine the Chatter Threshold [C]. Proc. 19th MTDR, 1978, 191-196

[5] Fu L Y, Yu J Y, Bao M. Study on the phase characteristics of cutting chatter [J]. Journal of vibration engineering, 2000, 13(4): 510-514

[6] Deng L S, Chen F. Analysis and prediction of fund index of nonlinear time series [J]. Journal of Tianjin University, 2004, 37(11): 1022-1025

[7] Lv J H, Lu J A, Chen S H. Chaotic Time Series Analysis and Applications [M]. Wuhan: Wuhan University Press, 2001.

Advanced Materials Research Vol. 819 (2013) pp 165-170
© (2013) Trans Tech Publications, Switzerland
doi:10.4028/www.scientific.net/AMR.819.165

Research on Agricultural Harvester Data Detection System Based On Remote Monitoring

Xiaojun Guo [1, a] Leilei Gao [2, b], Taiyong Wang [3, c], Zhennan Li[4, d]

[1]Tianjin University of Technology and Education, Tianjin 300222, China;

[2]Tianjin University of Technology and Education, Tianjin 300222, China;

[3]Tianjin University, Tianjin 300222, China

[4]Tianjin University of Technology and Education, Tianjin 300222, China;

[a]guoxiaojun126@126.com,[b]gaoleilei2010@163.com,[c]tywang@189.cn,[d]lzn906@163.com

Keyword: harvester, remote monitoring, data acquisition, UDP pattern, data processing

Abstract. On-line detection system of the harvester is an outcome from combination of modern computer technology and communication technology in harvester operations applications. With the help of the various sensors, the harvest yield, running routes, and threshing wheel speed, etc. are measured. These information and parameter are indicator of the harvesters working status. They are detected, processed, packed, and transmitted to the computer server in monitoring center via a wireless network. On the monitoring center server, the transmitted data is processed further, fault data are inspected, reliability data is calculated. Meanwhile, the harvester is controlled according to accepted data.

Introduction

In the ordinary condition, advanced harvesters are equipped with electronic auxiliary systems used to measure working status, inclusive of: quality of the harvest, walking route of the harvester, and threshing wheel speed, etc. Data is collected by the lower machine sensors situated in different electronic measuring apparatus, then uploaded to the host computer. The detection system of the host computer examine harvesters working parameters, process and pack data. By means with wireless network, data is uploaded to the monitoring center server. In the monitoring center server, data are further processed, and fault data are detected and further processed, and reliability data is calculated and accordingly the harvester are controlled. With help of a wireless telecommunication network, data transmission of the harvester from work end to monitoring center is realized, and the problem of the reliability and security of the harvester is solved, thus providing a basis for decision making of harvester.

System components framework

The overall system is composed of three parts: harvester data acquisition and processing unit, communication network, the monitoring center.

The overall system structure is shown in Fig.1.

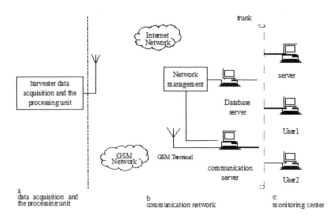

Fig 1. System Architecture

The structure of the various parts are stated as follows:

a. Harvester data acquisition and processing unit: data is collected by means with various types of sensors, then transmitted to the host computer, which is used to process data and judge system working state.

b. Communication network: pack the data processed by the host computer, and send data to the monitoring center through a communication network

c. Monitoring center: take responsibility of remote communication, harvesters monitoring and data management. The monitoring center is equipped with communications server and correspond software, database server and correspond software, data management software and users workstation terminal.

DATA acquisition and processing unit of harvester

Harvester data acquisition and processing unit are composed by variety of sensors. The GPS receive module, controller and memory hardware.

Each sensor and the GPS receive module, are referred to as lower machine .The main control system and the internal memory components are collectively referred to upper machine. Harvester data acquisition and processing unit diagram are as shown in Fig. 2:

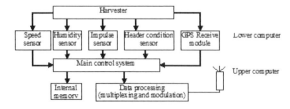

Fig2. Block diagram of data acquisition and processing unit

The function of various sensors are stated as follows: Speed sensor detects wheel revolutions per unit time for calculation of walking velocity, yield sensor of impulse type measures grain yield on-line; Capacitive humidity sensor measures grain moisture; Cutting table status sensor judges cutting operation by monitoring the lifting up and down. The GPS receive module receive satellite positioning signal, obtaining the position of the harvester, latitude data. The main controller receive each sensor as well as the GPS positioning information, calculating and processing data, storing parameter data in the internal memory, then packs processed data, transmit them to the monitoring center computer via a wireless telecommunication network.

The main technical parameters reflecting the harvester working states are of two types: (see Table 1, Table 2).

Table 1: Detection parameters of instrument in operation table

Simulation parameters	Engine cooling water temperature, fuel level, pulse parameters of engine speed, screw speed
Switch parameter	Low level input 1, oil pressure 2, S1 status 3,S2 status 4,S3 status 5,full storehouse 6, manual / automatic status 7, row 8, grass from the clutch switch High level input 1, manual switch 2, depth or manually drop switch 3, a charging indicator
Time parameter	1, the engine working hours 2, harvest time

Table 2 The detection module detection parameters

Simulation parameters	Hydraulic oil temperature
Pulse parameter	1, speed 2, From the roller speed 3, Cut tail speed
switch quantity parameter	Low level input 1, Cut off switch status 2, Cutting table drop switch state 3, Left switch state 4, The right switch state High level input 1, Left, right steering pressure switch status 2, Cutting condition of the pressure switch 3, Cutting table drop pressure switch state

Network data transfer

The data communication protocol between the user device and the GSM terminal is based on the communication 07.07 standard AT command set, which provide two short message format; text format and UDP format. Text format information is coded by the ASCII, can be directly read, the information in the UDP format short message is binary-coded, need to be decoded.

1.   Data package format: The result data, inclusive of processing, feasibility analysis and fault judgment is encapsulated into a data word. This data word reflects the harvester working condition. It has fixed content format which is stipulated by a certain agreement. In data packing agreement, each message content format is listed in Table 3:

2. Data packing and sending: Code word sequences encoded by the above content format, contains working state parameters of the harvester. In order to enhance transmission efficiency, multiple sets of working state parameters is multiplexed, digital modulated, frequency modulated. And the modulated information is emitted through the GSM terminal.

3. Example of data to be transmitted   TGY100000012621435131 23.45678912.345678AES902 070035040012. The specific meaning of data 345.61234.58512.3200100775 # 251 can be found out from table 3.

Finally after the GSM network transmission, data is received from the GSM terminal, after shunt, demodulation, decode steps, the original test information is recovered, then stored in the database server one by one, waiting for data processing by the monitoring center.

The Monitoring Center Data Processing

The monitoring center is composed of communication server equipped with communication software, database server, and user terminal workstation equipped with scheduling software, data management software and data analysis software. The client access database by the internet, thus improving the processing capacity of database, realizing multiple user's data sharing and coordination, etc.

Table 3 Content format

id	field	length	example	illustrate	id	field	Length	example	illustrate
1	Start character	3	TGY	Fixed character identifier	2	ID	8	10000001	Vehicle number
3	Data encoding	10	0000262143	Information for fixed string	4	Fault information	10	0000000513	Binary said fault information
5	Longitude	10	123.456789	Said longitude	6	Dimension	9	12.345678	Said latitude
7	Positioning state	1	A	A:orientation V: navigation	8	Longitude	1	E	E:east longitude, W:west longitude
9	Latitude direction	1	S	S:south latitude, N: latitude	10	Cooling water temperature	3	090	90℃
11	Fuel level	3	020	20%	12	Engine speed	4	0700	700RPM
13	#1auger speed	4	0350	350RPM	14	2#1auger speed	4	0400	400RPM
15	Working hours	7	12345.6	12345.6h	16	Harvest time	6	1234.5	1234.5h
17	Hydraulic oil temperature	3	085	85℃	18	Speed	4	12.3	12.3Km/h
19	Threshing cylinder speed	4	0200	200RPM	20	Cutting table tail speed	4	0100	100RPM
21	Status information table	4	0775	Binary said switch action	22	Terminator	1	#	Fixed character
23	Checksum	3	251	Consistent with valid data received checksum calculated checksum					

When the new data is transmitted to the monitoring center, communication software first receive data, conduct protocol analysis, restore original data stored in the database server, and then inform scheduling software new data arrive, scheduling software to receive new news arrival of news, query the database to get the latest data, conduct the corresponding data update.

Data management software and data analysis software process the latest data, executing fault-judgment and reliability-analysis the two big jobs.

1)   Fault judgment standard: data fault is judged according to the fault judgment, fault is divided into engine operating trouble and harvester working state failure.

(1) Engine run fault: when the engine is running (engine speed is > = idling speed, 500rpm), the oil pressure of the engine machine alarm, abnormal engine charging indicator, engine coolant temperature is >= 98°C (s) and stop processing, denoted failure once, while recording fault time, fault hours.

(2) Harvester working state fault:

a)The harvest working state minor fault: Lysimachus chain clogging alarm, 1 # input Valley auger shaft end speed <1 # auger standard value, 2 # input Valley auger shaft end speed <2 # standard value of the auger, work super load (working load> 99%) last for more than 5min, record minor fault once, at the same time record the fault time, fault hrs.

b) The harvest work state fault: 1 # input Valley auger shaft end speed is 0, which explains serious blockage or transmission system fault, the 2 # input Valley auger shaft end speed is 0, indicating severe blockage or transmission system failure, engine speed abnormal(working speed dips or flameout), abnormal active shutdown, respectively record fault once, at the same time record fault time, fault hours.

(3) The fault assessment standard:

a) Minor fault, such as stopping time or stopping time to repair is < 20 min, this is not counted as a fault in reliability calculation; Such as stopping and stopping time to repair is >= 20 min, this is counted as fault in reliability calculation.

b) if general fault appeared, the machine must stop to repair, after each repair, the harvest work status is resumed, record the fault-repairing time. Fig.3 is the fault standard management interface.

The fault standard list

id	Fault code	Fault name	Mark	Slight stop time	General	Shutdow time	Major	Major shutdown	Conditions 1	Conditions 2
1	BIT0	Low oil pressure	*	time=<0.5h	null	2h<=time>0.5h	■	time>2h		
2	BIT1	Charging indicator	*	time=<0.5h	○	2h<=time>0.5h	■	time>2h		
3	BIT2	Row grass	*	time=<0.5h	○	2h<=time>0.5h	■	time>2h		
4	BIT3	Full granary	*	time=<0.5h	○	2h<=time>0.5h	■	time>2h		
5	BIT4	Fuel level low	*	time=<0.5h	○	2h<=time>0.5h	■	time>2h		
6	BIT5	High engine temp	*	time=<0.5h	○	2h<=time>0.5h	■	time>2h		
7	BIT6	Emergency stop	*	time=<0.5h	○	2h<=time>0.5h	■	time>2h		
8	BIT7	1 screw speed alarm	*	time=<0.5h	○	2h<=time>0.5h	■	time>2h		
9	BIT8	2 screw speed alarm	*	time=<0.5h	○	2h<=time>0.5h	■	time>2h		
10	BIT9	High alarm	*	time=<0.5h	○	2h<=time>0.5h	■	time>2h		
11	BIT10	Cylinder stall			○	time=<1.0h	■	time>1h	Speed >=500rpm clutch switch=1 ,	
12	BIT11	Drive stall			○	time=<1.0h	■	time>1h	Speed >=500rpm clutch switch=1 !	

Fig.3. Fault Standard Management

2) Data reliability analysis

Data is obtained in real-time data from database server, examples of the specific data is seen in Fig. 4.

Test management	Test file	Test standard	Maintenance	Online help				

No. 1 Guangzhou, real-time monitoring data

The parameter name (code)	10:42:17	10:42:17	10:42:18	10:42:41	10:42:41	10:42:42	10:42:43	10:42:43
Temperature of cooling water	070	070	070	090	090	090	090	090
Fuel level	020	020	020	020	020	020	020	020
Engine speed	0900	0900	0900	0700	0700	0700	0700	0700
1# Screw speed	0350	0350	0350	0350	0350	0350	0350	0350
2# Screw speed	0400	0400	0400	0400	0400	0400	0400	0400
Hydraulic oil temperature	085	085	085	085	085	085	085	085
Speed	12.3	12.3	12.3	12.3	12.3	12.3	12.3	12.3
Threshing cylinder speed	0200	0200	0200	0200	0200	0200	0200	0200
Cutting table rail sp ed	0100	0100	0100	0100	0100	0100	0100	0100
Low oil pressure	1	1	1	1	1	1	1	1
Charging indicator	0	0	0	0	0	0	0	0
Row grass	0	0	0	0	0	0	0	0
Full granary	0	0	0	0	0	0	0	0
Fuel level low	0	0	0	0	0	0	0	0
High engine temperature	0	0	0	0	0	0	0	0
Emergency stop	0	0	0	0	0	0	0	0

Fig4. Specific data examples

The reliability analysis software for automatic analysis of the test data, statistical calculations and provide the results to provide objective data for ultimate reliability test report. Reliability related calculations, are carried out in accordance with the standard grain combine harvesters JB/T6287-2008 "Reliability Test Method for Evaluate the provisions of Section 9.2. The server receives data, and record accurately the exact time, including the year, month, day, hour, minute, seconds, the minor unit is second.

With the help of reliable analysis software, data is analyzed, and the characteristic curve of the various test parameters is calculated and drawn. Fig. 5 shows the characteristic curve for the engine speed changes with time. From the curve, Harvesters working parameters can be observed intuitively

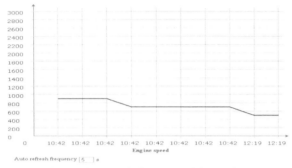

Fig. 5. Characteristic curve of the various test

Conclusion

In this system, the data transmission between the harvester and the monitoring center is realized, harvester working data is processed through the software, and working parameters state curve is drawn, and harvester working condition is monitored in real-time. The system can be extended further to embrace other functionalities. Fault expert processing system can be added to the monitoring center, when the harvester fault occurs, the system can give reason and fault treatment scheme. On the all, this system can be used in various field equipment monitoring, which shows a wide application prospects.

References

[1] Wang Xingliang. Principle and technology of digital communication[M] Xi'an Xi'an Electronic and Science University press 2003 10~13

[2] Jeong-Hyun Park. Wireless Internet access of the visited mobile ISP ubscriber on GPRS/UMTS network[J]. IEEE Transactions on Consumer Electronics, 2003, 49(1): 100-106

[3] Zheng Zuohui, Bao Zhiliang, Jing Ming, Zheng Lan, etc. Digital trunking mobile communication system[M]Beijing Publishing House of electronics industry 2002 35~36

[4] Dhawan Chander. Remote access networks and services: the Internet Access Companion. New York: McGraw-Hill, 1999, 4-6

[5] Li Hui. wireless communication system based on GSM short message[D] Nanjing Nanjing University of Science and Technology 2004

[6] Li Sixiang, Wang Lei, Zhang Feng, etc. A remote control device for earthmoving machines[J] Engineering machinery 2006（6）76~77

[7] Jiang Ping, Hu Wenwu, Sun Songlin, Luo Yahui, etc. Design of wireless remote control system for paddy field working machine. Journal of Agricultural Mechanization Research 2009（6）62~64

Advanced Materials Research Vol. 819 (2013) pp 171-175
© (2013) Trans Tech Publications, Switzerland
doi:10.4028/www.scientific.net/AMR.819.171

Research on Feature Extraction of Acoustic Emission Signals in Time-domain

Wang Wei[1,a], Li Qiang[2,b]

[1]School of Economy, Tianjin Polytechnic University, Tianjin 300387, China

[2]Technology and Equipment Department, Tianjin Port (Group) CO., LTD., Tianjin 300461, China

[a]wangweiuser@126.com, [b]liqiangmail@126.com

Keywords: acoustic emission, bearing fault detection, time-domain analysis, multi-channels.

Abstract. Acoustic emission detecting has been widely used in the diagnosis of bearing fault, but nearly all of these implements require that the transducer placed close to the source of acoustic emission. However, in actual industrial environment, the transducer couldn't be mounted very close to the bearings. In this paper, the time-domain wave and time-domain features based methods were analyzed and compared among four channels at different rotating speeds. And partial analysis and some conclusions drawn from the analysis were listed below.

Introduction

Rolling bearing is a key member for rotating machinery, and it is also one of a mechanical part prone to failure because of the structural characteristic of bearing. Therefore, fault detection of bearing has become a hot and difficult research focus. Acoustic emission (AE) is defined as transient and rapid release of the energy in the form of elasticwaves because of the deformation or damage within or on the surface of the material caused by stress[1-2].

Acoustic emission wave is a kind of mechanical wave, so it has the characteristics of fluctuation and attenuation. The AE wave generated in the material will propagate to the surface of the structure, and it is probed by the transducers mounted on the surface. From analyzing the information carried by the AE waves, we can get the information of the source of AE. Therefore, AE testing technology has been widely used in the detection of bearing faults.

D.Mba studied the implement of AE wave in detecting the early stage of the loss of mechanical integrity in low-speed rotating machinery [3]. C. James Li a and S.Y. Li b proved that AE was a better alternative to traditional bearing condition monitoring schemes by employing two normalized and dimensionless features in detecting the defects on the outer race and rollers of bearings [4]. L.D. Hall and D. Mba achieved the successful diagnosis of rubbing failure of bearings with the use of the KolmogorovSmirnov (KS) statistic method [5]. Abdullah M. et al. obtained the result that AE signal can detect and classify the early failure of bearings by studying the defects in outer race of deep groove ball bearing [6]. Catlin thought that high-frequency AE signal attenuated rapidly, and this high-frequency content can be used to distinguish the bearing faults with other disturbance when the sensor is placed close to the bearing [7]. The AE testing technology has been widely used in the detection of bearing fault, and nearly all of these implement require that the transducer is placed closed to the source of AE. However, in actual industrial environment, for example the bearing of machine tool spindle, the transducer couldn't be placed very close to the bearings. Therefore, one investigation in the processing method of AE signals is essential.

In this paper, the experiment setup was described in details in section 2; in section 3, several time-domain features, such as kurtosis, skewness, root mean square(RMS) etc., are extracted from the AE signals at different locations of the bearings, and the analysis and comparison are also listed in this part; the conclusions drawn from this paper are presented in section 4.

Experiment setup

To analyze the AE signal generated during the rotation of the bearings, several groups of experiments were conducted on a rotating machinery vibration fault test platform (Fig. 1). The one end of the spindle is driven by a control motor through a tooth belt, and the bearing is mounted at the other end of the spindle that can offer a radial load, and the bearing is N205E. The bearing was mounted on the platform which allowed the replacement of different bearings. They were (1) a good bearing, (2) a bearing with a single pitting on its outer race (Fig. 2).

Fig. 1 Rotating machinery vibration fault test platform and the mounting locations of the AE tranducers.

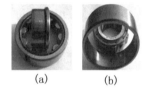

(a)   (b)

Fig. 2 Two kinds of bearings: (a) good bearing, (b) a pitting on the outer race.

For each of these two bearings, four AE transducers (R15a) were placed on the places emerged in Fig. 1. One first transducer was mounted on the top of the bearing seat (location ①), the second on the steel plate connected with the bearing seat (location ②), the third and the fourth on the steel plate of the base (locations ③ and ④). The horizontal distance between the second transducer and the third, the third and the fourth were 24cm respectively. The output signal of the transducers was firstly sent to an charge amplifier which was set 40dB, and then sent to a high-speed A/D converter installed on the computer for data collecting at a sampling rate of 3MHz. The experiment was conducted at different speed (300r/min, 600 r/min, 900 r/min, 1200 r/min, 1500 r/min) and the length of each sample was 1.2s. To study the characters of AE signal, some time-domain features, root mean square (RMS), skewness, kurtosis and the waveform of several revolutions picked up from the collected signal were analyzed.

Experiment results and analysis
Feature extraction. Fig. 3 and Fig. 4 are the time-domain waveforms of the AE signal of new bearing and failure bearing at 900rev/min. It is evident from these two figures that the characteristics of AE signal change with time. For example, there is a significant increase of the peak amplitude between these two figures, from which we can distinguish the normal bearing with the defective bearing. And in Fig. 4, for the amplitude of the waveform, there are considerable changes among these four channels. The amplitude is reduced from the first channel to the fourth channel, and the reduction between normal bearing and defective bearing at each channel is decreasing with the increase of the distance between the bearing and transducer. Therefore, from the time-domain waveform, it is difficult to distinguish the defective bearing when the transducer is far away from the bearing.

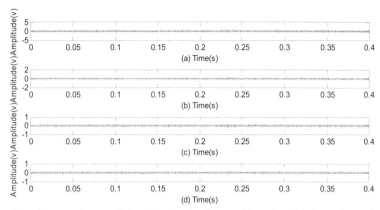

Fig. 3 Time-domain waveforms of the AE signal of normal bearing, (a) first channel, (b) second channel, (c) third channel, (d) fourth channel

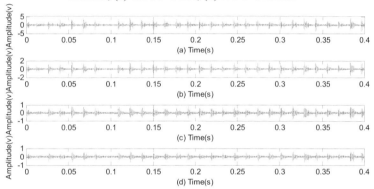

Fig. 4 Time-domain waveforms of the AE signal of failure bearing, (a) first channel, (b) second channel, (c) third channel, (d) fourth channel

Then some time-domain features are extracted from AE signal.

(1) Root mean square

Root mean square (RMS), which is the square root of the average of the squares, is a measure of signal intensity. In the case of given a series of input vectors $X = \{x_1, x_2, \cdots, x_n\}$, the RMS value calculated by the formula listed below:

$$x_{RMS} = \sqrt{\frac{1}{n} \sum_{i=1}^{n} x_i^2} \tag{1}$$

The RMS is one of the most commonly used variables to depict the energy of the signal. Great RMS value means that the energy of the signal is high.

(2) Skewness

The skewness α reflects the amplitude probability density function of the asymmetry to the vertical axis. The greater α is, the greater the asymmetry is. The skewness is given by,

$$\alpha = \frac{n \sum (x_i - \bar{x})^3}{(n-1)(n-2)\sigma^3} \tag{2}$$

Where σ and $\bar{x}$ are the standard deviation and mean value of the input sample.

(3) Kurtosis

Kurtosis is a physical quantity which is characterized by the steepness of the curve. When the defect appears, the value of kurtosis will increase. And it is more sensitive to the pulse signal than RMS, peak amplitude, etc., for which the kurtosis increases much faster when fault signal appears.

$$\beta = \frac{1}{N}\sum_{i=1}^{N}\left(\frac{x_i - \bar{x}}{\sigma}\right)^4 \tag{3}$$

Where σ and $\bar{x}$ are the standard deviation and mean value of the input sample.

In the test, we calculated these three features by six-cycle data picked up from the collected data. The rms and skewness were amplified by 100 times, and kurtosis was kept constant.

Results and analysis. From the three figures shown above, we can know that:

(1) The trends of the variation of these three features among four channels are different. RMS and kurtosis basically are decline whether the speed are low or high, but the variation of skewness is uncertain. Because that RMS and kurtosis have a great matter with the energy of the signal, and the skewness changes a lot for the reflection and refraction of AE wave.

(2) All of these features can used to judge whether the bearing has fault, whatever the speed are low or high. At low speed, RMS and kurtosis are more suitable than skewness. Because when these two values of the signal have a bigger difference between the fault bearing and normal bearing at farther distance. At high speed, on the contrary, the skewness is much more suitable. Because at high speed, the value of skewness of normal bearing almost keeps constant on all of these four channels. Thus, the value of skewness of fault bearing changes a lot among four channels. There aren't the same or better characters for the other two features.

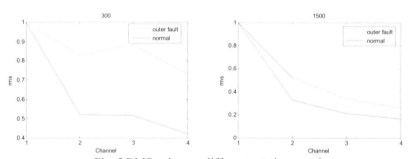

Fig. 5 RMS values at different rotation speed.

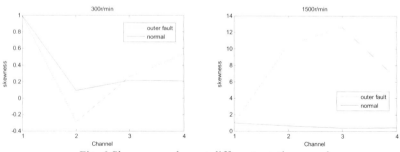

Fig. 6 Skewness values at different rotation speed.

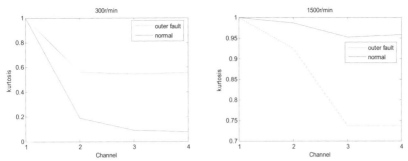

Fig. 7 Kurtosis values at different rotation speed.

Conclusions

Acoustic emission wave essentially is a kind of mechanical wave, then it has the characteristic of fluctuation and attenuation. When the transducer is mounted far away from the bearing, the fault bearing cannot be distinguished only relying on the time-domain waveform. The time-domain features are great supplements for the time-domain waveform. When RMS and kurtosis is more suitable than skewness at low speed (300 rev/min), on the contrary, the skewness has a better effect than RMS and kurtosis at high speed (1500 rev/min). This paper only study some simple processing method for AE signal when the transducer is far away from the AE source, a further investigation would be necessary in future work.

Acknowledgements

The support of the Planning of Philosophy and Social Science of Tianjin(Project No. TJYY11-2-042) is gratefully acknowledged.

References

[1] Mathews, J. R. Acoustic Emission, 1983, Gordon and Breach, New York.

[2] Pao, Y.-H., Gajewski, R. R. and Ceranoglu, A. N.Acoustic emission and transient waves in an elastic plate. J. Acoust.Soc. Am., 1979, 65(1), 96-102.

[3] D.Mba, R H Bannister, G E Findlay. Condition monitoring of low-speed rotating machinery using stress waves Part 1. Proc Instn Mech Engrs.1999, 213(E), 153-170.

[4] C. James Li a, S.Y. Li b., Acoustic emission analysis for bearing condition monitoring, Wear, 1995, 185, 67-74.

[5] L.D. Hall, D. Mba. Acoustic emissions diagnosis of rotor-stator rubs using the KS statistic. Mechanical Systems and Signal Processing. 2004, 18, 849-868.

[6] AbdullahM. Al-Ghamdi, ZhechkovD, MbaD. The use of acoustic emission for bearing defect identification and estimation of defect size. EWGAE2004Lecture38: 397-406.

[7] Catlin Jr, J. B. The use of ultrasonic diagnostic technique to detect rolling element bearing defects. In Proceedings of Machinery and Vibration Monitoring and Analysis Meeting, Vibration Institute, USA, April 1983, 123-130.

Advanced Materials Research Vol. 819 (2013) pp 176-180
© *(2013) Trans Tech Publications, Switzerland*
doi:10.4028/www.scientific.net/AMR.819.176

Research on the Embedded Nondestructive Testing System of Oil-well Tubing

Jingxiang Fang[1,a], Taiyong Wang[1,b], Luyang Jing[1] and Pan Zhang[1]

[1]Key Laboratory of Mechanism Theory and Equipment Design of Ministry of Education, Tianjin University, Tianjin300072, China

[a]zhenxiangfang@163.com, [b]tywang@189.com

Keywords: Magnetic flux leakage testing, Oil-well tubing, Signal analysis, Embedded Development, Qt/Embedded

Abstract. A Embedded Nondestructive Testing System of Oil-well Tubing based on the method of magnetic flux leakage (MFL) is designed. The function and structure of each part of the system are analyzed. The overall structure of the system and the function of each module are analyzed. The visual data processing and analysis program of magnetic flux leakage signal based on embedded Linux Operating System is designed, and the running interface of the system is given.

Introduction

Pumping tubing, also known as tubing, is one of the large-scale use oil extraction equipment in oil extraction operations. To ensure the security, reliability, efficiency and normal of oil operations, the detection of running and defective condition of the pumping tubing is necessary. The tubing is expensive and high-performance requirements, therefore non-destructive testing and quantitative analysis of the tubing can provide the basis for the safety of oil production and tubing reuse. The current PC-based tubing nondestructive testing system has powerful data file management and high-speed digital signal processing functions, but it's bulky and inconvenient to carry and subject to the restrictions of the space environment. Therefore, the development of the embedded non-destructive testing system which is small size, low power consumption and easy to use in the site is essential. Portable data acquisition and signal analysis instrument is similar to a regular PC with powerful data file management functions and network transport functions after installed embedded operating system[1]. And embedded nondestructive testing system can be integrated to tubing operation equipments, timely online testing the running and defective conditions of the tubing.

A embedded nondestructive testing system of oil-well tubing is built. The principle of magnetic flux leakage is used in the system, which can detect and analyse cracks, corrosion pits, holes and other defects on the inner or outer wall in the site or indoor. This can avoid a large number of the old pumping tubing being blind scrapped, reduce pumping tubing oil spills, guarantee the safety reuse of the old pumping tubing. At the same time, you can build the use file of in-service pumping tubing so as to manage the site production.

The principle of magnetic flux leakage testing

The magnetic flux leakage testing is a kind of nondestructive testing technology that finds defects through detecting magnetic flux leakage forming on the surface of the material due to the defects after the ferromagnetic material has attained saturation magnetization[2]. Tubing magnetic flux leakage testing is for such high magnetic permeability materials as the tubing (steel pipe). Defect lines of magnetic force will become bending deformation and many may leak out of the defect surface in the state of saturation magnetization. To detect the leakage magnetic field by using the magnetic sensing element (sensor) can determine whether there is a defect. As shown in Fig.1(a), if a tubing with no surface cracks, no internal defects and no inclusions is in saturation magnetization, all its magnetic lines of force through the magnetic circuit made of ferromagnetic materials in theory. If defects existing, as shown in Fig.1(b), the permeability of the medium is different from the ferromagnetic material and defect magnetoresistance is larger, it can be regarded as an obstacle in the magnetic circuit, then the leakage flux will occur in the defect.

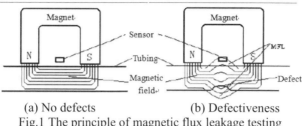

(a) No defects   (b) Defectiveness
Fig.1 The principle of magnetic flux leakage testing

The hardware design of the system

System architecture is displayed in the Fig.2, There are three parts in this system: probe body, signal conditioning module, ARM microprocessor control module. The ARM unit plays total control.

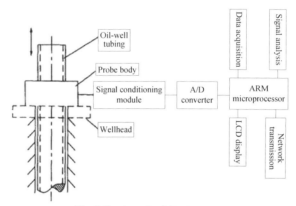

Fig.2 System Architecture

Probe body. It contains magnetized device and 28 Hall-effect element sensors. The Hall sensors are arranged along the tubing circumference evenly. It detects the axial component Φy of the defect leakage magnetic field. The scan of a individual sensor along the axial direction can only cover a certain region of the circumferential direction. It can't test the possible defects at any position along the tubing circumference. Therefore, in order to obtain the magnetic flux leakage testing signal with certain signal-to-noise ratio, using a plurality of sensors (Hall elements) arranged uniformly in the same section of the tubing along its periphery can achieve a scan test without undetected.

Signal conditioning module. Proper conditioning is necessary before the analog signal detected by probe body is supplied to the A/D converter, in order to meet the requirements of the sampling theorem[3]. To filter out the high frequency noise components of the detected signal, appropriate filtering through a specific filter is necessary. Its function is equivalent to the windowing processing in the frequency domain. In order to meet the requirements of the A/D sampling, appropriate amplification is needed through programmable amplification circuit, so that the range of the signal falls close to the voltage range of the A/D acquisition chip.

ARM microprocessor control module. ARM microprocessor is the core unit of the entire embedded devices, which are responsible for coordinating the work of the A/D acquisition, human-computer interaction, data storage, clock control, signal analysis, network communication and so on. The master module is the link combining other functional modules and the interface through which the user and the system exchange. The user can issue various commands to the computer through the main control module, to complete various operations required. It can also be used for providing the user with online acquisition, real-time help to guide you through the whole process of testing and data analysis.

The software design of the system

Online acquisition, defect real-time alarm and defect diagnosis through quantitative identification of the tubing magnetic flux leakage signal are required to complete in the software part during the testing process. Higher real-time performance is required in the software system, therefore acquisition module uses multi-threading technology. The software system is based on embedded Linux operating system in the ARM9 development board and developed in Qt/Embedded environments, in order to achieve the function of data acquisition, data analysis, data transfer and data management. The functional framework of the embedded application software designed and developed in this paper is shown in Fig.3.

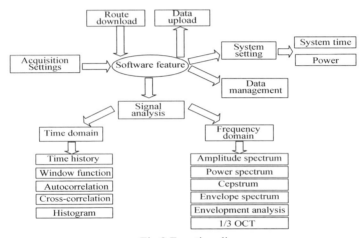

Fig.3 Function diagram

Data acquisition module. The system uses multi-thread technology to achieve high-speed continuous uninterrupted acquisition. Data acquisition thread is absolutely worker thread, which has no window graphic operations in the normal acquisition. In this way, when the users make any window operation, the thread will not be blocked. For the purpose of displaying the collected data progress on the screen, a sub-thread is needed.

Signal analysis module. After the acquisition of the tubing signal or selection of the stored tubing data file, the signal analysis module displays spectral array of the data to detect the fault of tubing. Then data extraction, pretreatment, features extraction and pattern recognition are achieved and classification and calculation are operated to obtain the defect size. [4]

Wavelet transform is operated to process the magnetic flux leakage signal of oil well tubing, distinguish the pit-like signal and partial wear signal, and remove the background noise of the multi-sensor signal, which lays a good foundation for the quantitative analysis of tubing defects. A wavelet neural network is used to predict tubing defect condition. Through the training of adjusting wavelet bases pan and telescopic coefficient, wavelet neural network provides good nonlinear approximation ability, faster convergence speed and a high accuracy, which changes the traditional estimation by experience and automatically find the nonlinear mathematical model based on the inherent laws[5]. Tests prove that it can achieve accurate quantitative identification of tubing defects and provide the basis for the maintenance[6].

Data transfer function. The module's main function is to back up the overall damage of the tubing and guide the transport of spare tubing, at the same time to establish use files of pumping tubing in service so as to combine with the ERP to manage the equipment. This function has two transmission modes, one is by U disk, the other is through the Internet directly to upload the magnetic flux leakage data to a remote server-side.

The running interface of the system

The software architecture is designed according to the requirements that all the modules should be displayed in the main interface as to illustrate the structure.

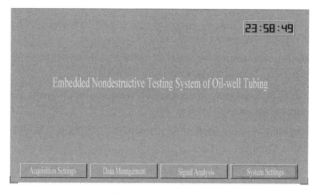

Fig.4 Main interface

There are three sub-blocks which are current settings, route settings and non-route settings in the acquisition settings module. The first sub-block displays all parameters of the last acquisition. If the task-route has been downloaded to the instrument, the operator can choose the channel, database and set the parameters in the route setting sub block, otherwise the acquisition has to be carried out in the non-route mode. The data management module consists of route-download, data-upload and data-delete. Both the task-download and data-upload can be completed between network and portable instrument through the serial, USB or Ethernet port. Signal analysis module accounts for nearly all of the signal analysis methods which are commonly used in engineering practice.

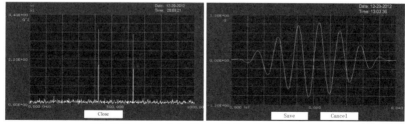

Fig.5 Running interface of signal analysis

Conclusion

A tubing embedded nondestructive testing system is successfully built in this paper. The system can be used in oil production site for quantitative detection of defects such as corrosion pits, corrosion hole and rod wear by pumping tubing magnetic flux leakage signal acquisition and analysis through multi-sensor. Compared with the existing domestic and international tubing damage detection device, the system has high detection sensitivity, powerful quantitative analysis capability, complying with the requirements of reliability, perfect site adaptability, good integration with the field devices.

Acknowledgment

This work was supported by National Science & Technology Pillar Program (2013BAF06B00), Doctoral Fund of Ministry of Education of China (20100032110006), Key Projects of Science and Technology Support of Tianjin(12ZCZDGX01600).

References

[1] Ganti, Ashwin, Plan 9 authentication in Linux, Operating Systems Review. 42 (2008) 27-33.

[2] Wang Taiyong, Yang Tao, Jiang Qi, The quantitative recognition for pipe pits on oil-gas pipe magnetic flux leakage inspection, Acta Metrologica Sinica. 25 (2004) 247-249.

[3] Wang Taiyong, Jiang Qi, Study on quantitative recognition technology of pipeline defect, Journal of Tianjin University. 36 (2003) 55-58.

[4] Yang Tao, Nondestructive testing of oil-well tubing and quantitative recognition of defects based on multi-sensor fusion, Tianjin University, Tianjin,2004.

[5] Jiang Qi, Quantitative technology and application research on magnetic flux leakage inspection of pipeline defects, Tianjin University, Tianjin, 2002.

[6] Jing Luyang, Key technology research of quantitative recognition of wire ropes injury, Qingdao Technological University, Qingdao, 2012.

Advanced Materials Research Vol. 819 (2013) pp 181-185
© *(2013) Trans Tech Publications, Switzerland*
doi:10.4028/www.scientific.net/AMR.819.181

Simulation Research of CNC Machine Servo System Based on Adaptive Fuzzy Control

Xian Wang[1,a], Zhe Wang[1,b], Taiyong Wang[1,c] and Jingchuan Dong[1,d]

[1]Tianjin Key Laboratory of Equipment Design and Manufacturing Technology, Tianjin University, Tianjin 300072, China

[a]wangxian320@163.com, [b]wzwhm@yahoo.com.cn, [c]tywang@189.cn,

[d]new_lightning@ sohu.com (corresponding author)

Keywords: fuzzy control, adaptive, CNC machine servo system, MATLAB simulation

Abstract. According to the characteristics and performance requirements of CNC machine, an adaptive fuzzy PID controller is designed. It uses fuzzy reasoning method to adjust the PID parameters online, and the system has good adaptive ability. The dynamic simulation of CNC machine servo system was carried out based on MATLAB. Simulation results show that the adaptive fuzzy control system has better control performance than ordinary PID control system. This method has less calculation and good dynamic quality. It is easy to implement and convenient for engineering application.

Introduction

CNC machine servo system affects the performance of the entire system of CNC system, and determines the efficiency and accuracy of the machine in a large extent. At present, the traditional PID control is still widely used in the control process of CNC machine servo system. Because of the instability of the input parameters, the conventional PID control method is difficult to consider both static and dynamic performance of the system. Poor anti-interference ability and large overshoot are not ideal for the robustness of the system [1,2].

Fuzzy control is a modern control theory established on fuzzy sets and based on language rules and fuzzy reasoning. It simulates logic inference rules, does not rely on the accurate mathematical model and is not sensitive to the change of the parameters of the object, so it can be used to overcome the effects of nonlinear, time-varying, coupling and other factors in the servo system. For the characteristics of CNC machine servo system, this paper designs an adaptive fuzzy PID controller. It has fuzzy control flexibility and strong anti-interference, and the characteristics of high steady precision, which can achieve good control effect of complex control systems and high-precision servo systems.

CNC machine servo system control model

CNC machine is a complex mechatronic system. The overall performance demands of the machine is high and requires high-speed, high-precision, high rigidity and smooth operation in the process. CNC machine servo system is an important part of CNC machine. It can precisely control the displacement, direction and speed of execution unit movement automatically according to the command signal. CNC machine servo system uses the three-loop control scheme that the position loop is outer loop and the speed loop and current loop are inner loop. The optimization of each link performance is the foundation to improve the performance of entire servo system [3].The diagram of CNC machine servo system is shown in Fig.1.

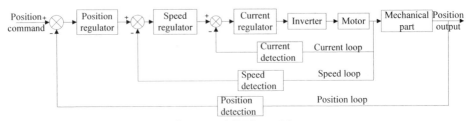

Fig.1 The diagram of CNC machine servo system

Three-loop control process is as follows: the position command and position feedback value are compared to obtain speed command value by the position controller, the speed command value and the speed feedback value are compared to obtain current command value by the speed controller, and then through current loop to the last of the angular displacement of the motor is converted to the linear displacement of the machine moving parts through the mechanical link [4].

The role of the position loop is to ensure the system static accuracy and dynamic tracking performance. It can make the servo system stable and efficient. Therefore, the position loop is the key to design the entire servo system. In this paper, the position loop uses a fuzzy PID controller, the current loop is designed to the P-type regulator and the speed loop is designed to the PI-type regulator, which is to satisfy the requirements for rapid response and no overshoot in the CNC servo system. The simplified dynamic model of servo system is shown in Fig.2.

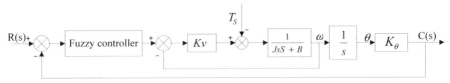

Fig.2 The simplified dynamic model of servo system

Design for adaptive fuzzy PID controller

Adaptive fuzzy PID controller structure. Adaptive fuzzy PID controller uses the basic theories and methods of fuzzy mathematics to express conditions and operations of the rule with fuzzy sets, and these fuzzy control rules and related information are stored in the computer knowledge base. The computer control system uses fuzzy reasoning to adjust PID parameters to the best automatically according to the actual response of the control system [5]. The structure of adaptive fuzzy PID controller is shown in Fig.3.

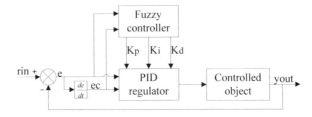

Fig.3 The structure of adaptive fuzzy PID controller

The controller is to seek the fuzzy relationship between PID three parameters and e, ec, test e and ec constantly in operation, and modify the three parameters online according to the principle of fuzzy control, thus the controlled object has a good dynamic and static performance.

According to the membership degree of fuzzy subsets, the assignment table of membership degree and the mode of fuzzy adjustment rules, we can use fuzzy synthesis reasoning to design PID parameters adjustment matrix table, which is the core of adaptive fuzzy control algorithm. The calculation formula for realizing the adaptive PID parameter is as follows:

$$K_p = K_p' + \Delta K_p \tag{1.1}$$

$$K_i = K_i' + \Delta K_i \tag{1.2}$$

$$K_d = K_d' + \Delta K_d \tag{1.3}$$

Complements: K_p', K_i' and K_d' are the initial values of PID parameters; ΔK_p, ΔK_i and ΔK_d are output of fuzzy controller; K_p, K_i and K_d are final output parameter values.

The establishment of fuzzy rules. The fuzzy control rule is a set which uses fuzzy language to express the mapping relationship between input and output. When designing these rules, according to the characteristics of the PID control, the below control laws need to follow:

When E is larger, K_p should be larger to accelerate the response speed of system, K_d should be smaller to avoid the differential overflow and K_i should be smaller to avoid a large overshoot.

When E is a medium size, K_p should be smaller to make the overshoot smaller and K_i should be appropriate to ensure the response speed of the system.

When E is smaller, K_p and K_i should be larger to ensure a good steady state and K_d should be appropriate to avoid the system oscillation at the equilibrium point.

According to the above parameters setting rules and accumulated practical experience, K_p, K_i and K_d fuzzy rules are shown in Table 1.

Table 1 Kp, Ki and K_d fuzzy rules

Kp/Ki/Kd \ EC \ E	NB	NM	NS	ZO	PS	PM	PB
NB	PB/NB/PS	PB/NB/NS	PM/NM/NB	PM/NM/NB	PS/NS/NB	ZO/ZO/NM	ZO/ZO/PS
NM	PB/NB/PS	PB/NB/NS	PM/NM/NB	PS/NS/NM	PS/NS/NM	ZO/ZO/NS	NS/ZO/ZO
NS	PM/NB/ZO	PM/NM/NS	PM/NS/NM	PS/NS/NM	ZO/ZO/NS	NS/PS/NS	NS/PS/ZO
ZO	PM/NM/ZO	PM/NM/NS	PS/NS/NS	ZO/ZO/NS	NS/PS/NS	NM/PM/NS	NM/PM/ZO
PS	PS/NM/ZO	PS/NS/ZO	ZO/ZO/PM	NS/PS/ZO	NS/PS/ZO	NM/PM/ZO	NM/PB/ZO
PM	PS/ZO/PB	ZO/ZO/NS	NS/NS/PS	NM/PS/PS	NM/PS/PS	NM/PB/PS	NB/PB/PS
PB	ZO/ZO/PB	ZO/ZO/PB	NM/ZO/PM	NM/PM/PM	NM/PM/PS	PB/PB/PS	NB/PB/PB

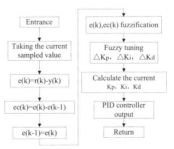

Fig.4 The workflow of self-tuning

In this paper, the change range of system error e and error change rate ec is defined as universe of fuzzy sets.Their range of change is the same, so e, ec={ -5, -4, -3, -2 , -1 , 0, 1, 2, 3, 4, 5},its fuzzy subset e, ec = {NB, NM, NS, ZO , PS, PM, PB}.The elements of subset are negative big, negative medium, negative small, zero, positive small, positive medium and positive big. The universe and membership function of fuzzy subsets can be Triangular Function and Gaussian Function according to the concrete situation.

In operation online, the control system deals with and computes the results of the fuzzy logic rules to accomplish self-tuning of PID parameters. Its workflow is shown in Fig.4.

System simulation and analysis

The controller of CNC machine servo system, for example, is designed by adaptive fuzzy PID method. The transfer function of closed-loop control is simplified:

$$G(s) = \frac{1}{s^2 + s + 1} \tag{1.4}$$

A step response is input for the system. We conduct the simulation design with MATLAB, which is shown in Fig.5,and the corresponding response result is shown in Fig.6.

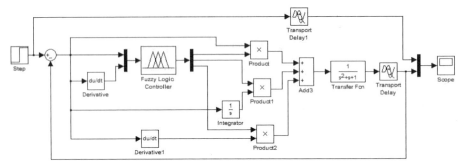

Fig.5 The simulation structure of fuzzy control system

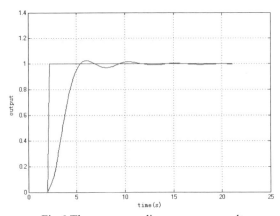

Fig.6 The corresponding response result

In Fig.6, the blue curve is the result of fuzzy control, while the red curve is the result of no treatment. We can see that the adaptive fuzzy PID control can improve the dynamic performance of system and make the step input signal transmission stable.

Conclusion

Because the traditional PID control can't meet the requirements on the performance of the servo system for a CNC machine tool, this paper uses the control strategy of adaptive fuzzy PID, which can improve the rapid response and anti-jamming ability. It can better meet the requirements of the control system and has a good engineering application value.

Acknowledgements

This work was financially supported by the National Science and Technology Support Project (2013BAF06B00), the Science and Technology Planning Project of Fujian (2012H1008), and Tianjin application foundation and frontier technology research program (12JCQNJC02500).

References

[1] Guozhi Li,Xiaohong Ren and Bin Ren: Modular Machine Tool&Automatic Manufacturing Technique,Vol.6(2012), p.67-70

[2] Shengjie Gu and Chunjuan Liu: Journal of Lanzhou Jiaotong University(Natural Sciences), Vol.23(2004), p.62-64

[3] Qidong Liu,Chunguang Xu,Juan Hao and Xifeng Song: Machine Tool & Hydraulics.Vol. 12 (2005), p.121-123

[4] Dasong Shu: Techniques of Automation and Applications. Vol. 29(2010), p.33-36

[5] Yanhai Zhang, Ruimin Li and Peigang Jiang: Journal of QingdaoTechnological University. Vol. 27(2006), p. 82-85

Advanced Materials Research Vol. 819 (2013) pp 186-191
© (2013) Trans Tech Publications, Switzerland
doi:10.4028/www.scientific.net/AMR.819.186

Stator Current-Based Locomotive Traction Motor Bearing Fault Detection

YANG Jiang-tian[1,a], ZHAO Wen-yu[2,b] and LEE Jay[2,c]

[1]School of Mechanical, Electronic and Control Engineering, Beijing Jiaotong University, Beijing 100044, China

[2]Center for Intelligent Maintenance Systems, University of Cincinnati, Cincinnati 45221, USA

[a]jtyang@bjtu.edu.cn (corresponding author), [b]zhaowy@mail.uc.edu, [c]jay.lee@uc.edu

Keywords: traction motor; bearing; fault detection; motor current signature analysis; wavelet packet transforms

Abstract. Rolling-element bearings are critical components in locomotive traction motors. A reliable online bearing fault-diagnostic technique is critically needed to prevent motor system's performance degradation and malfunction. Motor bearing failure induces vibration, resulting in the modulation of the stator current. Compared with conventional monitoring techniques such as vibration monitoring or temperature monitoring, stator current -based monitoring offers significant economic benefits and implementation advantages. In this paper, a novel approach to locomotive traction motor current signature analysis based on wavelet packet decomposition (WPD) of stator current is presented. The effectiveness and practicability of the proposed method is verified by locomotive running tests.

Introduction

The development of rail transportation is strongly dependent on the reliability of rail vehicles. All of the rail systems should be equipped with appropriate diagnostic tools to assure a safe transportation system [1]. Most of the diagnostic systems are equipped with additional sensors for safe operation. With extra sensors, the vehicle becomes more complicated for maintenance.

Typically, traction motors are equipped with vibration sensors [2]. The on-line vibration signals provide useful and reliable diagnostic information since they are sensitive to most of the faults or malfunctions. Motor bearing failure induces vibration, resulting in the modulation of the stator current. Motor-current signature analysis (MCSA) provides a nonintrusive way to assess the health of a machine [3].

Motor-current-signature analysis (MCSA) is a condition monitoring technique that has been widely used to diagnose problems in electrical motors. MCSA focuses its efforts on the spectral analysis of the stator current and has been successfully applied to detect broken rotor bars [4], abnormal levels of air-gap eccentricity [5], and shorted turns in stator windings [6], among other mechanical problems [7].

Compared to conventional vibration-based monitoring techniques, MCSA does not require additional sensors besides the existing transducers for electrical protection. As a result, current monitoring is non-invasive, and can be implemented in the motor control center remotely from the motors being monitored. Current monitoring therefore offers significant economic advantages. Another advantage of current monitoring is that a low cost, non-intrusive overall electric machine condition monitoring package is easy to be realized.

However, bearing faults directly affect the machine vibration, and the effects reflected into the stator current, so there they are typically very subtle [8]. Wavelet packet transform offers an efficient decomposition for signals containing both transient and non-stationary components. Fine frequency resolution may be achieved and fault feature can be extracted successfully by means of wavelet analysis for motor current. An online locomotive traction motor diagnostic method using wavelet packet analysis of stator current is presented. The proposed method was effectively applied to the operation tests of HX2D type locomotive, and the bearing incipient fault of traction motor was diagnosed successfully.

Motor current signature analysis

Current-based bearing fault detection has received more and more attention in the research literature and by industry. The occurrence of motor bearing faults usually results in an asymmetry in the windings and eccentricity of air gap, which lead to a change in the air-gap space harmonics distribution. This abnormality exhibits itself in the spectrum of the stator current as unusual harmonics. Therefore, by analyzing the stator current in search of current harmonics, the motor bearing faults can be detected. According to the effect the fault has on the stator current, traction motor bearing failures are categorized as either single-point defects or generalized roughness in this research.

Single-point defects. A single-point defect is defined here as a single, localized defect on an otherwise relatively undamaged bearing surface. A common example is a pit or spall. A single-point defect will cause certain characteristic fault frequencies to appear in machine vibration. The frequencies at which these components occur are predictable and depend on which surface of the bearing contains the fault; therefore, there is one characteristic fault frequency associated with each of the four parts (i.e. outer race, inner race, cage and ball) of the bearing. The majority of bearing-related fault detection schemes focus on these four characteristic fault frequencies.

Bearing faults result in the absolute motion (vibration) of the machine. The stator current is not affected by the absolute motion, but rather by the relative motion between the stator and the rotor (i.e., changes in the air gap). In the instance of a bearing fault, the characteristic fault frequencies are essentially modulated by the electrical supply frequency.

The most often quoted model that studies the influence of bearing damage on induction machine's stator current was proposed by Schoen et al. [9] The authors considered the generation of rotating eccentricities at bearing fault characteristic frequencies f_c, which leads to periodical changes in machine inductance. This should produce additional frequencies f_{bf} in the stator current, which is given by

$$f_{bf} = |f_s \pm k f_c| \tag{1}$$

where f_s is the electrical stator supply frequency, and $k = 1,2,3,\cdots$.

This model has been applied in a large amount of different works. However, this model does not consider torque variations as a consequence of the bearing fault. In certain applications including locomotive traction systems, the torque variations are incorporated [1]. Blodt [10] proposed a fine fault model, which considers fault-related air gap length variations and changes in the load torque. The comparisons of the two models are shown in Table.1

Table.1 Bearing fault-related frequencies in the motor stator current spectrum

	Schoen's model	Blodt's model	
		Eccentricity	Torque variations
Outer race	$f_s \pm k f_o$	$f_s \pm k f_o$	$f_s \pm k f_o$
Inner race	$f_s \pm k f_i$	$f_s \pm f_r \pm k f_i$	$f_s \pm k f_i$
Ball	$f_s \pm k f_b$	$f_s \pm f_{cg} \pm k f_b$	$f_s \pm k f_b$

Where f_s is the electrical stator supply frequency, f_r is the rotor frequency, f_{cg} is the rotation frequency of cage, f_o, f_i and f_b are the characteristic fault frequencies of outer race, inner race and ball respectively. $k = 1,2,3,\cdots$.

From Table.1, we can find that bearing faults produce the same additional frequencies in the stator current predicted by (1). The difference is, additional frequencies will emerge, according to Blodt's model. It should be pointed out, however, experimentation suggests that the presence of a characteristic fault frequency in the machine vibration does not guarantee its presence in the stator current.

Generalized roughness. Generalized roughness is a type of fault where the condition of a bearing surface has degraded considerably over a large area and become rough, irregular, or deformed. This type of failure is observed in a significant number of cases of failed bearings from various industrial applications. Since its impact is less than single point defects, they are often neglected in the research literature. In general, when a generalized roughness fault reaches a developed stage and the bearing is near failure, the fault can typically be detected via many techniques. An industrial condition monitoring scheme should be able to identify both types of faults while at incipient stages of development.

Experimental results obtained from former research suggest generalized roughness faults produce unpredictable (and often broadband) changes in the machine vibration and stator current. This is in contrast to the predictable frequency components produced by single-point defects. This type of fault feature can be extracted by means of multiple frequency bands analysis of stator current signal. In the light of frequency band division principle, the energy of the current signal in every frequency band is detected. If the energy in several frequency bands increased significantly, it indicates that generalized roughness fault may have occurred [11].

Above analysis shows that current-based bearing fault detection needs to employ powerful signal processing methods. Wavelet packet transform provides better analysis for nonstationary signals and permits tailoring of the frequency bands to cover the range of bearing-defect induced frequencies resulted from rotor speed variations. In this research, the root mean square (RMS) values for defect frequency bands are compared with kurtosis to determine any degradation in bearing health.

Wavelet packet transform (WPT)

Wavelet analysis is a time–frequency analysis, or more properly termed as a time-scaled analysis. It acts as bank of filters, i.e. it decomposes signal in the preceding level of decomposition into two sets of wavelet coefficients, which contain   information of a low-frequency band (which is the detailed coefficients) and a high-frequency band (which is the approximate coefficients). The detailed coefficients would be further decomposed into level 2 by going through a high-pass filter and low-pass filter, and the approximate coefficients would not further be decomposed.

Wavelet packet transform performs a similar operating mechanism as wavelet transform, except that the detailed coefficients would also be decomposed to equal band-width data.

Thus, wavelet packet system is a generalization of wavelet transform, in which at all stages both the low-pass and high-pass bands are split. Therefore, it can allow a finer adjustable resolution of frequencies at high frequencies. It also gives a rich structure that allows adaptation to particular signals or signal classes.

In order to study the frequency characteristic of a signal, wavelet with high frequency resolution is required. Among orthogonal wavelets, Shannon wavelet has the highest resolution theoretically, but sharp edges of their filters make them non-causal. In practice, their approximation,-Meyer wavelet is used more often [12]. This wavelet is a frequency bandlimited function whose Fourier transform is smooth, and cause a faster decay of wavelet coefficient in time domain.

Using Dmey wavelet, the separation of frequency band can be obtained with the best resolution. Therefore, in this research, Dmey is used as the mother wavelet function to analyze the current signal.

Experimental verification

HXD2 Electric Locomotive is an eight-axle AC drive electric locomotive, manufactured by CNR Datong Electric Locomotive Co., LTD. The rated power of this locomotive is 9,600 kW and the max speed is 120km/h. Since it is suitable for heavy haul train, this type of locomotive is running on Datong -Qinhuangdao railway for coal transport.

The locomotive is equipped with eight YJ90A six pole induction traction motors. The rated power of each motor is 1275 kW, the rated supply voltage is 1391V, and rated current is 620A. The shaft-end bearing is NU2322 (Cylindrical roller bearings single row) and the fan-end bearing is NJ2218 & HJ2218 (Cylindrical roller bearings single row with flange rings).

In order to inspect the assembly quality, running tests for a two-locomotive system were performed on the test line of Datong Electric Locomotive Co. In the test, one locomotive was towing and the other one braked as load. The running speed was 35 km/h. The typical current output waveforms of the forth axle traction motor is shown in Fig. 1. Fig. 2 is the corresponding power spectrum.

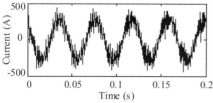

Fig. 1 Waveform of traction motor stator current

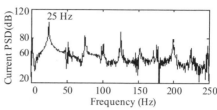

Fig. 2 Power spectrum of the current signal

From the figures, one can observe that there are many components in the traction motor current. The current spectrum is determined by the supply frequency, mechanical structures of the stator, rotor, bearing, winding, among others. Due to modulation effect, the sideband structure around the fundamental component (i.e. supply frequency 25Hz) can be found in Fig. 2. That is the basic characteristics of motor current spectrum.

This signal was decomposed into three levels via WPT algorithm using Dmey wavelet. The results are shown in Fig. 3. The wavelet packet coefficients of each node are on the left side, and their corresponding spectra are on the right side. Furthermore, the frequency range of wavelet packet nodes is shown in Table. 2.

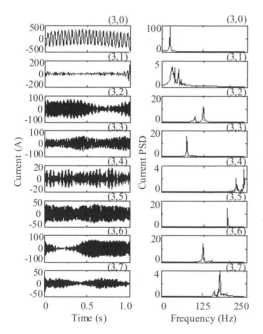

Fig. 3 Wavelet packet decomposition of traction motor stator current

Table.2 Frequency range of wavelet packet nodes

Node	Frequency (Hz)	Node	Frequency (Hz)
3,0	0 — 31.25	3,4	218.75 — 250.00
3,1	31.25 — 62.50	3,5	187.50 — 218.75
3,2	93.75 — 125. 0	3,6	125.00 — 156.25
3,3	62.50 — 93.75	3,7	156.25 — 187. 50

Table.3 Characteristic fault frequencies of traction motor bearings

Location	Frequency (Hz)	
	Shaft-end bearing	Fan-end bearing
Rotor rotation	12.9	12.9
Outer race fault	73	93.2
Inner race fault	108	126
Roller fault	64.8	83.7
Cage rotation	5.22	5.48

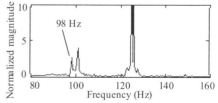

Fig. 4 Power spectrum of WP coefficients of node (3, 2)

For monitoring abnormal behavior of stator current, the root mean square (RMS) value and kurtosis of wavelet packet coefficients of nodes were calculated. The RMS value of each node reflects the amount of current induced in the corresponding frequency band. If RMS value of several nodes has increased, it indicates generalized roughness fault may have occurred. Kurtosis factor is sensitive to large magnitude of the signal. The increase of kurtosis indicates that there are new frequencies emerging in the corresponding frequency range.

The test locomotive was running at speed 35km/h, and the transmission ratio of HXD2 Electric Locomotive is 120:23. Therefore, the corresponding rotational speed of traction motor rotor is 775 r/min. The characteristic fault frequencies of shaft-end bearing and the fan-end bearing are shown in Table. 3.

The kurtosis of wavelet packet coefficients of node (3.2) has increased by 20% during the test. It indicates that there are new frequencies emerging in the corresponding frequency band 93.75-125.0 Hz. The power spectrum of WP coefficients of node (3, 2) is shown in Fig. 4. In the spectrum, the 98 Hz component can be found. According to Schoen's model, this component is the outer race fault frequency of shaft-end bearing reflected into the stator current (see Table. 1 and k =1). Fig. 5 is power spectrum of WP coefficients of node (3, 7). We can find 171-Hz component, which is another frequency mapping of bearing outer race fault frequency to stator current (see Table. 1 and k =2). Then, we came to a preliminary conclusion that a single point defect has occurred on the outer race of the shaft-end bearing.

The running speed of the locomotive was then raised to 46 km/h. The stator current of the traction motor was sampled, and the power spectrum is illustrated in Fig. 6. It can be found that the supply frequency 31.7 Hz is the fundamental component. The shaft-end bearing outer race fault frequency is 96 Hz at this speed. It is reflected into the stator current, and the abnormal harmonic frequency of 128Hz, 224 Hz, etc would emerge, in terms of Tabel.1.

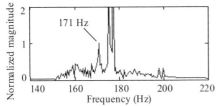

Fig. 5 Power spectrum of WP coefficients of node (3, 7)

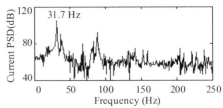

Fig. 6 Power spectrum of traction motor stator current (locomotive running speed 46 km/h)

This signal was decomposed into three levels using WPT algorithm. Fig.7 shows the power spectrum of WP coefficients of node (3, 6), in which the component of 128Hz can be found. Fig.8 illustrates the power spectrum of WP coefficients of node (3, 4), where there is the component of 224 Hz. These two frequencies demonstrate a single point defect occur on the outer race of the shaft-end bearing. We replaced the bearing with a new unit and consequently the failure disappeared. As a result, it is clear that the proposed algorithm is well designed to diagnose motor faults under the motor current signature analysis.

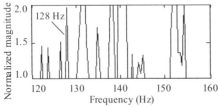

Fig. 7 Power spectrum of WP coefficients of node (3, 6)

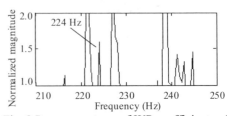

Fig. 8 Power spectrum of WP coefficients of node (3, 4)

Summary

Motor bearing failure induces vibration, resulting in the modulation of stator current. According to the effect the bearing fault imposes on the measurable machine parameters (e.g., vibration and current), this paper has categorized traction motor bearing failures as either single-point defects or generalized roughness. A single-point defect causes certain characteristic fault frequencies to appear in machine vibration. The characteristic fault frequencies are modulated by the electrical supply frequency and sideband components will emerge in motor current spectra. Generalized roughness faults produce unpredictable broadband changes in machine vibration and stator current. Motor current signature analysis provides a nonintrusive approach to assess the health status of a motor.

Wavelet packet transform offers an efficient decomposition for signals containing both transient and non-stationary components. Fine frequency resolution may be achieved and fault feature can be accurately extracted by means of wavelet analysis. This research investigates an online locomotive traction motor diagnostic method using wavelet packet analysis for stator current. The proposed method has been applied to the operation tests of HX2D type locomotive, and incipient bearing fault of traction motor was diagnosed successfully.

References

[1] H. Henao, S. Kia Hedayati, G. A Capolino, Torsional-vibration assessment and gear fault diagnosis in railway traction system, IEEE Trans on Industrial Electronics. Vol.58, No.5, (2011), p.1707-1717

[2] S. Nandi, H A. Toliyat, Xiaodong Li, Condition monitoring and fault diagnosis of electrical motors-a review, IEEE Trans on Energy Conversion. Vol.20, No.4, (2005), p.719-729

[3] J. H. Jung, J. Lee, B. H. Kwon, Online diagnosis of induction motors using MCSA, IEEE Trans on Industrial Electronics, Vol.53, No.6, (2006), p.1842–1852

[4] J. Milimonfared, H. M. Kelk, S. Nandi, A novel approach for broken-rotor-bar detection in cage induction motors, IEEE Trans On Industry Applications, Vol.35, No.5, (1999), p.1000–1006

[5] S. Nandi, T. C. Ilamparithi, Sangbin Lee, Detection of eccentricity faults in induction machines based on nameplate parameters. IEEE Trans on Industrial Electronics, Vol.58, No.5, (2011), p.1673–1683

[6] S.M.A. Cruz, A.J.M. Cardoso, Diagnosis of stator inter-turn short circuits in DTC induction motor drives, IEEE Trans on Industry Applications, Vol.40, No.5, (2004), p.1349-1360

[7] C. C. Lau Enzo, H. W. Ngan, Detection of motor bearing outer raceway defect by wavelet packet transformed motor current signature analysis, IEEE Trans on Instrumentation and Measurement, Vol.59, No.10, (2010), p.2683–2690.

[8] Fabio Immovilli, Alberto Bellini, Riccardo Rubini, Diagnosis of bearing faults in induction machines by vibration or current signals: a critical comparison, IEEE Trans on Industrial Applications, Vol.46, No.4, (2010), p.1350–1359

[9] R. R. Schoen, T. G. Habetler, F. Kamran, Motor bearing damage detection using stator current monitoring, IEEE Trans on Industrial Applications, Vol.31, No.6, (1995), p.1274–1279

[10] Martin Blodt, Pierre Granjon, Bertrand Raison, Models for bearing damage detection in induction motors using stator current monitoring, IEEE Trans on Industrial Electronics, Vol.45, No.4, (2008), p.1813–1822

[11] Fabio Immovilli, Marco Cocconcelli, Alberto Bellini, Detection of generalized-roughness bearing fault by spectral-kurtosis energy of vibration or current signals, IEEE Trans on Industrial Electronics, Vol.56, No.11, (2009), p.4710–4717

[12] J. O. Chapa, R. M. Rao, Algorithms for designing wavelets to match a specified signal, IEEE Trans on Signal Processing, Vol.48, No.12, (2000), p.3395–3406

Advanced Materials Research Vol. 819 (2013) pp 192-196
© (2013) Trans Tech Publications, Switzerland
doi:10.4028/www.scientific.net/AMR.819.192

Study on A type of Pneumatic Force Servo System

Nie Meng [a], Li Jian-yong[b], Shen Hai-kuo[c], Sun Hua-min[d]

School of Mechanical & Electronic Control Engineering   Beijing Jiaotong University, Beijing

[a]09116327@bjtu.edu.cn [b]jyli@bjtu.edu.cn, [c]shenhk@bjtu.edu.cn, [d]11121407@bjtu.edu.cn

Keywords: Pneumatic force servo system, Pneumatic system modeling, Cylinder system.

Abstract. A type of pneumatic force servo system is discussed in this paper. In this system, output pressure depends on pressure difference between two chambers of cylinder; by controlling pilot-operated reducing valve will get different cylinder chamber pressure. Dynamic mathematical model and control Strategy of this system is founded in this paper. By mathematical model and control Strategy, the digital simulation results show how system performance with different gas source pressure and rodless cavity size..

Introduction

Because the pneumatic force servo system has good flexibility and clean ability advanced advantage, it is used more and more to replace electromagnetic and hydraulic system in some way [1]. However, due to such reasons as air compressibility, friction and other nonlinear factors, pneumatic system is difficult to describe with precise mathematical model. How to build a suitable mathematic model for a pneumatic system and find out its parameter feature is always a research focus [2].

Theoretical research and engineering applications on the pneumatic system has been improved gradually after a few years of continuous development, and apply in different fields, 1965, T. Noritsugu [3] introduced the pressure feedback in the pneumatic position servo control system, and build pressure disturbance observer in order to speed up the system response speed to the external load force interference ; 1971 Sanville[4] seen the process of gas flow through the valve port approximately to the ideal gas through the one-dimensional isentropic flow of contraction nozzle to the valve port flow formula ; 1990 Araki[5] through the analysis of the advantage of compressed gas ,obtained a series of high-pressure gas flow pneumatic device characteristics; 1992, Xu Hongguang[6] adopted sliding mode variable structure control strategy to control pneumatic system of the position servo which used pressure proportional valve ; 1996, X.Lin[7] studied rodless cylinder electric / gas control system modeling ; 2004, Huang Chun[8] carried out cylinder crawling problems modeling and simulation to analyze the main factors of cylinder creeping phenomenon .

All of these research focuses on their own system and most of them put their research on pneumatic position control system. In this paper, a pneumatic force control system has been schemed out，and its mathematical model has been established .With the help of simulation some parameters how to influence the system have been studied.

Description of The Pneumatic System

This Pneumatic System is a cylinder force control system and the schematic diagram is shown in Figure. 1.

In this system, output pressure depends on pressure difference between two chambers of cylinder C1. By controlling Pilot-operated reducing valve A1 and A2 will get different cylinder chamber pressure. Proportional reducing valve A5 and A6 play the role as pilot valve of A1 and A2.In order to realize fast-uploading, set reversing valve A3 and A4 as fast-uploading valve.

The mathematical model

Actual gas subject to a number of factors, in order to facilitate analysis and research, during the modeling process to make the following assumptions:

a) Medium gas is an ideal gas.

b) The gas kinetic and potential energy is negligible.

c) The gas flowing state is the isentropic adiabatic process.

d) Ignore the cylinder inside and outside gas leak.

The mathematical model of the system consists the following components:

Cylinder piston force balance equation: According to Newton's second law, establish the piston force balance equation.

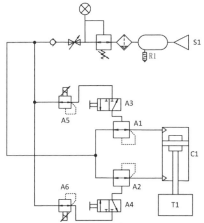

Fig.1 The pneumatic force servo system

S1: Gas source A1~A2: Pilot-operated reducing valve A3~A4: Fast-uploading valve

A5~A6: Proportional reducing valve C1: Cylinder T1: The pressure load

$$p_a A_a + Mg \cos \theta - Mg \sin \theta f - p_b A_b = F + m \frac{d^2 x}{dt^2} \tag{1}$$

Where A_a is work area of rodless chamber pressure, A_b is work area of the rod chamber pressure, p_a is rodless chamber side pressure , p_b is rod chamber side pressure value, M is quality of the load, θ is tilt angle of the cylinder, F is cylinder output pressure and x is position of cylinder piston.

Cylinder cavity flow continuity equation: According to the law of conservation of mass, flow into the accommodation chamber should be equal to the rate of change of the capacity of cavity quality.

$$\begin{cases} q_{ma} = \dfrac{dm_a}{dt} = \dfrac{d}{dt}(\rho_a V_a) \\[2mm] q_{mb} = \dfrac{dm_b}{dt} = \dfrac{d}{dt}(\rho_b V_b) \end{cases} \tag{2}$$

Where q_m is rodless cavity volume flow rate, m is quality of the corresponding capacity chamber, V is cylinder cavity volume and ρ is capacity cavity density.

Cylinder chamber pressure differential equation: According to the energy change values of the external work done is equal to the energy conservation of energy theorem into the accommodation chamber minus the capacity chamber cavity capacity.

$$\begin{cases} \dfrac{dE_a}{dt} = q_{ma}C_pT + \dfrac{dW_a}{dt} - p_a\dot{V}_a \\ \dfrac{dE_b}{dt} = q_{mb}C_pT + \dfrac{dW_b}{dt} - p_b\dot{V}_b \end{cases} \qquad (3)$$

Where q_mC_pT is unit-time input energy of chamber, dW/dt is unit-time input heat of chamber, $p\dot{V}$ is unit-time output power of chamber, dE/dt is control of the cavity of the total energy rate of change, W is tolerance for outside incoming heat chamber; C_p is pressure specific heat, C_v is gas and other capacity than hot, E is total energy capacity cavity, $E = C_v pV / R$, R is the gas constant.

Electric proportion valve governing equation:

$$m_b\dfrac{d^2x_{rb}}{dt^2} + b\dfrac{dx_{rb}}{dt} + k_a x_{rb} = k_u U_b - A_f P_b - k_g x_0 - F_c\,\mathrm{sgn}(\dfrac{dx_{rb}}{dt}) \qquad (4)$$

Where m_r is quality of valve spool , x_r is position of valve spool, b is viscous damping coefficient, k_a is equivalent stiffness, k_u is power coefficient gain, U_r is output control voltage, p_a is output pressure and F_c is coulomb friction.

According to the Sanville flow equation, the valve port flow equation:

$$q_m = \begin{cases} \dfrac{A(u)}{\sqrt{T_s}}\sqrt{\dfrac{k}{R}\dfrac{2}{k-1}}P_s\sqrt{(\dfrac{p}{P_s})^{2/k} - (\dfrac{p}{P_s})^{k+1/k}} \cdots\cdots\cdots \dfrac{p}{P_s} > C_t \\ \dfrac{A(u)}{\sqrt{T_s}}P_s\sqrt{\dfrac{k}{R}(\dfrac{2}{k+1})^{k+1/k-1}} \cdots\cdots\cdots \dfrac{p}{P_s} \le C_t \end{cases} \qquad (5)$$

Where $A(u)$ is opening area of the valve, p_s is gas source pressure , p is pressure of chamber side, k is gas adiabatic index and C_t is critical pressure ratio, $C_t = (2/(k+1))^{k/k-1}$.

Control Strategy

In this system, input signal is setting pressure F_{set} , and output signal is the real pressure F , including two controlled members: Cylinder rodless cavity control voltage U_{ra} and Cylinder rodless cavity control voltage U_{rb} , the Control block diagram is shown in Figure. 2.

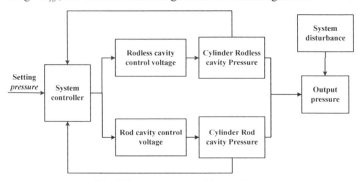

Fig.1 System control block diagram

As can be seen from block diagram, this system controller own three input singles and two output singles : it receive setting pressure single F_{set} and feedback pressure singles P_a and P_b, while cylinder rodless cavity control voltage U_{ra} and cylinder rod cavity control voltage U_{rb} are under the control of system controller. Control algorithm of this controller is based on PID Algorithm:

$$\begin{cases} U_{ra} = K_{pa}U_{ae} + K_{ia}\int U_{ae}dt + K_{da}\dfrac{dU_{ae}}{dt} \\ U_{rb} = K_{pb}U_{be} + K_{ib}\int U_{be}dt + K_{db}\dfrac{dU_{be}}{dt} \end{cases} \qquad (6)$$

Where K_p, K_i, K_d are PID control parameters and $U_{we} = K_w(p_{seta} - p_a)$; $U_{be} = K_b(P_{setb} - p_b)$.

Simulation Study

In order to design the system, some parameters how to influence the system need to be study. Such as gas source pressure and rodless cavity size. With mathematical model and control strategy, system simulation will have great helpful to such problem. According to the mathematical model, which the relevant parameters selected as shown in Table 1:

Table 1. Simulation of key parameters

Parameter	value	
Ambient temperature	293.14	[K]
Cylinder bore	125	[mm]
Load quality	100	[Kg]
Constant pressure specific heat	1.004	[J/(mol·K)]
Constant Volume Specific Heat	0.718	[J/(mol·K)]
Cylinder tilt angle	0	[°]
Polished target pressure	2	[KN]
Gas adiabatic index	1.41	
A1 A2 orifice area	5	[mm*2]

Gas source pressure p_s has great influence on this system. Select the same ratio to U_{ra} and U_{rb}, choose step signal with max 2KN and min 0KN as simulation input signal, and the simulation results are shown in Figure. 3; while choose sine signal with max 2.5KN, min 1.5KN and frequency 0.5Hz as simulation input signal, and the simulation results are shown in Figure. 4.

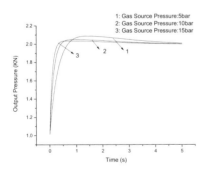

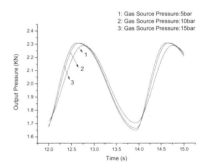

Fig.2 Step response simulation in different gas source pressure

Fig. 3 Sine response simulation in different gas source pressure

The above results show that in step response simulation, at gas source pressure 5bar, it takes more than 1s to achieve maximum value, however, it only takes less than 0.5s at gas source pressure 15bar. At the same time, higher gas source pressure brings less overshoot. But in sine response simulation, higher gas source pressure brings more overshoot and faster response. It means high gas source pressure will cause strong robust performance in step response, but it is not suitable in sine response.

Cavity size is composed by many factors, such as volume of air pipe, position of piston. In order to remove other influencing factors, take $U_{rb} = 0V$ and $p_s = 5bar$; Choose step signal and sine signal as same as before in different rodless cavity size. The simulation results are shown in Figure.5 and Figure. 6.

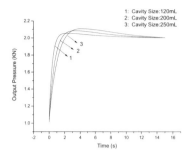

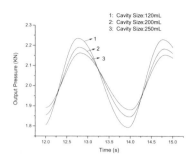

Fig.4 Step response simulation in different   Fig. 5 Sine response simulation in different
rodless cavity size   rodless cavity size

The above results show that oversize cavity will bring response rate reduce. As it is shown in Figure.5, cavity size at 250mL takes almost twice time to achieve maximum value as long as cavity size at 120mL and less cavity size brings less overshoot. But in sine response less cavity size brings more overshoot.

Summary

A type of pneumatic force control system is discussed in this paper, and the mathematical model is derived. With mathematical model and control strategy, system simulation about some parameters how to influence the system is studied. The results show that higher gas source pressure and less rodless cavity size will make the system get stronger robust performance in step response. At the same time, higher gas source pressure and less rodless cavity size brings more overshoot in sine response. The experiment system of pneumatic force control system will be constructed in Beijing Jiao Tong University; more experiments will be done to improve the performance of the system.

References

[1] ZHU Chunbo , BAO Gang, et al, Adaptive Neural Network Fuzzy Controller For Pneumatic Servo System, Journal of Mechanical Engineering. Vol.37, No.10,2001, 80-83.
[2] ZHANG Lin, LI Yanxi, Study on the Position Control of a Cylinder System Using MR Valve Actuators, Machine Tool & Hydraulics,Vol.12 ,No.12,2007, 26-27.
[3] T. Noritsugu , M. Takaiwa, Robust Positioning Control of Pneumatic Servo System with Pressure Control Loop, IEEE International Conference on Robotics and automation
 6(1965), pp95-96.
[4] F.E.Sanville ,A new method of specifying the flow capacity of pneumatic fluid power valves, Second Fluid power Symposium(1997).
[5] GUO Hao, YANG Gang. Characteristic Analysis on Large Volume Cushion Chamber of High-pressure Electro pneumatic Proportional Valve, MACHINE TOOL & HYDRAULICS. Vol.36, No.2,2008,p32-36
[6] XU Hongguang. Research on electric - pneumatic servo control system，Harbin Institute of Technology.1992,pp10-25.
[7] X.Lin, F.Spettel. Modeling and Test of an Electropneumatic Servo valve Controlled Long Rodless Actuator. Journal of Dynamic Systems, Measurement and Control.1996
[8] HUANG Jun, LI Xiaoning. Modeling and Simulation of Stick-slip Motion in Pneumatic Cylinder. Chinese Hydraulics &Pneumatics 6(2004), 20-24

Advanced Materials Research Vol. 819 (2013) pp 197-201
© (2013) Trans Tech Publications, Switzerland
doi:10.4028/www.scientific.net/AMR.819.197

Study on the Transmission Spectrum of Fiber Bragg Grating as Temperature Sensor

WANG Zheng[1,a], ZHU Pingyu[2, b], LIANG Hongqiang[1, c], PENG Wei[1,d]

[1] Provincial Key Lab of Hunan Province Health Maintenance for Mechanical Equipment,

Hunan University of Science and Technology, Xiangtan 411201, China

[2] School of mechanical and Electric Engineering, Guangzhou University, Guangzhou 510006, China

[a]uiopwz129@sina.com,[b]pyzhu@gzhu.edu.cn

Keywords: Fiber Bragg Grating (FBG); Optical Spectrum Analyzer (OSA); Transmission spectrum; Spectrum Width (SPEC-WD); Temperature sensing characteristic.

Abstract. The temperature sensing characteristic of Fiber Bragg Grating (FBG) has been studied. The temperature sensing equation of FBG is revised. The accurate analysis results are obtained. The center wavelength drifts of the tested FBG are displayed in the screen of Optical Spectrum Analyzer (OSA).The experimental results show that the center wavelength of FBG is thermally red-shift with the rise of the temperature and blue-shift with the decrease of the temperature. The results are agree with the revised temperature sensing equation. It is noticed that the transmission spectrum width (SPEC-WD) of FBG fluctuates with the change of temperature. Nevertheless, the trend of the fluctuation is demonstrably towards increase, which perhaps is some potential auxiliary sensing function in future.

Introduction

Fiber Bragg Gratings are capturing more and more attention of the engineer because of its distinguishing advantages such as electrically passive operation, immunity to electromagnetic interference, high resolution. K.O. Hill found the Photosensitivity effect in Ge-doped silica single mode fiber core in 1978 for the first time and started the application of FBG [1]. In 1989, United Technologies Research Center, G. Meltz use 244nm UV interference beams fringe side-exposure in Ge-doped silica single mode fibre core [2], make any work wavelength of phase grating written into the optical fiber. In 1993, K.O. Hill put forward phase mask manufacturing method that greatly reduces the complexity of the fiber grating fabrication system [3]. FBG based optical sensors becomes the mainstream, a variety of optical fiber grating sensors are designed by taking its axial strain and temperature sensing characteristics as the basic principle. With the development of research, people have found many new characteristics for FBG. For example, EI-Sherif discussed the FBG's lateral load sensitive characteristic in 2000, and found that FBG in lateral load effect reflection summit split into two parts [4].In 2002, Federico Bosia do research on optical fiber grating for lateral load test, it was found that the FBG reflection spectrum will find deformation with different stress load [5].Studies for the FBG on existing literatures are only recoded their center wavelength drifts with the changes in different environmental parameters(such as axial strain, temperature, bending and flow rate), some literatures are not give precision formulation for the theory temperature sense characteristic, one cannot obtain the accurate results from the formulations given by the existing literatures[6,7,8,9,10]. The paper revised the temperature theory Equation of FBG, The transmission notch temperature sensor characterization are studied using the OSA, and some different results are achieved.

Sensing principle of FBG

The shift of the Bragg wavelength due to strain and temperature is given by Eq.1 [11].

$$\Delta \lambda_B = 2(\Lambda \frac{\partial n_{eff}}{\partial l} + n_{eff} \frac{\partial \Lambda}{\partial l})\Delta l + 2(\Lambda \frac{\partial n_{eff}}{\partial T} + n_{eff} \frac{\partial \Lambda}{\partial T})\Delta T \qquad (1)$$

The first term of Eq.1 represents the strain effect on FBG fiber, this corresponds to a change in grating spacing and strain-optic induced change in the index of refraction. The second term represents the thermal effect on the FBG fiber, Only Regarding the thermal effect in equation1, then it can be written as Eq.2:

$$\Delta\lambda_{BT} = 2(\Delta n_{eff}\Lambda + n_{eff}\Delta\Lambda) \tag{2}$$

Where n_{eff} represents average refractive index, Λ represents the optical grating period, when the external temperature changes, one can rewrite Eq.2 as:

$$\Delta\lambda_{BT} = 2[\frac{\partial n_{eff}}{\partial T}\Delta T + (\Delta n_{eff})_{ep} + \frac{\partial n_{eff}}{\partial a}\Delta a] + 2n_{eff}\frac{\partial\Lambda}{\partial T}\Delta T \tag{3}$$

Where Δn_{eff} represents the change of the index of refraction, $\partial n_{eff}/\partial T$ represents the index thermal coefficient of FBG,$(\Delta n_{eff})_{ep}$ represents photoelastic effect due to the thermal expansion; $\partial n_{eff}/\partial a$ represents waveguide effect due to the thermal expansion that leads to changes the radial of the fiber, a is the fiber's radial; $\partial\Lambda/\partial T$ represents the linear thermal expansion coefficient; given that $\lambda_B=2n_{eff}\Lambda$, assuming:

$$a_n = (\frac{1}{n_{eff}})\xi, \quad a_\Lambda = (\frac{1}{\Lambda})\frac{\partial\Lambda}{\partial T} \tag{4}$$

Where a_n (approximately $6.8\times10^{-6}/°C$ for Ge-doped silica) represents thermal-optical coefficient, a_Λ (approximately $0.55\times10^{-6}/°C$ for silica) represents thermal expansion coefficient. As we know, the wavelength drift sensitivity coefficient K_ε caused by the photoelastic effect is determined by Eq.5 [12]:

$$K_\varepsilon = 1 - (\frac{n_{eff}^2}{2})[P_{12} - \upsilon(P_{11} + P_{12})] \tag{5}$$

Comprehensive considering the strain effect, the thermal expansion is described as Eq.6.

$$\begin{bmatrix} \varepsilon_{rr} \\ \varepsilon_{\phi\phi} \\ \varepsilon_{zz} \end{bmatrix} = \begin{bmatrix} a\Delta T \\ a\Delta T \\ a\Delta T \end{bmatrix} \tag{6}$$

The thermal term in equation 3 can be rewritten as Eq.7.,

$$\frac{\Delta\lambda_{BT}}{\lambda_B\Delta T} = \frac{1}{n_{eff}}[\xi - \frac{n_{eff}^3}{2}(P_{11} + 2P_{12})a_\Lambda + K_{wg}\frac{\Delta a}{\Delta T}] + a_\Lambda \tag{7}$$

Where K_{wg} represents the sensitivity coefficient of the wavelength drift caused by the waveguide effect, $K\varepsilon$ and K_{wg} can be determined by Eq.8.

$$K_{wg} = \frac{\partial neff}{\partial a} \tag{8}$$

When $K_{wg}=0$, according to Eq.4~Eq.8, Eq.7 can be rewritten as below,

$$\Delta\lambda_{BT} = \lambda_B K_T \Delta T \tag{9}$$

Where λ_B represents the center wavelength, K_T represents the amended thermal sensitivity coefficient, the above two parameters are given as Eq.10.

$$K_T = a_n + [1 - K_{\varepsilon T}]a_\Lambda, K_{\varepsilon T} = \frac{n_{eff}^2}{2}(P_{11} + 2P_{12}) \tag{10}$$

Where P_{11} and P_{12} represents the strain-optical tensor, Eq.10 is the amended theory formulation for FBG thermal sensitivity coefficient. The FBG's thermal sensitivity coefficient equation given on existing literatures is $K_T'=\alpha n+\alpha\Lambda$ [6-10]. Assume the parameters: $\alpha_n=6.8\times10\text{-}6/°C$, $\alpha_\Lambda=0.55\times10\text{-}6/°C$, $n=1.456$, $P_{12}=0.121$, $P_{12}=0.270$, $\upsilon=0.17$ [7,10,11]. It is difficult to obtain the accurate results as described in the literatures [6-10]. Nevertheless, the precision results can be calculated by Eq.10. Assuming $\lambda_B=1540nm$, $1545nm$, $1550nm$, we can calculate out the relative   thermal sensitivity coefficient is 10.725pm/°C, 10.76pm/°C and 10.80pm/°C, respectively.

Generally, it is a good linear relationship between the temperature and the axial center wavelength of FBG. And it has approximately same temperature sensitivity coefficient for FBG with different center wavelength. That is to say, the center wavelength of FBG is not the only important parameter as a thermal sensor

Spectral measuring principle of FBG

To know the characteristics of FBG, the reflection or transmission spectrum of FBG are generally measured using Optical Spectral Analyzer (OSA).

Experimental Setup. To monitor and control the change of the temperature, an experimental scheme is designed, as shown in Fig.1. A temperature monitoring control kit simulates the changes of the temperature.The instruments and devices are listed: ASE (C+L band), Optical Spectral Analyzer (OSA, AQ6317-YOKOGAWA), thermocouple (K type TP01), digital multi-meter, semiconductor refrigeration temperature controller (TC-10).

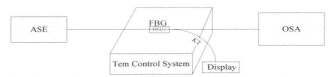

Fig.1 FBG Transmission Notch Monitor experimental system

Experimental Analysis. Connect FBG to the experiment system, we obtained the ASE-FBG transmission spectra, refer to Fig.2a.The measured FBG's transmission spectrum in 15 ℃, 25 ℃ and 35 ℃ are as shown in Fig. 2b.

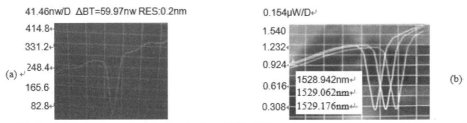

Fig.2 (a) FBG Transmission Notch λ_B =1529.050nm (b) Transmission spectral temperature

From Fig.2a, it shows the center wavelength of FBG is approximately 1529.050nm. From Fig.2b, it can be learned that the center wavelength of FBG on 15 ℃, 25 ℃ and 35 ℃ are 1528.942nm, 1529.062nm, 1529.172nm, respectively, the center wavelengths of FBG are drift with the increase of the temperature . At the same experimental condition, we recorded the center wavelength of FBG with different temperatures in three times.

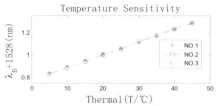

Fig.3 FBG Temperature sensitivity characteristic

Depict out the record datas as figure 3, where N0.1, N0.2, N0.3 represents the first, the second, and the third time experimental datas respectively. From Figure.3, it shows that all times experimental results fairly match the theoretical expectation. We can also learn from the Figure.3 that the total drift of wavelength of FBG is 0.472nm (equal to 472pm), conclusions can be observed that the center wavelength of FBG are red drift with the increase of the temperature, there are apparent linear relationship between the temperature and the center wavelength of FBG. Comprehensive analysis the recorded datas and the Fig.3, one can calculate out the approximate temperature sensitivity is 11.8pm/°C; which is very close to the amended Eq.10 has calculate out the results with the parameters: $\alpha_A=0.55\times10^{-6}/°C$, $\alpha_n=0.55\times10^{-6}/°C$, n=1.456, $P_{11}=0.121$, P12=0.270, $\upsilon=0.17$, verified the accuracy of the theory and experiment.

The Spectral width of FBG's transmission Notch also was recorded at the same experiment of monitoring the temperature changes, the recorded datas are depicted out as figure.4.

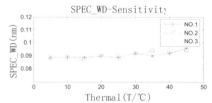

Fig.4 FBG-SPEC-WD Temperature

Where N0.1, N0.2, N0.3 represents the first, the second, and the third time experimental datas respectively. From Fig.4, it can be seen that the SPEC-WD of FBG fluctuates around its center value with the increase of the temperature. It has no obvious linear relationship between the SPEC-WD and the temperature. The spectrum width (SPEC_WD) value of FBG fluctuates with the change of the temperature. Nevertheless, the trend of the fluctuation is demonstrably towards increase.

Error analysis

From the experimental results, it is noticed that it is not always exactly coincident value between the center wavelength of FBG and the regulated temperature to the same point. Potential errors are mainly caused by the following three reasons. (1) The uncertainty of OSA and measuring accuracy error. From the manual of OSA, we learned that the wavelength accuracy range is from -0.02nm to +0.02nm (equals to+-20pm). Maximum of the wavelength resolution is 0.015nm (equals to15pm). (2) Temperature fluctuations during the measurement. (3) The coating of FBG products minor error.

Conclusion

The sensing Equation of FBG has been revised. The transmission spectrum of FBG has been measured by use of OSA in a controlled temperature system. The measurements demonstrated the effectiveness and accuracy of revised equation, Red-shift and blue-shift of FBG may be some potential use in real-world applications.

Acknowledgements

The research is supported by the National Natural Science Foundation of China (Grant No. 51105140)

References

[1] K.O. Hill, Y. Fujii, D.C. Johnson, et al. Photosensitivity in optical fiber waveguides: application to reflection filter fabrication. Appl. Phys. Lett, 1978, (32): 647-649.

[2] G. Meltz, M.M. Morey, W.H. Glenn. Formation of Bragg gratings in optical fibers by transverse holographic method. Opt. Lett, 1989, (14):823-825.

[3] K.O. Hill. Bragg gratings fabricated in mono-mode photosensitive optical fiber by UV exposure through a phase mask . Appl.Phs.Lett.1993, (62):1035-1037.

[4] R. Gafsi, M.A .EI-Sherif. Analysis of Induced-Birefringence Effects on Fiber Bragg Gratings [J]. Opt. Fiber Technol, 2000,(6): 229-322.

[5] F. Bosiaa, J. Botsisa, M. Facchinia, et al. Deformation characteristics of composite laminates part I: speckle interferometer and embedded Bragg grating sensor measurements [J]. Composites Science and Technology, 2002, 62:41-54.

[6] S.K. GUO, Z.K. YANG, et al. Sensing characters and experiments researches of the fiber bragg grating [J].CHINA MEASURMENT TECHNOLOGY, 2006, 32(20:71-73.

[7] Z. ZHOU, S.Z. TIAN, X.F. ZHAO, et al. Theoretical and experimental studies on the strain and temperature sensing performance of optical FBG [J]. Journal of functional material, 2002, 33 (5): 551-554.

[8] Y.J. ZHANG, H.B. WANG, Z.G. CHEN, W.H. BI. Study on FBG Sensor's Steel Capillary Packaging Technique and Sensing Properties [J]. LASER&INFRARED, 2009, 39(1):53-54, 66.

[9] Y.J. RAO, Y.P. WANG, T. ZHU .Principle and application of Fiber Grating [M].Beijing: science press, 2006.136-140.

[10]Z.X. WU, F. WU. Fiber grating sensing principle and application [M]. Beijing: National Defense Industry Press, 2011.

[11]Otghonos, K.Kalli.Fiber Bragg Gratings Fundamentals and Applications in Telecommunications and Sensing, Artech House, 1999 .

[12]M.LI, Y.B. LIAO. Optical fiber sensor and its application technology [M]. Wuhan: WU HAN University press, 2008.8.

Advanced Materials Research Vol. 819 (2013) pp 202-205
© *(2013) Trans Tech Publications, Switzerland*
doi:10.4028/www.scientific.net/AMR.819.202

The Construction of the Workpiece Coordinate in CNC System Based on Motion Control Card

Liu Hengli [1, 2, a], Hu Shiguang[3, b], Wang Taiyong[1, c] and Wang Dong [1, d]

[1]Key Laboratory of Mechanism Theory and Equipment Design of Ministry of Education, Tianjin University, Tianjin, 300072, China

[2] Tianjin Commerce University, Tianjin 300134, China

[3]Tianjin University, Tianjin, 300072, China

[a]lhl669@163.com, [b]sghu@yahoo.cn(corresponding author), [c]tywang@189.cn, [d]845073206@qq.com

Keywords: Motion control card; Workpiece coordinate; G54; CNC system

Abstract. Based on motion control card, it analyzed the methods and principles of the workpiece coordinate construction used the G54; created the G54 function modules, embedded CNC system of independent research and development and improved the system functions, processing efficiency and quality. The method described in the article can be directly applied to the machine tools controlled by numerical control system based on the card, and had a high practical value to the establish of other CNC machine tool workpiece coordinate system.

Introduction

The coordinates of CNC machine tools is machine coordinates. But in the actual processing, the NC code wrote based on the workpiece itself as the workpiece coordinates in order to facilitate the programming. So that it led to the deviation between machine coordinate origin and the origin of the workpiece coordinate, cannot be achieved on the normal processing. It must first establish workpiece coordinate system while write the CNC machining, and the coordinate values of the program based on the coordinate system [1]. The tools could arrived at the designated pot based on the coordinate system setting position to perform the normal processing, thus ensuring the quality and function of the workpiece.

At present, there are more study on the establishment of workpiece coordinate system, principles and methods is similar, but less Method of building workpiece coordinate system based on PMAC motion control card. This study will research for this problem, based on independent research and development CNC system, combined with the motion control card, developed the G54 functional modules to achieve automatic knife and improve processing efficiency and quality in the process.

CNC System NC Code Processing Functions Based on Motion Control Card

In recent years, with the rapid development of computer technology, the open CNC system came into being. The "PC + motion control card" mode is currently more popular. The NC code is delivered to the motion control card through the PC parallel bus, and then motion instructions issued to the drive by the motion control card, thus completing the CNC machining process [2]. The NC code is written by software or by hand, all based on the workpiece itself as the processing origin. In the actual processing, it called processing program stored in the system or artificial writing to meet the demand of processing.

In the "PC + motion controller" open CNC system, the NC code needs to be translated into the format for the card directly executed [3]. It means to complete functional identification of NC code, and extract the key word converted to corresponding movement r function parameters in motion controller. That is to say all of the commands must be converted to the language identified by the card, and then control the corresponding moving to the established action, shown in table 1.

Table 1 The translate between NC Code and PMAC language

	NC Code	Motion Control Card language
1	N1G91	N1M1012==78242M999==1G91
2	M03S500	M1015==39169M896==500M999==2M03S500
3	N1 G01 X-69.522 Y13.373 Z11.000 F200	N1M1012==78226M895==200M999==2 G01 X-69.522 Y13.373 Z11.000 F200

The Principles and Methods of G54 Workpiece Coordinate System Establishing

The workpiece coordinate system and the machine coordinate system are not only differences also linked. To some extent, the machine coordinate system is the foundation of establishing the workpiece coordinate system. It studied the principles and methods of establishing the workpiece coordinate system aimed at the motion control card, and ultimately achieved the G54 tool function.

The Differences and Relations Between the Workpiece and the Machine Coordinate System. CNC machine coordinate system is the origin of the device itself, usually fixed, and the machine back to the origin means to the origin. But the workpiece coordinate system is set based on the knife before machining, and the origin may be different with the different workpiece. There are some deviations between the two coordinate systems, and NC code is written in the workpiece coordinate system.

In actual processing, it needs re-processing of the knife while finishing the workpiece processing. In order to avoid the re setting the knife, the first setting knife number could be recorded, it means the difference between the workpiece and machine coordinate system, and set the workpiece coordinate system to achieve the automatically setting knife, and save time, improve processing efficiency and quality. So there are some relations between the two kinds of coordinate system, and the magnitude relation will change as the workpiece different, but it is remains the same in the logic, and it is also the principle of G54 functions.

The Method of G54 Setting Based on the Motion Control Card. In the NC machining, the tool motion controlled by the CNC system in the machine coordinate system, but the processing procedures and the tool motion are planned in the workpiece coordinate system. And it needs convert to coordinate form the workpiece coordinate system to the machine coordinate system after the processing procedures uploaded to the CNC system [4]. But the PSET command functions is to re-define the position of the current axis in the card, so the knife point coordinates need to be translated into workpiece coordinate from the machine coordinate.

Based on the above analysis and combined with PEST instruction, the current knife sites are read from the card variable based on the G54, and stored in the card corresponding variable. And then it will be converted to the coordinates in the workpiece coordinate system while compiling as the knife site coordinate in the coordinate system coordinates, as shown in Figure 1. It virtually determined the distance between the origin of the workpiece coordinate system and the machine reference point used the offset value. And once the workpiece coordinate system has been established, their position in the machine coordinate system fixed [5].

```
Tool position coordinates read from PMAC, and the coordinates of X,Y and Z
correspond board variations M1**,M2** and M3**
      COORX= DeviceGetVariable (m_ControlDevice.dwDevice, M_Varible, 1**, 0);
      COORY= DeviceGetVariable (m_ControlDevice.dwDevice, M_Varible, 2**, 0);
      COORZ= DeviceGetVariable (m_ControlDevice.dwDevice, M_Varible, 3**, 0);
Read G54 set coordinates stored in the variable and for compiling
      COORX=DeviceGetVariable (m_ControlDevice.dwDevice, M_Varible, 1**, 0);
      COORY=DeviceGetVariable (m_ControlDevice.dwDevice, M_Varible, 2**, 0);
      COORZ=DeviceGetVariable (m_ControlDevice.dwDevice, M_Varible, 3**, 0);
Compiling
      PSET X(M1**-M33) Y(M2**-M34) Z(M3**-M35)
```

Fig. 1 The setting of workpiece coordinate system origin

The Development of G54 Function Module Based on Visual C + +

Combing with Visual C + + visualization function, established the separate G54 functional unit, convenient for human-computer interaction and functional expansion; and based on the existing open CNC system, according to the actual processing, determined the functional logic relationship between G54 function modules and other modules; through debugging and experimental, achieved the automatically setting knife, avoided set the knife again when replacement parts, saving processing time.

The Construction of the G54 Function Module. It build G54 function modules basing on independent research and development CNC system, and embedded machine mode, as shown in Figure 2. The module includes editing and entry of five-coordinate value, as well as the man-machine interactive function, the right and bottom of the function keys. Through the right side key and the five-coordinate values edit field, it automatically captured the current position of the tool coordinate and displayed them, and set workpiece coordinate system; In addition, it used the bottom of the editing interface, can manually set the coordinates value of current workpiece coordinate system, also can correct the automatically the obtained coordinates. It also is able to realize switching between the two windows, easy to use and view based on the using of the bottom button to realize the switching between the other function modes.

The Implementation Process of G54 Function. First, the tool is at the zero point of the machine coordinate system after we start the machine, and move to the original point of the workpiece coordinate system manually. Then choose the G54 function module in the system, press the "F17 measurement" button, the coordinate value of the tool in the machine coordinate system will automatically be captured and displayed, as well as the origin of the workpiece coordinate system are saved. During processing, we can switch to G54 function module for browsing and viewing. At the end of processing, replace the same workpiece, it needs not setting the knife again, as long as the G54 instruction can be normally written in the NC program and run directly, and the tool can start processing from the origin of the workpiece coordinate system set originally. So it save time and improve processing efficiency.

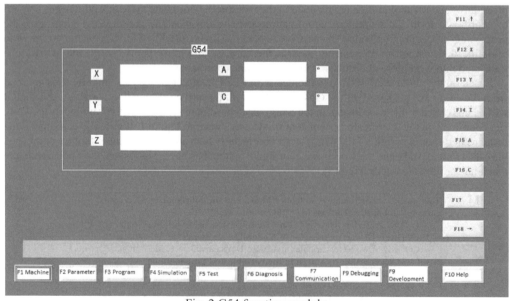

Fig. 2 G54 function module

Conclusion

In this paper, the relationship between workpiece coordinates and machine coordinates is researched, and discussed the method of establishing workpiece coordinate system, combing the motion control card's functions. And developed the G54 function module based on the Visual C + +, with good versatility and high open resistance; and embedded into the system independent research and development, realizing automatic tool setting functions, improve processing efficiency and quality in the six-axis CNC machining centers. The functional modules are apprized in other related projects, and proved that the module has a strong operability and high practical value.

Acknowledgment

The authors would like to thank the anonymous reviewers for their useful comments and suggestions. And this work was supported in part by National Science and Technology Support Program (Grant No. 2013BAF06B00), and Tianjin application foundation and frontier technology research program (Grant No. 12JCQNJC02500), and Fujian Science and Technology Project (Grant No. 2012H1008).

References

[1]   LI Yingping, HOU Wanming, SONG Yumei, ZHENG Wanjiang, Workpiece Coordinates Establishing and Tool Compensating Based on FUNUC 0 iT NC System, J. The process and equipment. 2(2008) 80-82.

[2]   ZHENG Hualin, MA Jianlu, PAN Shenghu, GUO Gaolei, Study of CNC System NC Code Compiler Techniques Based on PC Motion Control Panel, J. Machine Tool and Hydraulics. 20(2011) 94-96.

[3]   JIA Xv, LU Xiao-hong, WANG Xin-xin, JIA Zhen-yuan, Design and Realization of G-code Compiling for Micro-milling Machine NC System Based on PMAC, J. Modular Machine Tool & Automatic Manufacturing Technique. 3(2012)104-107.

[4]   NIU Lu feng, GAO X iu lan, WANG Bao, Some Questions about NC Machine Tool Program Technology and Correct Operation, J. Machine Tool and Hydraulics. 19(2006)49-51.

[5]   Chen Zhiqun, The application and analysis of the principle of establishing the workpiece coordinate system based on G50, J. Machine Tool and Hydraulics. 16(2011)124-126.

Advanced Materials Research Vol. 819 (2013) pp 206-211
© (2013) Trans Tech Publications, Switzerland
doi:10.4028/www.scientific.net/AMR.819.206

The feature extraction method of Gear magnetic memory signal

Yonggang Xu [a], Zhicong Xie, Linli Cui and Jing Wang

Key Laboratory of Advanced Manufacturing Technology, Beijing University of Technology, Beijing 100124, CHINA

[a]xyg_1975@163.com

Key words: Metal magnetic memory, Intrinsic time-scale decomposition, Proper rotation component, Potential gear fault , Feature extraction

Abstract. Magnetic memory test technology is a new nondestructive testing technique, which is able to detect of the stress concentration area and potential fault of low speed and heavy load gear. Because the magnetic memory signals are easy to be disturbed by various sources of noises, a new method based on the intrinsic time-scale decomposition (ITD) is proposed to achieve the extraction of magnetic memory signal. Firstly, the magnetic memory signals are decomposed into several proper rotation components (PRC) and a trend component by ITD. Then reconstruct the first four order PRCs to eliminate the low frequency cyclic composition of magnetic memory signal and magnetic noise. Finally, the magnetic signal strengths of each gear tooth root are extracted using cycle average and local statistic method. The results of Experiments show that the method is suitable to pick up effective ingredients of signal to extract signal feature and has important application value in potential fault diagnosis of low speed and heavy load gearbox.

Introduction

When the gear-box working in low speed and heavy load condition, it is difficult to extract useful information from weak vibration signals. However，the stress concentration and subtle deformation can be preserved in the form of magnetic memory. Therefore, the metal magnetic memory testing technology provides feasibility for early diagnosis of potential fault of low speed and heavy load equipments. Metal magnetic memory testing technology (MMM) is a new non-destructive testing techniques for early diagnosis of microscopic defects in ferromagnetic member, which is proposed by Russian scholars Doubov A in 1997[1-2].It has been widely used in the mechanical, aviation, chemical and other fields [3-4].

In this paper, magnetic memory testing technology is applied to determine the gear stress concentration area, then take advantage of the inherent time scale decomposition, the original magnetic memory signal noise is reduced and low-frequency cyclical trends is eliminated in order to extract the local features of each gear tooth stress concentration.

The principle of metal magnetic memory testing

Local stress and minor deformation can easily be produced in low-speed heavy-duty equipments. The magnetostrictive properties of the magnetic domain organization occurred which is directed and not reversible reorientation. It also has relationship with the maximum effect stress, the tangential component Hp (x) in the magnetic field has a maximum and the normal component Hp (y) has zero mean value point so that the stress concentration area of the workpiece can be inferred by measuring Hp (y) [5] .

The metal magnetic memory signal is a natural magnetic memory signal so it is susceptible to environmental magnetic interference in the actual work and determining the stress concentration area and metal defect location accurately might be affected. Therefore, it is extremely important to process magnetic memory signal.

The magnetic memory testing experiment of gear

The parameters of experimental gear. The parameters of experimental gear are as follows: gear material is 20CrMnTi carburizing steel; the modulus 5mm; the number of teeth 30, the reference diameter 150mm; the root diameter 137.5mm, the tip diameter 160mm. the physical map of gear is shown in Figure 1 (a) The instrument is stress concentrated magnetic detector TSC-1M-4 produced by the Russian power diagnostics company. The magnetic memory signal is measured by the sensors with a small round along the circumference of the tooth root. The position of Probe is shown in Figure 1 (b).

The Magnetic Memory testing of normal gear. The testing results of normal gear shown in Figure 2. Figure 2 (a) is the original magnetic memory signal measured for 5 circles along the inside circumference of the tooth root (sampling points 1950), then the number of sampling points in one circular should be 390. Figure 2 (b) is the average result of the five circles magnetic memory signals. In order to present the signal characteristics of t each tooth, a histogram is plotted in Figure 2 (c). The horizontal axis means the serial number of teeth, the vertical axis means magnetic memory signal of single tooth. We can see the overall trend of the gear magnetic memory signal is decreasing monotonically in the negative direction and increasing monotonically in the positive direction because the trend is associated with the circular shape of the gear.

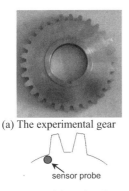

(a) The experimental gear

sensor probe

(b) The position of probe

Fig 1 The experimental gear and the scheme of probe position

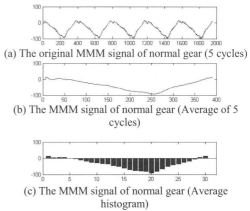

(a) The original MMM signal of normal gear (5 cycles)

(b) The MMM signal of normal gear (Average of 5 cycles)

(c) The MMM signal of normal gear (Average histogram)

Fig 2 The MMM signal of normal gear

The Magnetic Memory testing of gear crack fault & gear broken fault. Gear is mounted to the frequency fatigue testing machine loaded by tooth 7 and tooth26. After 20 minutes, the root of tooth 7 appears obvious crack and measure the MMM signal along the inside circumference of the tooth root. Similarly, mounted the gear again and loaded by tooth 11 and tooth 22 until tooth 22 brokes completely (as shown in Figure 3), and measure the MMM signal of gear broken fault.

The loaded tooth11

The broken tooth 22

The loaded tooth 26

The crack tooth 7

Fig 3 The testing gear with fault

The MMM signal of gear with crack fault is shown in Fig 4. Figure 4 (a) is the magnetic memory signal of gear measured 5 circles. Figure 4 (b) the averaged single circle signal, Figure 4 (c) the verage histogram of signal. Compared with Figure 2, a significant change occurs in the magnetic field strength near the position of tooth 7 and tooth 26. The magnetic field strength of tooth 7 is significantly higher because the root tooth cracks. As loaded tooth, tooth 26 also has stress concentration so that the magnetic memory signal has a significant change.

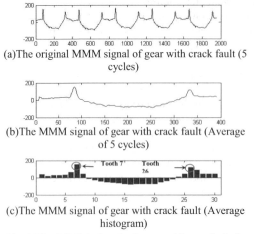

(a)The original MMM signal of gear with crack fault (5 cycles)

(b)The MMM signal of gear with crack fault (Average of 5 cycles)

(c)The MMM signal of gear with crack fault (Average histogram)

Fig 4 The MMM signal of gear with crack fault

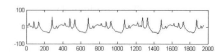

(a) The original MMM signal of gear with crack and broken fault (5 cycles)

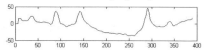

(b) The MMM signal of gear with crack and broken fault (Average of 5 cycles)

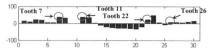

(c) The MMM signal of gear with crack and broken fault (Average histogram)

Fig 5 The MMM signal of gear with crack and broken fault

Figure 5 shows the MMM signal of gear with crack and broken fault, and the three graphs (a), (b), (c) have the same meaning as in Figure 2. Figure 5 (b) shows that magnetic memory signal mutation appears near the tooth7, tooth 11, tooth 11 and tooth 26, tooth root stress concentration, combined with the experiment, we know the root of tooth 7 cracks, tooth 22 broken completely, tooth 11 and tooth 26 are also stress concentration areas as loaded teeth. The average histogram in Figure 5 (c) also shows this situation, but the histogram of tooth 26 is not obvious because the slowly varying trend caused by the gear shape.

In summary, Magnetic memory testing technology can be used to detect gear tooth root stress concentration area, however, the signal is influenced by the ambient magnetic field noise and varying trend of gear. Based on this, we propose a method based on the intrinsic time-scale decomposition method (ITD) to extract the gear magnetic memory signal characteristics.

The basic principles of intrinsic time-scale decomposition

Intrinsic time scale decomposition method is a new non-linear non-stationary signal processing method proposed by Frei and Osorio in 2007, which is applied to biomedical signal processing, and achieved the ideal results [6]. Arbitrary complex signal can be decomposed into a number of proper rotation component (PRC) and a trend item based on ITD method, so that time-frequency distributions of the original signal can be obtained completely .

Given a signal, X_t, we define an operator L, which extracts a baseline signal from X_t in a manner that causes the residual to be a proper rotation. More specifically, X_t can be decomposed as

$$X_t = LX_t + (1-L)X_t = L_t + H_t \tag{1}$$

where $L_t = LX_t$ is the baseline signal and $H_t = (1-L)X_t$ is a proper rotation.

Suppose X_t is a real-valued signal, and let $\{\tau_k, k = 1,2,\cdots\}$ denote the local extrema of X_t, and for convenience define $\tau_0 = 0$. In the case of intervals on which X_t is constant, but which contain extrema due to neighbouring signal fluctuations, τ_k is chosen as the right endpoint of the interval. To simplify notation, let X_k and L_k denote $X(t_k)$ and $L(t_k)$, respectively.

Suppose that L_t and H_t have been defined on $[0, \tau_k]$ and that X_t is available for $[0, \tau_{k+2}]$. We can then define a (piece-wise linear) baseline-extracting operator L, on the interval $[\tau_k, \tau_{k+1}]$ between successive extrema as follows:

$$LX_t = L_k + \left(\frac{L_{k+1} - L_k}{X_{k+1} - X_k} \right)(X_t - X_k) \tag{2}$$

$$L_{k+1} = \alpha \left[X_k + \left(\frac{\tau_{k+1} - \tau_k}{\tau_{k+2} - \tau_k} \right)(X_{k+2} - X_k) \right] + (1 - \alpha)X_{k+1} \tag{3}$$

and $\alpha \in [0,1]$ is typically fixed with $\alpha = 0.5$ [8].then we can define the residual, proper-rotation-extracting operator H_t,

$$HX_t = (1 - L)X_t = H_t = X_t - L_t \tag{4}$$

Once the input signal has been decomposed into a baseline and a proper rotation component, with the latter representing the highest relative frequency 'riding wave' present in the input signal, the process can be re-applied using the baseline signals as input. This procedure can be iterated until a monotonic baseline signal (i.e. the 'trend') is obtained. This decomposes the raw signal into a sequence of proper rotations of successively decreasing instantaneous frequency at each subsequent level of the decomposition. More precisely,

$$X_t = HX_t + LX_t = HX_t + (H + L)LX_t = \left(H \sum_{k=0}^{p-1} L^k + L^p \right) X_t \tag{5}$$

Where $HL^k X_t$ is the $(k+1)$st level proper rotation and $L^p X_t$ is either the monotonic trend or the lowest frequency baseline extracted.

Analysis of experimental data

The ITD decomposed results of MMM signal with crack and broken fault is shown in Figure 6.The original signal is decomposed into 15 PRCs, the decomposition proceeds in the first eight order PRCs component and final trend is given. Original means magnetic memory signal, PRC1 ~ PRC8 mean the mutation ingredients of magnetic memory signal triggered by stress concentration, L means the tendency corresponding to low-frequency trend of magnetic memory signal. Retain the first four PRCs whose energy are larger and PRC5~PRC15 are ignored. Then reconstruct the first four order PRCs based on linear superposition to eliminate the big cycle composition of magnetic memory signal and magnetic noise, while the integrity of the mutations details of signal is retained.

Figure 7 shows the ITD decomposed results of normal gear. In figure 7 (a), the blue curve means the original magnetic memory signal and the red one means the reconstructed signal of the first four PRCs based on ITD method. Figure 7 (b) (c) means average signal and average histogram of signal. (similarly hereinafter)We can see that the average magnetic memory signal of normal gear and local mean is essentially zero and there is no obvious abnormalities projecting ingredient. Compared with Figure 2, magnetic memory signal local characteristics eliminates the impact of the low-frequency trend after ITD processing whose physical meaning is more apparent.

Figure 8 shows the ITD decomposed results of gear with crack.

Seen from the figure 8 (a), low-frequency magnetic memory signal trend term and part of the noise is removed after ITD processing. The magnetic memory signal of tooth 7 and tooth 26 tooth is significantly higher than other parts of the gear. What's more, signal intensity in tooth 7 is greater than the tooth 26 and section 7 of the signal strength of the tooth, as shown in figure (b) and figure (c) which is fully consistent with the actual stress concentration. The above characteristics can't be fully reflected in Figure 4 so that the effectiveness of ITD method is further illustrated in gear magnetic memory signal feature extraction.

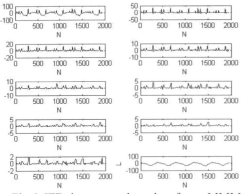

Fig 6. ITD decomposed results of gear MMM signal

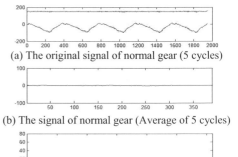

(a) The original signal of normal gear (5 cycles)

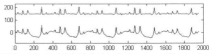

(b) The signal of normal gear (Average of 5 cycles)

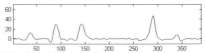

(c) The signal of normal gear (Statistics histogram)

Fig 7. ITD decomposed results of normal gear signal

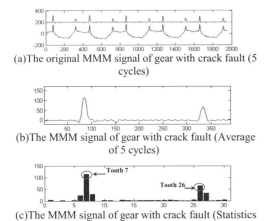

(a)The original MMM signal of gear with crack fault (5 cycles)

(b)The MMM signal of gear with crack fault (Average of 5 cycles)

(c)The MMM signal of gear with crack fault (Statistics histogram)

Fig 8. ITD decomposed results of crack gear MMM signal

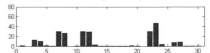

(a) The original MMM signal of gear with crack and broken fault (5 cycles)

(b) The MMM signal of gear with crack and broken fault (Average of 5 cycles)

(c) The MMM signal of gear with crack and broken fault (Statistics histogram)

Fig 9. ITD decomposed results of crack & broken gear MMM signal

Figure 9 shows the ITD decomposed results of gear with crack and broken fault. Seen from the figure, four roots of gear teeth have visible magnetic memory signal strength who are tooth 7 (cracked tooth), tooth 11 (loaded tooth), tooth 22 (broken tooth) and 26 teeth (loaded teeth) What's more, tooth 7 has the highest intensity, followed by cracked tooth, the crack experiment loaded tooth is lower in the two loaded teeth which is also consistent with the actual stress concentration In addition, tooth 3 also have a small amount of highlighting which is presumed to be the caused in the latter part of the experiment. Due to stress concentration is not limited to the center position of the tooth root in a single tooth, it is normal that both the adjacent teeth have a magnetic field the highlight in figure(c).

In summary, it is significantly better to extract local features based on the magnetic memory signal after ITD processing ,stress concentration can be judged at each position of the gear size the extent and results are fully consistent with the actual situation which is not only can detect dominant failure such cracks or broken teeth, but also effective to detect the no obvious gear detects just with stress concentration of two loaded teeth , which provides a practical technology support for early diagnosis of low speed and heavy gear latent failure.

Conclusions

(1) Magnetic memory method can effectively detect the ferromagnetic component failure defects, especially for obvious defects and stress concentration area which has obvious advantages compared to conventional detection method such as vibration signals and acoustic emission signals.

(2) There are obvious trend of low-frequency items and noise in magnetic memory signal caused by gear shape and the surrounding environment, but we can extract the deep-seated the characteristic information with the help of modern signal processing method .

(3) The method applied to the characteristics of magnetic memory signal extraction, The experiments show that the ITD method is able to extract the signal active ingredients for gear composite failure to make a more accurate judgment of the areas of stress concentration and size which has a good prospect of application in magnetic memory.

Acknowledgment

This work is partially supported by Nature Science Foundation of China (51075009), and Beijing talents training subsidy scheme(2011D005015000006).

References

[1] Dubov A A. Express method of quality control of a spot resistance welding with usage of metal magnetic memory[J]. Welding in the World，2002, 46(6): 317-320.

[2] Doubov AA. Diagnostics of equipment and constructions strength with usage of magnetic memory[J]. Inspection Diagnostics, 2001,35(6):19-29.

[3] Zhang Weimin,Dong Shaoping.State-of-the-art of metal magnetic memory testing technique[J]. China Mechanical Engineering, 2003,14(10):892-896(in Chinese).

[4] Tang Dedong, Zhou Peng. Research on device for inspection of defects of metal structures[J]. Chinese Journal of Scientific Instrument, 2005,25(4): 254-256 (in Chinese).

[5] Wen Wei-gang,Sa Shu-li. Mechanism and implementation of magnetic memory metal diagnostic technique [J]. Journal of Northern Jiaotong University, 2002，26(4):67-70(in Chinese).

[6] Mark G Frei and Ivan Osorio. Intrinsic time-scale decomposition: time-frequency-energy analysis and realtime filtering of non-stationary signals[J]. Proceedings of the Royal Society A, 2007, 463: 321-342.

Advanced Materials Research Vol. 819 (2013) pp 212-215
© *(2013) Trans Tech Publications, Switzerland*
doi:10.4028/www.scientific.net/AMR.819.212

Application of DIC System in the Tensile Test of Medium and Heavy Plate

Yu Aiwu[1, a], Zhu Chuanmin[2,b]

[1]ZIBO Vocational Institute   Zibo, 255314, Shandong, P. R. C
[2]School of Mechanical Engineering, Tongji University, Shanghai, 200092, P. R. C
[a]yaw8456@sina.com, [b]01065@tongji.edu.cn

Keywords: DIC   Medium and heavy plate   Material performance tests   Full range strain

Abstract. Digital image correlation (DIC) is a kind of non-contact strain measurement method. It could provide deformation information of a specimen by processing two digital images that are captured before and after the deformation. the application of DIC method in performance testing and evaluation of new materials has great superiority over the traditional methods in the aspects that help to explore the features of new materials, such as mechanical performance, deformation mechanism and dynamic and static performance[1-4]. Tensile test of 8mm steel plate based on DIC is introduced in this paper, this test is easy operated and high accuracy. It can not only get all the data that traditional method done, but also reflect the full range strain, strain history and strain rate. It can also transform the result among different type strain. The experiments based on DIC method of medium and heavy plate can help to learn the properties of material like this kind much more comprehensively, reduce testing time and promote the wide application of new material in manufacturing.

Introduction

Material performance test is very important before the new material applied into manufacturing. It needs many test to comprehensively realize the properties of material. traditional test method generally is hard operated and single output, so it is hard to completely realize the properties of new materials. The application of Digital Image Correlation method (DIC) in material performance tests simplifies the tests, makes the test results more accurate. DIC method is a kind of photometric test method which can quantitative analysis the full range displacement and strain of the specimen[1]. This method has many advantages such as non-contact, non-destructive testing, high accuracy, large measuring range [2]. DIC has been successfully used in the study on material's mechanical behavior explore, dynamic measurement and fracture mechanics [3,4], etc. In this paper, DIC method is used in material mechanics experiment of medium and heavy plate, and illustrate a tensile test of 8 mm steel plate which use DIC method, compared the difference between measured data and traditional theoretical calculated method.

Basic method and principle of DIC

The basic consist of DIC system. DIC system is mainly consists of CCD camera, constant light source, 1394 data acquisition card and computer that install image acquisition software (Fig.1), accessories is for fixed installation and data transmission device, such as tripod, beams, data transmission, etc.

Work steps of DIC system. Firstly, specimen for DIC test will be simple processing, Because of DIC method is to deal with speckle digital images in order to get information of displacement and strain of the specimens (Fig. 2), in order to make digital image data collected by DIC acquisition system easy to be identified [5]. Secondly, to ensure that the environment light intensity is relatively stable during test process, a constant light source is needed. It is mainly used to eliminate the color difference caused by light intensity, so as to reduce the calculation error. Thirdly, make sure that it is synchronized between the time CCD began filming and the time test instrument began work. CCD camera continuously shooting the surface speckle pattern of specimen, and   transmit and storage speckle images to the computer by data acquisition card. Finally, to calculate and contrast speckle images through software that integrate DIC algorithm, extracting data such as strain, strain rate, displacement; Test instrument is used to record force that load on the specimen during tests. To ensure the synchronization between stress and strain, it will make data processing accuracy and effective.

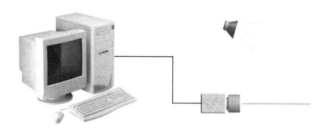 

Fig. 1 Constitution of DIC system   Fig. 2 Speckle image

Work principle of DIC. DIC system view the grey value of discrete digital images' pixel as density of the point [4], the data is stored in a two-dimensional array. Assume that discrete pixels and their gray value of two before and after deformation digital images in the calculation process are corresponded. It is based on pixel set in images processing, while calculation of deformation in a set is completed, the set offset a few steps (less than set the width of the pixels) to continue to calculate. In order to ensure the approximate linear strain relationship, the set of pixels should be small [5].

 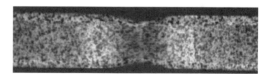

Fig. 3 Set of image pixels   Fig. 4 DIC calculation result

There are several DIC calculation algorithm used to calculate deformation in pixel set. DIC method calculates the displacement field in the different area by matching two speckle image (deformation image before and after images) (Fig. 3), finally obtain strain field by the numerical calculation [6,7].

Digital image correlation method (DIC) has already been integrated into computer software successfully, computer software extracts pixel gray value of images collected during the material specimen deformation process, gray level information of each pixel in the pictures will be marked in the form of numerical matrix and participate in calculation [6]. Fig. 4 is the principal strain color picture obtained by dealing with the tension specimen.

Medium and heavy steel plate tensile test based on DIC method

Experimental conditions and preparation. Firstly, to cut out tension specimens from the test plate due to tensile test standard [8]. Secondly, to burnish the parallel section surface of specimen equably with sand paper, wearing off defects of parallel section such as burr in order to prevent stress concentration from affecting test result; polishing out oxide layer on the surface of the parallel section, because the nature of the oxide layer is much different from original material and will reduce the adhesion ability of matte paint on the surface of the specimen. Thirdly, clean off the stains on the surface of the specimens, to ensure the quality of speckle paint. Finally, spray a layer of speckle matte paint on the surface of specimen parallel section carefully.

Experimental process. Clamp the specimen in the tensile machine, adjusting the position of the CCD camera and the cross of camera, to make sure it is at a comfortable place relative to specimens. To adjust focal length and aperture, making computer achieve the most clear state images. At the same time confirm data acquisition channel of tensile machine is unobstructed, start data collection of tensile machine and DIC at the same time, to ensure data synchronization. After tensile is finished, save and process data.

The experimental data analysis. DIC method has advantages of non-contact, high precision, wide measurement range, could calculation strain and displacement of specimen in full range test surface [7], could show deformation strain, strain rate of specimen by the visual form clouds at the same time, can extracted data of arbitrary point, line, face in the strain range. Traditional material tensile test using extensometer as the tool to measure and record deformation, while the measuring accuracy of extensometer is high, the measuring output of extensometer is the mean value of the deformation within the range, could not reflect local strain and three-phase strain. Medium and thick steel plate specimen cannot be considered as plane strain during tensile deformation, for the necking period of specimen is very long, after the specimen is stretched over dispersive instability points, necking phenomenon occurs.

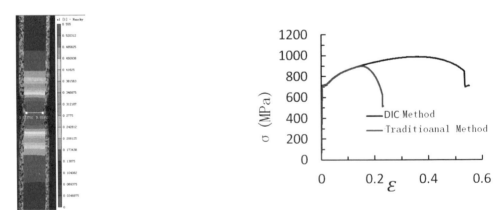

Fig. 5 Extract data   Fig. 6 Comparison of traditional method and the DIC method

After necking occurs, the concentration of deformation occurs at necking part, the deformation in necking area could be far greater than in other areas of specimen. While deformation developed to concentration instability point, it will produce obvious deformation in the thickness direction of the medium and thick steel plate specimen, so the deformation cannot be simplified as plane strain deformation. Traditional tensile experiments can only get engineering strain by means of extensometer, to get a true strain, with the aid of mathematical model:

$$\varepsilon = \ln l/l_0 \tag{1}$$

The mathematical model is based on assuming constant volume:

$$lA = l_0 A_0 \tag{2}$$

But after dispersive instability point, the cross-sectional of specimen parallel section area is always changes, it will reduce shrink sharply at necking section, the real strain obtained by Eq.1 began to appear deviation. DIC processing methods of strain is to extract the strain of fracture location (as is show in Fig. 5), it is the true deformation strain of the specimen fracture in the process of fracture. As shown in Fig. 6, the calculated shrink neck true strain value is much smaller than the actual value which is got by DIC means, it is not exactly the true strain of material process. On the other hand, DIC method analysis the full range strain, can extract the local point, line or plane strain and calculate the strain rate. What's more, DIC system can display the strain process dynamically, help the experimenter intuitive understanding of material deformation process.

Conclusion

By contrast of the results of two kinds data processing in tensile experiments of 8 mm steel plate, obviously, the traditional test method has certain limitation in performance test of medium and heavy plate material, the DIC system application in medium and heavy steel material performance test can help to deep and thorough realize the comprehensive performance of new materials, the mechanism of deformation and complement the disadvantages of traditional test method.

Acknowledgement: the project is supported by the National Key Technologies R&D Program of MOST (No. 2011 ZX04016-021) and the National Basic Research Program of China National Science and Technology of the MOST (No. 2011BAC10B08)

References

[1] Peters W.H, Ranson W.F. Digital imaging technique in experimental stress analysis [J]. Opt. Eng., 1982, 21(3):427-431.

[2] Chu T.C, Ranson W.F, Sutton M.A, et al. Applications of digital image correlation techniques to experimental mechanics [J]. Exp Mech, 1985, 25 (3):232-244.

[3] Schreier Hubert W. Investigation of two and three-dimensional image correlation techniques with applications in experimental mechanics [D]. Doctor dissertation, University of South Carolina, 2003.1~4.

[4] Hreier Hubert W. Investigation of two and three-dimensional image correlation techniques with applications in experimental mechanics [D]. Doctor dissertation, University of South Carolina, 2003:1—4.

[5] Bruck H.A, McNeill S.R, Sutton M.A, Peters W.H. Digital image correlation using Newton–Raphson method of partial differential correction. Exp Mech 1989; 29:261-7.

[6] Choi D, Thorpe J.L, Cote W.A, et al. Quantification of compression failure propagation in v wood using digital image pattern recognition [J]. Forest Prod, 1996, 46:87.

[7] Li E.B, Tieu A.K, Yuen W.Y.D. Application of digital image correlation technique measurement of the velocity field in the deformation zone in cold rolling[J]. Optics and Lasers in Engineering, 2003, 39(4):479—488.

[8] JIS Z 2241-1998

Advanced Materials Research Vol. 819 (2013) pp 216-221
© (2013) Trans Tech Publications, Switzerland
doi:10.4028/www.scientific.net/AMR.819.216

Approach to weak signal extraction based on Empirical Mode Decomposition and Stochastic Resonance

Pan Zhang[a], Taiyong Wang, Lu Liu[b], Luyang Jin, Jinxiang Fang

Tianjin Key Laboratory of Equipment Design and Manufacturing Technology, Tianjin University, Tianjin 300072, China

[a]zhangpan2003@126.com, [b]lordman1982@aliyun.com

Keywords: SR, EMD, Weak signal

Abstract. The empirical mode decomposition (EMD) of weak signals submerged in a heavy noise was conducted and a method of stochastic resonance (SR) used for noisy EMD was presented. This method used SR as pre-treatment of EMD to remove noise and detect weak signals. The experiment result prove that this method, compared with that using EMD directly, not only improve SNR, enhance weak signals, but also improve the decomposition performance and reduce the decomposition layers.

Introduction

In the quest of accurate time and frequency localization, Huang et al. [1] proposed the empirical mode decomposition (EMD) scheme which offers a different approach in time-series processing. This method can decompose the signal into a set of oscillatory modes by taking advantage of the characteristic time scales embedded in the data. So there is no need for a basis function and no need for transformation. But EMD method can not eliminate boundary problem because it uses cubic splines method to obtain signal's instantaneous average. If EMD is used in strong noise condition it will affect the decomposition performance, especially increase decomposition layers and lower efficiency of arithmetic, even may lead EMD lose definitude significance. So denoising process must be performed before the EMD decomposition. References [4] and [5] use wavelet and SVD to denoise and get good effect. But these methods are not good at weak signals submerged in heavy noise. At the meantime, stochastic resonance (SR), which is put forward by Benzi wherein he address the problem of the primary cycle of recurrent ice ages, has a strong development for its advantages in weak signal's enhancement and detection. With the cooperation effect of signal and noise in non-liner system, SR can put noise power into lower-frequency signal which can enhance weak signal and reduce noise at the same time. In a word, this paper proposes a new method of weak signals detection which based on SR theory and EMD. Decomposition. Firstly, do denoising procession with SR and then do EMD decomposition. Experiments proved that this method is better than direct EMD decomposition method.

Empirical Mode Decomposition

Empirical Mode Decomposition (EMD) is a novel method for adaptive of non-linear and non-stationary signals. It can decompose any non-linear signal into several Intrinsic Mode Functions (IMFs), and a residue.

The components resulting from EMD, called Intrinsic Mode Functions (IMFs), each admit an unambiguous definition of instantaneous frequency. By definition an Intrinsic Mode Function(IMF) satisfies two conditions

1). the number of extreme and the number of zero crossing may differ by no more than one, and

2). the local average is zero

Where the local average is defined by the average of the maximum and minimum envelopes discussed in the following section. These properties of IMFs allow for instantaneous frequency and amplitude to be defined unambiguously.

In order to obtain the separate components called IMFs, we perform a sifting process. The goal of sifting is to subtract away the large-scale features of the signal repeatedly until only the fine-scale features remain. A signal $x(t)$ is then divided into the fine-scale detail, $h(t)$ and a residual, $m(t)$

so $x(t) = m(t) + h(t)$. This detail becomes the first IMF and the sifting process is repeated on the residual, $m(t) = x(t) - d(t)$.

The sifting process requires that a local average of the function be defined. If we knew the components before, we would naturally define the local average to be the lowest frequency component. Since the goal of EMD is to discover these components, we must approximate the local average of the signal. Huang's solution to finding a local average creates maximum and minimum envelopes around the signal using natural cubic splines through the respective local extreme. The local average is approximated as the mean of the two envelopes.

The first IMF, $c_1(t)$ of a signal, $x(t)$, is found by iterating through the following loop.

1. Find the local extreme of $x(t)$.

2. Find the maximum envelope $e_+(t)$ of $x(t)$ by passing a natural cubic spline through the local maxima. Similarly, find the minimum envelope $e_-(t)$ with the local minima.

3. Compute an approximation to the local average, $m(t) = (e_+(t) + e_-(t))/2$

4. Find the proto-mode function $h_1(t) = x(t) - m(t)$

5. Check whether $h_1(t)$ is an IMF. If $h_1(t)$ is not an IMF, repeat the loop on $h_1(t)$. If $h_1(t)$ is an IMF then set $c_1(t) = h_1(t)$.

The sifting indicates the process of removing the lowest frequency information until only the highest frequency remains. The sifting procedure performed $x(t)$ on can then be performed on the residual $r_1(t) = x(t) - c_1(t)$ to obtain $r_2(t)$ and $c_2(t)$, Repeat the process as described above for n time, then n IMFs of signal $x(t)$ could be got. Then

$$
\begin{cases}
r_2(t) = r_1(t) - c_2(t) \\
r_3(t) = r_2(t) - c_3(t) \\
\quad \vdots \\
r_n(t) = r_{n-1}(t) - c_n(t)
\end{cases}
\tag{1}
$$

At last, the signal $x(t)$ is decomposed into several Intrinsic Mode Functions (IMFs), and a residue.

$$x(t) = \sum_{i=1}^{n} c_i(t) + r_n(t) \tag{2}$$

Residue $r_n(t)$ is the mean trend of $x(t)$, The IMFs $c_1, c_2, \ldots c_n$ include different frequency bands ranging from high to low. The frequency components contained in each frequency band are different and they change with the combination signal $x(t)$, while $r_n(t)$ represents the central tendency of signal $x(t)$.

In order to illustrate the performance of EMD, we use a combination of two pure sine waves, which are added together as

$$x(t) = a_1 \sin(2\pi f_1 t) + a_2 \sin(2\pi f_2 t) \tag{3}$$

The values of f_1 and f_2 are respective 10Hz and 30Hz, amplitudes are both 1.5.

As shown in Fig.1, mixed signal $x(t)$ is decomposed into two IMFs: c_1, c_2, and the residue $r_2 \cdot c_1$ is corresponding with sine wave of 30Hz , c_2 is the other sine wave of 10Hz. It can be seen from this example that, with the EMD method, signal can be decomposed into some different time-scales IMFs by which the characteristics of the signal can be presented in different resolution ratio.

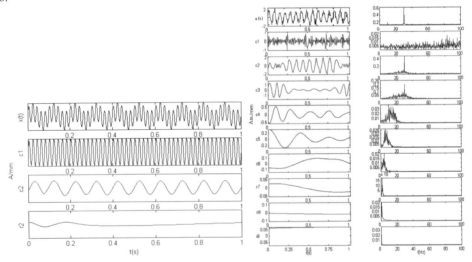

Fig.1 The EMD decomposed of mixed signal   Fig.2 The decomposed of noisy and weak signal

However, signals are not always pure in practice. They usually mix much noise which can affect EMD performance. In Fig2, the mixed signal is still composed   by two sine waves 30Hz and 10Hz, but mixed   noise of intensity D=0.2. While the amplitude of 30Hz is still 1.5, but the one of 10Hz changes into 0.1, belongs to weak signals. As shown in Fig.2, $c_1 \sim c_7$ are IMFs decomposed form EMD, and r_7 is the residue. It can be seen from this figure that the first IMF c_1 is mainly composed of high frequency noise, and for the noise existence, the sine wave of 30Hz is decomposed into c_2 and c_3, the wave is serious distortion especially in time domain. $c_4 \sim c_6$ are supposed to depict sine wave of 10Hz, but due to small amplitude, they cannot   be distinguished in despite of in frequency spectrum.   Therefore, when signal is very weak and mixed with noise, direct EMD performance is bad. It will result distorted IMFs and cannot detect the weak signal. So before EMD operation denoising process is necessary.

Stochastic resonance

Stochastic resonance (SR) is a phenomenon in which a nonlinear system heightens the sensitivity to a weak signal input and when noise with an optimal intensity is presented simultaneously. So SR doesn't remove but make use of noise to gain the optimal and desirable signal-to-noise ration (SNR) of output signals.

SR is most often considered by the example of the motion equation of a light particle in a bistable potential field. It is disturbed by a small periodic signal and additive white noise:

$$\dot{x} + f(x) = A\sin(\omega t) + n(t) \tag{4}$$

Where is x the particle displacement, $A\sin \omega t$ is a weak periodic signal at frequency ω, $f(x) = dU(x)/dx$, $U(x)$ is a symmetric double-well potential (we will consider the simplest case of such a potential, namely, $U(x) = -ax^2/2 + bx^4/4$), $n(t)$ is white noise of intensity .

As it follows from [7], under the condition of adiabatic elimination and small parameters (frequency, amplitude and intensity of noise are less than 1), the power spectrum of Eq.(4) is described that high frequency noise is weakened, and spectra energy is center around in the low frequency area, especially, there is a peak at the frequency of ω. However, the adiabatic elimination SR theory in small parameters cannot meet larger signals in practice. So the method of twice sampling stochastic resonance is recommended. According to this method, we first compress the measurement signals to low frequency ones in a linear mode to meet the requirement of small parameters. Secondly, analyses the compressed data spectrum to acquire characteristics of signals. At last we reconstruct the compressed data to origin according the linear mode above.

Fig.3 is a given example to illustrate the implementation of stochastic resonance under larger parameters signal condition. As shown in fig.3 the f_0 value is much more than 1, belongs to large parameter signals, so the twice sampling stochastic resonance method is adopted to compress it to small parameter signal.

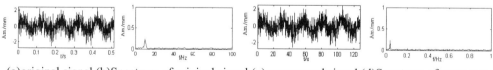

(a)original signal (b)Spectrum of original signal (c)compressed signal (d)Spectrum of compressed

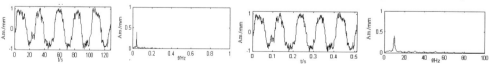

(e) SR output   (f) Spectrum of SR   (g) reconstructed signal   (h) Spectrum of reconstructed
Fig.3 Implement of SR of large parameters single, where $f_0 = 10$Hz, $f_s = 2000$Hz, and data length is 2048, Parameter values are $a=0.1$, $b=1$, $A=0.5$, and $D=0.6$.

Twice sampling frequency is selected 8Hz, then the compressed frequency is calculated 0.04Hz. In the figure, (c) and (d) graphs display waveform and spectrum of compressed signal, (c) and (d) is the SR output of compressed signal. Compared with graph (c) and (e), because high-frequency noise in original signal is weakened, the waveform of SR output become smooth, at the same time the amplitude of signal is heightened,, as shown in graph there is a peak at the frequency of f_0 .

Test of EMD base on SR denosing

We still use the 10Hz and 20Hz sine wave signals. Sampling frequency is 2KHz and data length is 2048. Amplitude of 30Hz and 10Hz signal is 1.5 and 0.1. The noise of intensity is 0.2 Parameters of SR are set as: a=0.1, b=1,twice sampling frequency is 8Hz.

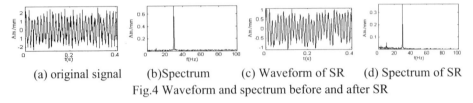

(a) original signal   (b)Spectrum   (c) Waveform of SR   (d) Spectrum of SR

Fig.4 Waveform and spectrum before and after SR

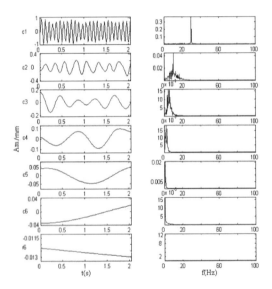

Fig.5 EMD decomposition result of SR output

Fig 4 shows signal's waveform and spectrum before and after SR process. The 10Hz weak signal nearly submergence in noise seen in graph (b), but after SR system the weak signal is enhanced as show in grph (d). Fig5 gives the EMD decomposition results of signal which is processed by SR. It is clear that the high frequency noise is filtering by SR system and 10Hz weak signal is enhanced. It can be seen from frequency domain, the first IMF c_1 is 30Hz frequency component, and it is smooth and obvious. IMF c_2 is correspondence with 10Hz signal. IMF $c_3 \sim c_6$, for their small amplitudes, are consider as residue.

Fig2 and Fig5 use the same original signal. In Fig2, we decompose signal with EMD directly, for the reason of noise the first IMF c_1 is mainly high-frequency noise component. 30Hz component is decomposed into IMF c_2, c_3 and is distortion. 10Hz correspondence with $c_4 \sim c_6$, for the low amplitude it is hard to detect. Compared with Fig2 and Fig6, we can conclude that SR+EMD method is better than only EMD method. For large signal the former performance is better than the latter and for weak signal only the former can be detected and decomposed. Otherwise, SR+EMD method can reduce EMD decomposition's layer, as shown in figures, There is 8 layers in Fig.5 but only 6 ones in Fig.2.

Conclusion

In order to use EMD decomposition in weak signals under noisy background, this paper puts forward a new method: using SR as pre-treatment and then performing EMD decomposition. The experiment result proved that this method, Compared with EMD directly, not only improve SNR, enhance weak signals, but also improve the decomposition performance and reduce the decomposition layers of signals.

Acknowledgements

This work was financially supported by Ministry of Education Doctoral Fund of 2010 (20100032110006).

References

[1] Huang N E, Shen Z, Long S R, et al. The empirical mode decomposition and the Hilbert spectrum for nonlinear and non-stationary time series analysis [J]. Proc R Soc Lond A , 1998, 454: 903∼ 995.

[2] Huang N E. Computer implicated empirical mode decomposition method, apparatus, and articale of manufacture [P].U.S.Patent Pending, 1996.

[3] Deng Yongjun, Comment and modification on EMD and Hilbert transform method[J].Chinese Science Bulletin,2001，46(3):257-263.

[4] Dai Gui-ping,Liu Bin, Instantaneous Parameters Extraction Based on Wavelet Denoising and EMD[J].Acta Metrological Sinica,2007,28(2):158-162.

[5] Wang Taiyong,Wang Zhengying,Xu Yonggang, Empirical mode decomposition and its engineering applications based on SVD denoising[J].Journal of Vibration and Shock, 2005,24(4): 96-98.

[6] R.Benzi, A.Sutera, A.Vulpiana. The Mechanism of Stochastic Resonance. J. Phys. A, 1981, 14 (11):L453∼4572

[7] Leng Yong-Gang,Wang Tai-Yong,etal. Power spectrum research of twice sampling stochastic resonance response in a bistable system [J]. Acta Physica Sinica, 2004, 53 (3)：717-723.

[8] Leng Yong-gang, Wang Tai-Yong, Numerical research of twice sampling stochastic resonance for the detection of a weak signal submerged in a heavy Noise[J].Acta Physica Sinca, 2003, 52(10)：2432-2437.

Advanced Materials Research Vol. 819 (2013) pp 222-228
© *(2013) Trans Tech Publications, Switzerland*
doi:10.4028/www.scientific.net/AMR.819.222

Neural Networks Based Attitude Decoupling Control for AUV with X-Shaped Fins

Xiujun Sun[1, a], Jian Shi[2, b] and Yan Yang[2, c]

[1]National Ocean Technology Center, No.219 Jieyuanxi Road, Nankai District, Tianjin, China 300112

[2]Tianjin Institute of Urban Construction, No.26 Jinjing Road, Xiqing District, Tianjin, China 300384

[a]sunxiujun@yahoo.com, [b]shijian15153420811@163.com, [c]yangyan_0103@163.com

(Corresponding Author: Yan Yang)

Keywords: AUV, X-shaped fins, Dynamics modeling, Hydrodynamics, Decoupling control, Neural networks.

Abstract. Attitude control in three-dimensional space for AUV (autonomous underwater vehicle) with x-shaped fins is complicated but advantageous. Yaw, pitch and roll angles of the vehicle are all associated with deflection angle of each fin while navigating underwater. In this paper, a spatial motion mathematic model of the vehicle is built by using theorem of momentum and angular momentum, and the hydrodynamic forces acting on x-shaped fins and three-blade propeller are investigated to clarify complex principle of the vehicle motion. In addition, the nonlinear dynamics equation which indicates the coupling relationship between attitude angles of vehicle and rotation angles of x-shaped fins is derived by detailed deduction. Moreover, a decoupling controller based on artificial neural networks is developed to address the coupling issue exposed in attitude control. The neural networks based controller periodically calculates and outputs deflection angles of fins according to the attitude angles measured with magnetic compass, thus the vehicle's orientation can be maintained. By on-line training, twenty four weights in this controller converged according to index function.

Introduction

There are many advantages for autonomous underwater vehicles to equip with x-shaped fins comparing to vehicles with cross shaped fins. Mechanical configuration becomes more compact since height and width of vehicle decreases. Performance of hydrodynamics increase 10% more because of the increment of steering plane. Many countries in the world (Swiss Holland, Norway and Germany etc) have adopted the x-shaped fins for autonomous vehicles, because it features high maneuverability thanks to high lift and lateral force, enables convenient recovery and launch owe to smooth berthing and storage, and ensures safe operation due to the system backup in case of one wing failure[1, 2]. To conquer inconvenient manipulation brought about by the strong coupling and attain simultaneous operation of somersault and inclination for vehicle with x-shaped fins, this paper firstly builds motion model of vehicle and analyze hydrodynamics acting on x-shaped fins and thruster to find the coupling principle between vehicle attitude and deflection angles of fins, then develops a neural network based algorithm to self-adaptively decouple the attitude control.

Kinetic equations of winged underwater vehicle

Mathematic model of vehicle is generally utilized for description of space motion state of underwater vehicles and serves as reference for motion numerical simulation, performance analysis and engineering design. The earth-fixed frame of reference and body-fixed frame of reference are assigned as shown in Fig.1. For earth-fixed frame E–ξηζ, coordinate axis E-ξ is located on the surface of the sea, and heads to the magnetic north, coordinate axis E-η is on the same surface with coordinate axis E-ξ and rotates for 90 degrees by right-hand rule, coordinate axis E-ζ is perpendicular to the plane ξ-E-η and points to the center of earth, origin of frame E is a arbitrary point for vehicle launch on the sea surface. For body-fixed frame, axis B-x, B-y and B-z arranged by right hand rule respectively point to vehicle stem, vehicle starboard and vehicle spine, frame origin B is set on the buoyancy center of vehicle.

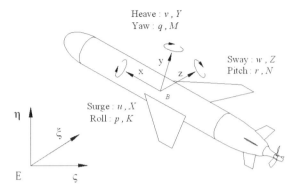

Fig.1 Illustration of motion for autonomous underwater vehicle

Generalized position vector $P=[\xi\ \eta\ \zeta\ \varphi\ \theta\ \psi]^T$ is used for the description of position and attitude of vehicle with respect to earth-fixed frame. Note that φ, θ and ψ are Cardan angles which can express the attitude angles. Generalized velocity vector $V=[u\ v\ w\ p\ q\ r]^T$ includes linear and angular rates of vehicle with respect to body-fixed frame. Generalized force vector $F=[X\ Y\ Z\ K\ M\ N]^T$ is designed to denote force and moment acting on vehicle with respect to body-fixed frame. Simultaneously, the control input of vehicle can be defined as vector $I=[n\ \delta_{al}\ \delta_{ar}\ \delta_{bl}\ \delta_{br}]^T$. The elements in it respectively mean prop speed, deflection angles of four fins.

According to dynamics of torpedo [4, 5], nonlinear equations of motion in matrix form which denotes the interactive relation between forces applied to vehicle and motion state of vehicle can be expressed as

$$M\dot{V} + A\dot{V} + V^T CV - D(V) - G(P) = F(I) \tag{1}$$

In Equ.1, M is mass matrix of inertia of vehicle, A denotes the add mass matrix of vehicle submerged in water , D(V) denotes water-damping vector counteracting on vehicle relative with body-fixed frame. Note that L is the whole length of vehicle, S is the cross section area of vehicle and V_B is the resultant velocity of vehicle. G(P) means the gravity and buoyancy vector expressed in body frame. For detailed illumination, refer to [6, 7].

The deflection angles of fins are defined as positive if the fins generate a yaw torque to steer the vehicle starboard [8]. As is shown in Fig.2, all four fins turn anticlockwise in A and B view to have positive angles. Note that the top-left fin and bottom-right one in the A view are A pair, the top-right fin and bottom-left fin in the B view are defined as B pair.

Denote deflection angles, lift forces and drag forces of right and left fins in A pairs as $\delta_{ar}\ \delta_{al}$, $\tau_{ar}\ \tau_{al}$ and $f_{ar}\ f_{al}$. In the same way, denote deflection angles, lift forces and drag forces of right and left fins in B pairs as $\delta_{br}\ \delta_{bl}$, $\tau_{br}\ \tau_{bl}$ and $f_{br}\ f_{bl}$. Hydrodynamic steer force acting on fins while vehicle cruises at a constant speed in water can be expressed as

$$F_{fin} = \begin{bmatrix} X_{fin} & Y_{fin} & Z_{fin} & K_{fin} & M_{fin} & N_{fin} \end{bmatrix}^T \tag{2}$$

where l_1 and l_2 are the corresponding force arms for lift and drag to generate moments.

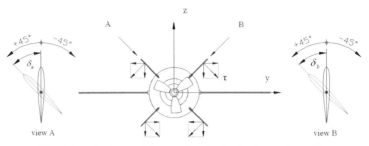

Fig.2 Regulation and denotation for hydrodynamics of fins

There will be three types of hydrodynamic forces acting on the fins when underwater vehicles navigate in the uncompressed viscous water. They are forces and moments of idea fluid, viscous position forces and moments, and viscous damping forces and moments. According to [9], the formula is written as

$$\begin{cases} \tau_{ar} = k_1\delta_{ar} \\ \tau_{al} = k_1\delta_{al} \\ \tau_{br} = k_1\delta_{br} \\ \tau_{bl} = k_1\delta_{bl} \end{cases}, \quad \begin{cases} f_{ar} = k_2\delta_{ar}^{\;2} \\ f_{al} = k_2\delta_{al}^{\;2} \\ f_{br} = k_2\delta_{br}^{\;2} \\ f_{bl} = k_2\delta_{bl}^{\;2} \end{cases}$$

where k_1 and k_2 are force co-efficiencies.

Generalized thrust force is expressed as

$$\mathbf{F}_{pro} = \begin{bmatrix} X_{pro} & Y_{pro} & Z_{pro} & K_{pro} & M_{pro} & N_{pro} \end{bmatrix}^T \tag{3}$$

where $X_{pro}=X'_n\rho n^3 D^4$, $Y_{pro}=Z_{pro}=0$, $K_{pro}=K'_n\rho n^3 D^5$, $M_{pro}=N_{pro}=0$. According transformation from body-fixed to earth-fixed frame, the kinematic equations can be denoted as

$$\dot{\mathbf{P}} = \mathbf{T}(\mathbf{P})\mathbf{V} \tag{4}$$

For more detailed description on transformation matrix, look up for reference [10].

In light of the investigation above, the dynamic model is strong nonlinear system with 5 inputs and 6 outputs. The control principle of vehicle can be illustrated by using schematic diagram as shown in Fig.3. The generalized control force F is a linear function of the control input I by approximating the real model. The generalized velocity V is obtained through integral function of generalized control force F, generalized position P and generalized velocity V. And the generalized position P is an integral function of generalized velocity V. Neural network has strong nonlinear mapping capability [11, 12], hence we attempt to introduce attitude controller with neural network algorithm. Without loss of generality, we simplify the dynamic model according to reasonable assumption and actual operation, and develop a tailored BP neural network for it. Here we chose the typical simplified dynamic model with pitch and yaw as outputs and angles of two pairs of fins as inputs.

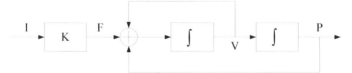

Fig.3 Schematic diagram of the dynamic model of x-shaped-fin vehicle

Prop speed is assumed to be constant and the surge velocity can be considered as uniform. Roll balance is achieved with two configurations: gravity center is below the buoyancy center, and the roll moment caused by prop is counteracted by roll mass in vehicle. Each pair of fins is operated together in the same direction and magnitude, because the vehicle is stable in roll and angle of Jenckel rudder is not needed. We denote the control angles of two pairs of fins as δ_a, δ_b. Note that $\delta_a>0$ will create a positive pitch moment (nose-up) and a positive yaw moment (turn-right), $\delta_b>0$ will produce a negative pitch moment (nose-down) and a positive yaw moment (turn-left). The gravity center is vertically below the buoyancy center, namely, the gravity center is on the axis B-y. The assumptions and actual configurations can be expressed as follows: u=0, n=const, p=0, K=0, $\delta_a=\delta_{ar}+\delta_{al}$, $\delta_b=\delta_{br}+\delta_{bl}$, $\delta_d=0$, K'n=0, $x_G=0$, $z_G=0$.

By substituting the equations above into dynamic equations of vehicle (Equ.1 and Equ.4), we can derive the simplified dynamic model for AUV with x-shaped fins, viz.

$$
\begin{cases}
(m+a_{vv})\dot{v}+a_{vr}\dot{r}+mur-my_Gr^2 = \dfrac{1}{2}\rho V_B^2 S\left(Y_\alpha'\alpha+Y_r'r'+Y_{\delta a}'\delta_a+Y_{\delta b}'\delta_b\right) \\[2mm]
(m+a_{rr})\dot{w}-a_{wq}\dot{q}-muq+my_Grq = \dfrac{1}{2}\rho V_B^2 S\left(Z_\beta'\beta+Z_q'q'+Z_{\delta a}'\delta_a+Z_{\delta b}'\delta_b\right) \\[2mm]
(J_y+a_{qq})\dot{q}+a_{wq}\dot{w}=\dfrac{1}{2}\rho V_B^2 SL\left(M_\beta'\beta+M_q'q'+M_{\delta a}'\delta_a+M_{\delta b}'\delta_b\right) \\[2mm]
(J_z+a_{rr})\dot{r}+a_{vr}\dot{v}-my_G\left(wq-vr\right)=Gy_G\sin\theta+\dfrac{1}{2}\rho V_B^2 SL\left(N_\alpha'\alpha+N_r'r'+N_{\delta a}'\delta_a+N_{\delta b}'\delta_b\right) \\[2mm]
\dot{\psi}=q/\cos\theta \\[2mm]
\dot{\theta}=r
\end{cases}
\tag{5}
$$

where $\{X_{(.)}'\,Y_{(.)}'\,Z_{(.)}'\,K_{(.)}'\,M_{(.)}'\,N_{(.)}'\}$ is the force co-efficiencies relative to control variables, $\{X_{(.)}'\,Y_{(.)}'\,Z_{(.)}'\,K_{(.)}'\,M_{(.)}'\,N_{(.)}'\}$ is the force co-efficiencies with respective to generalized velocity, $(x_G\ y_G\ z_G)$ is the coordinate of gravity center of vehicle expressed in body-fixed frame, and $(J_x\ J_y\ J_z)$ is the moment of inertia of vehicle about three axes of body-fixed frame.

Specification of vehicle and parameters of dynamic model

The physical specification of winged vehicle is listed in Table.1. The control variables are showed in Table.2. The hydrodynamics coefficients are obtained both experimentally and theoretically and are illustrated in Table.3.

Table.1 Geometric and physical specification of x-shaped-fin vehicle

notation	value	unit	Notation	value	unit
m	130	kg	y_G	0.02	m
L	3.19	m	J_x	1.1739	$kg{\cdot}m^2$
S	0.06	m^2	J_y	87.324	$kg{\cdot}m^2$
D	0.18	m	J_z	86.655	$kg{\cdot}m^2$

Table.2 Control variables of the vehicle

variable	minimum	maximum	unit	description
δ_a	$-\pi/4$	$\pi/4$	rad	Deflection angle of A pair
δ_b	$-\pi/4$	$\pi/4$	rad	Deflection angle of B pair

Attitude decoupling controller construction

The decoupling controller for attitude maintaining of vehicle is made up of three-layer feed-forward networks which use generalized subsection function as computing cell, and can perform nonlinear mapping between the input and output. The requirement of decoupling control can be satisfied only if the learning process of back propagation algorithm is completed according to index function and the nonlinear mapping with decoupling relationship is built. Actually attitude control can also work without identifying the controlled plant. Motion modeling and hydrodynamic analysis before used for deriving the decoupling function is just done to facilitate the validation of feasibility, because the decoupling function of plant is necessary for computer simulation.

As is shown in Fig.4, the plant means simplified dynamic model to be controlled, it is a strong coupled system with deflection angles of A and B pairs fins as inputs, and pitch and yaw angle as outputs. The neural network controller is consisted of two parts, one is for pitch approximation and the other is for yaw approximation. The two parts is independent in the input layer and correlated in the output layer. θ^- and ψ^- are respectively the desired pitch and yaw angle.

Table.3 Hydrodynamic parameters of the vehicle

coef	Value	description	Coef	value	description
a_{uu}	12 kg		$Y_{\delta b}$	-0.34	
a_{vv}	170.49 kg	Added mass	M_β	0.4	
a_{ww}	125.79 kg		M_q	-0.145	Torque coefficient about y axis
a_{vr}	-46 kg·m	Added moment	$M_{\delta a}$	0.12	
a_{wq}	31 kg·m		$M_{\delta b}$	0.12	
a_{pp}	6 kg·m^2		Z_β	3	
a_{qq}	46 kg·m^2	Added moment of inertia	Z_q	-0.58	Force coefficient along z axis
a_{rr}	52 kg·m^2		$Z_{\delta a}$	0.34	
X_n	0.21	Force coefficient along x axis	$Z_{\delta b}$	0.34	
X_0	-0.34		N_α	-0.61	
Y_a	8		N_r	-0.157	Torque coefficient about z axis
Y_r	2	Force coefficient along y axis	$N_{\delta a}$	0.11	
$Y_{\delta a}$	-0.34		$N_{\delta b}$	0.11	

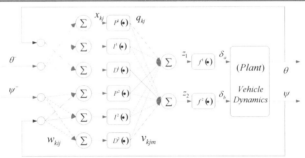

Fig.4 Attitude decoupling controller with neural networks

The input of controller is $r_1=[\theta\ \theta^-]^T$ and $r_2=[\psi\ \psi^-]^T$, the output of controller is $y=[\theta\ \psi]^T$, and the control quantity of plant is $u=[\delta_a\ \delta_b]^T$. The weights in input layer are w_{kij} (k=1,2; i=1,2; j=1,2,3), the weights in output layer are v_{kjm} (j=1,2,3; m=1,2). The variables in the net can be calculated as follows

$$x_{kj} = \sum_i r_{ki} \cdot w_{kij}, \quad z_m = \sum_{k,j} q_{kj} \cdot v_{kjm} \tag{6}$$

Together with controlled plant, attitude decoupling controller becomes a generalized network. Backpropgation algorithm is utilized to update the synaptic weights through online learning according to index function as follow

$$J = \frac{\left(\theta^- - \theta\right)^2 + \left(\psi^- - \psi\right)^2}{2} \le \varepsilon$$

To get weights in output layer proper and right, its increment ratio should be inversely proportional to the partial derivative of index function

$$\frac{dv_{kjm}}{dt} = -\lambda \frac{\partial J}{\partial v_{kjm}} \tag{7}$$

where

$$\frac{\partial J}{\partial v_{kj1}} = \frac{\partial J}{\partial \theta} \cdot \frac{\partial \theta}{\partial \delta_a} \cdot \frac{\partial \delta_a}{\partial z_1} \cdot \frac{\partial z_1}{\partial q_{kj}} + \frac{\partial J}{\partial \psi} \cdot \frac{\partial \psi}{\partial \delta_a} \cdot \frac{\partial \delta_a}{\partial z_1} \cdot \frac{\partial z_1}{\partial q_{kj}}$$

$$\frac{\partial J}{\partial v_{kj2}} = \frac{\partial J}{\partial \theta} \cdot \frac{\partial \theta}{\partial \delta_b} \cdot \frac{\partial \delta_b}{\partial z_2} \cdot \frac{\partial z_2}{\partial q_{kj}} + \frac{\partial J}{\partial \psi} \cdot \frac{\partial \psi}{\partial \delta_b} \cdot \frac{\partial \delta_b}{\partial z_2} \cdot \frac{\partial z_2}{\partial q_{kj}}$$

note that λ is study step size.

The weights in input layer can also be gotten in the same way as weights in output layer.

$$\frac{dw_{kij}}{dt} = -\eta \frac{\partial J}{\partial w_{kij}} \qquad (8)$$

where

$$\frac{\partial J}{\partial w_{11j}} = \sum_{j,m} \left(\frac{\partial J}{\partial v_{1jm}} \cdot \frac{\partial q_{1j}}{\partial x_{1j}} \cdot \frac{\partial x_{1j}}{\partial \theta} \right), \quad \frac{\partial J}{\partial w_{12j}} = \sum_{j,m} \left(\frac{\partial J}{\partial v_{1jm}} \cdot \frac{\partial q_{1j}}{\partial x_{1j}} \cdot \frac{\partial x_{1j}}{\partial \theta^-} \right)$$

$$\frac{\partial J}{\partial w_{21j}} = \sum_{j,m} \left(\frac{\partial J}{\partial v_{1jm}} \cdot \frac{\partial q_{1j}}{\partial x_{1j}} \cdot \frac{\partial x_{1j}}{\partial \psi^-} \right), \quad \frac{\partial J}{\partial w_{22j}} = \sum_{j,m} \left(\frac{\partial J}{\partial v_{1jm}} \cdot \frac{\partial q_{1j}}{\partial x_{1j}} \cdot \frac{\partial x_{1j}}{\partial \psi} \right)$$

note that η is step size for learning.

Advanced learning algorithm for weights update can be used to accelerate the process of convergence and avoid local minimum.

Simulation of attitude decoupling control

To get started with the simulation, the function of plant on vehicle motion is needed. The simplified dynamics of vehicle with x-shaped fins expressed in Equ.5 is used as plant function, which features a strong coupling relationship between deflection angles of fins and attitude angles of vehicle.

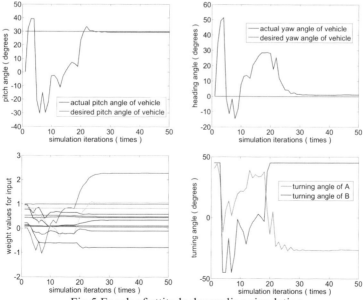

Fig.5 Epoch of attitude decoupling simulation

Many times of simulations have been performed, the initial synaptic weights is selected randomly, so the simulations are different with each other, some iterations curves converged quickly and some takes more time to converge. If the weights converged in the last simulation were chosen for the new initial weights of new simulation, then the curve will converged quickly. Taking one simulation (Fig.5) for an instance, its initial weights is chosen randomly, its initial yaw angle and pitch angle are all 0, and the appointed yaw and pitch angle are respectively 0 and 30 degrees. After around twenty five seconds of iterations, the weights which include 24 numbers start to fix, nearly at the same time, the control deflection angles get fixed too, the A pair turn to the maximum 45 degrees and the B pair turn to about 25 degrees. The actual heading angle and actual pitch angle become approximate to the corresponding desired values in less than thirty seconds of iterations. In this simulation, the step size η and λ are all set to be 0.03. This simulation shows that coupling relation between control quantity and output result can be mapped and the attitude of vehicle can be approximated quickly only if the weight is properly chosen, even the synaptic weights are not the optimal ones.

Summary

Attitude control for autonomous underwater vehicle with x-shaped fins is meaningful and the decoupling of control system is critical. Dynamic model of vehicle is analyzed and decoupling controller based on neural networks is developed. In this paper, yaw and pitch of vehicle are taken into consideration (the roll is balanced by inner mass redistribution) and four fins were divided into A and B pair (there is no Jenckel rudder). To increase the flexibility and maneuverability, three attitudes (pitch, roll and yaw) should be controlled together and the four detached (all-movable) fins should be operated respectively. Thus the controlled plant of vehicle with x-shaped fins will become much stronger coupling system. A new decoupling control model with four fin-angle inputs and three attitude-angle outputs needs to be constructed and its feasibility needs to be confirmed.

References

[1] Yamamoto I. "Research and development of past, present, and future AUV technologies." in Masterclass in AUV Technology for Polar Science, Southampton, UK, 2006.

[2] Li Y, Pang Y J. "Stability analysis on speed control System of autonomous underwater vehicle." China Ocean Engineering, 2009, 23(2): 15~18

[3] Wu J G, Cheng C Y, Wang S X. "Hydrodynamic Effects of a Shroud Design For a Hybrid-Driven Underwater Glider." Sea Technology, 2010, 51(1):16~20

[4] Yan Weisheng. "Torpedo Sail Dynamics[M]." Xi'an: Northwestern Polytechnical University Publishing Company, 2005.

[5] Wang B, Su Y M. "Modeling and motion control system research of a mini underwater vehicle." Proceedings of 6th International Symposium on Underwater Technology, Wuxi China, 2009: 71~75

[6] Wang X. M. "Dynamical behavior and control strategies of the hybrid autonomous underwater vehicle." Ph. D. Thesis, Tianjin University, 2009

[7] Wang X M, Zhang H W. "Analysis on the landing strategy of autonomous underwater vehicle based on fuzzy control." Chinese Journal of Mechanical Engineering, 2009, 45(3): 71~75

[8] Hu K, Xu Y F. "X rudder submarine's mathematical model of space motion simulation." Transaction of Computer Simulation, 2005, 22(4): 50~52 (In Chinese)

[9] Hu K, Xu Y F. "Simulation analysis of effect of submarine rudder type on maneuvering characteristics." Transaction of Ship Engineering, 2005, 27(1): 41~45 (In Chinese)

[10] Song J D. "Analysis and design of performance of underwater launched capsules." Beijing: National Defence Industrial Press, 2008.1: 22~25

[11] Ham F M, Kostanic I. "Principles of neurocomputing for science and engineering." Beijing: China Machine Press, 2003, 7

[12] Xu L L. "Artificial neural networks control." Beijing: Publishing House of Electronics Industry, 2003

Advanced Materials Research Vol. 819 (2013) pp 229-233
© (2013) Trans Tech Publications, Switzerland
doi:10.4028/www.scientific.net/AMR.819.229

New Hydraulic Synchronization System Based on Fuzzy PID Control Strategy

Zhong Liu[1,2,a], Jia Chen[2,b] and Kai Zhang[3,c]

[1]Changshu Institute of Technology, School of Mechanical Engineering, Changshu, Jiangsu, China, 215500;

[2]Soochow University, School of Mechanical Engineering, Suzhou, Jiangsu, China, 215502;

[3]China University of Mining and Technology, School of Mechanical and Electrical Engineering, Xuzhou, Jiangsu, China, 221116

[a]liuzhong678@sina.com [b]chenjia601303@163.com [c]zhangkai2211445@163.com

Keywords: Synchronization control，High-speed switching valve，Transfer function，Fuzzy PID control

Abstract. Proposition of a high-speed switching valve pilot control of two-cylinder two-way electro-hydraulic synchronous drive system, the establishment of a mathematical model of the system, and using fuzzy PID control strategy designed controller, at the same time building a electro-hydraulic synchronization system simulation model based on fuzzy PID controller . Simulation results show that ,when using the fuzzy PID control strategy, slave cylinder of the synchronization system follow the initiative cylinder movement well, the peak-to-average speed of the slave cylinder is 20.3mm / s. Fuzzy PID control process according to the operating conditions change error and error change, by which it has automatic adjustment of PID parameters of the synchronization system. Therefore, fuzzy PID control has better adaptive ability, and the synchronization error is 0.04 mm, achieving high synchronization accuracy. Verifying that high-speed switching valve pilot control of the synchronous drive system and its control strategy is feasible.

Schematic Analysis Based on High-speed Switching Valve Electro-hydraulic Synchronization System

Hydraulic synchronization control system in modern engineering equipment has being used more and more widely. Compared with other drivers, hydraulic synchronous drive has a simple structure, easy to control, and suitable for overloaded high-power occasions.

High-speed switching valve can be directly controlled with the PWM digital signal without D / A conversion part. The power of high-speed switching valve is small, so it can be used for pilot control occasions. Therefore, synchronization control system made up of high-speed switching valve is simple, reliable, and easy to control. For the advantages shown as above, this paper designed a two-way hydraulic synchronization control loop based on the high-speed switching valve pilot driving cartridge cone valve, the principle is shown in Fig.1 below.

As is shown in Fig.1, the output port of the high-speed switching valve is connected to the cone valve control port to control the flow of two working ports in cone valve to achieve the larger proportion of flow control. It works as follows: the load acting on the piston rod end of the hydraulic cylinder 9,10, 3 and 4 solenoid valve is energized, the left station controller output PWM signal role in the high-speed switching valve 7,8 electromagnetic coils by amplifier, high-speed switching valve to open, the hydraulic oil in the cone valve 5,6 back into the tank via the high-speed switching valve7,8, due to the role of the orifice, so that the pressure in cone valve 5,6 pressure drops, two cone valves open at the same time, the cylinder rod chamber oil return 9,10, two cylinders at the same time drive the load lifting. On the contrary, the cone valve is closed, the cylinder rod chamber oil return, the two cylinders maintain the current position unchanged. When the two cylinders in motion are not synchronized by various reasons, the two shift sensors 11,12 respectively measuring the displacement values of the two cylinders and transmission through the detection and processing of the displacement signal to the controller. Through the comparison and calculation of displacement error , the controller outputs a certain value of the modulation rate of the PWM signal, acting on the speed switching valve via the amplifier, then high-speed switching valve opens to act on the cone valve control port to change the flow through the cone valve, controlling the changes in the return flow of the hydraulic cylinder, thereby getting different speed of movement of the cylinder in order to reach the two-cylinder synchronization.

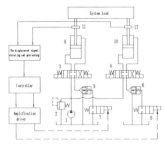

1-Hydraulic pump 2-Relief valve 3,4-The three position four way electromagnetic valve   5,6-Two-way cartridge valves 7,8-High-speed switching valve 9-Active cylinder 10-Slave cylinder 11,12-Displacement sensor

Fig.1 New hydraulic synchronization system

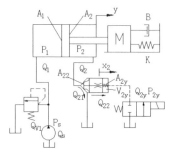

Fig.2 Asymmetric cylinder model of high-speed switching valve poppet valve

The program is actually a master-slave closed loop synchronization control system, I.e. motion displacement of cylinder 9 is applied as a reference, the cylinder 10 follows cylinder 9, and the displacement difference of the two-cylinder is an input control signal of the high-speed switching valve 8, in order to control the flow of cone valve 8 through adjusting the modulation rate of the high speed switching valve 8. So it can achieve precise control of speed, displacement; master-slave circuit has completely symmetrical form, that the parameters of each link is basically the same, thereby greatly reduce the negative impact on the accuracy of synchronous control from the system asymmetry; realize a bidirectional synchronous control of the system.

Mathematical Modeling of the System

The system uses high-speed switching valve pilot to drive cone valve, the system employ return oil flow to control valve control cylinder system, and two-cylinder circuit has a completely symmetrical type, therefore, we can use the hydraulic cylinder 9 as active cylinder to control circuit to establish the mathematical model, simplified physical structure is as shown in Fig.2.

Change the cone valve opening degree through controlling the flow Q_{2y} of the high-speed switching valve, thereby adjusting the return flow Q_2, and then control the movement of the hydraulic cylinder displacement y. Therefore, the transfer function of the active cylinder portion includes three parts: ①The transfer function of the high-speed switching valve control poppet valve; ②The transfer function of the valve control cylinder; ③ The transfer function of the non-symmetric hydraulic cylinder.

The transfer function reflecting relationship of the flow in the high-speed switching valve and displacement of cylinder piston is finally available through the mathematical derivation.

$$G(s) = \frac{Y(s)}{Q_{2y}(s)} = \frac{\frac{vm_1}{\beta_e}s^3 + (\frac{vc_f}{\beta_e} + m_1 k_{22})s^2 + (\frac{vk}{\beta_e} + c_f k_{22} + 2A_{2y}^{\ 2})s + kk_{22} + A_{2y}k_{21}}{\frac{vm_1}{\beta_e}s^3 + (\frac{vc_f}{\beta_e} + m_1 k_{22})s^2 + (\frac{vk}{\beta_e} + c_f k_{22} + A_{2y}^{\ 2})s + kk_{22}}$$

$$\times \frac{C_1 A_2}{C_1 C_2 m s^3 + C_1 C_2 B s^2 + (C_1 C_2 K_1 + C_2 A_1^2 + C_1 A_2^2)s} \qquad (1)$$

In the formula: A_{2y}-cartridge valves spool control the cavity area (m²); k_{21}-coefficient

$(C_d \pi d \sin a \sqrt{\dfrac{2}{\rho} p_2})$; k_{22}-coefficient $(\dfrac{\pi d_{22}{}^4}{128 \mu l})$; v-cone valve control chamber volume (m³); the m_1-poppet

quality (kg); β_ε-fluid bulk modulus of elasticity (N/m²), k-cone valve spring (N / m); c_f-cone valve of the viscous damping coefficient (Ns / m); C_1, C_2-fluid capacitance of piping into the oil chamber, piping back to the oil chamber; wherein $C_1 = \dfrac{V_1}{K}$, $C_2 = \dfrac{V_2}{K}$, wherein V_1, V_2 -the volume of the rod chamber, no rod chamber, K is the elastic modulus of the fluid volume; A_1-the active area of the hydraulic cylinder rod chamber(m²); A_2-role in the area of the hydraulic cylinder rod chamber (m²); m-hydraulic cylinder load weight (kg); B-viscous friction coefficient; K_1-load elastic modulus;

The Sync Control Strategy Based on Fuzzy PID Control

Fuzzy PID parameters self-tuning idea is based on the controlled object response error E and error change EC to determine the parameters to adjust the amount of size and polarity. The algorithm process is using set of rules related to control the conditions of fuzzification, then match it with the knowledge base of the fuzzy rule (that is, to determine whether control conditions and the conditions of the rule set is the same), if the rule is matched, then the implementation of the results portion of the rules, thereby to obtain a parameter corresponding to the adjustment amount. Structure of fuzzy PID controller block diagram shown in Fig.3.In the synchronization system, the control system, according to the reference input and the feedback signal of the position of the slave cylinder, calculated the active cylinder position and slave cylinder position error E and error change EC. Expert knowledge is concluded on the basis of the process analysis and operational experience, rules are taken shaped in the form of "if … Then", and after the fuzzy inference output of the PID controller integral coefficient K_I proportional coefficient K_P and the gain of the differential coefficient K_D fuzzy query matrix to form them as current controller reference PID regulator. If response curve with the desired curve in a sampling time of observation, can be based on the E and EC through real-time adjustments in the knowledge base for the corresponding fuzzy inference matrix parameter adjustment, until its output reaches the desired response mode.

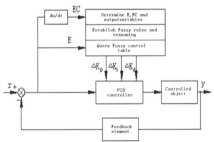

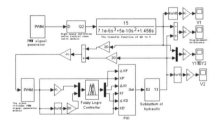

Fig.3 Fuzzy PID control block diagram   Fig.4 Fuzzy PID control simulation block diagram of hydraulically driven synchronization system

Synchronization System Fuzzy PID Controller Design

Fuzzy PID controller employs two-input and single-output type; fuzzy controller is applied with two-dimension input (E, EC) and three-dimension output (ΔK_P, ΔK_I ΔK_D) type. The input variables are error E and error change EC between the slave cylinder and active cylinder position. And E and EC changes in the scope is defined as fuzzy set theory domain {-5, -4, -3, -2, -1 , 0, 1, 2, 3, 4, 5}, the fuzzy subset {NB, NM, NS, ZO, PS, PM, PB}; output variables ΔK_P, ΔK_I, ΔK_D are defined as a fuzzy set on the universe {-3, -2, -1,0,1,2,3}, the fuzzy subset {NB, NM, NS, ZO, PS, PM, PB}, and a function of the triangular establish their membership function. The core of fuzzy control design is a summary of the technical knowledge and practical experience of the designers, to establish a suitable fuzzy rule table, three parameters for ΔK_P, ΔK_I, ΔK_D tuning fuzzy control table. For the synchronous system with reference to the general experience, ΔK_P, ΔK_I, ΔK_D parameter tuning fuzzy control rule table can be created.

The computer control system correct parameter by detecting a sampling timing corresponding to the above table, and substituted into the following formula:

$$K_P = K_{P0} + \{E_i, EC_i\}\Delta K_P$$
$$K_I = K_{I0} + \{E_i, EC_i\}\Delta K_I$$
$$K_D = K_{D0} + \{E_i, EC_i\}\Delta K_D$$

(2)

Completion of fuzzy control reasoning , realize the on-line adjustment of PID parameter.

Synchronous Drive System Simulation Analysis

In MATLAB window, type fuzzy to start fuzzy Editor, based on the principle of the fuzzy controller, build a two-dimensional input, three-output fuzzy system to set the relevant parameters, based on the design in the previous section: If the membership function editor enter E, EC, Δ KP, Δ KI, Δ KD membership function and the universe, and then in the rule editor shape fuzzy control rules in the form of IF E is ·· and EC is ·· then Δ K$_P$ is ·· and Δ K$_I$ is ·· and Δ K$_D$ is ··. This established a dual input and three output fuzzy controller based FSI system files; Simulink Model file structures PID controller module and create subsystem fuzzy controller combination which can establish the synchronization system of fuzzy PID The control simulation block diagram shown in Fig.4.

Create the above model, input the relevant parameters and the control amount of the system, through simulation we can obtained the speed of the two cylinders, the displacement error response, and the curve is shown in Fig.5 respectively.

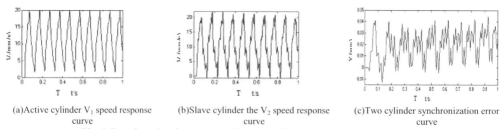

(a)Active cylinder V$_1$ speed response curve   (b)Slave cylinder the V$_2$ speed response curve   (c)Two cylinder synchronization error curve

Fig.5 Synchronization system based on fuzzy PID control response cure

As can be seen from Fig.5 (a) and (b), the slave cylinder can be a good follower of the active cylinder, the peak average rate of the active cylinder motion is 20.3mm / s. The speed of the slave cylinder during movement varies with system error, and this is because the fuzzy PID control can make automatically PID parameter adjustment according to errors and the error change in operating conditions, and thus it has good adaptive ability to achieve high synchronization accuracy. Seen from the figure(c), the error of two-cylinder synchronization is 0.04mm when employing fuzzy PID control in the synchronized system. Seen from the simulation results, the fuzzy PID controller has a good control effect.

Conclusion

Propose the program that making high-speed switching valve as the pilot valve to control cone valve in a two-cylinder synchronous closed loop circuit, large traffic cone valve is controlled through the small flow of high-speed switching valve, closed-loop control system using master-slave loop error control, the application of computer makes large flow, high response synchronous system digital control come true. Through the synchronous drive system modeling and simulation analysis, it shows that: the fuzzy PID controller has good control of the system, which both improves the speed of the system tracking accuracy, but also improves the synchronization accuracy of displacement.

This study is to solve problems for cylinder synchronous, whether the synchronization system control program has a good synchronization control effect pending further study, and it is even more meaningful. And we will concentrate on the point next step.

Acknowledgements

This article is funded by National Natural Science Foundation of China (51275060) and Jiangsu Province Blue Project funded (young and middle-aged academic leaders)

References

[1] LEI Jin-zhu .YANG Fei, JIA Jian-tao. Simulation and analysis of hydraulic synchronous system based on MATLAB and AMEsim[J]. China Water Transport,2008(3):175-176.(In Chinese)

[2] LUO Yan-lei. The methods of the synchronous hydraulic circuit and control system[J]. Chinese Hydraulic & Pneumatics，2004(4):65-67. (In Chinese)

[3] HE Qian,LIU Zhong. Design of the hydraulic synchronization control system based on high speed on-off valve[J].Manufacturing Technology & Machine Tool，2009(1): 142-143. (In Chinese)

[4] DING Yi, ZHAO Ke-ding et al. Research on synchronization control of bi-cylinder systems [J]. Fluid Power Transmission and Control,2007 (2): 24-26. (In Chinese)

[5] Takahashi K, Yamada T. Application of an immune feedback mechanism to control systems. JSME International Journal , Series C ,1998 , 41 (2) :184~191.

Advanced Materials Research Vol. 819 (2013) pp 234-237
© (2013) Trans Tech Publications, Switzerland
doi:10.4028/www.scientific.net/AMR.819.234

Remote monitoring and intelligent fault diagnosis technology research Based on open CNC system

Zhili Lu[1, a], Shiguang Hu[1, b] Taiyong Wang[1,c], Dongxiang Chen[1,d] Qingjian Liu[1,e]

[1]Key Laboratory of Mechanism Theory and Equipment Design of Ministry of Education Tianjin University, Tianjin 300072, China.

[a]812270432@qq.com, [b] sghu@yahoo.cn, [c]tywang@189.com, [d]dxchen@tju.edu.cn

[e]lklqj-5759@163.com

Keywords: open CNC system; remote monitoring; intelligent fault diagnosis ;Fault Tree Analysis Method;

Abstract. In modem manufacturing, the various and complex requirement of industry makes the CNC machine tools more and more automatic and networking. While a remote monitoring and intelligent fault diagnosis system is the basic and indispensable unit for automatic and networking machine tools. This paper is focused on open CNC system, the condition monitoring, and fault diagnosis technology are researched of open CNC system. Integration achieved the CNCmachine tools' status remote monitoring and intelligent fault diagnosis, and detailed analysis of the key technologies for the components of the system. Through effectively integration of the computer technology, Fault Tree Analysis method, or other technologies to enhance the automation, networking and intelligent level of the open CNC system.

Introduction

In modem manufacturing, the various and complex requirement of industry makes the CNC machine tools more and more automatic and networking.The economic losses could be caused due to CNC machine tool failure.While a remote monitoring and maintenance system with the module of intelligent fault-diagnosis is the basic and indispensable unit for automatic and networking machine tools. This can not only increase the efficiency of productivity,but also improve the degree of automation and flexibility.This paper study the status information acquisition system in intelligent fault diagnosis-- remote monitoring system .On one hand ,this message can used to ensure the status of the CNCmachine tools[1].On the other hand, it can be used to extract feature information In order to through the Fault Tree Analysis module to reason the causes of failure .Which can detect faults accurately, reduce mechanical losses and casualties in the production process.At the same time this paper mainly introduces the Fault Tree Analysis Method in CNC equipment intelligent fault diagnosis .

The Realizing Scheme For Remote Monitoring and Controlling Module.

The whole structure of remote monitoring and controlling system based on open CNC system as Fig .1.the realizing scheme based on the module of C/S.The following contents are included[3]: the open CNC system is used as the Client-side,the centern of the remote controlling is used as the server-side.Run in laboratory LAN network environment,The remote monitoring and Controlling center server is listening.The worker on the site opened the Client-side,so that the Client-side can send a connect command to the server-side,and make a connection with it. The UDP and TCP network as the base of transport protocol,so that it can send the viedo of machining scenre to the remote monitoring and controlling center with high speed.At the same time,the remote monitoring and controlling center can send the control instruction to the CNC system to control the CNC tools.

Open CNC system installed in the Host computer,Connected with a control system through the bus,gatering the motion status of the CNC tools.The following contents are included: 64 I/O staus of the machine's key part,the status of the Spindle load etc.The CNC system send this message to the remote monitoring and controlling center through the internet.At the same time,it also can receive the feedback message from the remote monitoring and controlling center .According to this message,the CNC tools can execute some related action.

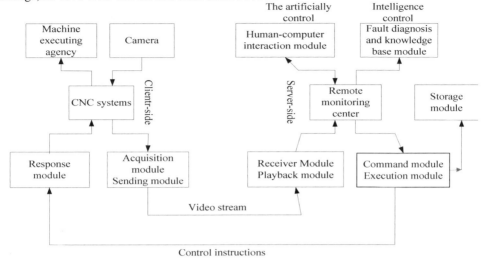

Fig. 1 Process monitoring module structure

The remote monitoring and controlling center, it main provide three parts service:Fistly, monitoring and controlling,according to the status from the CNC system,the worker in the controlling center can remote control the CNC tools.Secondly,Information sharing,it can keep all the message in the Database. Thirdly, Fault Diagnosis. Under the infomation of the Database,the center can diagnosis the fault.

The Database is used to record,inquiry,delete,add information . the Remote Monitoring and Controlling infomation ,the infomation are included Operating time,contents etc[2].The Technology of the database is employ the Method of ADO Execute the SQL Language to coduct the ACCESS table.the ACCESS table make a connection with a list control in the VC++.The Realizing flow chart as follow,

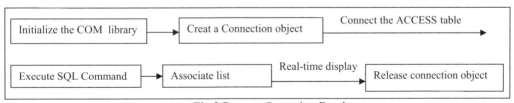

Fig.2 Process Operating Database

Send and receive streaming video: The client-side send the video data of the scene to the remote monitoring and controlling center , adopt the Directshow technology to realize it, a specialized filter —NetSender Filte should be set up. it's function mainly to receive the media data from the video data gather filter, checked the media type, and packed Media data into a UDP packet. According to the UDP protocol to send to the specified address [4].

Transport network is the Information transmission channel of the Remote Monitoring and Controlling center and the CNC tools. Mainly by uses of Internet.

The remote monitoring and controlling module monitor the site of CNC tools and control it.

Fault Diagnosis Module based on Fault Tree Analysis

The whole structure of Fault Diagnosis Module based on Fault Tree Analysis as figure 3, a group of sensor gathering the dates from the CNC tools,then,the gathered dates send to the remote Fault Diagnosis center,the center Intelligence analysis those date,judge and predict the type of the fault,and decide to remote conduct the CNC tools.

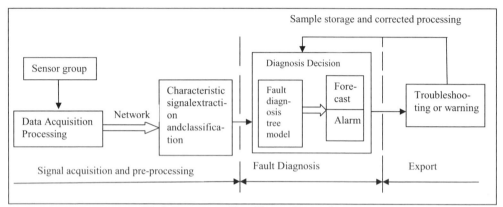

Fig.3 Structure of Intelligent diagnosis system

The Fault Tree Analysis, is a analytical methods of formed a system failure from the general to the part of the step by step according to dendritic refinement [5].
Set minimal cut set expression for Lj (X), the minimal cut set structure function is

$$\theta(X) = \sum_{J=1}^{n} k_j(X) \tag{1}$$

Where, k is the minimum number of cut sets, Lj (X) is defined as

$$L_J(X) = \prod_{I \subseteq K_J} X \tag{2}$$

Seeking top event probability, if $\theta(X) = 1$ is the probability, Take the mathematical expectation on both sides, the left end is the probability of occurrence of the top event

$$G = P_r\{\sum_{J=1}^{n} L_j(X) = 1\} \tag{3}$$

Suppose E_i is the occurrence event of all events of the end which belong to the minimum cut k_j, then the top event will be occur is

$$G = P_r\{\sum_{i=1}^{k} E_i\} \tag{4}$$

If the event and the probability which Write as

$$F_j = \sum_{1 < j_1 < j_2, \dots < K} P_r\{E_{i1} \cap E_{i2} \cap E_{i3} \dots \cap E_{ij}\} \tag{5}$$

Then can figure out the probability of the top event

$$g = \sum_{j=1}^{k}(-1)^{j-1}F_i = \sum_{1<i<j<k}P_r\{E_i \cap E_J\}+....(-1)^{K-1}P_r\{\bigcap_{r=1}^{k}E_r\} \tag{6}$$

The Fault Diagnosis Module based on Fault Tree Analysis can intelligent analysis the data which acquired from the site, and generated the corresponding fault diagnosis report which could be auto-transmitted to the interface.

Summary

Remote monitoring and intelligent fault diagnosis technology is a combiniton of computer science,communication technology and virtual reality technology.it realized A technological innovation of" Mobile data instead of people".The Remote monitoring and intelligent fault diagnosis system is not only very convenient and efficient,but also the degree of inteligent is very high.what'more,it commendably conduct on-line control and reduce the operation cost.But, Remote monitoring and intelligent fault diagnosis technology is still in the stage of development in china at present.There still have many technology crux problem need to solve.Whereas , no Matter from which point of view,with the help of internet it. Increase the efficiency of productivity to sovle problem on CNC machine tools.

Acknowledgements

This work is financially supported by the CNC generation of mechanical product innovation demonstration project of Tianjin (2013BAF06B00), The recovery method of weak information and study of algorithm hardened design based on generalized parameter adjustment stochastic resonance (20100032110006) and High-end CNC machining chatter online monitoring and optimization control technology research (12JCQNJC02500). Please communicate with the corresponding author Zhili Lu, if there are any questions in this paper.

Reference

[1] Lin Jinzhou, Jiang Dayong, Geng Bo, et al. Research on the remote monitoring and fault diagnosis of CNC system based on network, 2012 2th International Conference on Functional Manufacturing and Mechanical Dynamics, China: Hangzhou, 2012, 141(1): 465~470

[2] Zhang Y, Wang T Y, Leng Y G.etal. Application of stochastic resonance signal recovery[J]. Chinese Journal of Mechanical Engineering, 2009,22(4): 542-549.

[3] Xiang J, Ogata K, Futatsugi K. Formal fault tree analysis of state transition systems. In: Proc. of Fifth International Conference on Quality Software, 2005: 124—131.

[4] Xiang J, Futatsugi K. Fault tree and formal methods in system safety nalysis, In: Proc. of the fourth International Conference on Computer and Information Technology, 2004: 1 128—1 1 15.

[5] Steyer J, Lardon L, Bernard Q. Sensors network diagnosis in anaerobic digestion processes using evidence meory[J]. Water Science and Technology, 2004, 1 1(50): 2 1. 29.

Advanced Materials Research Vol. 819 (2013) pp 238-243
© (2013) Trans Tech Publications, Switzerland
doi:10.4028/www.scientific.net/AMR.819.238

Research of Control Strategy based on the Electro-hydraulic Proportion Position Control System

Zhu Yinfa[1,a] ,Chen Bingbing[2,b]

[1]College of Technology, Lishui University, Lishui, 323000, China

[2]College of Mechanical Engineering, Donghua University, Shanghai, 201620, China

[a]lszhuyinfa@163 com, [b]cbb214@163 com

Keywords: electro-hydraulic proportion positional control; PID control; self-tuning fuzzy PID control

Abstract. The self-tuning fuzzy PID controller of the electro-hydraulic proportion position control system is designed and researched. Compared the self-tuning fuzzy PID control with the traditional PID control through experiments for the track effect on sinusoidal signals, the results show that the self-tuning fuzzy PID controller has higher accuracy and better stability. It is a more excellent performance controller.

Introduction

Now in the field of national defense science and technology, industrial automation and control, with the advantages of low-cost and anti-pollution ability, the electro-hydraulic proportional directional valve generally replaces the electro-hydraulic servo valve in many industrial areas. The electro-hydraulic proportional directional valve is a kind of device in which proportional solenoid generates the corresponding action according to the input voltage signal, make the spool produce displacement, and complete the pressure, flow output proportional to the input voltage. The electro-hydraulic proportional position control system controlled by computer has simple structure and good performance, but the requirements to the electro-hydraulic proportional position control system are also very high. While traditional PID control is difficult to achieve satisfactory results, in order to solve this problem, it appeared the intelligent control such as fuzzy control and neural network control. The electro-hydraulic proportional position control system combines the advantages of PID controller and fuzzy controller, modify the three parameters of PID controller in real time by fuzzy control. After the experiment, it shows that the position control accuracy and stability of the electro-hydraulic proportional position control system are improved greatly and the system can be widely used in the industrial areas.[1]

Composition and principle of the electro-hydraulic proportional position control system

The electro-hydraulic proportional position control system is composed of displacement sensor, electro-hydraulic proportional valve, pressure gauge, pressure sensor, overflow valve, vane pump, amplifier, hydraulic cylinder and so on. The structure of system is shown in figure 1. The working principle of system is listed below: when the system works, the displacement sensor detects the current position signal of the hydraulic cylinder piston rod and feeds back to the computer through A/D transformation, and then compares to the specified inputs, at last the system gets the deviation. After D/A transformation and amplification of the power amplifier, it changes the orifice opening of electro-hydraulic proportional direction valve, and makes the piston rod reach to the required position.

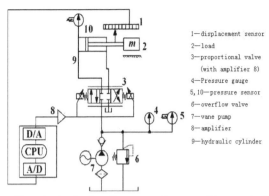

Fig.1   The electro-hydraulic proportioan position control sytem

Study of the control strategy

The electro-hydraulic proportional position control system select single-variable & two-dimensional fuzzy controller, which includes two inputs: error e and error rate of change ec and a control output: u. the self-tuning fuzzy PID controller tests the e and ec constantly during the working and then output the adjustment of three parameters of PID controller: ΔKp, ΔKi, ΔKd accord to the fuzzy control rules. The correction algorithm is as follows:

$$K_p = K_p' + \Delta K_p, \quad K_i = K_i' + \Delta K_i, \quad K_d = K_d' + \Delta K_d.$$

Among them, Kp′, Ki′ and Kd′ are predetermined, ΔKp, ΔKi and ΔKd are outputs of the fuzzy controller, Kp, Ki and Kd are outputs of the PID controller. The flow chart of self-tuning fuzzy PID controller is shown in Figure 2. [2] [3]

For the fuzzy controller, whether inputs are deviation or change rate of deviation, they're all accurate inputs. It must transform them into fuzzy date and express with the corresponding fuzzy sets. The domain of them (include e, ec, Kp, Ki and Kd) is {-6 -5 -4 -3 -2 -1 0 1 2 3 4 5 6}, and their fuzzy subset is {PB, PM, PS, ZO, NS, NM, NB}, the elements in the subset represent positive big, positive middle, positive small, zero, negative small, negative middle and negative big. We suppose the basic domain of deviation e is [-75, 75]mm, the basic domain of change rate of deviation ec is [-1,1]mm/s. The basic domains of Kp, Ki and Kd are [-0.6, 0.6], [-0.1, 0.1], [-2.5, 2.5]. The membership function is shown in Figure 3.

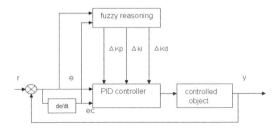

Fig.2   Flow chart of the self-tuning fuzzy PID controller

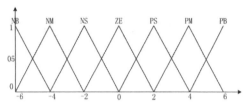

Fig.3   The triangular membership function

After fuzzying the exact amount of system, according to the theoretical knowledge, practical experience and design ideas of self-tuning fuzzy PID control, we can draw the fuzzy control rule table of output variable Kp, Ki and Kd, as shown in table 1, table 2 and table 3.

Table1   Fuzzy control rule table of Kp

e	ec						
	NB	NM	NS	ZO	PS	PM	PB
NB	PB	PB	PM	PM	PS	ZO	ZO
NM	PB	PB	PM	PS	PS	ZO	NS
NS	PM	PM	PM	PS	ZO	NS	NS
ZO	PM	PM	PS	ZO	NS	NM	NM
PS	PS	PS	ZO	NS	NS	NM	NM
PM	PS	ZO	NS	NM	NM	NM	NB
PB	ZO	ZO	NM	NM	NM	NB	NB

Table 2 Fuzzy control rule table of Ki

e	ec						
	NB	NM	NS	ZO	PS	PM	PB
NB	NB	NB	NM	NM	NS	ZO	ZO
NM	NB	NB	NM	NS	NS	ZO	ZO
NS	NB	NM	NM	NS	ZO	PS	PS
ZO	NM	NM	NS	ZO	PS	PM	PM
PS	NM	NS	ZO	PS	PS	PM	PB
PM	ZO	ZO	PS	PS	PM	PB	PB
PB	ZO	ZO	PS	PM	PM	PB	PB

Table 3 Fuzzy control rule table of Kd

e	ec						
	NB	NM	NS	ZO	PS	PM	PB
NB	PS	NS	NB	NB	NB	NS	PS
NM	PS	NS	NB	NM	NM	NS	ZO
NS	ZO	NS	NM	NM	NS	N	ZO
ZO	ZO	NS	NS	NS	NS	NS	ZO
PS	ZO	ZO	ZO	ZO	ZO	ZO	ZO
PM	PB	NS	PS	PS	PS	PM	PB
PB	PB	PM	PM	PM	PS	PS	PB

According to the fuzzy control rules, the fuzzy output can be inferred, after anti-blurring, it can get the actual exact amount of three PID parameters, and at last realise the real-time adjustments of three PID parameters. In order to achieve better control, it take the median anti-fuzzy approach.

Study of the system experiments

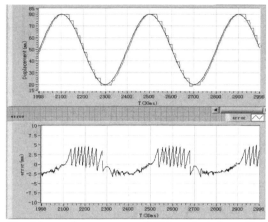

Fig.4   Tracking and error curve of PID control sinusoidal signal

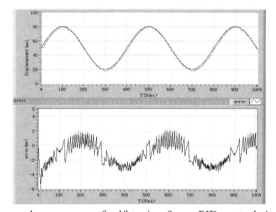

Fig.5   Tracking and error curve of self-tuning fuzzy-PID control sinusoidal signal

The control system realizes the real-time control by adopting programming of MATLAB. By use of the fuzzy logic toolbox of Matlab/Simulink, electing Mandani mode, the controller structure is shown in figure 6, where e and ec are two input parameters, Kp, Ki and Kd are three output parameters. Setting up the fuzzy PID simulation model shown in figure 7, figure 8 is the structure of its sub-system.[4][5]

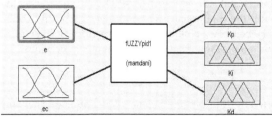

Fig.6   The structure of fuzzy controller

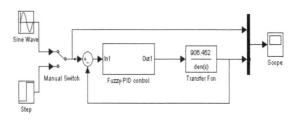

Fig.7   The fuzzy PID simulation model

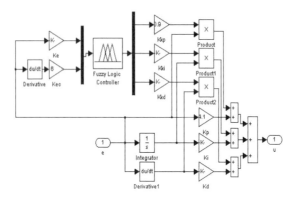

Fig.8   Structure of the fuzzy PID sub-system

If input the sinusoidal signal with amplitude 30mm and cycle 4s, the tracking results of PID control and self-setting fuzzy PID control are shown in the figure 4 and 5. According to the experimental results above, we know that when the system adopts the traditional PID control, the overall error is larger and control accuracy and stability of the system is poor. When the system uses the self-tuning fuzzy PID control, the overall error of the self-tuning fuzzy PID controller is small, and displacement error can be controlled within ±2 mm. The control accuracy and stability of the system are improved greatly. Compared with the PID controller, the self-tuning fuzzy PID controller has better precision and stability. It's an excellent performance controller.[6]

Conclusion

The experimental research is made in the electro-hydraulic proportional position control system by adopting the PID controller and self-tuning fuzzy-PID controller respectively. Experimental results show that compared with the traditional PID controller, fuzzy-PID controller is improved greatly whether in the control accuracy or stability, and provide   reference to the application of this electro-hydraulic proportion position control system on other occasions.

Acknowledgments

The paper is Supported by 2012 Natural Science Foundation of Zhejiang province (LY12E05011)

References

[1] Parnichku M, Nagecharoenkul C. Nonlinear observer-based fault detection technique for electro-hydraulic positioning systems. 2005 SIE.

[2] Yang Yong Mei, Chen Ning. The Design and Simulation of MATLAB-based Fuzzy Self-tuning PID Parameters [J]. Computer Cnformation. 2005(24).

[3] H.Khan,S.Abou and N.Sepehri. Fuzzy control of a new type of piezoelectric direct drive electro-hydraulic servo valve.2002 IEEE.

[4] Zhang Guo Liang, Deng Fan Ling. Fuzzy Control and Application of MATLAB [M]. Xi'an: Xi 'an jiaotong university press.

[5] Yuan Feng Lian. Fuzzy self-tuning PID Controller Design and Simulation of MATLAB [J]. Aviation industry journal 2006(1).

[6] Tan Jia You, Tan Guan Jun, Wu Wei. Fuzzy and self-adaptive fuzzy-PID composite control of electro-hydraulic proportional pressure valve [J]. Moderm manufacturing engineering 2010(9).

Advanced Materials Research Vol. 819 (2013) pp 244-248
© *(2013) Trans Tech Publications, Switzerland*
doi:10.4028/www.scientific.net/AMR.819.244

Research on Acoustic Source Localization Technology Based on

Cross-correlation

Xie zhijiang[1,a], Zeng hai[1,b] and Chen ping[1,c]

[1] Department of Mechanical Engineering, Chongqing University, Chongqing 400044,

People's Republic of China

[a]email: xie@cqu.edu.cn, [b] email:330821416@qq.com, [c]email: 68103331@qq.com

Keywords: Acoustic source localization; Cross-correlation; Microphone array; Time delay; NI CompactRIO

Abstract. In order to realize acoustic source localization, we proposed to use microphone array[1,4] and time delay based on cross-correlation algorithm, and analyzed the factors to influence the locating accuracy, and established a real-time acoustic source localization system on the NI CompactRIO system. By experiments, cross-correlation algorithm for wideband signal(sweep signal, noise signal) localization is more accurate, for narrowband signal (organ signal) localization is less significant. After analyzed factors to influence the localization accuracy, we improved sound source system, and the factors that affected the organ signal localization is verified by experiment, and improved accuracy of narrowband signal localization .

Introductions

Determine the localization of the sound source in space is a broad background. In the civilian aspects, acoustic source location has wide applications in video conference, monitoring systems, intelligent robots, robust speech recognition and other fields.

1   A cross-correlation positioning principle

1.1 The time delay of signal

According to the trigonometric functions of Fourier series, the periodic signal can be mixed by one or several, or even infinite plurality of different frequency harmonics, so multi-frequency signal can be disintegrated as follows:

$$x(t) = a_0 + \sum_{n=1}^{\infty} B_n \sin(nw_0 t + \varphi_0(n)) \qquad y(t) = b_0 + \sum_{m=1}^{\infty} A_m \sin(mw_0(t+\tau) + \varphi_1(m)) \qquad (1)$$

The cross-correlation function of random single-frequency signal x (t) and y (t) of two erotic processes $R_{xy}(\tau)$ is defined:

$$R_{xy}(\tau) = \lim_{T \to \infty} \frac{1}{T} \int_0^T x(t) y(t+\tau) dt$$

$$= \frac{1}{T_0} \int_0^{T_0} \left[a_0 + \sum_{n=1}^{\infty} B_n \sin(nw_0 t + \varphi_0(n)) \right] \left[b_0 + \sum_{m=1}^{\infty} A_m \sin(mw_0(t+\tau) + \varphi_1(m)) \right] dt$$

$$=a_0b_0+\frac{1}{2}\cdot\sum_{n=1}^{\infty}A_nB_n\cos(\varphi_0(n)-nw_0\tau-\varphi_1(m))-\frac{1}{nw_0}\left[\cos(2\pi n+nw_0\tau+\varphi_0(n)+\varphi_1(m))\cdot\sin(2\pi n)\right]$$

(2)

$$=a_0b_0+\frac{1}{2}\cdot\sum_{n=1}^{\infty}A_nB_n\cos(\varphi_0(n)-nw_0\tau-\varphi_1(m))$$

Theoretically speaking, when we know the maximum value of the cross-correlation function of two signals, it could indicate its horizontal coordinate; the horizontal coordinate is the time delay[2, 5]. When the sound source signal is a multi-frequency signal, such as organ signal, cross-correlation[3] is the superposition of the cross-correlation with the same frequency signal, so it adapts the multi-frequency signal.

1.2 Cross-Correlation localization algorithm

As shown in Fig.1, r_0, r_1, r_2 are the distance from the target to the acoustic emission sensor 1, sensor 2 and sensor 3. Distance difference is Δr_i, i = 1,2, then the positioning equations:

$$r_0^2=(x-x_0)^2+(y-y_0)^2 \quad r_i^2=(x-x_i)^2+(y-y_i)^2 \quad c\Delta t_i=c\cdot(t_i-t_0)=r_i-r_0 \quad (i=1,2)$$

(3)

As follows:

$$k_i+c\Delta t_i r_0=x(x_0-x_i)+y(y_0-y_i)$$

(4)

"C" stands for the speed of light. $k_i=\frac{1}{2}\left[(c\cdot\Delta t_i)^2+(x_0^2+y_0^2)-(x_i^2+y_i^2)\right],(i=1,2)$

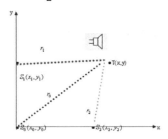

Fig.1 The schematic diagram of TDOA location

According to (4), we can constitute non-linear equations, written in matrix as:

$$AX=F$$

(5)

$$A=\begin{bmatrix}x_0-x_1 & y_0-y_1\\x_0-x_2 & y_0-y_2\end{bmatrix},X=\begin{bmatrix}x\\y\end{bmatrix},F=\begin{bmatrix}k_1+r_0\Delta r_1\\k_2+r_0\Delta r_2\end{bmatrix}$$

(6)

$$\hat{X}=A^{-1}F$$

(7)

$$x=\frac{(k_1+r_0\Delta r_1)(y_0-y_2)-(k_2+r_0\Delta r_2)(y_0-y_1)}{(x_0-x_1)(y_0-y_2)-(x_0-x_2)(y_0-y_1)} \quad y=\frac{(k_2+r_0\Delta r_2)(x_0-x_1)-(k_1+r_0\Delta r_1)(x_0-x_2)}{(x_0-x_1)(y_0-y_2)-(x_0-x_2)(y_0-y_1)}$$

(8)

$$A^{-1}=\frac{\begin{bmatrix}y_0-y_2 & -(y_0-y_1)\\-(x_0-x_2) & (x_0-x_1)\end{bmatrix}}{(x_0-x_1)(y_0-y_2)-(x_0-x_2)(y_0-y_1)}=\begin{bmatrix}a_{11} & a_{12}\\a_{21} & a_{22}\end{bmatrix}$$

(9)

Among: $m_i=\sum_{j=1}^{2}a_{ij}\cdot k_j,n_i=\sum_{j=1}^{2}a_{ij}\cdot\Delta r_j,i=1,2$ .The estimated value of target localization:

$$\hat{x}=m_1+n_1\cdot r_0 \quad \hat{y}=m_2+n_2\cdot r_0$$

(10)

$$a \cdot r_0^2 + 2b \cdot r_0 + c = 0 \tag{11}$$

Among: $a = n_1^2 + n_2^2 - 1$ $b = (m_1 - x_0)n_1 + (m_2 - y_0)n_2$ $c = (m_1 - x_0)^2 + (m_2 - y_0)^2$

The two values are as follows by the equations (11): $r_0 = \dfrac{-b \pm \sqrt{b^2 - ac}}{a}$ (12)

Take (12) into (10) can be obtained the coordinates of x, y.

2   The analysis of experimental results

2.1 The layout of experimental site

The experiment is processed in the room which has 8 meters long and 6 meters wide, 3 meters high, the layout of sensor is in the length of 100 cm, width 50 cm, high 67 cm of space, it is used a tripod to support microphone, and sound as a sound source, the use of NI 9234 as a data acquisition card, data via NI CompactRIO system using cable sent to the PC. Moderate temperature and humidity indoor.

2.2   Broadband signals

2.2.1 Sweep signal

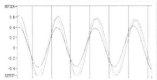 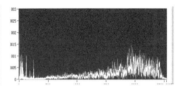 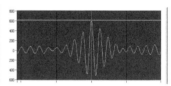

Fig.2 The time-domain wave-   Fig.3 The frequency domain   Fig.4 The cross- correlation
form diagram of sweep signal   waveform of sweep signal   waveform of sweep signal

According to Fig.2, it can be found that the time-domain waveform of sweep signal is relatively regular, and can be seen relatively accurate between time delay of different channels, less interfere information. The frequency domain waveform diagram in Fig.3 can be seen, the amplitude of the sweep signal is relatively small, and the frequency from 0 to 2500Hz and the high band is less. By Fig.4 can be found that the peak of the cross-correlation waveform diagram of the channels 1 and 2 is obvious, therefore, the sweep signal has the higher accuracy of sound source localization.

2.2.2 Noise signal

According to Fig.5, the overall time-domain waveform of the noise signal is relatively regular, the time delay of the signal between the different channels is obvious. Fig.6 is a frequency domain waveform diagram。 the amplitude of the noise signal is relatively small. By Fig.7, it can be found that the peak of the cross-correlation waveform diagram of channels 1 and 2 is obvious; therefore, the sound source of the noise signal has a higher positioning accuracy.

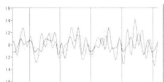  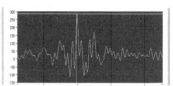

Fig.5 The time-domain wave-   Fig.6 The frequency domain   Fig.7 The cross- correlation
form diagram of noise signal   waveform of noise signal   waveform of noise signal

2.3   Narrowband signal
2.3.1 Organ signal
(1) Organ signal experimental waveforms

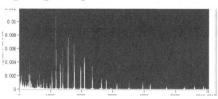
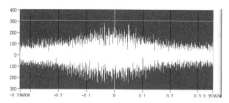

Fig.8 The domain waveform of organ signal   Fig.9 The cross-correlation waveform

According to Fig.9, the peak of cross-correlation function of the organ signal is not obvious, so time-delay determination is inaccurate. By observing Fig.8, the frequency of organ signal ranges from 0 to 4000Hz, but the continuous frequency bandwidth is relatively narrow, the distribution of band is more dispersed and the amplitude of this band is relatively large. The amplitude of organ signal is very small when the frequency ranges from 400 to 700 Hz.

(2) Waveform of filtered organ signal

By analyzing the spectrum of the signal of organ, the use of the digital IIR filter in the band pass filtering between 450 to 700 Hz, the cross-correlation waveform is shown in Fig.10, using a digital IIR filter of the low pass filter in the range of 200Hz, cross-correlation waveform as shown in Fig.11.

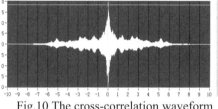
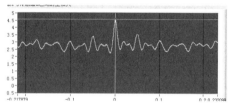

Fig.10 The cross-correlation waveform   Fig.11 The cross-correlation waveform
of band-pass filtering   of low-pass filtering

By observing Fig.10 and Fig.11, using a filter in a digital IIR filter, the peak of cross-correlation function waveform of organ signal is highlighted, the corresponding time delay will be able to judge accurately, while a high-pass filtering by filtering, the positioning of the sound source signal is inaccurate. Resulting in the organ signal inaccurate positioning because the influence of the high-band signal. High frequency, short wave length, short cycle, and therefore positioning distance is short.

3 Experimental results

Table 1 Experimental results of sound source localization

Theory of sound source coordinates	The type of signal		
	Sweep signal	Noise signal	Organ signal
(0.6,0.25)	(0.5953,0.2543)	(0.5998,0.2629)	(0.5909,0.2457)
(0.8,0.35)	(0.7885,0.3328)	(0.7927,0.3413)	(0.7984,0.3426)
(0.67,0.35)	(0.6410,0.3188)	(0.6453,0.3214)	(0.6527,0.3481)

Summary

Whether to be able to locate is related to the frequency of the signal. In the case of a relative fixed position of the sensor and the sound, the frequency is high to a certain extent, it can not locate. However, by filtering, the high frequency of the signal is filtered out and retains the positioning of the low-frequency signal; the positioning accuracy can be improved. High frequency has a negative effect on the positioning, so the filter should use low-pass, and useful low frequency is preserved. Very wide frequency band of the sweep and noise, but they are close to 0Hz, the low-frequency nearby strong, which play a major role in the positioning; the low frequency of organ is very low, middle-frequency positioning play a major role. The experiment use noise signal, sweep signal,organ signal to locate, and experimental results showed that there are two ways to locate signal sound source: First, use the noise carried by the signal itself and the low-frequency of signal, that is to say, the low-pass filtering. Second, use a band-pass filter to set the frequency corresponding band filtering, the determination of band select required by the signal spectrum, which is the frequency range of small amplitude.

References

[1]   Cao hulin, Zhou lingling, Based on sound source localization system of microphone array in hardware design and algorithm [D].Shanghai: Shang hai jiao tong University, 2011

[2]   Ji qinghua, Wang hongyuan, Research about the location algorithm vibration plane based on time difference of arrival (TDOA) technology [D].Shenyang: Shen yang li gong University, 2009

[3]   Dong dai, Liu rong, Sun minglei,Automatic focusing method based on the cross correlation [J]. Journal of Beijing University of Aeronautics and Astronauticon, 2006, 32(3)

[4]   Guo juncheng, Gu hongbin,Sound source localization technology based on microphone array [D].Nanjing: Nanjing University of Aeronautics and Astronautics, 2007

[5]   Wang yi, Wu changqi, Hu shuangxi, The Several comparison of the delay estimation algorithm on TDOA [D]. Hebei Province: Yan shan University, 2008

Advanced Materials Research Vol. 819 (2013) pp 249-253
© *(2013) Trans Tech Publications, Switzerland*
doi:10.4028/www.scientific.net/AMR.819.249

Systematical Signal Unified Model and Intelligent Unit of MEMS Sensor

Chuande Zhou[a], Gaofa He, Zelun Li, Xianlin Deng, Li Lai

College of Mechanical Engineering, Chongqing University of Science and Technology, Chongqing,China

[a]zcd0013@163.com

Keywords: MEMS, Systematical Signal Unified Model, Intelligent Unit, Plug and Play

Abstract: Based on the studying of IEEE145 international standard and achievement of hardware and software of TEDS intelligent sensor, it suggested designing an intelligent unit that won't change the input / output interface of the sensor but enables device plug-and-play. The system still can operate normally when the intelligent unit was removed. The intelligent unit includes various modules or functions, such as the logo of IEEE1415 standard device, humiture measurement, nonlinear data table, unified model as well as MCU achievement, normalized treatment and synthesis of signals, aiming to realize the normalization of sensor's output signals and achieve the quick connection of test system as well as plug-and-play.

Introduction

Various MEMS devices of aerospace system like inertial navigation and acceleration transducer achieved rapid development recently and their performances were improved significantly. However, it is inevitable for MEMS to produce nonlinearity of analog signal and conformity error due to its manufacture errors, isomerism of material structure, impact from external environment, which will cause accumulative errors after signal transformation and transition control. It can avoid error accumulation effectively and guarantee the accuracy as well as uniformity of MEMS sensor or execution components by applying amended, normalized and standardized plug-and-play through the intelligent perception and execution unit.

Intelligent perception and intelligent execution were summarized as the important development direction of intelligent manufacture by "China's Mechanical Engineering Technology Roadmap in 2030" compiled by nearly one hundred experts from China's Mechanical Engineering Technology Association. It is expected to develop standard of plug-and-play technology, extensive perception technology and driving technology that oriented at intelligent sensor, intelligent instrument and intelligent terminal by 2030, which supports the plug-and-play and system reconfiguration of mechanical (manufacturing) system as well as intelligent holographic man-machine interactive system. Currently, the IEEE1415 intelligent sensor, or generally known as the TEDS intelligent sensor, is the most advanced technical standard that representing the development direction, such as the TEDS acceleration transducers of B&K, PCB and KISTLER. TEDS provides the standard for achieving intelligent sensor, mainly including TEDS achievement based on hardware and virtual TEDS method based on pure software.

Hardware achievement implements TEDS through hardware, as shown in Fig.1. Sensor of mixed model needs to pack the sensor and module containing TEDS information together, while the intelligent wire separates the hardware module containing TEDS information from the sensor, satisfying the intelligent reconstruction requirement of the traditional sensor. The hardware-based TEDS achievement makes the sensor and control system independent mutually, thus possessing excellent interchangeability. Such sensor has complicated structure, which will surely develop lower reliability. Since many users have already equipped with relative complete traditional test system, they are more concerned on how to use advanced technology to better develop the efficiency of the existing test system.

Fig.1 The achievement of hardware TEDS

Virtual TEDS isn't in hardware, but it stores relevant information of sensor in the software system of computer as a database, as shown in Fig.2. It just needs to update the content of the TEDS file after periodic calibration in order to update the parameters of the intelligent sensor. Virtual TEDS achievement only simplifies the operation in setting the control system, only requiring inputting the TEDS information related with sensor from the database. However, depending on the establishment of test system, it needs application programs to accomplish the setting and invoking of all relevant parameters, and failed to identify the sensor automatically.

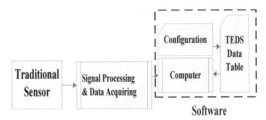

Fig.2 The achievement of virtual TEDS

Since TEDS achievement is restricted by the establishment of test system and influenced by various aspects like cost for achieving TEDS, TEDS intelligent sensor was less applied, especially in the existing complete traditional test system. This paper put forward the sensor normalization based on the unified model of signal treatment and intelligent unit, which makes the connection and application of sensor compatible with the traditional sensor, convenient for adding and removing and thus enabling to be applied in both newly established and traditional test systems flexibly.

Implementation plan of signal normalized intelligent unit

An intelligent unit was developed in view of the nonlinearity, temperature drift and inconsistency existed in traditional perception devices (sensor), as shown in Fig.3. Intelligent unit (IU) includes various modules or function, such as the logo of IEEE1415 standard device, humiture measurement, nonlinear data table, unified model as well as MCU achievement, normalized treatment and synthesis of signals, etc. The IU-based test system enables to the signal output by MEMS sensor normalized, thus achieving the quick connection and plug-and-play of the sensor.

The module of IEEE1415 standard device logo mainly stores the device type, unique numbers and range of the device, and other basic information for the sake of plug-and-play, intelligent management and compatibility with traditional TEDS. The humiture measurement module is an inbuilt humiture MEMS-sensing test device, which provides parameters for the environmental drift compensation. The nonlinear data table is the calibration data of the device, that is, the output value of the sensor when it was driven by standard 5-point or 7-point physical signal source. The unified model as well as MCU module achievement calculates the deviant through the unified model according to environmental temperature, humidity and a signal value and then superposes the deviant into signal through the synthesis treatment module.

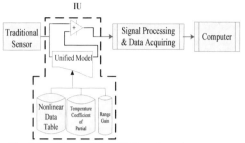

Fig.3 Intelligent Unit (IU) and testing system

Compared to the traditional test system, the IU-based test system serial connects an intelligent unit module without changing the interfaces of the sensor and subsequent test control system. The system can work normally when the IU was removed, only causing some accuracy deviation. The operating principle of IU-based test system was shown in Fig.4 and the physical model of IU-based test system was shown in Fig.5.

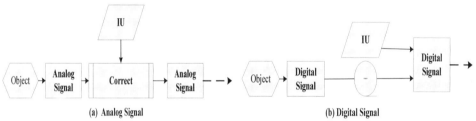

Fig.4 Framework of testing system based on IU

Fig.5 Sample of testing system based on IU

Unified model of signal

The unified model creates function based on the close-open and redundant theorem to unify calculations under different conditions into a single formula. For instance, the Fourier transform, short-time Fourier transform (STFT), Gabor transform, wavelet transform (WT), Wigner-Ville distribution, Wigner bispectrum are unified through 1-4 control parameters, thus making the calculation, handover and management easier and more convenient.

For the convenience of IU input, the proportionality coefficient, inconsistency and nonlinear data are unified into an individual nonlinear table which can be obtained through calibration (Fig.6). The length of the table can be extended dynamically, where t_0 is the temperature reference point; b_1 and b_2 are temperature calibration coefficients; N is the length of the table; Unit is the data unit; x_i and y_i are the theoretical output and practical output. If the simplification only takes the temperature offset into account, the established unified model function is shown as equation (1), where a_i is the nonlinear compensation coefficient of the sensor that calculated dynamically through x and x_i.

$$f(x) = x + (\sum_{i=1}^{N} a_i y_i + b_1(t - t_0) + b_2(t - t_0)^2 - x) \tag{1}$$

$$y_i = F(x_i) \tag{2}$$

t_0	
b_1	
b_2	
N	
Unit	
x_1	y_1
x_2	y_2
......	......
x_N	y_N

Fig.6 Table header for IU

Conclusion

(1) Plug-and-play of IU. IU doesn't change the input/output interface and its interface electrical characteristics are in consistent with that of sensor without IU, which can update quickly and increase the accuracy as well as performance without changing the existing test system structure.

(2) Reliability and maintainability design. The IU corrects errors through the unified model. The system can work normally when the IU is removed and won't be failed due to the IU breakdown. The system possesses higher reliability and maintainability.

(3) TEDS standard compatibility. The first part and second part of IU header are designed in accordance with IEEE145 international standard, which are compatible with hardware TEDS sensor. The test systems supporting TEDS sensor also can use sensor system with IU.

(4) This paper carries out various tasks, including designing and testing the IU as well as its parts, correcting and calibrating input/output errors, refining and verifying the unified model, determining the IU standard, etc.

Acknowledgment

This work was supported by the National Natural Science Foundation of China (No.51205431) and Chongqing Scientific and Technological Program (No.CSTC2012gg-yyjs70012).

References

[1] Reina Nieves Álvaro, Martnez Madrid Natividad, Seepold Ralf and etc. UPnP Service to Control and Manage IEEE 1451 Transducers in Control Networks[J].IEEE Transactions on Instrumentation & Measurement.2012,61(3):791-800

[2] Ye Xiangbin,LI Wen,YANG Xue,and etc. Design of Plug-and-play Smart Transducer Interface Module[J].Instrument Technique and Sensor,2009,10:28-30

[3] Wu Zhongjie,Lin Jun,Li Ye, and etc. Design of networked capable application processor for IEEE1451 standard smart sensor [J].Transducer and Micro-system Technology, 2006, 25(6):85-88.

[4] Sun Haijun. Study on Transforming Method from Traditional Transducer to Intelligent Transducer[J].MECHANICAL ENGINEERING & AUTOMATION,2007,145(6):182-184

[5] Kim Jeong-Do, Lee Jung-Hwan, Ham Yu-Kyung and etc. Sensor-Ball system based on IEEE 1451 for monitoring the condition of power transmission lines [J].Sensors & Actuators A: Physical, 2009, 154(1):157-168.

[6] Song Guangming, Song Aiguo, Huang Weiyi. Distributed measurement system based on networked smart sensors with standardized interfaces [J].Sensors & Actuators A: Physical, 2005, 120(1):147-153.

Advanced Materials Research Vol. 819 (2013) pp 254-258
© *(2013) Trans Tech Publications, Switzerland*
doi:10.4028/www.scientific.net/AMR.819.254

The Error Caused By Order Analysis for Rolling Bearing Fault Signal

Weidong Cheng[1,a], Robert X. Gao[2], Jinjiang Wang[2], Tianyang Wang[1], Weigang Wen[1], Jianyong Li[1]

[1]School of Mechanical Electronic and Control Engineering, Beijing Jiaotong University, Beijing 100044, China

[2]Department of Mechanical Engineering, University of Connecticut, Storrs, CT 06269, USA

[a]wdcheng@bjtu.edu.cn (corresponding author)

Keywords: rolling element bearing, varying rotational speed, order analysis, spectral error

Abstract. Defect diagnosis of rolling element bearings operating under time-varying rotational speeds entails order tracking and analysis techniques that convert a vibration signal from the time domain to the angle domain to eliminate the effect of speed variations. When a signal is resampled at a constant angular increment, the amount of data padded into each data segment will vary, depending on the rate of change in the rotational speeds. This leads to changes in the distance between the adjacent impulse peaks, and consequently, the result of order analysis. This paper presents a quantitative analysis of key factors affecting the accuracy of order analysis on rolling element bearings under variable speeds. An analytical model is established and simulated. The effects of speed variation, instantaneous speed, angular interval between impulses, and the rising time of impulse are specified. It is concluded that the results of order analysis will be smaller as the rotational speed increases, and becomes larger when the speed decreases. Furthermore, the error is larger under low speeds than high speed.

Introduction

There widely exist the equipments which operate under intermittent state in the industry of machinery, energy, petrochemical engineering, transport, national defense and so on. Electric shovel, a kind of large scale excavator for mining, is a proper example, and it works periodically under intermittent mode with its bucket. The whole working cycle consists of filling-up, rotating to the unloading place，unloading and turning back to the digging location. The detail working procedure means that the lifting gear cooperates with the crowding mechanism to finish the loading action, and the swing mechanism coordinates with the running gear to complete the unloading task. Transmission gear in every working portion operates intermittently, and the rotational speed of the axles and the rolling element bearings in the gear box changes all the time, in a pattern of keeping static – increasing speed- continuous speed fluctuating - reducing speed - keeping static. The rolling element bearings in this kind of equipment almost never run at a constant rotational speed, and they work at a large range of speed variation. Intermittent machine exists universally in a huge quantity and covers extensively in lots of industry. Many of them need to work reliably, safely and continuously, to be monitored working condition. For diagnosing the fault of rolling element bearings in this kind of machine, it needs to extract the fault characteristics from the vibration signal with changing rotational speed on a large scale [1].

The fault diagnosis of rolling element bearing is widely studied by considerable researchers, and many algorithms have been proposed [2,3]. Among them, the envelope analysis is the most basic and effective one. The essence steps of this algorithm are doing band filter to the vibration signal at higher frequency, demodulating the signal to obtain the envelope, then doing spectral analysis based on the envelope. The fault information carried on high frequency will be appeared in the spectrum of envelope. The envelope analysis is a practical and effective technique to detect the faults of bearings when the speed is fixed. Although this technique has been presented for almost 40 years, most of new approaches are still improvement and enhancement version according to it.

Under changing rotational speed, order tracking and order analysis can be used to eliminate the influence of speed variation [4]. Order tracking is the algorithm to resample the vibration signal at a constant angle increment according to a rate proportional to the shaft speed, and order analysis is doing spectral analysis based on the resampled single. To realize order tracking technique, the problems of cost and mounting space are two key points. Computed Order Tracking (COT) which makes use of tachometer and interpolation algorithm to realize order tracking is proposed to solve this problem. The development of COT will help order tracking to lower the dependency to hardware [5,6]. In real application, higher sampling frequency of keyphasor pulse [7] and cubic spline fitting [8] are used to improve the accuracy of resampling signal.

However, using order tracking to deal with the fault signal of rolling element bearing is of some particularity, and different from other applications of order tracking. The method to realize resample will apply padding data points to the data segments under relatively high rotational speed, and not reducing data points from segment under relatively low speed. But the method of padding sampling data has its own problem which is padding data in different rotation will deform the shape of envelope. The deformation will elongate or shorten the distance between adjacent impulse peaks on the envelope. Most of all, this envelope deformation will further affect the envelope specturm.

This paper mainly discusses the key factors which result in the deformation of envelope, and how much they affect outcome of the order analysis.

Waveform deformation caused by angular resampling

In this section, we will discuss effects of resampling algorithm, and assume that the acceleration is constant in each revolution.

The impulses generated by tachometer can be looked on as the coordinate reference in angular domain. The data between two impulses are the vibration signal in one rotation. At first, these data are transformed into angular signal based on equation (1).

$$\phi = \int_0^t \frac{n(t)}{60} dt \tag{1}$$

Where, ϕ represents the angle the rolling bearing has rotated pass at corresponding time t, which can be calculated by the integration of rotational speed by time. $n(t)$ is the instantaneous rotational speed of bearing with the unit of rpm. Then construct Cubic spline curve according to the angular signal. At last, resample the whole curve by sampling frequency $f_{s\phi}$,

$$f_{s\phi} = \frac{f_{st}}{n_{min}/60} \tag{2}$$

Where, f_{st} is the sampling rate whose unit is Hz, and n_{min} is the minimum of rotational speed.

Just as the signal in time domain, the interval between adjacent sampling data of signal in angular domain is constant. To keep things simple, a triangular signal is used to explain the wave deformation during the transformation.

Constant speed situation (a=0)

Fig. 1a shows the first situation in which the rotational speed does not change with time (acceleration equals to zero). Equation (1) is used to transform the horizontal axis of signal from time domain to angle domain. In here, there exist a constant ratio between angle and time because of fixed speed.

The relationship between angle and time is displayed in Fig.1b. Supposing that there exist 5 impulses in one revolution, these impulses are uniform distribution in time domain (Fig.1c). The impulses in angle domain are uniform distribution too (Fig. 1d), and waveforms have not changed.

Speed-up situation (a>0)

Fig.1e shows the situation whose rotational speed goes up with time (acceleration equals to 30 r/s^2 and beginning rotational speed equals to 60 rpm). According to equation (1), the horizontal axis can be transformed from time to angle, and the relationship between angle and time during speed-up period is displayed in Fig.1f. We can see the nonlinear relationship between two factors because of the integral operation.

The interval between starting points of adjacent impulse is only determined by the angle passed through. If the speed increases, the interval will change from longer to shorter just like Fig.1g. From Fig.1h, the resampled version of Fig.1g, we can see that the starting points become uniform distribution in each rotation, but the lengths of the impulses have been expanded with high rotational speed. The expansion will be larger as the rotational speed increases. As a consequence, the peak points of impulse are shifted a little to the right too.

Speed-down situation (a<0)

Fig.1i shows the situation of speed-down in which acceleration equals to -30 r/s^2 and the initial rotational speed equals to 2100 rpm. Fig.1j shows that the angle changes with time, and the curve is convex which is just opposite direction compared to Fig.1f. If the speed decreased, the interval between starting points will change from shorter to longer shown in Fig.1k. From Fig.1m, the resampled version of Fig.1k, we can see that, not only the starting points become uniform again, but also the lengths of the impulses have been contracted at lower rotational speed. The contraction will be larger when the rotational speed decreases. As a consequence, the peak points of impulses are shifted a little to the left too.

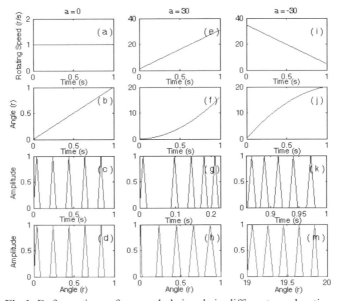

Fig.1. Deformations of resampled signals in different acceleration.

Quantitative analysis of the change of peak points interval

Suppose the rotational speed of rolling bearing changes with constant acceleration a, and the peak points of adjacent impulses equals to each other, the rising time t_r of them are the same too. As shown in Fig. 2.

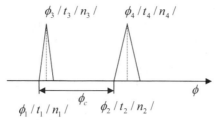

Fig.2. Relationship between adjacent impulses.

We can see that the angular interval between the start points of two adjacent impulses and the one between the peak points of the two adjacent impulses are different. And this difference is named as $\Delta\phi_c$ which is actual the angular interval offset.

$$\Delta\phi_c = (\phi_4 - \phi_3) - (\phi_2 - \phi_1) = (\sqrt{2a\phi_c + n_1^2} - n_1)t_r \qquad (3)$$

Following equation calculate the ratio of this difference towards the total angle interval.

$$\frac{\Delta\phi_c}{\phi_c} = \frac{(\sqrt{2a\phi_c + n_1^2} - n_1)t_r}{\phi_c} \times 100\% \qquad (4)$$

When the acceleration changes from -100 to 100 r/s^2, Fig. 3 represents how the acceleration affect the peak points interval, in which instantaneous rotational speed n_1, the starting point interval ϕ_c and rising time t_r are respectively $10r/s^2$, $1r$ and $0.01s$. From the Fig. 3, we can see that the expansion percentage deviation increases with the acceleration whose value is positive, and the contraction percentage deviation increases when the acceleration increases in the opposite direction.

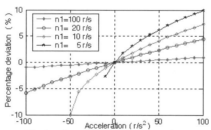

Fig.3. The effect of acceleration.

Based on the simulation result shown in Fig. 3, the error increases when acceleration becomes larger; the instantaneous rotational speed of the impulse affects the error more when the instantaneous rotational speed is relatively low.

Conclusion

In this paper, we have discussed that the angle domain resampling may lead to the deformation of envelope signal, and this deformation will cause the variation of angular interval between the peak points of impulses, which then affects the envelope spectrum. It has been proved that the main factors which determine the deviation contain the acceleration of rolling element bearing, the corresponding instantaneous rotational speed of impulse, the angular interval between the adjacent impulses and the rising time of impulse.

The magnitude and direction of acceleration affect on the deformation of envelope signal. The deformation of envelope will be larger as the acceleration increases, and it will be smaller on the contrary. The outcome of order analysis will be smaller than that in the acceleration equaling to zero. In addition, if the acceleration is negative, the deformation of envelope will be bigger as absolute value of the acceleration increases, and the result of order analysis will be larger than that in the acceleration equaling to zero.

The corresponding instantaneous rotational speed of impulse will affect on the deformation of envelope. The deformation in the period of low speed is larger than the one in high speed. The machinery worked under intermittent situation always runs in low rotational speed, so the deviation introduced by this paper should be given more attention in real signal processing.

The interval between adjacent impulses will affect on the deformation of envelope. For rolling element bearing, the angular interval of fault in outer race is larger, and it is smaller in inner race fault, so the effect on outcome of envelope spectrum is different among different faults.

The rising time affects on the deformation of envelope. It is important to accept the existence of the rising time, and do not assume t_r as zero in processing signal. In time domain, the rising time won't change with rotational speed, but its corresponding angular version will change. The deformation of envelope will be larger as the rising time increases, and it will be smaller on the contrary.

Acknowledgment

The authors gratefully acknowledge funding provided to this research by the National Science Foundation of China under Grant No. 51275030. Experimental support from EMS Lab in department of mechanical engineering, University of Connecticut (USA) is appreciated. The authors would also like to thank the reviews for their valuable suggestions and comments that have helped improve this paper.

Reference

[1]Weidong Cheng, Tianyang Wang, et al., Anomaly detection for equipment condition via frequency spectrum entropy, Advanced Materials Research, 2012, 433-440 (3753): 3753-3758

[2]Ruqiang Yan, Robert X. Gao, Approximate entropy as a diagnostic tool for machine health monitoring, Mechanical Systems and Signal Processing, 21(2007): 824-839

[3]Jinjiang Wang, Robert X. Gao, et al., A hybrid approach to bearing defect diagnosis in rotary machines, CIRP Journal of Manufacturing Science and Technology, 5(2012): 357-365

[4]Robert B.Randall, Jerome Antoni, Rolling element bearing diagnostics - A tutorial, Mechanical Systems and Signal Processing, 25(2011): 485-520

[5]K.R.Fyfe, E.D.S.Munck, Analysis of computed order tracking, Mechanical Systems and Signal Processing, 11(1997): 187-205

[6]K.M.Bossley, R.J.Mckendrick, Hybrid computed order tracking, Mechanical Systems and Signal Processing, 13(1999): 627-641

[7]Anders Brandt, Thomas Lagö, et al., Main principles and limitations of current order tracking methods, Sound and Vibration, March 2005, 19-22

[8]P.N.Saavedra, C.G.Rodriguez, Accurate assessment of computed order tracking, Shock and Vibration, 13 (2006): 13-32

Advanced Materials Research Vol. 819 (2013) pp 259-265
© (2013) Trans Tech Publications, Switzerland
doi:10.4028/www.scientific.net/AMR.819.259

Longitudinal Control for AUV with Self-adaptive Learning Law

Xiujun Sun[1, a] and Yan Yang[2,b]

[1]National Ocean Technology Center, No.219 Jieyuanxi Road, Nankai District, Tianjin, China 300112

[2]Tianjin Institute of Urban Construction, No.26 Jinjing Road, Xiqing District, Tianjin, China 300384

[a]sunxiujun@yahoo.com, [b]yangyan_0103@163.com, (Corresponding Author: Yan Yang)

Keywords: Mini AUV, BP Neural Network, Pitch Attitude Tracking, Depth Control.

Abstract. A mini AUV (Autonomous Underwater Vehicle) with cross shaped rudders and one single thruster is presented, which features high maneuverability due to the intelligent control algorithm. A single variable PID neural network controller is also proposed, which is utilized to maintain attitude for the vehicle. In order to testify feasibility of the control methodology, a spatial motion mathematic model is constructed and linear equations that indicate the relation between attitude angles of vehicle and deflection angles of rudders is deduced firstly. Subsequently, the neural network PID controller is developed according to the deduced equations and the attitude control simulation of the vehicle with this controller is conducted. Taking actual and desired attitude angles of the vehicle as input and deflection angles of the rudders as output, this controller performs self-adaptive update for 9 synaptic weights through back-propagation algorithm and employs the converged weights to calculate the appropriate deflection angle of each rudder.

Introduction

For most autonomous underwater vehicles in the world, the basic configuration is consisted of rudders and thrusters. Generally, rudders are mounted by the stern or/and bow of vehicle for vehicle steering and attitude keeping, while some other rudders evolve to be fixed ones (here called fins) just to keep sTab. during cruising voyage. Sometimes, there is no rudders even fins configured on the vehicle body but several thrusters mounted on vehicle in orthogonal directions and different positions to maneuver the vehicle. One innovative method is to use one single vectored thruster to have the vehicle actuated. The vectored thruster can implement either vehicle driving or attitude maintaining [1,2]. Generally speaking, rudders (including fins) and thrusters are crucial components for motion control of autonomous underwater vehicles.

Spatial Motion Modeling

Mathematic model for depicting spatial motion of vehicle plays a key role in numerical simulation and performance analysis [3]. Fixed reference frame and dynamic reference frame is assigned as Fig. 1. For fixed frame E-ξηζ, ξη-plane superposes on the surface of sea, ξ-axis points to the magnetic north, ζ-axis is orthogonal to ξη-plane and points to the center of earth, η-axis rotates for 90 degrees by right-hand rule, E -origin is a point for vehicle launch. For dynamic frame O-xyz, x-axis aligns along the longitudinal axis of vehicle and points to the stern, y-axis heads to the starboard of vehicle, z-axis arranged by right hand rule pointing to the bottom, and O-origin coincide with buoyancy center of vehicle [4,5].

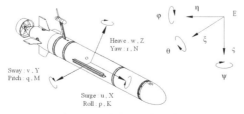

Fig. 1 Fixed reference frame and dynamic reference frame assignment

Generalized position vector $L=[\xi\ \eta\ \zeta\ \varphi\ \theta\ \psi]^T$ is defined for depiction of position and orientation with respect to fixed frame, generalized velocity vector $V=[u\ v\ w\ p\ q\ r]^T$ for expression of linear and angular velocity with respect to fixed frame and generalized force vector $F=[X\ Y\ Z\ K\ M\ N]^T$ denotation of force and moment with respect to dynamic frame. According to moment theory and frame transformation, the dynamic model of vehicle can be deduced.

For simplification in dynamics model construction, the motions in longitudinal and horizontal plane are assumed to be decoupled. With detailed deduction, we derive the longitudinal dynamics model as

$$AS = B \tag{1}$$

where,

$$A = \begin{pmatrix} m+\lambda_{11} & 0 & 0 & 0 & 0 & 0 \\ 0 & m+\lambda_{33} & \lambda_{35} & 0 & 0 & 0 \\ 0 & \lambda_{35} & J_Y+\lambda_{55} & 0 & 0 & 0 \\ 0 & 0 & 0 & 1 & 0 & 0 \\ 0 & 0 & 0 & 0 & 1 & 0 \\ 0 & 0 & 0 & 0 & 0 & 1 \end{pmatrix} \tag{2}$$

and

$$B = \begin{bmatrix} K_T\rho D^4 n^2 + \dfrac{1}{2}\rho(u^2+w^2)A_D C_X(0) \\[2mm] -mup + \dfrac{1}{2}\rho(u^2+w^2)A_D\left(C_Z^a \arctan\left(-\dfrac{w}{u}\right) + C_Z^p\dfrac{pL}{\sqrt{u^2+w^2}} + C_Z^{\delta_h}\delta_h\right) \\[2mm] mgz_G\sin(\theta) + \dfrac{1}{2}\rho(u^2+w^2)A_D L\left(C_M^a \arctan\left(-\dfrac{w}{u}\right) + C_M^p\dfrac{pL}{\sqrt{u^2+w^2}} + C_M^{\delta_h}\delta_h\right) \\[2mm] u\cos(\theta)-w\sin(\theta) \\[1mm] u\sin(\theta)+w\cos(\theta) \\[1mm] p \end{bmatrix} \tag{3}$$

In addition, $S=[u\ v\ q\ \xi\ \zeta\ \theta]^T$ is critical status variables in longitudinal plane.

The parameters can be found in reference [6], and the values for these parameters are listed in Tab. 1 and Tab. 2.

<table>
<tr><th colspan="2">Tab. 1 Physical parameters</th></tr>
<tr><td>Mass of Vehicle, m</td><td>120 kg</td></tr>
<tr><td>Length of Vehicle, L</td><td>3.2 m</td></tr>
<tr><td>Sectional area, A_D</td><td>0.05 m^2</td></tr>
<tr><td>Moment of Inertia, J_Y</td><td>80 kg·m^2</td></tr>
<tr><td>Diameter of Prop, D</td><td>0.2 m</td></tr>
<tr><td>Thrust Coefficient, K_T</td><td>0.023</td></tr>
<tr><td>Gravity Center, z_G</td><td>-0.02 m</td></tr>
<tr><td>Water Density, ρ</td><td>1 kg/m^3</td></tr>
<tr><td>Rotation Velocity, n</td><td>300 r/m</td></tr>
</table>

<table>
<tr><th colspan="2">Tab. 2 Hydrodynamic coefficients</th></tr>
<tr><td>Added Mass along x Axis, λ_{11}</td><td>12 kg</td></tr>
<tr><td>Added Mass along z Axis, λ_{33}</td><td>180 kg</td></tr>
<tr><td>Added Moment of Inertia, λ_{35}</td><td>8.6 kg·m^2</td></tr>
<tr><td>Added Moment of Inertia, λ_{55}</td><td>110 kg·m^2</td></tr>
<tr><td>Main Drag Coefficient, C_X</td><td>-0.24</td></tr>
<tr><td>Attack angle Related Coefficient, C_Z^a</td><td>2.8</td></tr>
<tr><td>Pitch Related Coefficient, C_Z^p</td><td>0.324</td></tr>
<tr><td>Rudder Related Coefficient, $C_Z^{\delta h}$</td><td>1.01</td></tr>
<tr><td>Attack Angle Related Coefficient, C_M^a</td><td>0.42</td></tr>
<tr><td>Pitch Related Coefficient, C_M^p</td><td>-0.11</td></tr>
<tr><td>Rudder Related Coefficient, $C_M^{\delta h}$</td><td>-1.2</td></tr>
</table>

Note that, the chosen control input is the velocity of rotation for propeller n and the deflection angle of horizontal rudders δ_h, and the control output is the depth ζ and pitch angle of vehicle θ.

Pitch Controller with Neural Network

Attitude control like pitch angle and yaw angle approximation is one-input and one-output control system, it can be controlled by PID neural network and get a good control result without identifying the complex non-linear controlled plant [7~9].

For AUV, pitch is controlled by turning the two horizontal rudders, and yaw is adjusted by the two perpendicular rudders while the vehicle is sailing. PID neural network based pitch control model aims to minimize the disparity between the actual output of control plant (pitch angle from magnetic compass TCM3) and the desired output, and the PID neural network controller will change and calculate the control quantity of plant (deflection angle of two horizontal rudders). During the training phase of PID neural network, the synaptic weights are adjusted to calculate a proper control quantity and complete the actual output of plant to approach the desired output.

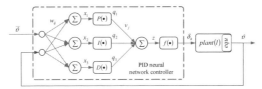

Fig. 2 Pitch control model with PID neural network

Set $\bar{\theta}$ as desired pitch angle of vehicle, and θ as actual output of pitch angle. Correspondingly, deflection angle of horizontal rudders δ_h acts as control quantity. The speed of velocity is assumed to be a constant 300r/min.

As shown in Fig. 2, there are 2 inputs, 3 neurons in the hidden layer and only one neuron in output layer. Input of the ith neurons of hidden layer.

$$x_i(k) = \bar{\theta}w_{1i} + \theta w_{2i}, \ i = 1,2,3 \tag{4}$$

The outputs of proportional neuron, integral neuron and derivative neuron in hidden layer can be calculated.

PID neural network controller combines with the controlled plant to be a generalized network. Back-propagation algorithms are used for online learning to minimize the index function defined as follows.

$$E(k) = \frac{[\theta(k) - \bar{\theta}(k)]^2}{2} = \frac{e(k)^2}{2} \tag{5}$$

The training algorithms are comprised of feed forward and back propagation phases. Set the weights as random nonzero small numbers at the beginning of training PID neural network. The weights of network should be adjusted to develop a proper nonlinear mapping between input and output of the network as follows.

After the k times of calculation in training process, the weights between hidden layer and output layer can be represented as

$$v_i(k+1) = v_i(k) - \eta_2 \frac{\partial E(k)}{\partial v_i(k)} \tag{6}$$

where, η_2 is the study step-size. And

$$\frac{\partial E}{\partial v_i} = \frac{\partial E}{\partial \theta} \cdot \frac{\partial \theta}{\partial \delta_h} \cdot \frac{\partial \delta_h}{\partial z} \cdot \frac{\partial z}{\partial v_i} \tag{7}$$

after k times of adjustments, the weights between input layer and hidden layer can be written as

$$w_{ij}(k+1) = w_{ij}(k) - \eta_1 \frac{\partial E}{\partial w_{ij}}, \ j = 1,2 \tag{8}$$

where, η_1 is learning step length, and

$$\frac{\partial E}{\partial w_{ij}} = \frac{\partial E}{\partial \theta} \cdot \frac{\partial \theta}{\partial \delta_h} \cdot \frac{\partial \delta_h}{\partial w_{ij}} \tag{9}$$

where

$$\frac{\partial \delta_h}{\partial w_{ij}} = \frac{\partial \delta_h}{\partial z} \cdot \frac{\partial z}{\partial q_i} \cdot \frac{\partial q_i}{\partial x_i} \cdot \frac{\partial x_i}{\partial w_{ij}} \tag{10}$$

Depth Controller with Neural Networks Controller

To have a gradual change of the pitch angle while cruising from original depth to predetermined depth, the pith angle is regulated by the following function.

$$\theta = \begin{cases} \dfrac{|\theta_{min}|}{2}\cos\left(2\pi\cdot\dfrac{(d-\bar{d})}{|d_1-\bar{d}|}\right)-\dfrac{|\theta_{min}|}{2} & d_1 < d \le \bar{d} \\[3mm] \dfrac{|\theta_{max}|}{2}-\dfrac{|\theta_{max}|}{2}\cos\left(2\pi\cdot\dfrac{(d-\bar{d})}{|d_1-\bar{d}|}\right) & \bar{d} < d < d_2 \end{cases}$$

(11)

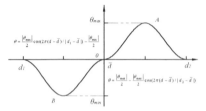

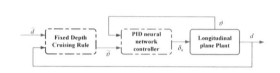

Fig. 3 Fixed depth control rule while cruising from original depth to specified depth

Fig. 4 Depth control model with PID neural network

As is shown in Fig. 3, $\bar{d}$ represents the specified depth, d_1 and d_2 are the original depth of vehicle, θ_{max} and θ_{min} are respectively the maximum and minimum pitch angle the vehicle can take. If the depth difference between the original depth and specified depth is much larger, we can prolong the range in A and B point in that the vehicle can profile the water upward or downward rapidly. For an instance, the vehicle departures from depth d_2, arrives at preset depth d and cruises forward with little oscillation of depth. In this procedure, the pitch angle of vehicle is changed according to the actual depth and the preset depth.

As is shown in Fig 4, depth control model with neural network is constituted with two feedback loops. One is for pitch angle control and the other for depth approximation. The depth loop and pitch loop coordinates each other to fulfill the depth control mission. Depth control rule calculates the desired pitch angle and pitch loop outputs the deflection angle of horizontal rudder.

Simulation with Pitch Control Model

By presetting the initial conditions as listed in Tab. 3, the simulation can be conducted.

Tab. 3 The initial values for simulation with neural networks

Step size, η_1	0.02
Step size, η_2	0.02
Given pitch angle, $\bar{\theta}$	$\pi/9$ rad
Initial pitch angle, θ_0	0 rad
Initial surge velocity u_0	0.01 m/s
Initial dive velocity, w_0	0 m/s
Initial pitch velocity, p_0	0 rad/s

It is of important to select the original weights for rapid convergence. To find the proper initial weights for rapid convergence and moreover the convergence law, a set of simulations was conducted with random initial synaptic weights. By analyzing varied simulating cases, we could derive the elecTab. scope for initial weights. The simulations can be divided into three categories, one is converged, the second is non-converged and the third is locally converged.

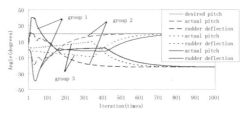

Fig. 5 Convergence iterations in pitch control simulation in longitudinal plane

As shown in Fig. 5 is three converged simulations, the approximation velocities is different and the group 2 has the fastest speed. The common character for three simulations is that they initially counterturn the horizontal rudder and finally deflect the appropriate rudder angles through self-adaptive weights update.

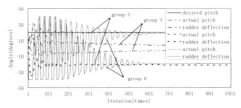

Fig. 6 Iterations of oscillation convergence and local convergence in pitch control simulation

As shown in Fig. 6 is three groups of simulation curves. For group 4 and 6, the controller gives the deflection angle of horizontal rudders to cause the vehicle to approach the desired pitch angle initially, then overshoot that, finally get to the desired angle after many times of oscillation. However, simulation such as group 5 gets trapped in local minimum, and no longer achieves desired pitch maintenance.

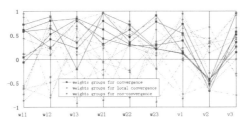

Fig. 7 Statistic figure of trained synaptic weights groups in convergence, local convergence and non-convergence simulation

Parts of trained weights groups of non-convergence, local convergence and convergence are listed in different color and line style as shown in Fig.7. This statistic figure indicates that we can choose v2 as a negative number at about -0.5, and others for positive numbers to obtain a converged control simulation. But the converged weights in this figure are only instructional for your selections and do not hold all possible synaptic weights for convergence simulation. For an instance, the weights in group 7 in Fig. 7 will result in a rapid convergence. This is the attributes of neural network that there are lots of revolutions for the same control system

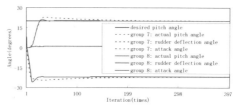

Fig. 8 Simulation iteration in pitch control by using the trained weights as initial weights

The best approach to get a high performance control in pitch maintenance is to employ the well trained synaptic weights group in the convergence simulation for a new simulation. As shown in Fig. 8, the desired pitch angle is achieved after about 100 times of iteration for group 7 and 40 times for group 8. The initial weights used in this simulation are listed in Tab. 3. These two groups of weights are converged weights and can not self-adaptively update any more. However, there is an obvious distinction in convergence performance of pitch control for these two groups' weights.

Tab. 4 The initial weights for rapid convergence in pitch simulation

Group 7	w_{11}	0.06	w_{21}	-0.78	v_1	-0.15
	w_{12}	1.02	w_{22}	-0.76	v_2	-0.42
	w_{13}	-0.53	w_{23}	-0.86	v_3	-0.27
Group 8	w_{11}	0.61	w_{21}	0.43	v_1	0.39
	w_{12}	0.24	w_{22}	0.27	v_2	-0.52
	w_{13}	0.79	w_{23}	0.29	v_3	0.35

Simulation with Depth Control Model

To escape the singular solution, we set the original depth d_1 is 0 meter and the original pitch angle 1 degree. Assuming that the specified depth is 8 meters and the minimum pitch angle θ_{min} is 20 degrees, we can derive the simulations as follows (Fig 9). The converged weights of the simulation are listed in Tab 4.

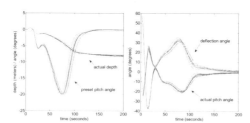

Fig. 9 Simulation of depth transition from sea surface to the depth of 8 meters

Tab. 5 The converged weights in 8 meters depth control simulation

w_{11}	0.88	w_{21}	0.89	v_1	-0.83
w_{12}	0.77	w_{22}	0.25	v_2	-0.21
w_{13}	-0.52	w_{23}	-0.59	v_3	-0.93

Summary

The developed attitude and depth control model with neural network controller are validated to be feasible and reliable by the simulation and experiment, and the approach to select initial synaptic weights and the rough regime are presented. Unlike traditional PID control rule, experiments should be done many times to acquire a set of proper proportional, integral, derivative values. PID single variable neural network based control algorithms is self-adaptive and can search the proper parameters of network according to the changefully conditions.

References

[1] P. Antonio, O. Paulo, S. Carlos, B. Anders. "MARIUS: an autonomous underwater vehicle for coastal oceanography," Int. J Robotics and Automation Magazine, vol 3, no 1, pp. 46-59, 1997.

[2] G. Ayela, A. Bjerrum, S. Bruun, A. Pascoal. "Development of a self-organizing underwater vehicle – SOUV," Proc 21st MAST-Days and Euromar Conference, Sorrento, Neapolitan, 1995, vol 3, pp. 23-26.

[3] J. Yuh. "Design and control of autonomous underwater robots: a survey," J Autonomous Robots, vol 8, pp. 7-24.

[4] S. D. McPhail, M. Pebody. "Navigation and control of an autonomous underwater vehicle using a distributed networked control architecture," Int. J Underwater Technology, 1988, vol 23, pp. 19-30.

[5] J. Anthony, D.L. Healey. "Multivariable sliding mode control for autonomous diving and steering of unmanned underwater vehicles," J Oceanic Engineering, vol 18, pp 12-16, 1993.

[6] K. Y. Pettersen, O. Egeland. "Time-varying exponential stabilization of the position and attitude of an underactuated autonomous underwater vehicle," J Transactions on Automatic Control, 1999, vol. 44, pp. 25-29.

[7] G. J. Bellingham, C. A. Goudey, T. R. Consi, J. W. Bales, D. K. Atwood. "A second generation survey AUV," Symp. Autonomous Underwater Vehicle Technology, 1994, pp. 148-155.

[8] V. Riqaud, J. M. Laframboise. "First steps in Ifremerís autonomous underwater vehicle program- a 3000m depth operational survey AUV for environmental monitoring", Proceedings of the Fourteenth International Offshore and Polar, 2004, pp. 201-206.

[9] J. Yuh, S. K. Choi, C. Ikehara. "Design of a semi-autonomous underwater vehicle for intervention missions (SAUVIM)", pp 63-68.

Advanced Materials Research Vol. 819 (2013) pp 266-270
© *(2013) Trans Tech Publications, Switzerland*
doi:10.4028/www.scientific.net/AMR.819.266

Approach to extraction of incipient fault features on unstable rotating rolling bearings based on time-frequency order tracking and SPWVD

Song Longlong[1, a], Song Degang[2, b], Cheng Weidong[1, c], Wang Taiyong[3, d], Su Kaikai[1, e]

[1] School of Mechanical, Electronic and Control engineering, Beijing Jiaotong University, Beijing, 100044, PR China

[2] CSR Qingdao Sifang Locomotive & Rolling Stock Co., Ltd, Qingdao, 266111, PR China

[3] School of Mechanical Engineering, Tianjin University, Tianjin, 300072, PR China

[a]song_bjtu@163.com, [b]songdegang@cqsf.com, [c]wdcheng@bjtu.edu.cn, [d]tywang@189.cn, [e]12116313@bjtu.edu.cn

Keywords: fault diagnose, rolling bearing, time-frequency order tracking, SPWVD, instantaneous frequency estimation

Abstract. Frequency-smear phenomenon caused by fierce rotating speed variation made it difficult to extract the fault features of the rolling bearing. An approach based on time-frequency order tracking and SPWVD was proposed in this paper. The influence of speed variation was reduced by resampling the time-domain non-stationary signal at constant angle increments with order tracking analysis. The precision of instantaneous frequency estimation (IFE) in time-frequency order tracking was improved by the use of Smoothed Pseudo Wigner-Ville Distribution (SPWVD). The simulation signal and experiment on test-rig revealed that the proposed method was more effective than the traditional order tracking in clarifying incipient fault.

Introduction

Rotary machinery condition monitoring is typically based on vibration analysis and shaft run-ups and run-downs are of particular interest as these modes of operation can highly previously unobservable system defects[1]. The spectral contents of emitted vibration signals are traditionally analyzed to ascertain the current condition of the monitored process. Vibration signals produced from rotating machinery are speed-dependent. During periods of fast acceleration and deceleration, frequency-smear phenomenon may happen, thus the traditional FFT analysis become invalid. Order track analysis was proposed to avoid the adverse influence of the speed variation. K.M Bossley et al. proved the effectiveness of the hybrid computed order tracking in testing speed-related vibrations[1]. Qin Shuren et al. developed a virtual instrument in characteristic analysis of rotating machinery based on time frequency estimation[2]. Pan M Ch and Lin Y F researched the use of Void-Kalman-filtering in order track analysis to extract fault features of rolling bearing[3].

Order tracking analysis provided an approach to diagnose rolling bearings whose rotating speed fluctuated, but the precision was limited because the instantaneous frequency estimation (IFE) was usually based on STFT. Because of the Heisenberg uncertainty principle, the frequency resolution of STFT is limited which influences the accuracy of IFE directly. Due to high frequency resolution and energy focus, this paper improved the accuracy of time-frequency order tracking by SPWVD. The simulation and diagnosing example analysis revealed that the proposed method is more effective in clarifying incipient fault.

Time-frequency order tracking analysis and the limitation of STFT in IFE

Order tracking analysis is a new approach in detecting rotating machines. One of the biggest advantages of order tracking analysis is that it provides successful results in non-stationary signals which will vary in frequency and amplitude with the rotation of a shaft[4].

The order could be defined as the frequency normalized by the shaft speed.

$$O = \frac{60f}{n} \tag{1}$$

Where O is the order, f (Hz) represents the frequency of observed vibration, and $n(rev/\min)$ denotes the rotating speed of the reference shaft.

The main steps of the time-frequency order tracking analysis include:

(1) Sampling the original vibration signals of the reference shaft by traditional synchronous sampling approach such as vibration acceleration transducers. The original vibration signal was signed as $x(t)$.

(2) Estimating the instantaneous frequency of $x(t)$ with time-frequency methods such as STFT, WVD et al.

(3) Calculating the speed curve of the reference shaft according to the relation between the instantaneous frequency $f(x)$ and the instantaneous rotating speed n.

(4) Obtaining the sampling-time marks of the even-angle sampling on the basis of the speed curve and to perform the even-angle sampling from the sampled vibration signals.

$$2\pi \int_{T_0}^{T_n} f(x)dt = m\Delta\theta \quad m = 1, 2, \cdots, L \tag{2}$$

Where T_n is the even-angle sampling-time marks, T_0 is the first sampling time mark, m denotes the serial number of the re-sampling, $\Delta\theta$ represents the angle increment and L is the number of re-sampling signals.

$$\Delta\theta = \frac{1}{2} \cdot \frac{2\pi}{O_{max}} \tag{3}$$

$$x(T_n) = x(t_i) + \frac{x(t_{i+1}) - x(t_i)}{t_{i+1} - t_i}(T_n - t_i) \quad t_i \leq T_n \leq t_{i+1} \tag{4}$$

Where O_{max} represents the maximum of orders and $x(T_n)$ denotes the even-angle re-sampled cyclostationary signal.

(5) Data processing approaches such as order spectrum analysis, envelop analysis et al. would be applied to the even-angle re-sampling signals to extract the fault features.

IFE is the most important step in time-frequency order tracking analysis which is usually based on STFT in the past. STFT has advantages of fast computation speed and clear principle, but it was limited by the Heisenberg uncertainty principle. STFT could not reach high time resolution and frequency resolution at the same time. For a given time-domain signal $x(t)$, if $\lim_{t\to\infty}\sqrt{t}x(t) = 0$, $\Delta_t\Delta_\Omega \geq 0.5$. The equal sign is correct if and only if $x(t)$ was a Gaussian signal. Where Δ_t represents the effective time width and Δ_Ω denotes the effective frequency bandwidth. So the product of Δ_t and Δ_Ω is a fixed value when the signal is given. On the other hand, STFT analyses time-varying vibration signals by cutting it into short segments with windows. Short segments in each window are still analyzed as stationary signals. Actually, STFT only suits for the processing of vibration signals whose rotating speed fluctuated slowly. Due to its high temporal and frequency resolution and high energy focus of Wigner Ville distribution (WVD), this paper improved the accuracy of instantaneous frequency estimation by WVD and SPWVD was also imported to decrease the cross-term interference in WVD.

The improvement of instantaneous frequency estimating with SPWVD

The Wigner-Ville distribution (WVD) is part of the Cohen class of distribution. Some good properties assured it has wide interest in non-stationary signal analyzing. WVD is a time-frequency energy density computed by correlating $f(t)$ with a time and frequency translation of itself. This avoids any loss of time-frequency resolution. The discrete WVD could be described as followed:

$$WVD_x(t, \omega) = \frac{1}{2\pi}\int_{-\infty}^{+\infty} x^*(t - \frac{1}{2}\tau)x(t + \frac{1}{2}\tau)e^{-i\tau\omega}d\tau \lim_{x\to\infty} \tag{5}$$

Where $x(t)$ is the obtained signal, $x^*(t)$ denotes the conjugate complex of $x(t)$. It is highly related with the data front and back. The non-locality assures that the time-frequency characteristic of WVD is much better than that of STFT. A multi-component chirp signal $x(t)$ consists of two mono-component chirp signals.

$$x(t) = \cos(2\pi f_0 \times \left(\frac{f_1}{f_0}\right)^{\frac{1}{t_1}} t + \varphi_0) + \cos(2\pi \omega_0 \times \left(\frac{\omega_1}{\omega_0}\right)^{\frac{1}{t_1}} t + \psi_0) \tag{6}$$

Where $f_0 = 10Hz$, $f_1 = 400Hz$, $\omega_0 = 50Hz$, $\omega_1 = 500Hz$, the initial phase $\varphi_0 = 0$, $\psi_0 = 0$, $t_0 = 0$, $t_1 = 5s$.

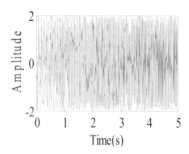

Fig.1 Time-domain waveform of $x(t)$

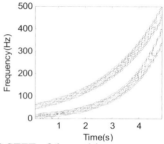

Fig.2 STFT of the multi-component chirp signal $x(t)$

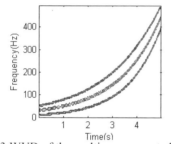

Fig.3 WVD of the multi-component chirp signal $x(t)$

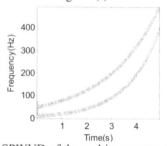

Fig.4 SPWVD of the multi-component chirp signal $x(t)$

Fig.2 showed the STFT of $x(t)$. Where a Hamming window length 256 was chosen, the overlapping ratio between segments was 50% and the sampling frequency was 1000Hz. STFT showed the frequency contents in $x(t)$, but it couldn't obtain the precise frequency at one point. WVD (Fig.3) of this signal was calculated to compare with STFT. The frequency bandwidth of WVD is narrower and more focused, which made it more conductive to estimate the trend of each component's frequency. But another problem of cross term interference occurred because $WVD[x(t)]$ was not the summation of $WVD[x_1(t)]$ and $WVD[x_2(t)]$.

$$WVD(x) = WVD(x_1) + WVD(x_2) + 2\,\mathrm{Re}\left[WVD(x_1 x_2)\right] \tag{7}$$

The IFE in time-frequency order tracking analysis was relied on searching the peaks of the frequency lines. So the searching result would deviate from the right course because of the cross term. The cross term interference must be restrained first.

Firstly, a window was added at time-axis to highlight the segments that near time t. Thus the segments that were far from time t were restrained effectively. The cross term interference was greatly restrained and the restraining extent could be adjusted by changing the length of the windows. In a similar way, windows could also be added at frequency-axis to restrain the cross term interference in frequency domain. When windows were added at both time-axis and frequency-axis, WVD was called Smoothed Pseudo Wigner-Ville Distribution (SPWVD).

$$SPWVD_x(t,\omega) = \frac{1}{2\pi}\int_{-\infty}^{+\infty}\int_{-\infty}^{+\infty} h(\tau)g(s-t)R[s,\tau]e^{-i\tau\omega}dsd\tau \tag{8}$$

The interference caused by cross term was restrained perfectly. At the same time, the high temporal and frequency resolution and high energy focus retained well in SPWVD (Fig.4).

Experiment on test-rig and analysis of the result

To verify the proposed approach for the signals collected in the real word, an experiment on a test-rig was performed and introduced as follows. The physical map of the whole experimental system was shown in Fig.5.

Fig.5 Multi-functional rotor test-rig

The fault rolling bearing was mounted in the bearing pedestal at left of the coupling for the inner race faults. The test conditions included a fault rolling bearing type 6000 SKF deep groove ball bearing (the number of rolling elements Z=7, the inner diameter d=10mm, the outer diameter D=26mm, the thickness is 8mm). The speed of the reference shaft varies in the coast-up progress from starting. The data acquisition card is SINOCERA YE6231 (four channels synchronous sampling) with the sampling rate 96 kHz, a piezoelectric accelerometer CA-YD-103(sensitivity 20 pC/g, frequency response range 0.5~12000 Hz) mounted on the bearing pedestal. The original time-domain waveform and its FFT spectrum were shown in Fig.6 and Fig.7.

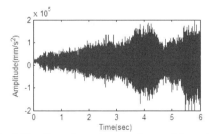

Fig.6 The time-domain waveform of the original signal

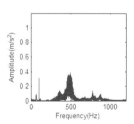

Fig.7 FFT spectrum of the original signal

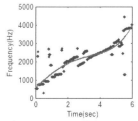

Fig.8 The instantaneous frequency estimation with STFT

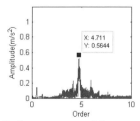

Fig.9 Order spectrum of the re-sampling signal with STFT

Frequency-smear phenomenon was obvious in Fig.7. It was difficult to obtain the characteristic frequency because the peak of the spectrum was smeared. Fig.8 and Fig. 9 depicted the instantaneous frequency curve and order spectrum processed with STFT.

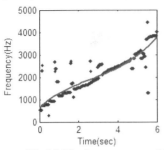

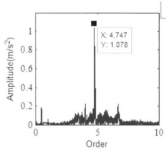

Fig.10 The instantaneous frequency estimation with SPWVD

Fig.11 Order spectrum of the re-sampling signal with SPWVD

Fig.10 and Fig.11 showed the corresponding results to compared with the traditional approach and verify the proposed method in this paper.

The contrast effective before and after was remarkable. It was obvious that the frequency-smear phenomenon was partly weakened in the order spectrum with STFT, but the energy was still dispersive in a relatively wide band. The frequency resolution was much higher and the energy was more focused in the order spectrum shown in Fig.11, and the accuracy of the frequency curve (Fig.10) was more precise than that pre-processed by STFT (Fig.8). The fault order characteristic $O_{fault} = 4.747$ was clear and centralized in the order spectrum that was improved with SPWVD.

Summary

This paper proposed an approach of improving the accuracy of time-frequency order tracking. The effects of multi-component signals analyzing respectively with STFT, WVD and SPWVD were compared first. Then SPWVD was used to restrain the cross-term interference and estimate the instantaneous frequency. The simulation and experiment on test-rig clearly showed that the time-frequency order tracking analysis with the proposed improving approach was more effective and accurate in the varying-speed running condition of rotating machinery.

Acknowledgment

This research is supported by the National Natural Science Foundation of China (No. 51275030); the National Science & Technology Pillar Program (No. 2013BAF06B00). We also thank the anonymous reviewers for their constructive comments.

References

[1] K.M Bossley, R.J.Mackendrick, Hybrid computed order tracking, J. Mechanical Systems and Signal Processing. 13(1999) 627-641.

[2] Yang Jiongming, Qin Shuren, Ji Zhong, Guo Yu, Development of virtual instrument in characteristic analysis of rotating machinery based on time frequency estimation, J. Chinese Journal of Construction Machinery. 17(2004) 490-493.

[3] Pan M Ch, Lin Y F, Further exploration of Void-Kalman-filtering order tracking with shaft-speed information-II Engineering applications, J. Mechanical Systems and Signal Processing. 20(2006) 1134-1154.

[4] J.R. Blough, Development and analysis of time variant discrete Fourier transform order tracking, J. Signal Process. 17 (2003) 1185–1199.

Advanced Materials Research Vol. 819 (2013) pp 271-276
© *(2013) Trans Tech Publications, Switzerland*
doi:10.4028/www.scientific.net/AMR.819.271

Fault Diagnosis of Rolling Bearing Based on Dual-tree Complex Wavelet Transform and AR Power Spectrum

Zhipeng Meng , Yonggang Xu [a], Guoliang Zhao and Sheng Fu

Key Laboratory of Advanced Manufacturing Technology, Beijing University of Technology, Beijing 100124, CHINA

[a]xyg_1975@163.com

Keywords: Dual-tree complex wavelet transform, AR power spectrum, Hilbert envelope, Fault diagnosis

Abstract. Aiming at the strong background noise involved in the signals of rolling bearing and the difficulty to extract fault feature in practice, a new fault diagnosis method is proposed based on Dual-tree Complex Wavelet Transform (DT-CWT) and AR power spectrum. Firstly, the non-stationary and complex vibration signal is decomposed into several different frequency band components through dual-tree complex wavelet decomposition; Secondly, Hilbert envelope is formed from the components which contains the fault information. Finally, the auto-power spectrum can be obtained by auto-regressive (AR) spectrum. The noise interference was eliminated effectively, and the effective signal information was retained at the same time. Thus, the fault feature information was extracted. In this paper, the fault test and the engineering practical fault data of rolling bearing were analyzed by dual-tree complex wavelet transform and AR power spectrum. The results show that the noise of the vibration signal was eliminated effectively, and the fault feature were extracted. The feasibility and effectiveness of the method were verified.

Introduction

The operation states of rolling bearings which are the most common and important parts in the mechanical equipment, will affect the whole machine running condition directly. It is very important to study the rolling bearing fault diagnosis, because the main reason of the mechanical equipment fault is caused by the rolling bearing fault. Due to the working environment of rolling bearing is complicated, the fault vibration signal of rolling bearing is usually non-stationary, and the strong noise interference is contained in the vibration signals at the same time. The main research of fault diagnosis is the direction that the noise should be eliminated and the fault feature information should be extracted effectively [1].

Dual-tree complex wavelet transform (DT-CWT) [2-3] was proposed by Kingsbury firstly, then Selesnick researched algorithm of DT-CWT. DT-CWT retained the good properties of the complex wavelet transform, and the perfect reconstruction of signal was insured by using dual-tree filters. DT-CWT is a kind of wavelet transform that has many good characteristics, for example, approximate shift invariance, good directional selectivity, perfect reconstruction, limited data redundancy and efficient computational efficiency. At present, DT-CWT has been successfully applied to the field of image processing [4-5], speech recognition [6], signal noise reduction processing [7] and fault diagnosis [8-9]and so on.

AR model is a kind of time series analysis method [10], the autoregressive parameters is very effective to the change of state, the objective laws of dynamic system can be expressed profoundly and intensively. The autoregressive parameter of AR model is regarded as a feature vector to analyze the state of the system change effectively. The AR power spectrum is only applicable to stationary signal analysis, but the fault vibration signal of rolling bearing is usually non-stationary, so the result of rolling bearing fault vibration signal which was processed by AR power spectrum directly is not ideal.

In view of the above situation, a new fault diagnosis method was proposed based on dual-tree complex wavelet transform and AR power spectrum, and applied to fault diagnosis of rolling bearing successfully. The results of the experiments and engineering application show that the noise of signal can be eliminated, the fault feature of rolling bearing can be extracted effectively, the feasibility and effectiveness of the method were verified.

Dual-tree Complex Wavelet Transform

The Dual-tree Complex Wavelet Transform (DT-CWT) uses two parallel real wavelet transform to realize signal decomposition and reconstruction，they called the real part tree and imaginary part tree respectively, and the decomposition and reconstruction process of DT-CWT will be showed in Fig.1[2-3]. In the process of signal decomposition and reconstruction, the location of imaginary part tree should be kept in the middle of the real part tree. The DT-CWT can effectively utilize wavelet decomposition coefficients of real part tree and imaginary part tree，so as to realize information complementary between the real part tree and imaginary part tree. This kind of wavelet decomposition method makes DT-CWT has approximate translation invariance, and to reduce the loss of useful information. In each layer of the decomposition process of DT-CWT, the use of wavelet coefficient dichotomy reduces redundant calculation, and increase the calculation speed. According to the double tree complex wavelet construction method, Complex wavelet can be expressed as:

$$\varphi(t) = \varphi_h(t) + i\varphi_g(t) \tag{1}$$

Where $\varphi_h(t)$，$\varphi_g(t)$ are two real wavelet，i is unit of complex.

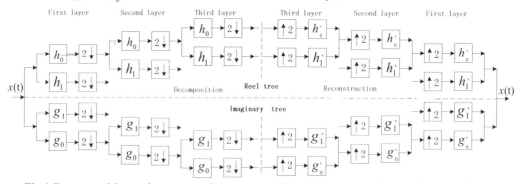

Fig.1 Decomposition and reconstruction process using dual-tree complex wavelet transform

Due to the DT-CWT consists of two parallel wavelet transform composition, Therefore, according to the wavelet theory, the real part tree wavelet transform of wavelet coefficient and scale coefficient can be calculated by the formula (2) and formula (3):

$$dI_j^{Re}(n) = 2^{j/2} \int_{-\infty}^{+\infty} x(t)\varphi_h(2^j t - n)dt \quad j = 1,2,......, J \tag{2}$$

$$cI_j^{Re}(n) = 2^{J/2} \int_{-\infty}^{+\infty} x(t)\varphi_h(2^J t - n)dt \tag{3}$$

By the same, the imaginary part tree wavelet transform of wavelet coefficient and scale coefficient can be calculated by the formula (4) and formula (5):

$$dI_j^{Im}(n) = 2^{j/2} \int_{-\infty}^{+\infty} x(t)\varphi_g(2^j t - n)dt \quad j = 1,2,......, J \tag{4}$$

$$cI_j^{Im}(n) = 2^{J/2} \int_{-\infty}^{+\infty} x(t)\varphi_g(2^J t - n)dt \tag{5}$$

So the wavelet coefficient and scale coefficient of DT-CWT can be obtained:

$$d_j^{\varphi}(n) = d_j^{Re}(n) + id_j^{Im}(n) \quad j = 1,2,......, J \tag{6}$$

$$c_j^{\varphi}(n) = c_J^{Re}(n) + ic_J^{Im}(n) \tag{7}$$

Finally, wavelet coefficient and scale coefficient of DT-CWT can be reconstituted by formula (8) and (9) :

$$d_j(t) = 2^{(j-1)/2}[\sum_{-\infty}^{+\infty} d_j^{Re}(n)\varphi_h(2^j t-n) + \sum_{-\infty}^{+\infty} d_j^{Im}(n)\varphi_g(2^j t-k)] \tag{8}$$

$$c_J(t) = 2^{(J-1)/2}[\sum_{-\infty}^{+\infty} c_J^{Re}(n)\varphi_h(2^J t-n) + \sum_{-\infty}^{+\infty} c_J^{Im}(n)\varphi_g(2^J t-k)] \tag{9}$$

The reconstruction signal of DT-CWT can be expressed as:

$$x(t) = d_j(t) + c_J(t) \tag{10}$$

AR power spectrum

AR power spectrum is used to modern power spectrum estimation commonly. AR model is a full pole model, and the difference equation of AR model is:

$$x(n) = -\sum_{i=1}^{p} a_i x(n-i) + u(n) \tag{11}$$

In which, $u(n)$ is sequence of white noise, p is the order of AR model, a_i is model parameters of AR model ($i = 1,2,..., p$). It is easy to get the transfer function form of AR model:

$$H(z) = \frac{1}{1 + \sum_{i=1}^{p} a_i z^{-i}} \tag{12}$$

According to the spectrum of definition, estimation of auto-power spectrum can be obtained:

$$S(e^{jw}) = |H(e^{jw})|^2 S_w(e^{jw}) = \sigma_w^2 |H(e^{jw})|^2 = \frac{\sigma_w^2}{|1 + \sum_{i=1}^{p} a_i e^{-jwi}|^2} \tag{13}$$

Where σ_w^2 is the power spectrum density of white noise .

Diagnosis method of DT-CWT and AR auto-power spectrum

Firstly, the signal is decomposed by DT-CWT to get several components of different frequency band. Secondly, Hilbert envelope is formed from the component which contains the fault information. Finally, the AR auto-power spectrum can be obtained. The noise interference was eliminated effectively, and the effective signal information was retained at the same time, and the fault frequency will be found, the accuracy of fault identification will be improved. The process flow diagram of this method is showed in Fig. 2.

Fig.2 Diagnosis method based on DT-CWT and AR auto-power spectrum

Analysis of experiment

Fig.3 shows experimental system consisting bearing test rig, piezoelectric acceleration sensor, data acquisition instrument and laptop. The normal and faulty bearing installed in the bearing test rig in turn that gather the data of experiments. Then data acquisition instrument made data to laptop and analyzed it. In the experiment, the type of rolling bearing is 6307, speed of motor is 1496rpm, sampling frequency is 15360Hz, which simulated outer ring fault of bearing at end bearing of the bearing fault test rig. The fault frequency of outer ring is 76.728Hz.

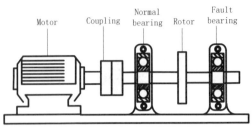

Fig.3 Bearing fault test rig

Waveform and spectrum of bearing outer ring fault is showed in Fig. 4.The impact appeared in waveform obviously, but fault feature is not clear. The sideband also can easily be found in the spectrum, at the same time there is some strong noise interference composition in the signal. So the original signal was decomposed to 4 layers by DT-CWT, then the every layer was reconstructed to 5 different frequency bands which are a_4, d_4, d_3, d_2 and d_1 as shown in Fig.5. There is obvious periodical impact in the third component d_3.

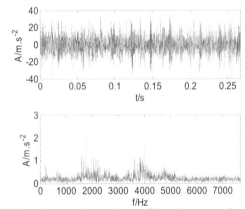

Fig.4 Waveform and spectrum of bearing outer ring fault

Due to the method of DT-CWT and AR auto-power spectrum, Hilbert envelope of d_3 can be obtained , then the AR auto-power spectrum can be got in Fig. 6, from which we can clearly see the frequency of 75 Hz, 153.8Hz,228.8Hz and so on. They are close to harmonics feature of outer race. Thus it can be concluded that it exists bearing outer race fault. Fig.7 is Hilbert envelope spectrum of original signal directly, it was clear that the signal has a strong interference composition.

Engineering application

The gear boxes of 27 finishing mills of a steel plan were bashed to broken in an accident. The location of fault was the 12th bearing of the II axis. The whole bearing was damaged seriously so that the machine had to stop for 4 hours. When the fault occurred, the rotating speed of motor was 1178 r/min, the sampling frequency of the system was 12000Hz, the number of the sampling points was 2048, the characteristic frequency of the system is 117.1875 Hz. The waveform and amplitude spectrum eight days before the fault happened was shown in Fig.8. The time domain waveform can display some impact composition, but the corresponding characteristic frequency of the faulty bearing can't be identified accurately according to the spectrum.

According to the method proposed in this paper, Fig.9 is the results of fault signal that was decomposed to 4 layers by DT-CWT. Fig.10 is shown AR auto-power spectrum of d_3 .It is easy to get the frequency of 117.2 Hz, 345.7Hz ,457Hz and 574.2Hz. They are close to harmonics feature of outer race(117.1875 Hz). Fig.11 is Hilbert envelope spectrum of original signal directly, it was clear that the signal has a strong noise interference composition, it is difficult to identify fault.

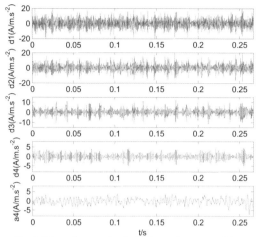

Fig.5 Waveform of DT-CWPT decomposition

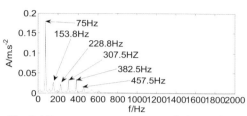

Fig.6 AR auto power spectrum of d_3 envelope

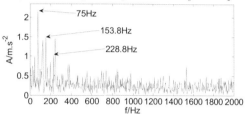

Fig.7 Hilbert envelope spectrum of original signal

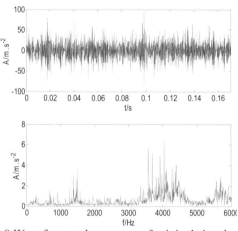

Fig.8 Waveform and spectrum of original signal

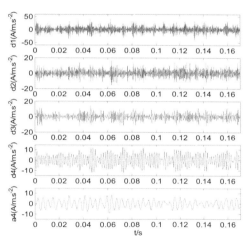

Fig.9 Waveform of DT-CWT decomposition

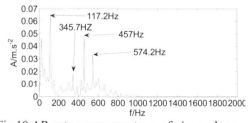

Fig.10 AR auto power spectrum of d_3 envelope

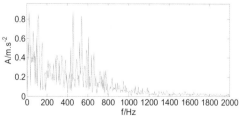

Fig.11 Hilbert envelope spectrum of original signal

Conclusions

This paper studied the method of fault diagnosis based on DT-CWT and AR power spectrum. The validity of the proposed method was verified through the rolling bearing fault experiment and engineering case.

1) Due to the approximate shift-invariant, avoiding frequency aliasing and de-noising effectively of DT-CWT, components of different frequency band were obtained by DT-CWT for the bearing fault vibration signal.

2) The noise which is contained in signal can be eliminated effectively, the useful component is reserved, it is useful to the fault identification.

3) The method based on DT-CWT and AR power spectrum can be used in the field of fault diagnosis, the noise of the vibration signal was eliminated effectively, and the fault feature was extracted.

Acknowledgment

This work is partially supported by Nature Science Foundation of China (51075009), and Beijing talents training subsidy scheme(2011D005015000006).

References

[1] Peng Zhike, Chu Fulei, He Yongyong. Vibration signal analysis and feature extraction based on reassigned wavelet scalogram[J]. Journal of Sound and Vibration, 2002, 253(5):1087-1100.

[2] Kingsbury N G.The dual-tree complex wavelet transform: a new technique for shift invariance and directional filters[J]. IEEE Digital Signal Processing Workshop, 1998,98(1):2-5.

[3] Selesnick I W, Baraniuk R G, Kingsbury N G. The dual-tree complex wavelet transform[J].IEEE Digital Signal Processing Magazine, 2005, 22(6): 123-151.

[4] Edward H S L, Pickering M R, Frater M R, et al. Image segmentation from scale and rotation invariant texture features from the double dyadic dual-tree complex wavelet transform[J]. Image and Vision Computing,2011,9(1): 15-28.

[5] Priyaa K J, Rajeshb R S. Local fusion of complex dual-tree wavelet coefficients based face recognition for single sample problem[J]. Procedia Computer Science,2010,2(1): 94-100.

[6] Wang Na, Zheng Dezhong, Liu Yonghong. New method for speech enhancement based on dual tree complex wavelet packet transform[J]. Journal of Sensore and Actuators, 2009,22(7):983-987.

[7] Wang Yanxue, He Zhengjia, Zi Yanyang. Enhancement of signal denoising and multiple fault signatures detecting in rotating machinery using dual-tree complex wavelet transform[J]. Mechanical Systems and Signal Processing, 2010, 24(1):119-13.

[8] Chen Zhixin, Xu Jinwu, Yang Debin. Denoising method of block thresholding based on DT-CWT and its application in mechanical fault diagnosis[J]. Journal of Mechanical Engineering, 2007, 43(6): 200-204.

[9] Su Wensheng, Wang Fengtao, Zhu Hong, et al. Denoising method based on hidden Markov tree model in dual tree complex wavelet domain and its application in mechanical fault diagnosis[J].Journal of Vibration and Shock, 2011, 30(6):47-52.

[10] Cheng Junsheng, Yu Dejie, Ynag Yu. A fault diagnosis approach for roller bearings based on EMD method and AR model[J]. Mechanical Systems and Signal Processing, 2006,20(2):350-362.

Advanced Materials Research Vol. 819 (2013) pp 277-280
© *(2013) Trans Tech Publications, Switzerland*
doi:10.4028/www.scientific.net/AMR.819.277

The Design of IPC Chassis Structure Based on CNC System

Ruoyu Liang[1,a] , Dongxiang Chen[1,b] , Taiyong Wang[1,c] , Kaifa Wu[1,d] ,
Fuxun Lin[1,e] , Yichao Li[2,f] and Xiaofeng Xu[2,g]

[1]Key Laboratory of Mechanism Theory and Equipment Design of Ministry of Education, Tianjin University, Tianjin 300072, China

[2]College of Architecture & Art, Hefei University of Technology, Hefei, 230002, China

[a]lryasa@163.com, [b]dxchen@tju.educn, [c]tywang@139.com, [d]sdwkf.good@163.com, [e]lfx3107@163.com, [f]lyc20073733@sina.cn, [g]xiaotianfacai@yahoo.cn

Keywords: Structural design, Electromagnetic shielding, Thermal design.

Abstract. The article introduces the new technologies and methods in chassis of IPC design. It elaborates the process from following four parts: electromagnetic shielding, anti-libration and anti-impact design, thermal design and other protective design. The last part summarizes the design procedure and looks ahead to the future of this field.

Introduction

Today, as a collaborative automation control equipment, IPC plays an irreplaceable role in the industrial areas. The chassis of the IPC stores a variety of electronic devices, it's a powerful guarantee to let the IPC achieve various functions stably. With the continuous development of science and technology, the development of the IPC chassis is turning towards the direction of small, modular and humane. The CNC machine tools' working environment is often complex, so the chassis requirements are also higher than other types.

Structure Planning

The proper plan often plays a very important role in the design process, so an accurate planning should be formulated in the early stage of the chassis structure design.

In order to radiate better and optimize the overall electromagnetic compatibility, the chassis utilizes sub-cavity structure. There are two cavities in the chassis, board cavity and aid cavity. The main function of former is to store the motherboard, PMAC, I/O board, adapter board and collection board. The aid cavity is the container of the power supply, hardware, cooling fans and some other auxiliary devices.

The main part of the chassis is made of aluminum alloy 5052, this label material has good forming properties, corrosion resistance, blanch resistance, fatigue strength and the medium static strength.

Electromagnetic Design

In order to ensure that the chassis achieves various functions in the complex electromagnetic environment, the electromagnetic shielding performance should be considered at the beginning of the chassis design.

There are several technologies in this design to ensure the chassis have a good electromagnetic compatibility: individually furnished strong sources of interference and set shielding boxes for them, set shield plates in the cavities to prevent electromagnetic interference, the contact surface between the chassis frame and cover should be smooth, with as many screws clamping, thereby forming a reliable electrical connection. [1]

Power Interference Shielding. To the whole internal electromagnetic environment of the chassis, the power supply is a strong source of interference, so it should be shielded. The mains should be the

IPC dedicated power, this type has good electromagnetic properties and mounting dimensions. In addition, the power supply should be disposed in a separate cavity, and a power filter is necessary. The power should be set close to the wall of the chassis, the filter should be set near the power cable, the input line and output line should be isolated, there should have a good electrical connection between the filter and the chassis.

Shielding of the gap. There are gaps often between the chassis frame and the cover, they're the main passage of electromagnetic leakage. Therefore, it is necessary to do something useful to deal with the gaps in order to reduce electromagnetic leakage.

There're several factors affect the performance of the gap shield, such as the maximum size of the gap, the depth of the gap, the characteristic of the shielding materials, etc. The effective methods to limit the maximum size of the gap include the following parts, reduce the distance of the fastening points, strengthen the rigidity of the frame and the cover, improve the surface accuracy of the combination of surface, etc. Increase the depth of the joint surface gap, and select the reasonable shielding material to make the cover and frame of chassis are also the effective methods to reduce electromagnetic leakage.

Shielding of the Holes. In order to achieve the cooling, I/O functions, chassis surface often set a large number of holes, these holes is one of the important channels of electromagnetic leakage, therefore, researchers must take measures to eliminate the affect of the holes to the whole electromagnetic performance.

It's an effective way to bore holes on metal plate to radiating. The maximum size of the hole and the depth of the hole are the major factors affecting its shielding effectiveness, so when the shielding effectiveness of the ventilation holes and heating contradiction, increase the depth, reduce the diameter of the hole and increase the density and number of the holes can solve the problem. Trying to find out the equilibrium point between the shield and radiation. Perforated metal plate structure is simple, inexpensive, and can be applied to most kinds of the chassis.

The other holes for the cables can set shield box, in order to achieve the shielding function.

Cable Shielding. Through the adapter plate, the cables in the chassis can contact with external port, the adapter plate and the interfaces are handled in accordance with the standards specified by NEMA, can effectively ensure the shielding effect. As shown in Fig.1.

Anti-libration and Anti-impact Design

The technologies such as isolation technology, decoupling techniques, damping technology, rigid technology, etc can be used to achieve the function of anti-libration and anti-impact. [2]

Most chassis anti-vibration and impact of the measures is to install the chassis in a framework, this way can effectively alleviate the impact from outside, however, due to the low degree of freedom of the framework, it is often unable to buffer the impact of the plurality of directions at the same time, so it is possible to damage the internal chassis components. If take some measures inside the chassis to achieve anti- impact function, then can reduce the impact during the machine working and transportation.

The measures in the chassis that can anti-impact include install boards layering and use spring screws as fasteners. Board batten generally is set at the top of the motherboard PCI slot(or above the passive motherboard PCI slot), rigid connection to the chassis wall, blew the layering is a L-shaped metal plate, it's height is adjustable, at the bottom of the plate is soft insulating material gasket, it contacts the top of the board, this design can effectively prevent the board pulled out due to shock or vibration from PCI slots, protect the board won't be damaged. As shown in Fig.2.

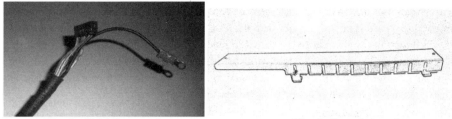

Fig. 1. Shielding conductor   Fig. 2. Boards layering

The components are more vulnerable and need to be installed on the wall of the chassis like motherboards and graphics cards should be fixed by the spring screws, it has buffer trip, can prevent components deformed by rigid impact, also can guarantee that each board to maintain normal operation under vibration environment.

Thermal Design

When the electronic devices work, they will also have a greater heat, if it does not dissipate in time, it will result in the operating temperature of the power components rising, thus affecting the performance of electronic devices. [3]

The thermal structure of the chassis is designed to minimize the thermal resistance, to accelerate the speed of heat dissipation and reduce the rate of temperature rise. The cooling process as shown in Fig.3.

Fig. 3. Cooling process

The measures of radiating include: natural heat, forced air cooling, direct liquid cooling and evaporative cooling (Heat pipe cooling).

Assume that the internal chassis components heat flux is q, $q=Q/(S*t)$, which Q represents the amount of heat (W), t represents the cooling time (s), S represents the surface area of the cooling element (m^2). Assuming the thermal resistance is R, $R=L/KA$, L represents the length (m), K represents the thermal conductivity, A represents the cross-sectional area.

With the concept of thermal resistance, substituted into the basic equation of the thermal conductivity: $Q=KA\Delta t/L$, Q represents power dissipation (W), t represents the rise temperature (°C), $\Delta t=QR$. At room temperature (25°C), when requires the thermal resistance between 1°C/W and 0.05°C/W, the forced air cooling can be used, when requires blew 0.02°C/W, the heat pipe cooling can be used.

According to the thermal density of the heating elements and the requires of the thermal resistance, we chose the forced air cooling as the radiating method. The radiating process in the chassis as shown in Fig.4.

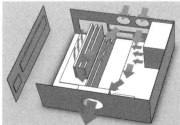

Fig. 4. The radiating process in the chassis

The chassis uses the frame structure, sets the cooling holes on front and rear cover. The position of the front cover vents is slightly higher than the motherboard, set a positive pressure fan and filter on the inside. The vents on the rear cover are set in the middle of the body, set double negative pressure fans on the inside. The positive / negative pressure fans formed airflow channel inside the chassis, thus exhaust the heat that released by the elements. The component which exudes large amount of heat can be installed separately heatsink to improve the cooling efficiency.

Other Protective Design

Hot flashes, salt spray and mold, these three environmental factors also affect electronic components, the technical precautions for them are called defense technology. At present, the defense technology achieved through the following measures, structural protection, material protection, surface treatment and professional maintenance.

The main part of the chassis is made of aluminum alloy 5052, this label material has good corrosion resistance, the external of the chassis adopt plating, using conductive tape sealed seams, to do so not only to meet the requirements of electromagnetic shielding but also improve the performance of the defense. In addition, at the vents installed a multi-purpose filter can also effectively improve the chassis performance of the defense.

Conclusions

The design is based on the proven technology of the current domestic, take advantage of the commercialization of parts, and develop the chassis for CNC system by optimal combination. It satisfies several functions and ergonomic requirements. The chassis can be an effective carrier to assist the IPC achieve control functions. After further design procession can develop series of products, to meet the needs of different CNC system.

Acknowledgements

This work is financially supported by the CNC generation of mechanical product innovation demonstration project of Tianjin (2013BAF06B00), The recovery method of weak information and study of algorithm hardened design based on generalized parameter adjustment stochastic resonance (20100032110006) and High-end CNC machining chatter online monitoring and optimization control technology research (12JCQNJC02500). Please communicate with the corresponding author Fuxun Lin, if there are any questions in this paper.

References

[1] L. Yuan, J. F. Huang, The structure and defense design of a portable reinforcement machine: Modern Manufacturing Engineering, 2012, v7, p123

[2] L. Z. Dong, D. Y. Zhang, J. Lin, Structural design of reinforce cases: Infrared and Laser Engineering, 2006, v35 Supplement, p195

[3] J. H. Liu, Study on Cooling Structure of Sealed Case of a Radar: Electro − Mechanical Engineering, 2012, v28, p36

Advanced Materials Research Vol. 819 (2013) pp 281-285
© *(2013) Trans Tech Publications, Switzerland*
doi:10.4028/www.scientific.net/AMR.819.281

Dynamic Fuzzy Reliability Analysis of Mechanical Components with Small Samples of Load

Peng Gao [1,2,a], Shaoze Yan [1,b], Jianing Wu [1,c], Tianfu Yang [1,d]

[1]State Key Laboratory of Tribology, Department of Mechanical Engineering, Tsinghua University, Beijing 100084, China

[2]Department of Chemical Mechanical Engineering, Liaoning Shihua University, Liaoning 113001, China

[a]gaogaopeng@163.com, [b]yansz@tsinghua.edu.cn, [c]wujn09@mails.tsinghua.edu.cn, [d]ytf10@mails.tsinghua.edu.cn

Keywords: Dynamic Fuzzy Reliability, Small Samples, Failure rate.

Abstract. Dynamic fuzzy reliability models of mechanical components are proposed in this paper in the case of small samples of load. The proposed models are established based on the distribution of load obtained from a load history, which reflects the relationship among the occurrence frequency of loads with different magnitude in a given time period, rather than the distribution at each load application. To deal with the uncertainty of the distribution of load from a load history due to the randomness of the magnitude of load at each load application, the statistical parameters involved in the distribution of load are modeled as fuzzy variables. Explosive bolts are chosen as representative examples to illustrate the proposed models. The results show that reliability increases with the increase in the level of cut set. Besides, the dispersion of strength has great influences on the reliability and failure rate of mechanical components.

Introduction

The load–strength interference (LSI) model is the most important approach in the reliability assessment of mechanical components, which is only suitable for static reliability estimate[1]. However, from the definition of reliability, it can be learned that reliability is the function of time. During the last two decades, many investigations have been involved in developing dynamic reliability models with strength and load modeled as two stochastic processes[2]. Pourgharibshahi calculated the dynamic reliability of structures by using two stochastic performance evaluation processes[3]. Valdebenito proposed a method for reliability sensitivity estimation of linear structural systems under the load that follows the Gaussian process[4]. Lewis analyzed the dynamic behavior of a redundant system based on a Markov model[5]. In these models, the probability density function (pdf) of load and the pdf of strength are obtained from sufficient experimental data. However, in the aerospace engineering, mechanical components always operate in complicated random environments and only limited load history samples can be obtained. It is difficult or impossible to acquire the accurate distribution of load at each time instant though experiment.

In this paper, dynamic reliability models are developed in the case of limited load history samples. The proposed reliability models provide a basis for dynamic reliability of mechanical components with small samples of load and can be used to analyze the impacts of the variation in statistical parameters of strength on the dynamic characteristics of reliability and failure rate of mechanical components.

The remainder of this paper is organized as follows. In Section 2, dynamic fuzzy reliability models are developed. In Section 3, explosive bolts are chosen as representative examples to illustrate the proposed models. Besides, the influences of the variation in the dispersion of strength on the fuzzy reliability and fuzzy failure rate are analyzed through the numerical examples.

Dynamic fuzzy reliability models

The methods for reliability analysis based on fuzzy set theory have been widely investigated since the concept of fuzzy probability was proposed by Zadeh. The triangular fuzzy number is denoted by a triplet (a,b,c) in this paper. b is the center point, whose degree of membership is one. a and c are, respectively, the left parameter and right parameter, which determine the range of the possible value of a fuzzy variable.

According to the definition of fuzzy probability proposed by Zadeh, the α-cut of a fuzzy set provides a basis for fuzzy reliability assessment in the context of probability. Therefore, the reliability can be written by the integral of the fuzzy reliability at different levels of α in the following form [6].

$$P(A) = \int_0^1 \int_{\tilde{X}_\alpha^L}^{\tilde{X}_\alpha^U} f(k) dk d\alpha \tag{1}$$

In fact, the distribution of load obtained from a given load history sample reflects a possible relationship among the occurrence frequency of load with different magnitude in a given time period. Therefore, in the case of a given distribution of load obtained from a load history sample with the mean value of m and the standard deviation of σ, the application times of load with the magnitude of L_i, with the total application times of n, can be expressed as

$$n_i = \frac{n}{\sqrt{2\pi}\sigma} \exp[-\frac{1}{2}(\frac{l_i - m}{\sigma})^2] \Delta l_i \tag{2}$$

The conditional probability that a component survives the application of load L_i for n_i times can be given by

$$P = \left[\int_{l_i}^\infty f_r(r) dr\right]^{n_i} = \left[\int_{l_i}^\infty f_r(r) dr\right]^{\frac{n}{\sqrt{2\pi}\sigma} \exp[-\frac{1}{2}(\frac{l_i - m}{\sigma})^2] \Delta l_i} \tag{3}$$

where $f_r(r)$ represents the pdf of strength. Therefore, the reliability of components under the application of random load for n times can be calculated as follows

$$R(n) = \prod_i \left[\int_{l_i}^\infty f_r(r) dr\right]^{\frac{n}{\sqrt{2\pi}\sigma} \exp[-\frac{1}{2}(\frac{l_i - m}{\sigma})^2] \Delta l_i}$$

$$= \exp\left\{\sum_i \frac{n}{\sqrt{2\pi}\sigma} \exp[-\frac{1}{2}(\frac{l_i - m}{\sigma})^2]\{\ln[\int_{l_i}^\infty f_r(r) dr]\} \Delta l_i\right\} \tag{4}$$

Let $\Delta l_i \to 0$, then the reliability of components can be calculated as follows

$$R(n) = \exp\left\{\frac{n}{\sqrt{2\pi}\sigma} \int_{-\infty}^\infty \exp[-\frac{1}{2}(\frac{l - m}{\sigma})^2]\ln[\int_l^\infty f_r(r) dr] dl\right\} \tag{5}$$

To consider the fuzziness of the statistical parameters involved in the distribution of load, denote the distributions of the mean value and standard deviation of load on the interval of the α-cut by $f_{ma}(m)$ and $f_{\sigma a}(\sigma)$, respectively. Therefore, the fuzzy reliability of components at the level of α can be given by

$$R_\alpha(n) = \int_{\tilde{m}_\alpha^L}^{\tilde{m}_\alpha^U} f_{ma}(m) \int_{\tilde{\sigma}_\alpha^L}^{\tilde{\sigma}_\alpha^U} f_{\sigma a}(\sigma) \exp\left\{\frac{n}{\sqrt{2\pi}\sigma}\times\right.$$

$$\left. \int_{-\infty}^\infty \exp[-\frac{1}{2}(\frac{l - m}{\sigma})^2]\ln[\int_l^\infty f_r(r) dr] dl\right\} d\sigma dm \tag{6}$$

From the definition of failure rate, the fuzzy failure rate of components at the level of α can be written as

$$h(n) = \frac{F(n+1)-F(n)}{R(n)}$$

$$= \left\{ \int_{\bar{m}_\alpha^L}^{\bar{m}_\alpha^U} f_{m\alpha}(m) \int_{\bar{\sigma}_\alpha^L}^{\bar{\sigma}_\alpha^U} f_{\sigma\alpha}(\sigma) \exp\left\{ \frac{n}{\sqrt{2\pi}\sigma} \int_{-\infty}^{\infty} \exp[-\frac{1}{2}\left(\frac{l-m}{\sigma}\right)^2] \times \right. \right.$$

$$\ln\left[\int_l^\infty f_r(r)dr \right] dl \left\} \left\{ 1 - \exp\left\{ \frac{1}{\sqrt{2\pi}\sigma} \exp[-\frac{1}{2}\left(\frac{l-m}{\sigma}\right)^2] \right\} \times \right. \quad (7)$$

$$\ln\left[\int_l^\infty f_r(r)dr \right] dl \right\} dodm \right\} / \left\{ \int_{\bar{m}_\alpha^L}^{\bar{m}_\alpha^U} f_{m\alpha}(m) \int_{\bar{\sigma}_\alpha^L}^{\bar{\sigma}_\alpha^U} f_{\sigma\alpha}(\sigma) \exp\left\{ \frac{n}{\sqrt{2\pi}\sigma} \times \right. \right.$$

$$\int_{-\infty}^{\infty} \exp[-\frac{1}{2}\left(\frac{l-m}{\sigma}\right)^2] \ln\left[\int_l^\infty f_r(r)dr \right] dl \right\} dodm \right\}$$

The fuzzy reliability of components can be expressed as follows

$$R(n) = \int_0^1 \int_{-\infty}^\infty f_{m\alpha}(m) \int_{-\infty}^\infty f_{\sigma\alpha}(\sigma) \exp\left\{ \frac{n}{\sqrt{2\pi}\sigma} \times \right.$$

$$\int_{-\infty}^{\infty} \exp[-\frac{1}{2}\left(\frac{l-m}{\sigma}\right)^2] \ln\left[\int_l^\infty f_r(r)dr \right] dl \right\} dodmd\alpha \quad (8)$$

Correspondingly, the fuzzy failure rate of components is given by

$$h(n) = \frac{F(n+1)-F(n)}{R(n)}$$

$$= \left\{ \int_0^1 \int_{-\infty}^\infty f_{m\alpha}(m) \int_{-\infty}^\infty f_{\sigma\alpha}(\sigma) \exp\left\{ \frac{n}{\sqrt{2\pi}\sigma} \int_{-\infty}^{\infty} \exp[-\frac{1}{2}\left(\frac{l-m}{\sigma}\right)^2] \times \right. \right.$$

$$\ln\left[\int_l^\infty f_r(r)dr \right] dl \left\} \left\{ 1 - \exp\left\{ \frac{1}{\sqrt{2\pi}\sigma} \exp[-\frac{1}{2}\left(\frac{l-m}{\sigma}\right)^2] \right\} \times \right. \quad (9)$$

$$\ln\left[\int_l^\infty f_r(r)dr \right] dl \right\} dodmd\alpha \right\} / \left\{ \int_0^1 \int_{-\infty}^\infty f_{m\alpha}(m) \int_{-\infty}^\infty f_{\sigma\alpha}(\sigma) \exp\left\{ \frac{n}{\sqrt{2\pi}\sigma} \times \right. \right.$$

$$\int_{-\infty}^{\infty} \exp[-\frac{1}{2}\left(\frac{l-m}{\sigma}\right)^2] \ln\left[\int_l^\infty f_r(r)dr \right] dl \right\} dodmd\alpha \right\}$$

Numerical examples

In this section, explosive bolts are chosen as representative examples to illustrate the proposed reliability models. In the launch process of satellites, the explosive bolts are used for the connection between the launch vehicle and the satellite. When considering the stiffness k_1 of the explosive bolts and the stiffness k_2 of the components connected by the explosive bolts, the stress on the explosive bolts can be calculated as follows

$$s = \frac{k_1}{A(k_1+k_2)} F + \frac{F_0}{A} \quad (10)$$

where A and F_0 stand for cross-sectional area and preload, respectively. Therefore, the distribution of stress can be obtained by using the distribution of environmental load F from Eq.(10).

Suppose that the strength follows the normal distribution with the mean value of $m(r_0)$ and the standard deviation of $\sigma(r_0)$. The mean value and standard deviation of load are (650,700,750)MPa and (20,30,40)MPa, respectively. Besides, the mean value and standard deviation of load follow the uniform distribution on the interval of their α-cut. To consider the influences of the dispersion of strength on the dynamic fuzzy reliability and fuzzy failure rate of explosive bolts, consider the following three cases with the statistical parameters of strength in each case listed in Table 1. The reliability and failure rate of explosive bolts with different standard deviation of strength are shown in Fig.1 and Fig.2, respectively.

Table.1. Statistical Parameters of Strength of Explosive Bolts

	$m(r_0)$ (MPa)	$\sigma(r_0)$ (MPa)
Case 1	800	10
Case 2	800	20
Case 3	800	30

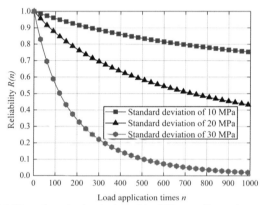

Fig.1. Reliability of explosive bolts with different dispersion of strength.

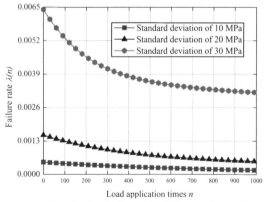

Fig.2. Failure rate of explosive bolts with different dispersion of strength.

From Fig. 1and Fig. 2 it can be seen that the dispersion of strength has great influences on the dynamic fuzzy reliability and fuzzy failure rate. Reliability decreases and failure rate increases with the increase in the dispersion of strength. Besides, the slope of the failure rate of explosive bolts decreases with the increase in the load application times, which is consistent with the bathtub curve theory when strength degradation is not taken into consideration.

Conclusions

Dynamic fuzzy reliability models of mechanical components in the case of small samples of load is proposed in this paper. The statistical parameters involved in the distribution of load are modeled as fuzzy variables to deal with the uncertainty of the distribution of load from a load history sample. The proposed reliability models can be used to analyze the impacts of the variation in statistical parameters of strength on the fuzzy reliability and fuzzy failure rate of explosive bolts. The results show that the dispersion of strength and the α-cut level have great influences on reliability. Reliability increases with the increase in the α-cut level and the decrease in the dispersion of strength.

Acknowledgments

This work was supported by the National Science Foundation of China under Contract No. 11072123, the National High Technology Research and Development Program of China (863 Program) under Contract No. 2009AA04Z401, the Major State Basic Research Development Program of China (973 Program), and the Project-sponsored by SRF for ROCS, SEM.

References

[1] J.F. Castet, J.H. Saleh, Satellite and satellite subsystems reliability: Statistical data analysis and modeling, Reliab Engineering and System Safety, 94 (2009) 1718–1728.

[2] J.N. Wu, S.Z. Yan, L.Y. Xie, Reliability analysis method of a solar array by using fault tree analysis and fuzzy reasoning Petri net, Acta Astronautica, 69(2011) 960-968.

[3] Ali Pourgharibshahi, Touraj Taghikhany, Reliability-based assessment of deteriorating steel moment resisting frames, Journal of Constructional Steel Research, 71(2012) 219-230.

[4] M.A. Valdebenito, H.A. Jensen, G.I. Schuëller, F.E. Caro, Reliability sensitivity estimation of linear systems under stochastic excitation, Computers and Structures, 92-93 (2012) 257-268.

[5] E. E. Levis, A Load-capacity interference model for common-mode failures in 1-out-of-2: G systems, IEEE Transactions on Reliability, 50 (2001) 47-51.

[6] B. Li, M. Zhu, K. Xu, A practical engineering method for fuzzy reliability analysis of mechanical structures, Reliability Engineering and System Safety, 67 (2000) 311-315.

Advanced Materials Research Vol. 819 (2013) pp 286-291
© (2013) Trans Tech Publications, Switzerland
doi:10.4028/www.scientific.net/AMR.819.286

Method of Plugging Cylinder Selection for Leak Detection Equipment Design Based on Differential Pressure Decay

Xiao Xinhua[1,2, a], Wang Taiyong[1, b], Cheng Bing[1, c]

[1]Key Lab of Mechanism Theory and Equipment Design of Ministry of Education, Tianjin University, Tianjin, China

[2]School of Mechanical Engineering, Tianjin Polytechnic University, Tianjin, China

[a]xxinhua@126.com, [b]tywang@139.com, [c]chengbing200082@yahoo.com.cn

Keywords: leak detection; line seal; face seal; plugging cylinder selection; subsection curve fitting-integration; CAD

Abstract. In auto leak detection equipment based on differential pressure decay, plugging quality directly impact on its accuracy and reliability. Since the cylinder is usually used to provide plugging force, plugging cylinder selection is an important part the detection equipment design. The main mechanical structure of the detection equipment was described. The equilibrium equations of plug were established and the bore size calculation formulas were derived in line seal and in face seal. To acquire the parameters such as circumference, area, etc. of the irregular seal surface, the subsection curve fitting-integration method and distilling from CAD model method were studied.

Introduction

The leak detection equipments used for cylinder block, cylinder head and clutch leak testing play important role in automobile production line [1-2]. In order to improve the precision and efficiency, differential pressure detection method is wildly used [3]. Positioning, clamping, plugging, pressurization, balance, detection and exhaustion are the basic detection stages [4-5].However, plugging quality is greatly impact detecting accuracy and reliability. The cylinder used to block tested cavity is known as plugging cylinder. How to select plugging cylinder reasonably is a problem need to be solved during the detection equipment design.

Main Structure of Detection Equipment

Detection equipment consists of main frame, lower clamp, upper clamp, leak tester, hydraulic & pneumatic system, electrical control system, etc. Fig.1 shows the main structure of the equipment. Its working principle is as follows:

1) Locate the workpiece (12) on the lower fixture board (2) by the location elements.

2) Push the lower fixture board (12) with the workpiece (2) into the bottom board (1) along the chute, and locate them in the equipment.

3) The clamping cylinder (7) extends out and push the middle motherboard (5) down, the upper fixture board（9）fixed on it takes the clamping device (11) and the upper plugging move down. Therefore the workpiece is fixed and the upper seal surface is sealed.

4) The side plugging cylinders (3, 13) extend out and the side surface of the cavity are sealed.

5) After the workpiece has been sealed, the test cavity is filled with gas until introduce the test pressure to the entire system.

6) Fill valve closes to isolate the pressure source from the workpiece. The workpiece and the master are pneumatically connected inside the tester for stabilization.

7) Equalization valve closes to isolate the work and the master in order to detect pressure difference between them.

8) After a certain period of time, small leak is detected and displayed on the leak tester.

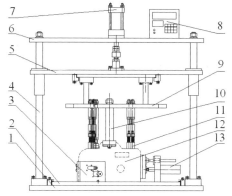

1-bottom board 2-lower fixture board 3-side plugging 1 4-guaid shaft 5-middle
motherboard 6-upper board 7-clamping cylinder 8-leak tester 9-upper fixture board
10-upper plugging 11- clamping device 12-workpiece 13-side plugging 2

Fig.1 Main structure of detection equipment

According to the contact size of seal element and seal surface, there are two seal types as line seal and face seal. The calculation method in the two cases is different.

Bore Size Determine Method of Plugging Cylinder Diameter

Bore Size Determine Method in Line Seal. When the middle of the seal element is empty and the seal area is relatively small, there is a linear relationship between the seal element's compressive force and its length. This seal type is named as line seal [6]. To improve efficiency and simplify of the plugging mechanism, the mechanism usually designed as seal element imbedded in the slot of plugging head, while the plugging cylinder driving the plugging head directly. Assuming the seal surface is circular. Taking the plugging head as force analysis object, forces acting on the object shows as Fig.2.

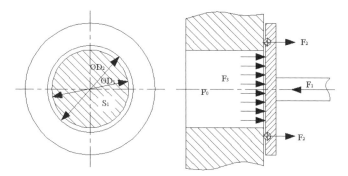

D_1-diameter of seal surface D_2-inner diameter of O ring P_0-detect pressure
F_3-force of seal gas F_2-compressive force F_1-plugging force

Fig.2 Forces acting on the plugging head in line seal

Statics shows that F_1, F_2, F_3 meet equilibrium equation, their relationship can be formulated as follow:

$$F_1 - F_2 - F_3 = 0 \tag{1}$$

Set inner diameter of O ring as D, diameter of plugging cylinder rod as d, detection pressure as P, the plugging force F_1 can be expressed as follow:

$$F_1 = \frac{1}{4}\pi(D^2 - d^2)P \tag{2}$$

Set diameter of O ring as d_o, the compressive force at unit length as λ, circumference of O ring as L, the compressive force F_2 can be formulated as follow:

$$F_2 = \pi\lambda(D_2 + d_o) \tag{3}$$

Set sealing area as S, testing pressure as P_0, the force of the test gas F_3 may be expressed as follow:

$$F_3 = \frac{1}{4}\pi D_2^2 P_0 \tag{4}$$

Substituting Eqs.(2)-(4) into Eqs.(1) and solving it , the diameter of plugging cylinder can be written as follow:

$$D = \sqrt{d^2 + \frac{4(D_2 + d_0)\lambda + D_2^2 P_0}{P}} \tag{5}$$

In order to ensure the sealing performance, the determine method of plugging diameter has to take safety factor into consideration. Therefore Eqs.(1) will be written as the following formula:

$$F_1 - \eta(F_2 + F_3) = 0 \tag{6}$$

The corresponding formula of the diameter of plugging cylinder will be follow:

$$D = \sqrt{d^2 + \frac{4\eta(D_2 + d_0)\lambda + \eta D_2^2 P_0}{P}} \tag{7}$$

Under normal circumstances, the sealing surface is irregular. Eqs.(7) needs to be extended to general form. Set the circumference of sealing surface boundary as L, the valid sealing area as S, the general form of the bore size of plugging cylinder in line seal can be shown as

$$D = \sqrt{d^2 + \frac{4\eta(\lambda L + P_0 S)}{\pi P}} \tag{8}$$

Bore Size Determine Method in Face Seal. When the seal element is a whole plane and the seal area is relatively big, there is a linear relationship between the seal element's compressive force and its area [7].This seal type is named as face seal. Assuming the seal surface is circular. Taking the plugging head as force analysis object, forces acting on the object shows as Fig.3.
 In the face seal case, F_1, F_2 and F_3 still meet Eqs. (1). F_1 and F_3 calculation method is the same as line seal, the compressive force F_2 can be formulated as follow:

$$F_2 = \kappa S_2 = \frac{1}{4}\kappa\pi(D_2^2 - D_1^2) \tag{9}$$

In this case, the diameter of plugging cylinder can be formulated as follow:

$$D = \sqrt{d^2 + \frac{D_1^2 P_0 + \kappa(D_2^2 - D_1^2)}{P}} \tag{10}$$

Taking safety factor into consideration,F_1,F_2,F_3 meet Eqs.(6).The diameter of plugging cylinder can be formulated as follow:

$$D = \sqrt{d^2 + \frac{\eta D_1^2 P_0 + \eta(D_2^2 - D_1^2)}{P}}$$ (11)

Under normal circumstances, set the compression area as S_2, the valid sealing area as S_1, the general form of the bore size of plugging cylinder in face seal can be shown as

$$D = \sqrt{d^2 + \frac{4\eta(kS_2 + S_1 P_0)}{\pi P}}$$ (12)

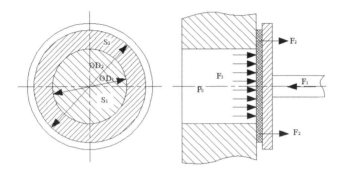

D₁-diameter of seal surface D₂-inner diameter of O ring P₀-detect pressure
F₃-force of seal gas F₂-compressive force F₁-plugging force

Fig.3 Forces acting on the plugging head in face

Acquisition Methods for Irregular Seating Surface Parameters

Subsection Curve Fitting-Integration Method. In most cases, the actual blocking surface is usually irregular plane.Fig.4 is the sketch for seating surface parameters acquisition. In Fig.4, the length of the centre line of seal ring is the circumference of seal ring (L); the area enclosed by the inside boundary of seal ring is approximately the valid area in line seal(S); the area enclosed by the boundary of the tested cavity is the plugging area (S_1); the area between the boundary of seal gasket and the boundary of tested cavity is the compressive area (S_2).

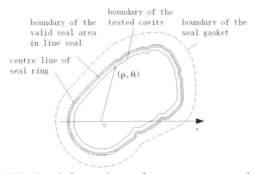

Fig.4 The sketch for seating surface parameters acquisition

If the workpiece's CAD model is not available, the subsection curve fitting-integration method can be employed. Its main process is:
(1) Select point O and create the polar coordinate in sealing plane.

(2) Set data point $P_i(\rho_i,\theta_i)(i=1,2,\ldots\ldots n)$ on the boundary of the tested cavity and measure its value in the polar coordinate. The density of data points should be adjusted according to the curvature of the boundary of the tested cavity. Increase data point appropriately on the large curvature section, so that the connection between data points is approximately a straight line.

(3) Divide the boundary to m sections.

(4) Curve fit for each section using the method of least squares [8]. Set the fitting curve equation as $\varphi_i(\theta)$ $(i=1,2\ldots\ldots m)$. The calculation method of S_1 is shown as follow:

$$S_1 = \sum_{i=1}^{m} \int_0^{2\pi} \frac{1}{2}[\varphi_i(\theta)]^2 \, d\theta \tag{13}$$

(5) By the same method, We can obtain the fitting curve equation $\alpha_i(\theta)$ $(i=1,2\ldots\ldots k)$ for the centre line of seal ring, the fitting curve equation $\beta_i(\theta)$ $(i=1,2\ldots\ldots t)$ for the boundary of the valid seal area, the fitting curve equation $\gamma_i(\theta)$ $(i=1,2\ldots\ldots w)$ for the boundary of the seal gasket. The calculation method of parameters L,S and S_2 are shown as follow:

$$L = \sum_{i=1}^{k} \int_l \alpha_i(\theta) \, ds \tag{14}$$

$$S = \sum_{i=1}^{t} \int_0^{2\pi} \frac{1}{2}[\beta_i(\theta)]^2 \, d\theta \tag{15}$$

$$S_2 = \sum_{i=1}^{w} \int_0^{2\pi} \frac{1}{2}[\gamma_i(\theta)]^2 \, d\theta - \sum_{i=1}^{m} \int_0^{2\pi} \frac{1}{2}[\varphi_i(\theta)]^2 \, d\theta \tag{16}$$

Distill parameters from CAD model. If the workpiece's CAD model is available, draw the reference lines and planes on the CAD model according to Fig.5, and the parameters L, S, S1 and S_2 can be distilled from its model. The distillation process based on SolidWorks 2012 as follows:

(1) Invoking CreateObject("SldWorks.Application") to create SolidWorks session.

(2) Invoking ISldWorks::OpenDoc7 method to open the workpiece's CAD model and ISldWorks:: ActiveDoc method to obtain the document object.

(3) Select the centre line of seal ring and invoking IModelDoc2:: SelectionManager method to obtain the selection manager object. Using SelectionManager:: GetSelectedObject5 method to obtain the object of the centre line of seal ring.

(4) Invoking IEdge:: GetCurve method to obtain the curve of the centre line object.

(5) Invoking ICurve::GetLength method to obtain the circumference of the curve (parameter L).

(6) Select the plane of the valid seal area and invoking SelectionManager:: GetSelectedObject5 method to obtain the plane object.

(7) Invoking IFace2::GetArea method to obtain the area of the valid seal plane (parameter S).

(8) By the same method,We can obtain parameters S1 and S2.

Summary

This paper established the method of born size calculation in line seal and in face seal respectively, and presented the subsection curve fitting-integration method and CAD integration method to acquire the parameters of plugging surface. Cases show that the plugging quality and efficiency is raised and the design period is shorted.

Acknowledgement

Project supported by the national science & technology support program, China(No. 2013BAF06B00),the research fund for doctoral program of higher education,China(No. 20100032110006),the major project supported by Tianjin science & technology support program, China(No. 12ZCZDGX01600),and the Fujian science & technology program, China(No. 2012H1008).

References

[1] Janez Tusek, Bozidar Bajcer,Zlato Kampus,On line leak testing of welded water heaters, Journal of Materials Processing Technology, 3(2004) 1164-1170.

[2] Zhang Yong,Yang Mengtao, Control system reconstruction of leak tightness detection equipment for automobile water pump, Modern Manufacturing Engineering,8(2012)112-115,132.

[3] Pascall, M.A, Evaluation of a laboratory-scale pressure differential (force/decay) system for non-destructive leak detection of flexible and semi-rigid packaging, Packaging Technology and Science,4(2002), 197-208.

[4] Changsoo Jang, Byeng Dong Youn, Ping F. Wang,etc, Forward-stepwise regression analysis for fine leak batch testing of wafer-level hermetic MEMS packages, Microelectronics Reliability,4(2010),507-513.

[5] HU Hao,ZHONG Liqiong,YIN Cunhong,Simulation Research on the Leak Detection of Differential Pressure Leak Detector, MACHINE TOOL & HYDRAULICS,17(2011),37-41.

[6] Ha, Tae-Woong,Lee, Yong-Bok; Kim, Chang-Ho,Leakage and rotordynamic analysis of a high pressure floating ring seal in the turbo pump unit of a liquid rocket engine,Tribology International,3(2002), 153-161.

[7] Liu, Zhong, Liu, Ying, Liu, XiangFeng, Optimization design of main parameters for double spiral grooves face seal,Science in China, Series E: Technological Sciences,4(2007), 448-453.

[8] De Levie, Robert, Curve fitting with least squares, Critical Reviews in Analytical Chemistry, 1(2000), 59-74.

Advanced Materials Research Vol. 819 (2013) pp 292-296
© *(2013) Trans Tech Publications, Switzerland*
doi:10.4028/www.scientific.net/AMR.819.292

The Delayed Correlation Envelope Analysis Technique based on Sparse Signal Decomposition Method and Its Application to Bearing Early Fault Diagnosis

Daiyi Mo[a], Lingli Cui , Jin Wang ,Yonggang Xu

College of Mechanical Engineering and Applied Electronics Technology, Beijing University of Technology, Beijing 100124, China

[a]fendoumdy@163.com

Keywords: Sparse-Decomposition, Delayed Correlation Envelope, Gabor-Atom, D-value of Adjacent Residual Energy

Abstract： In order to extract the early weak fault information submerged in strong background noise of the bearing vibration signal, a delayed correlation envelope technique based on sparse signal decomposition method is proposed. This method can improve the signal to noise and extract the fault information efficiently. For the strong noise problem in the early fault, based on D-value of adjacent residual energy as the termination condition of the iterative method, reducing the noise effectively, and combine it with delayed correlation to enhance the denoising effect. The analysis results of roller bearing experimental data confirm the feasibility and validity of this method.

Induction

Rolling element bearings are one of the most important and frequently encountered components in a wide variety of rotating machinery, Its running state is normal or not often directly affect the performance of the whole machine, As a result, the condition monitoring and fault diagnosis of rolling element bearings has important practical significance. In the early fault diagnosis of bearing, due to the vibration signal is very complicated, and it often mixed with a large amount of background noise, which makes the diagnostic work becomes quite difficult[1]. To get reliable analysis conclusion, it should extract fault characteristic signal effectively from the noise signal. Resonance demodulation technique is based on that the fault bearing vibration signal appears modulation phenomenon, selecting the modulation signal which carry fault information by the band-pass filter, then separating the fault signal from the modulation signal by envelope demodulation technique, and diagnose the bearing became faulted or not accord to whether its spectrum is contain obvious bearing fault characteristic frequency[2]. However, the traditional resonance demodulation had the band-pass filter parameter (center frequency and filter bandwidth) need to be human determine and fixed filter band defects. Aiming at the defect, domestic and overseas scholars have been studied extensively. Literature [3] used the wavelet packet coefficient entropy threshold to rolling bearing vibration signal demodulation analysis, putting forward the enhanced resonance demodulation method, providing a new idea that overcome the traditional resonance demodulation problem in determining the resonance, extracting the resonance zone obviously, however, as the extraction effect of resonance has big relation to the scale N and entropy threshold h_E, so its parameter optimization problem need to be studied further. Literature [4] use the short time Fourier transform to rolling bearing vibration signal demodulation analysis, which determine the optimal band-pass filter band accord to the kurtosis of different frequency band demodulation signal, but it has only signal resolution and a large amount of calculation, can be used for general stationary signal, but for rolling bearing vibration signals, which are non-stationary signals, and when the signal waveform change sharply, the main frequency is high frequency, which require higher time resolution, when the signal waveform change gently, the main frequency is low frequency, which require higher frequency resolution, apparently, STFT cannot give attention to the both. In order to achieve the more flexible, concise and adaptive expression of signals, Mallat and Zhang summed up previous research achievements based on wavelet analysis and raised the idea of decomposing signals on the over-complete dictionary in 1993: the basic function was replaced by the over-complete redundant function which was known as the atom dictionary, and the elements in the atom dictionary were known as atoms; atoms with the optimal linear combination were picked out from the atom dictionary to represent a signal, which was known as the Sparse Approximation of the signal[5]. Besides, a matching pursuit (MP) algorithm based on the time-frequency atom dictionary was proposed. The algorithm adopted a strategy of obtaining the sparse expression of signals by gradual approximation. A group of primitive functions, i.e., atoms, were selected from the atom dictionary to calculate a linear expansion of signals and to achieve the successive approximation of signals by solving the rectangular projection of the signals on each atom. The proposal of the sparse decomposition based on the MP algorithm aroused extensive interest among researchers who did a lot of work on the optimization and improvement of the algorithm as well as the expansion of its application fields [6-7].

Using the advantages of signal sparse decomposition, a delayed correlation envelope technique based on sparse signal decomposition method is proposed, matching the impact component of signals gradually, and the impact components in resonance zone are more obvious than other zones. Sparse decomposition has good noise reduction effect, and reconstructed signals can secondary denoise effectively by delayed correlation, so it could extract the resonance zone of weak fault signal from strong noise by choosing appropriate iterative termination conditions. The algorithm uses the Gabor atomic to decompose the original fault signal, and uses delayed correlation to reconstructed signals, extracting the fault characteristic frequency, providing a powerful tool to bearing early faults. The analysis results of roller bearing experimental data confirm the feasibility and validity of this method.

The basic principle of MP based on Gabor atomic
The consist of Gabor atomic [8]

A Gabor atom consists of a cosine-modulated Gaussian window function:

$$g_\gamma(t) = \frac{1}{\sqrt{s}} g(\frac{t-u}{s}) \cos(vt + w)$$

where, $g(t)=exp(-\pi t^2)$ is the Gaussian window function, $\gamma(s, u, v, w)$ are the time-frequency parameters. Where, s is scale factor, u is shift factor, v is frequency factor, w is phase factor. The space of time-frequency parameters can be discretized as $\gamma=(a^j, pa^j\Delta u, ka^{-j}\Delta v, i\Delta w)$, with $a=2$, $\Delta u=1/2$, $\Delta v=\pi$, $\Delta w=\pi/6$, $0<j\leq log_2 N$, $0\leq p\leq N2^{-j+1}$, $0\leq k\leq 2^{i+1}$, $0\leq i\leq 12$.

The choice of iterative termination conditions

Generally speaking, the set of MP iterative termination conditions has the following two methods:

(1)Set hard threshold. It means to set the iterative termination conditions to the upper limit of iteration, and use the linear combination of m atoms to reconstruct the approximation of noise-signal, and the residual signal as noise. But this method cannot determine the upper limit of iteration m accurately, too small m will influence the accuracy of signal reconstruction, too big will weaken the denoising effect, similarly, will influence the accuracy of signal reconstruction.

(2)Set the residual signal energy less than a certain threshold value as iterative termination conditions. For large SNR signal, this method cannot influence the accuracy of signal reconstruction obviously. However, if the SNR of signal is too small, the residual threshold is not easy to set.

As the research object is the signal of rolling bearing early fault, whose SNR is smaller , therefore , in order to further enhance the robustness of iterative termination conditions, choose adjacent residual energy difference value is less than a critical value as modified iterative termination conditions, when adjacent residual energy difference value change gently, residual signal is noise basically. Therefore, it could choose adjacent residual energy difference value flexibly based on different needs.

The method to improve the speed of signal decomposition

As to norm length signal, the atomic number is a large number with the Gabor atom, for example, if the signal length N is 256, the atomic number is 119756, such a great atomic library will make it difficult to decomposition process, and this paper use the following method to improve the decomposition rate.

(1) Not form atomic, generating atom on one side, selecting the best atom and delete atoms on the other side.

(2) Use the global optimization performance of genetic algorithm, the Gabor atomic expression as the parameters of solving parameters, the inner product of signal and the unknown parameters of the Gabor atom as the adaptive function, realize the best match analysis result.

The Delayed Correlation

Although it could reduce the noise of the signal effectively base on Gabor atomic signal sparse decomposition method, but in bearing early weak fault information, due to the strong background noise influence, the SNR is very low, it may contain some noise interference in reconstruction signal. For broadband random noise signal, whose correlation function $R_x(\tau)$ quickly to zero with enlargement of τ, correlation function has the characteristic of noise reduction, therefore, to the reconstruction signal, it could enhance the denoising effect with delayed correlation, as a result, increase the robustness of the algorithm[9-10].

Bearing experiment signal analysis

The experiment data was obtained through experiment system, which is made of bearing test ben ch, data acquisition instrument named HG3528A and computer. Where, test bench (Fig.1) contained three phase asynchronous motor through the coupling which equipped with rotor shaft connection, s haft is supported by two types of 6307 deep groove ball bearing. Bearing point corrosion failure size is 0.1 mm, motor speed $n=1496r/min$, bearing large diameter D=80 mm, inside diameter d=35mm , Rolling element number Z=8, contact angle a=0 degree. Take the above parameters to the correspo nding fault characteristic frequency calculation formula, calculating that the bearing's outer fault cha racteristic frequency is 76.7Hz, the bearing's inner fault characteristic frequency is 122.7Hz, the be aring's ball fault characteristic frequency is 51.7Hz, sampling points is 8192 and sample frequency i s 15360Hz.

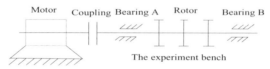

Fig.1 Experiment system

Selecting the inner ring of the single point early pitting fault test data to analysis, the original signal waveform and spectrum diagram is shown in Fig.2.

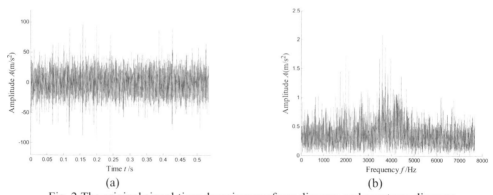

(a)   (b)

Fig. 2 The original signal time domain waveform diagram and spectrum diagram

Processing the original signal by sparse decomposition based on Gabor atoms, selecting different adjacent residual energy difference, automatic extraction of resonance zone as shown in Fig. 3:

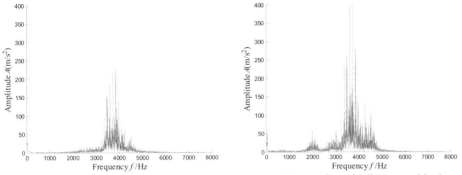

(a) D-value of adjacent residual energy less than 200　　(b) D-value of adjacent residual energy less than 160

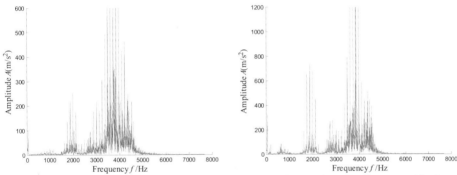

(c) D-value of adjacent residual energy less than 120　　(d) D-value of adjacent residual energy less than 40

Fig.3 The resonance zone extracted base on D-value of adjacent residual energy

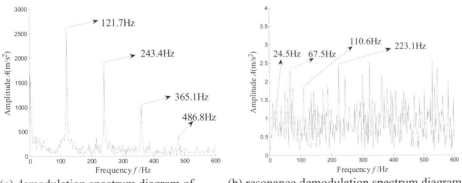

(a) demodulation spectrum diagram of the resonance zone　　(b) resonance demodulation spectrum diagram of the original signal

Fig.4　Spectrum diagram

It can be seen from Fig.3, extracted two order resonance zone which are 1620 ~ 2240 Hz and 3250 ~ 4450 Hz successfully by using the proposed method. And the selection of adjacent residual energy difference has great influence on extraction of resonance, when it change between 200-160, extraction of resonance could not change obviously, all extract the resonance zone close to 3250 ~ 4450Hz, when adjacent residual energy difference changed from 120 to 40, further extract the resonance zone of 1620~2240Hz, but adjacent residual energy difference can not too small, otherwise it cannot reflect the essence of the sparse decomposition. Take the resonance zone which were extracted from the adjacent residual energy difference less than forty for demodulation analysis, get the demodulation spectrum, as shown in Fig.4(a), where, the fault characteristic frequency of 121.7 Hz (close to the inner fault characteristic frequency) and its all order harmonic frequency is quite clear, and we can draw a correct diagnosis, confirming the feasibility and validity of this method. While, use the traditional resonance demodulation to the original signal, obtain the traditional resonance demodulation spectrum diagram, as shown in Fig.4(b), rotational frequency and other frequency exist, will affect the accurate judgement of the fault element.

Conclusion

(1) Aimed at the problem of traditional signal sparse decomposition algorithm, whose atomic library is too big, signal decomposition speed is slow and decomposition is not sparse, not structure atomic library sparse decomposition method is adopted in this paper, and combining it with genetic algorithm, thus enhancing the signal decomposition and reconstruction speed greatly, as a result, enhance the usefulness of the algorithm.

(2) Direct at early failure fault problems, putting forward a modified iterative termination conditions based on adjacent residual energy difference value, making the sparse decomposition has not only the advantage of adaptability and sparse etc, and can more reduce the noise in the original signal effectively, improve SNR. Combining the sparse decomposition and delay correlation, can detect and diagnose the rolling element bearing early weak traumatic failure better.

Acknowledgment

This work is supported by National Natural Science Foundation of China (51175007).

Reference

[1] Mei Hongbin. Vibration monitoring and diagnosis of rolling bear[M]. Beijing: China Machine Press, 1995.

[2] Ding Fang, Gao Lixin, Cui Lingli. Application of resonance demodulation technology in equipment's fault diagnosis[J]. Machinery Design & Manufacture, 2007(11)：178-179.

[3] Cui Lingli, Kang Chenhui, Zhang Jianyu.Enhanced Resonance Demodulation Based on the Delayed Correlation and Entropy Threshold of Wavelet Packet Coefficients[J]. Chinese Journal of Mechanical Engineering, 2010, 46(20): 53-58.

[4] Ding Xiawan, Liu Bao, Liu Jinzhao. Fault Diagnosis of Freight Car Rolling Element Bearings with Adaptive Short Time Fourier Transform[J]. Chinese Journal of Railway Science, 2005, 26(6) :24-27.

[5] Mallat S, Zhang Z. Matching pursuit with time-frequency dictionaries[J]. IEEE Trans. On Signal Processing, 1993, 41(12):3397-3415.

[6] Remi Gribonval. Fast Matching Pursuit with a multiscale dictionary[J]. IEEE Transactions on Signal Processing, 2001: 994-1001.

[7] McClureM.R, Carin L. Matching pursuits with a wave-based dictionary[J]. IEEE transactions on .Signal Processing, 1997, 45(12): 2912-2927.

[8] A. Lobo and P. Loizou, Voiced/unvoiced speech discrimination in noise using Gabor atomic decomposition [J]. IEEE Proc Int Conf Acoust Speech Signal Process (ICASSP), 2003, 1: 820–823.

[9] Sun Hui, Zhu Shanan. Hilbert-Huang transform demodulation based on delayed auto correlation pretreatment[J]. Journal of Zhe jiang University, 2005, 39 (12):1998-2001

[10] Meng Tao, Liao Mingfu. Detection and diagnosis of the rolling element bearing fault by the delayed correlation envelope technique[J]. Acta Aeronautica et Astronautica Sinica, 2004, 25(1)：41-44.

CHAPTER 3:

Advanced Design Technology, Optimization and Modelling

Advanced Materials Research Vol. 819 (2013) pp 299-303
© (2013) Trans Tech Publications, Switzerland
doi:10.4028/www.scientific.net/AMR.819.299

An Improved Parallel Computation Method for Delaunay Triangulation

Zhiyu Chen[a], Jianzhong Fu, Hongyao Shen and Wenfeng Gan

The State Key Lab of Fluid Power and Transmission and control, Zhejiang University, Hangzhou, China

[a]chenzy713@163.com (Corresponding author)

Keywords: Delaunay Triangulation; compound algorithm; parallel computation; uniform grid

Abstract: Amongst the flourishing Delaunay Triangulation methods, growth algorithm has been widely accepted because of its reputation of being simple and elegant. However, the parallelization of growth algorithm has not been fully exploited. In this work, a novel Growth algorithm of Delaunay Triangulation is proposed. The point cloud is first divided into two parts by a suitable curve and the separated areas are calculated by incremental algorithm. Triangles which cross with the curve are generated by a growth algorithm associated with uniform grid. At the process of merging, these grew triangles are used to detect incorrect triangles of the incremental algorithm areas. Method about generating triangles on curve is elaborated and a simple way to detect interferential triangles is also explained. With above method, triangulation calculation can be parallelized. Unlike the traditional divide-and-conquer method, no flip operation is needed in the proposed methodology. Thus, three dimensional applications are also made possible. A comparative research between tradition incremental algorithm and the proposed method has been conducted. Results show, the algorithm has a higher performance with less computation time.

Introduction

Triangulation over an arbitrary point cloud is an inevitable technique in the realm of reverse engineering (RE), finite element analysis and CAD/CAM. Amongst all triangulation methods, Delaunay Triangulation (DT) is most widely used. Two unique properties distinguish DT from any other methods, i.e. the empty circum-circle property and the largest minimum interior angle property. Empty circum-circle property means that, for an arbitrary triangle in the triangulated mesh, there will be no point located inside its circum-circle. The largest minimum interior angle property guarantees that the minimum interior angle of DT is the largest when comparing with the minimum interior angle among all triangulation methods[1]. Due to such characteristics, narrow triangles are diminished during DT operation, resulting in a high quality mesh[2].

In all the flourishing DT methods, growth algorithm, divide-and-conquer algorithm, and incremental algorithm have been widely accepted. Growth algorithm came into existence in 1980's[3] making advantage of the empty circum-circle property. The basic procedure of growth algorithm starts from a line. Then find the third point to construct a triangle which has no other points in its circum-circle. This procedure is repeated on the newly generated edges to create more triangles. The iteration stops till all points are used up. Incremental algorithm employs another strategy of triangle construction. A giant triangle containing the whole point cloud is created firstly. Then points are inserted inside this triangle one by one to split it into smaller triangles. During the insertion, flip operation of edges according to Bowyer-Watson principle[4]is performed When all points have been inserted, the desired mesh will be obtained after deleting the triangles which have relationship with the initial giant triangle. Divide-and-conquer algorithm[5] first divides the point clouds into several sets. Each of them is triangulated to mesh patches by using growth algorithm. Then all patches are combined to get the final mesh. Flip operation are performed along boundaries between the patches. Often than not, edges inside each patch should be flipped again because of the changes on the boundaries[6]. In a word, algorithms including flip operation are computational complex.

Moreover, flip operation fails in three-dimensional triangulation if two adjacent tetrahedrons violate the empty circum-circle property[7]. In order to solve the problem, P Cignoit proposed a divide-and-conquer algorithm called Dewall algorithm[8]. Point cloud is separated by divider lines/faces iteratively thus through growth algorithm triangles/tetrahedrons can be generated on these lines/faces in parallel. Flip operation is excluded from the algorithm so it is implementable in both 2D and 3D. However, all triangles generated by just growth algorithm will cause high complexity, it makes the goal of parallel computing fail.

In this work, a novel algorithm feasible for both 2D and 3D is proposed. It combines the merits of both Dewall algorithm and incremental algorithm. Point cloud is still separated by divider lines, but not all triangles are generated on them. In the Areas with no divider lines, triangles are calculated by incremental algorithm, and triangles on divides lines are obtained growth algorithm.

The rest of the paper is organized as follows. Section 2 presents a low complexity uniform grid based growth algorithm. In section 3 the flow path of triangles generating on curve is provided and the process of merging separated areas is explained. Besides, a 3D example is also given here. To verify the utility, the compare test between incremental algorithm and the proposed algorithm is carried out in section 4. Section 5 draws the conclusion.

Uniform Grid Based Growth Algorithm

The complexity of growth algorithm is $O(n^2)$ for the search of the corresponding point and the judge for validity of searched point are both n times. The search and judge times can be reduced by using uniform grid to just verify the vicinity points[9]. Since one edge can only belong to two different triangles, the complexity can be reduced more by adopting a strategy to avoid checking points that lie on the triangulated side. The three steps in uniform grid based growth algorithm are points dividing, first line construction and triangle growth, respectively. It is shown in Fig. 1.

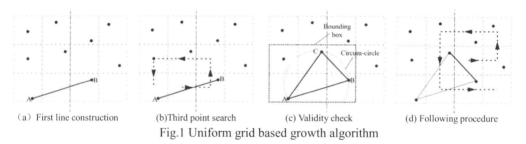

（a）First line construction   (b)Third point search   (c) Validity check   (d) Following procedure

Fig.1 Uniform grid based growth algorithm

In points dividing, the extreme values of points are found and defined as $X_{min}, X_{max}, Y_{min}, Y_{max}$. Tolerance T is set so two points within this tolerance are considered coincidence. The starting point of the grid is set as $(X_{min} - T, Y_{min} - T)$. For the purpose to make the average points number in each grid is one, the length and height of the grid is set as $(X_{max} - X_{min})/\sqrt{n}$ and $(Y_{max} - Y_{min})/\sqrt{n}$ respectively. In first line construction, due to the lines of convex hull are parts of the Delaunay mesh, so in convex hull the line which crosses with divider line can be taken for the first line. Particularly when the divider line is vertical, the first line is just the connecting line of the lowest point at the two sides, as Fig. 1(a) shows. For a points set which already has the grid, search of the lowest row of the grid is usually enough to obtain the first line.

After the first line is constructed, the left endpoint defines as point A and the right endpoint defines as point B. Starting from the grid which contains the middle point of the line AB, a third point is searched for generating the target triangle in reverse spiral, see Fig. 1(b). Defining the checking point as point C, the cross product $\overrightarrow{AB} \times \overrightarrow{AC}$ is calculated. If the result is negative, the checking point C must be at the triangulated side of line AB and it is an invalid point. The empty circum-circum test can be significantly reduced through this judgment, but the right choosing line direction is necessary. When the cross product is positive, the circum-circle of triangle $\triangle ABC$ is calculated firstly, and then the empty circum-circle test is done in the bounding box. If all the points in the bounding box meet the property of the empty circum-circle, point C is the desired point. By this method, the low relevancy points can be eliminated, this operation is shown as Fig. 1(c).

Parallel DT Computation

Grow triangle along the curve. A curve which passes through a triangular mesh will cross a series of consecutive triangles. These triangles will divide the mesh into two different parts. In a similar way, the consecutive triangles and the separated areas can be calculating simultaneously and then be combined into one. Generating triangles on curve is mostly using the uniform grid based algorithm, what differs is the choosing of new starting line. Every time when a new triangle is generated, there will be two new line segments. Choose the segment which crosses with the curve as the new starting line and repeat the procedure until no DT point can be found.

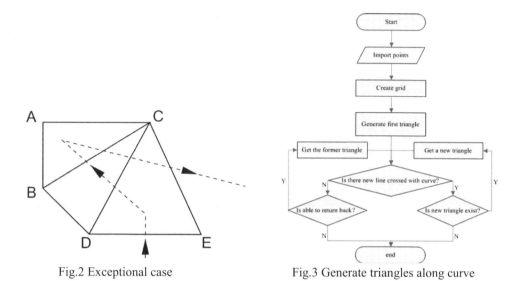

Fig.2 Exceptional case   Fig.3 Generate triangles along curve

There will be two kind exceptional cases in the algorithm. Exception 1 is that both two new lines are not crossed with the curve, $\triangle ABC$ in Fig. 2 is the example. At this time, a return strategy is proposed. If there is a line of triangles can generate new triangle, this line will be used as a new starting line, see line CE of $\triangle CDE$ in Fig. 2. Exception 2 is triangle recreating, see $\triangle BCD$ in Fig. 2. When using the return strategy, the first starting line obtained will be either BC or CD. There will be recreated triangle $\triangle BCD$ or $\triangle CDE$ if BC or CD is used to generate triangles. This can be eliminated by using a data structure which stores the status of the line. If one line already has two triangles associated with it, this line would not create triangle again. The flow chart is shown in Fig. 3.

Splicing method. There would be interference at the vicinity of the divider line if two patches merge directly, it is shown in Fig. 4. Triangles on the divider lines are always correct because they are calculated on the global concern, while some incorrect triangles at the left side got by incremental algorithm need to be deleted. There is a rule to verify the triangle validity in the incremental side, that is if two lines have a counterclockwise angle which exceed $180°$ at the divider line side, the triangle comprising these two lines is a valid one. For example, the counterclockwise angle of line BC and line CD is more than $180°$ so the $\triangle BCD$ of left side is a valid one. However, there is too much possible invalid triangles which contains the point near the divider line, but not all of them do exists. It is prohibitive to mark all the invalid triangles because only valid triangle can be found through the above rule. If simply adding a deleting flag on certain points to mark the vertices of incorrect triangles, it erroneously deletes the correct one. For example, if point C,D,E,F are marked as incorrect triangle vertex, while the incorrect triangle $\triangle BCD$ $\triangle CDF$ $\triangle DEF$ are deleted, the correct triangle $\triangle CDE$ (if it exists) will also be deleted. To solve this problem, a new way to mark the points is proposed.

With the concept of digit, a new property is added to the points of the triangles on the divider line. Suppose there are only four digits of added value. Before mark the points, a value called tag is set as 0001. Each time when a correct triangle is found, the tag value is left shifted and then an OR operation is done with corresponding points digit value. When in the procedure of deleting triangles, a AND operation is done among three triangle points. Only the triangle whose result is 0001 will be deleted. To explain this there is a examples in the Fig. 4. $\triangle BCD$ is a correct one, so tag value left shifts to 0010 and a OR operation is done with B,C,D, the digit value of them becomes 0010. Then $\triangle CDE$ is judge as a correct one so the tag left shifts again to 0100, and after the OR operation, the digit value of C,D,E are 0111,0111 and 0101 respectively. After all these operation, the incorrect $\triangle ACF$ & $\triangle CDF$ etc. can be detected easily.

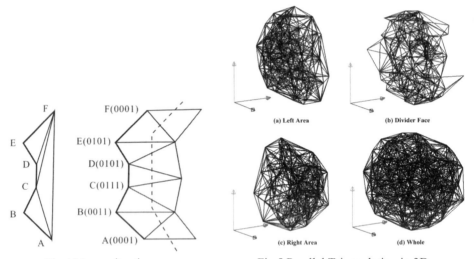

Fig.4 Merge situation   Fig.5 Parallel Triangulation in 3D

This method is not confined in 2D triangulation. In 3D triangulation a divider face was constructed firstly to obtain the tetrahedrons that separate the points. Points of different areas are triangulated individually and then be combined into one. A 3D triangulation case of 500 points is shown in Fig. 5. **Degeneracy.** Using this divide and conquer algorithm will meet two kinds of degeneracies. Degeneracy 1 is the point coincidence, which will prevent the incremental algorithm proceeding when coping with the coincident point. In this paper coincident point is eliminated when the point is adding into the grid. The adding point is going to be checked with other points of same grid whether they are under tolerance T. Point whose check result is under the tolerance will be considered as a invalid points. Degeneracy 2 is that multiple points have the same one circum-circle. When multiple points have a same circum-circle, the DT mesh of these points is not unique. But the largest minimum interior angle is unchanged though degeneracy exists, so set a proper tolerance will be enough to generate the right mesh.

Implementation and Analysis

The algorithm is implementation by C++ in visual studio 2008. Comparing with other algorithms, the $O(n^2)$ complexity of growth algorithm is higher,. Even the strategy of uniform grid is adopted, the complexity is still $O(n^{3/2})$. Nevertheless, the mesh generated by this algorithm is very small part of the total. The proportion of triangles generated by growth reduces progressively when total point number grows. In the test, the proportion is always under 1% when the total point number is beyond 10000 and reaches 3‰ when the number is 100000.

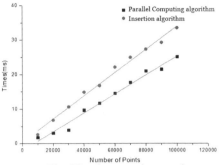

Fig.6 Implementation result

Fig. 6 shows the relationship between point number and triangulation time. The test shows the proportion of the triangles which are generated by growth algorithm or need to be deleted is always under 2%. This proportion decreases while the point number grows. Because the algorithm can use two CPU to calculate, the calculating time decreased significantly.

Conclusion

In this paper, the merits of growth algorithm and incremental algorithm are combined by using a curve to divide the point cloud into separated parts. These parts are calculated in incremental algorithm and merging together by growth algorithm to obtain the final mesh. This algorithm is not only easily achievable but also stable when the degeneracies exist. No flip operation make it can be extended to three-dimensional, which meets the trend ofparallel computation.

Acknowledgments

This work was financially supported by the Program for Zhejiang Leading Team of S&T Innovation (No.2009R50008), National Nature Science Foundation of China (No. 51105335).

Reference

[1]   C.J. Du, An algorithm for automatic Delaunay triangulation of arbitrary planar domains. Advances in Engineering Software[J], 1996. 27(1-2): p. 21-26.

[2]   J.Yu, J. LU, C.W. ZHENG, A Comparative Research on Methods of Delaunay Triangulation. Journal of Image and Graphics[J] in Chinese, 2010(08): p. 1158-1167.

[3]   D.T. Lee, B.J. Schachter, Two algorithms for constructing a Delaunay triangulation. International Journal of Computer & Information Sciences[J], 1980. 9(3): p. 219-242.

[4]   M. Zadravec, B. Zalik, An almost distribution-independent incremental Delaunay triangulation algorithm. Visual Computer[J], 2005. 21(6): p. 384-396.

[5]   S.W. Yang, Y. Choi, C.K. Jung, A divide-and-conquer Delaunay triangulation algorithm with a vertex array and flip operations in two-dimensional space. International Journal of Precision Engineering and Manufacturing[J], 2011. 12(3): p. 435-442.

[6]   M.B. Chen, T.R. Chuang, J.J. Wu, Parallel divide-and-conquer scheme for 2D Delaunay triangulation. Concurrency and Computation-Practice & Experience[J], 2006. 18(12): p. 1595-1612.

[7]   Z. Gao, Z. Yu, M. Holst, Quality tetrahedral mesh smoothing via boundary-optimized Delaunay triangulation. Computer Aided Geometric Design[J], 2012. 29(9): p. 707-721.

[8]   P. Cignoni, C. Montani, R. Scopigno, DeWall: A fast divide and conquer Delaunay triangulation algorithm in E-d. Computer-Aided Design[J], 1998. 30(5): p. 333-341.

[9]   T.P. Fang, L.A. Piegl, Delaunay triangulation using a uniform grid. IEEE Computer Graphics and Applications[J], 1993. 13(3): p. 36-47.

Advanced Materials Research Vol. 819 (2013) pp 304-310
© (2013) Trans Tech Publications, Switzerland
doi:10.4028/www.scientific.net/AMR.819.304

Case-based Reasoning Rapid Design Approach for CNC Turret

Wang Haiqiao[1, 2, a], Sun Beibei[1] and Shen Xianfa[2, 3]

[1]School of Mechanical Engineering, Southeast University, Nanjing 211189

[2]School of Mechanical Engineering, Sanjiang University, Nanjing 210012

[3]College of Engineering, Nanjing Agricultural University, Nanjing 210031, China

[a]wanghaiqiao_000@163.com

Keywords: case-based reasoning (CBR); case retrieval; similarity; combination weight; CNC turret

Abstract: In order to provide more efficient knowledge services in the CNC turret design process, a rapid design method of a case-based reasoning is proposed. Firstly, according to different types of demand in case retrieval, the similarity measurement models for crisp and fuzzy attribute type demands are constructed respectively. Secondly, in the weights assignment, this paper utilized the deviation information of similarity values to calculate objective weights, and then combined the objective weights and subjective weights to form synthesis weights. Finally, the similarity measurement and weights coefficient assignment methods were applied in a CNC turret design CBR system, and using the calculation function of MATLAB. It was demonstrated that this method could improve the accuracy of case retrieval.

CBR Composition

CBR overview. CBR is one of the emerging paradigms for designing intelligent systems. In essence, it is the use of past experiences as a basis for dealing with novel problems. In CBR, each case is stored alongside its known solution in a database, referred to as the case-base (CB)[1]. This CB is reviewed for each new problem being solved. This methodology differs from other learning problem-solving paradigms in that the cases themselves are used as the basis for coping with a new situation, and not some derivative representation, as found in adaptive neural-nets, adaptive methods in fuzzy control or in other inductive methods. As shown in Figure 1, the case-based reasoning system can be finished by four steps[2].

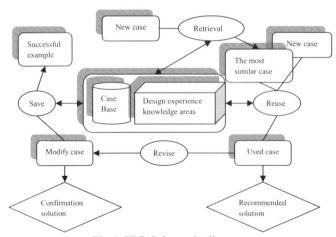

Fig.1 CBR Schematic diagram

Overall frameworks for similarity calculation. Case retrieval in CBR systems often come with a certain degree of ambiguity, because the same instance with new problems rarely found in the instance library, and more likely to find similar instances with new problems. Figure 2 shows the overall framework of similarity. First, according to the precise attributes and fuzzy attribute similarity calculation model to calculate a new instance of the instance library instance attribute similarity matrix; then combined with subjective weights and similarity-based objective weight deviation calculated using the multiplication synthesized instance each attribute combination weight, based on the similarity matrix and the right combination of re-calculate the global similarity; Finally, solving the new instance based on the most similar instance[3].

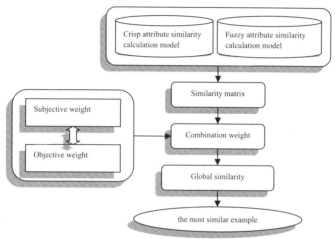

Fig.2 Overall similarity calculation framework

Similarity measures for CBR

CNC turret attribute value description is incomplete determine the type, there is a fuzzy attribute values, concrete can be subdivided into five categories: deterministic numerical Crisp numeric values (CN), determine the type symbol Crisp symbolic values (CS), fuzzy type value Fuzzy number values (FN), fuzzy interval Fuzzy interval values (FI), and fuzzy symbols fuzzy linguistic values (FL). If we are still in accordance to the general similarity calculation method, the fuzzy attribute values retrieval will fail, so here we study similarity calculation model for all types of property values[4].

Similarity Measure for Crisp Sets. Commonly, CBR systems use the inverse of weighted normalized Euclidian distance or Hamming distance as the similarity measure. That is use Euclidian distance and Hamming distance respectively.

$$SIM(X,Y) = 1 - DIST(X,Y) = 1 - \sqrt{\sum_i w_i^2 dist^2(x_i, y_i)} \tag{1a}$$

$$SIM(X,Y) = 1 - DIST(X,Y) = 1 - \sum_i w_i dist(x_i, y_i) \tag{1b}$$

Where X, Y represent the two cases whose similarity is to be measured. Weight w_i is normalized to denote the importance of the i th attribute, and i takes a value from 1 to n, where n is the number of attributes. The normalized dist (x_i, y_i) is often represented as:

$$dist(x_i, y_i) = \frac{|x_i - y_i|}{|max_i - min_i|} \tag{2}$$

Where x_i, y_i are the ith attribute values in the two cases. For crisp numerical attributes, "max_i" and "min_i" denote the maximum and minimum values of the ith attributes, respectively. For crisp symbolic attributes, dist (x_i, y_i) is equal 0 if $x_i = y_i$; otherwise, dist $(x_i, y_i) = 1$. Eq. (1) shows that when DIST=0, SIM attains the maximum value 1, meaning that the two cases being evaluated are identical. When DIST is 1, SIM has the minimum value 0, meaning that they are completely different. **Similarity Measure for Fuzzy Sets.** Similarity Measure for Crisp Sets is not suitable for Fuzzy Sets either. Accuracy and simplicity are the other major considerations. Our area ratio method has advantages in both accuracy and simplicity. The area ratio method calculates the similarity between two fuzzy values as the ratio of the overlapping area of two corresponding membership functions to the total area as follows:

$$sim(x_i, y_i) = \frac{A(x_i \cap y_i)}{A(x_i \cup y_i)} = \frac{A(x_i \cap y_i)}{A(x_i) + A(y_i) - A(x_i \cap y_i)}$$

(3)

Where A represents the area of the corresponding membership function and $x_i \cap y_i$ is the intersection of two fuzzy sets. For fuzzy number and interval attributes, five types of intersection can be classified as depicted in Figure 3. For fuzzy linguistic attributes, only types (a), (b) and (c) apply. The following algorithm is developed to calculate the fuzzy similarity sim(x, y) for these types [5, 6].

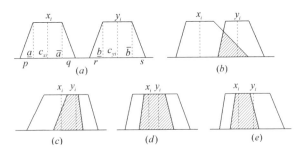

Fig.3 Five similarity types of two fuzzy sets x_i and y_i.

Algorithm FSM (Fuzzy similarity method):

$$c_{xi} = \frac{(a + \bar{a})}{2}, c_{yi} = \frac{(b + \bar{b})}{2} \qquad \text{//calculate centers of two fuzzy sets.}$$

if $c_{xi} > c_{yi}$, then exchange x_i and y_i, endif

$$x_i^* = \frac{(q\underline{b} + r\bar{a})}{r + q}, \; y_i^* = 1 - \frac{x_i^* - \bar{a}}{q} \qquad \text{//calculate intersection point } (x_i^*, y_i^*).$$

if $y_i^* \le 0$, then $A(x_i \cap y_i) \leftarrow 0$, $sim(x_i, y_i) \leftarrow 0$ // it belongs to type(a).

else $A(x_i) \leftarrow \dfrac{(2\bar{a} + q - 2\underline{a} + p)}{2}$ $A(y_i) \leftarrow \dfrac{(2\bar{b} + s - 2\underline{b} + r)}{2}$ // calculate $A(x_i)$, $A(y_i)$.

if $0 < y_i^* < 1$, then $A(x_i \cap y_i) \leftarrow \dfrac{(\bar{a} + q - \underline{b} + r)y_i^*}{2}$ // it belongs to type(b).

else   // it belongs to type(c),(d)or(e).

if $\bar{a} + q < \bar{b} + s$ and $\underline{a} - p < \underline{b} - r$ // it belongs to type(c).

$A(x_i \cap y_i) \leftarrow \dfrac{(2\bar{a} + q - 2\underline{b} + r)}{2}$

else $A(x_i \cap y_i) \leftarrow \min(A(x_i), A(y_i))$ // it belongs to type(d)or(e).

endif

endif

$$sim(x_i, y_i) \leftarrow \frac{A(x_i \cap y_i)}{A(x_i) + A(y_i) - A(x_i \cap y_i)}$$

endif

end_of_FSM

In our study, the similarities of two fuzzy linguistic attributes for all possible combinations of fuzzy sets are calculated and stored in the system in advance. When matching cases, these values are retrieved from the data-base if needed to save computing time. This is sensible because all fuzzy sets are predefined for each fuzzy linguistic attribute. For the similarity between two fuzzy number or interval attributes, the distance between centers c_{xk} and c_{xk} also considered, and the similarity is computed when two cases are matched. The weighted average operator is used to aggregate the similarities of all attributes.

In summary, a hybrid similarity measure proposed for matching two cases X and Y is:

$$SIM(X,Y) = \sum_i w_i sim_{CS}(x_i, y_i) + \sum_j w_j sim_{CN}(x_j, y_j)$$
$$+ \sum_k w_k (\varepsilon_{k1} sim1_{FNI}(x_k, y_k) + \varepsilon_{k2} sim2_{FNI}(x_k, y_k)) + \sum_l w_l sim_{FL}(x_l, y_l) \qquad (4)$$

Where Eq. (3) is used to calculate $sim1_{FNI}(x_k, y_k)$ and $sim_{FL}(x_l, y_l)$, and

$$sim_{CS}(x_i, y_i) = 1 \quad x_i = y_i$$
$$sim_{CS}(x_i, y_i) = 0 \quad \text{otherwise}$$

$$sim_{CN}(x_j, y_j) = 1 - dist(x_j, y_j) = 1 - \frac{|x_j - y_j|}{|\max_j - \min_j|}$$

$$sim2_{FNI}(x_k, y_k) = 1 - dist(c_{xk}, c_{yk}) = 1 - \frac{|c_{xk}, c_{yk}|}{|\max_k - \min_k|}$$

Where w_i , w_j , w_k and w_l are the weights for each attribute and $\sum w_i + \sum w_j + \sum w_k + \sum w_l = 1$; $i = 1,...,n1$; $j = n1+1,...,n1+n2$; $k = n1+n2+1,...,n1+n2+n3$; $l = n1+n2+n3+1,...n$, $n1$, $n2$, $n3$ and n are the number of fuzzy number and interval, fuzzy linguistic, crisp symbolic, and total attributes, respectively.

Properties weight calculation

Combination weight. The attribute weights essence property in a subjective assessment of the relative importance of the decision-making process and objectively reflect the comprehensive measure, by nature are divided into the following two categories: First subjective rights reflect the degree of importance of the property itself or the decision makers on the preferences of each attribute, experts direct weighting method and binary comparison method using said; Second objective weight, reflecting the impact of the attributes contained in the amount of information on the decision result, such weight and its attribute value of the program is directly proportional to the ability to distinguish between the attributes of each program distinguish ability should be given a relatively large weight, irrespective how, with the degree of importance of the attribute itself[7, 8].

The essence of attribute weights, the weight should be a combination of the above two types of weight, that is. The following describes the second objective weight calculation method.

Calculate objective weights based on similarity deviation information. New examples, for instance in the Case Base, the new instance library and attribute similarity attribute of all instances of similarity a similarity matrix:

$$S = \begin{bmatrix} s_{11} & s_{12} & \cdots & s_{1m} \\ s_{21} & s_{22} & \cdots & s_{2m} \\ \vdots & \vdots & & \vdots \\ s_{n1} & s_{n2} & \cdots & s_{nm} \end{bmatrix}$$

By the nature of the objective weight shows that the difference between the similarity matrix can determine attributes influence the search results. For the first attribute, the similarity matrix ($i = 1, 2, \ldots, n$) mutually smaller differences, should be given a smaller weight coefficient, because the influence of the attribute instance retrieved is small; if large difference between its influence, and that this property should be given a larger weighting coefficient, regardless of the subjective weighting coefficient[9].

Based on the above analysis of the similarity matrix in information related to the assignment of attribute objective weight coefficient can be calculated based on the similarity deviation information objective weight, the calculation expression for:

$$w_j^{(2)} = \frac{\sum_{i=1}^{n}\sum_{k=i+1}^{n}(s_{ij}-s_{kj})^2}{\sqrt{\sum_{j=1}^{m}[\sum_{i=1}^{n}\sum_{k=i+1}^{n}(s_{ij}-s_{kj})^2]^2}} \tag{5}$$

Where: $\sum_{i=1}^{n}\sum_{k=i+1}^{n}(s_{ij}-s_{kj})^2$ each instance of the j-th attribute similarity squares of deviations.

In this paper, multiplication synthesis calculated combination weight:

$$w_i = \frac{w_i^{(1)}w_i^{(2)}}{\sum_{j=1}^{m}w_j^{(1)}w_j^{(2)}} \qquad i = 1, 2, 3, \ldots, m. \tag{6}$$

Table 1 Example of the CNC turret attribute instance library

Attribute	type	subjective weights	CASE1	CASE2	CASE3	CASE4	CASE5	CASE6	New case
Turret Model	CS	0.05	ELT	ELT	ELT	HLT	HLT	SLT	SLT
Center height (mm)	CN	0.15	63	80	100	80	125	80	125
Knife-digit (N)	CN	0.15	8	8	12	8	12	12	12
Turn 30 ° and tighten the (S)	FN	0.15	0.41	0.48	0.65	0.5	0.6	0.3	0.3
Turn 180 ° and lock (S)	FN	0.15	1.16	1.38	1.83	1.3	1.5	0.6	0.6
Repeat positioning accuracy (mm)	FI	0.15	<0.005	<0.005	<0.005	<0.005	<0.005	<0.003	<0.003
Maximum unbalanced torque (Nm)	CN	0.05	10	12	40	20	50	20	50
Maximum tangential torque (KNm)	CN	0.05	0.75	1.6	3.5	1.5	3.6	1.5	3.6
Maximum axial torque (KNm)	CN	0.05	0.85	1.9	5.2	2	4	1.6	3.4
Net Weight (without knives disk)	FL	0.05	Light	Light	Medium	Light	Heavy	Medium	Heavy

The combination of the weight of properties taking into account its own characteristics as well as the influence of the attribute information contained on the instance of the search results. Shows, the combination weight is more conducive to the similarity of the overall calculation example, in order to ensure the accuracy and reliability of the case retrieval results.

Designing a fuzzy case-based reasoning system of CNC turret

The following example of CNC turret shows how the proposed hybrid measure works. Two cases X (New Case) and Y (Case1) of failure analysis are used; n1, n2, n3, and n are 1, 2, 3 and 10, respectively.

Based on the similarity calculation method and with used the matrix calculation of Matlab software, the similarity matrix can be calculated as follows:

$$S = \begin{bmatrix} 0 & 0.6125 & 0.6667 & 0.4450 & 0.4067 & 0.7000 & 0.6000 & 0.7150 & 0.7450 & 0 \\ 0 & 0.7188 & 0.6667 & 0.4100 & 0.3700 & 0.7000 & 0.6200 & 0.8000 & 0.8500 & 0 \\ 0 & 0.8438 & 1.0000 & 0.3250 & 0.2950 & 0.7000 & 0.9000 & 0.9900 & 0.8200 & 0.1032 \\ 0 & 0.7188 & 0.6667 & 0.4000 & 0.3833 & 0.7000 & 0.7000 & 0.7900 & 0.8600 & 0 \\ 0 & 1.0000 & 1.0000 & 0.3500 & 0.3500 & 0.7000 & 1.0000 & 1.0000 & 0.9400 & 1.0000 \\ 1.000 & 0.7188 & 1.0000 & 1.0000 & 1.0000 & 1.0000 & 0.7000 & 0.7900 & 0.8200 & 0.1032 \end{bmatrix}$$

According to the formula (5) of objective weight, and the combination of the Matlab programming, we can available:

$$w_j^{(2)} = \{0.6609 \ 0.0722 \ 0.1322 \ 0.2565 \ 0.2755 \ 0.0595 \ 0.1026 \ 0.0555 \ 0.0161 \ 0.6176\}$$ From

Table 1, we can obtained subjective weight $w_i^{(1)}$:

$$w_i^{(1)} = \{0.05 \ 0.15 \ 0.15 \ 0.15 \ 0.15 \ 0.15 \ 0.05 \ 0.05 \ 0.05 \ 0.05\}$$

Based on the formulas (6) of combined weight, using Matlab programming:

$$w_i = \{0.1721 \ 0.0564 \ 0.1033 \ 0.2004 \ 0.2152 \ 0.0465 \ 0.0267 \ 0.0144 \ 0.0042 \ 0.1608\}$$ Above

combination weights calculated results show that the original turret Model and net weight to the subjective right of weight is relatively small, but after considering the property itself large amount of information on the similarity of the results given greater weight, which is in line with the actual situation.

Finally, based on the attribute similarity weighted sum obtained new instance and instance library instance similarity:

$$sim(x, y) = w_i \cdot s_{ij}^{\ T} = \{0.3421 \ \ 0.3354 \ \ 0.3704 \ \ 0.3383 \ \ 0.5436 \ \ 0.8281\}$$

Can be calculated with the new instance is the most similar to Example 6, for Examples of the retrieved certain modifications strategy adjustment finally obtained a new instance of solving program, so as to achieve the purpose of rapid design.

Summary

Property values for the five types exist in the CNC turret instance retrieval process, given the overall similarity calculation model to unify standard measure of similarity between the various types of properties. Considering the nature of subjective and objective weights, the combination weights reflect the impact on retrieval results instance attributes. The CNC turret instance instance-based reasoning that this overall similarity calculation model to retrieve the most similar instance, can be applied to a new instance of design experience and knowledge accumulated by the previous processing instance solving, greatly improved the CNC the turret design efficiency, the promotion and application of CNC turret has very important significance.

References

[1] B. Bouchon-Meunier, M. Ramdani, L.Valverde, Fuzzy logic, inductive and analogical reasoning, Proc. Fuzzy Logic in Artificial Intelligence Workshop, IJCAI'93, Springer, Berlin, 1994, 38–50.

[2] Mantaras R L, Plaza E. Case-based reasoning: an overview. AI Communications,1997,10: 21$\sim$29

[3] LI Jun-jun, QI Jin, HU Jie, et al. Similarity measurement method based on membership function and its application. Application Research of Computers, 2010, 27(3):891-893.

[4] Chen S M, Yeh M S, Hsiao P Y. A comparison of similarity measures of fuzzy values. Fuzzy Sets and Systems, 1995, 72(1):79-89.

[5] Liao T W, Zhang Z M, Mount C R. Similarity measures for retrieval in case-based reasoning systems. Applied Artificial Intelligence.1998, 12(4), 267-288.

[6] SLONIM T Y, SCHNEIDER M. Schneider, Design issues in fuzzy case-based reasoning. Fuzzy Sets and Systems, 2001, 117(2):251-267.

[7] Li D. Fuzzy multiattribute decision-making models and methods with incomplete preference in formation. Fuzzy Sets and Systems, 1999, 106(2):113-119.

[8] FIN N IE G,S U N Z. Similarity and metrics in case-based reasoning. International Journal of Intelligent Systems, 2002, 17(3):273-287.

[9] JIANG Zhan-si, CHEN Li-ping, LUO Nian-meng. Similarity analysis in nearest- neighbor case retrieval. Computer Integrated Manufacturing Systems, 2007, 13(6):1165-1168.

Advanced Materials Research Vol. 819 (2013) pp 311-316
© *(2013) Trans Tech Publications, Switzerland*
doi:10.4028/www.scientific.net/AMR.819.311

Design and fabrication of a piezoelectric bend mode drop-on-demand inkjet printhead with interchangeable nozzle

Senyang Wu[1, a], Yong He[1], Jianzhong Fu[1], Huifeng Shao[1]

[1]The State Key Lab of Fluid Power Transmission and Control, Zhejiang University, Hangzhou 310027, China

[a]wusenyangzju@gmail.com

Keywords: Drop-on-demand, inkjet printing, interchangeable nozzle, piezoelectric droplet generator

Abstract. The drop-on-demand (DOD) inkjet printing technology has been widely used in many fields and several types of droplet generators are developed. This paper presents the design, fabrication and tests of a piezoelectric bend mode drop-on-demand inkjet printhead with interchangeable nozzle. A disk-type PZT is actuated to push the liquid out of inkjet printhead by a function generator, and a droplet is formed because of surface tension. The interchangeable nozzle design enables the same printhead to be fitted with nozzles of different orifice size, thus a clogged nozzle can be easily removed for cleaning or replacement. An experimental platform for micro-droplet jetting is built in this paper. The droplet formation is recorded by a CCD camera as pictures, which can be used to measure the droplet dimension. The experiments are carried out by using the self-developed bend mode piezoelectric inkjet printing system. The influence of the drive parameters on the droplet quality is also studied by dispensing water.

Introduction

The drop-on-demand (DOD) inkjet printing technology has been widely used in many fields, such as rapid prototyping[1], fabrication of integrated circuits[2, 3],MEMS ,LED[4], cell printing[5] and drug delivery[6] ,due to its advantages in automation, low cost, non-contact and ease of material handling. Accordingly, the dispensed liquids have been expanded from the conventional pigmented ink (or standard dye-based ink) to polymers , gels, cell ink or other materials which often have higher viscosities or even contain large particles or cells[7].

When using the traditional inkjet printhead to dispense complex liquids, nozzle clogging may lead to unreliable or failed dispensing. Fluids containing particles, or cells, can easily block the nozzle orifice, resulting in time-consuming nozzle cleaning or even damage of the entire conventional printhead. The easiest way to solve the problem is to use a nozzle with a bigger orifice, as bigger orifices are less likely to clog. However, bigger nozzles result in bigger droplets and lower printing resolution, which is often not desirable in inkjet printing.

There are several types of droplet generators depending on four piezoelectric material deformation modes, namely, the squeeze mode[7], the shear mode, the bend mode[8] and the push mode[9]. In this paper, we will present an in-house-developed piezoelectric bend mode drop-on-demand inkjet printhead with an interchangeable nozzle.

Printhead fabrication

In order to design a simple structure of a low cost inkjet printhead, we use a disk-type piezoelectric buzzer as driving source and the PZT motion is equivalent to the bend mode of the printhead. The interchangeable nozzle design allows one to easily clean or change the clogged or damaged nozzle, avoiding the destruction of the whole printhead assembly. Several development steps are described in the followings.

Printhead chamber. The design of the printhead chamber is illustrated in Fig.1a. The piezoelectric buzzer is made up of a thin piezoceramic layer and a vibration diaphragm which sticks on the layer. The piezoceramic layer is about 0.2 mm thick, and the brass diaphragm is 27 mm in diameter. The piezoelectric buzzer is fixed within the main body and bends if a voltage pulse is applied. The pressure is generated in the liquid flow channel and to pushes the liquid out of the glass nozzle when a voltage pulse is applied. The structure of the proposed droplet generator consists of five main parts: (1) main body with liquid flow channel and upper gland, (2) piezoelectric buzzer, (3) cap, (4) gasket and glass nozzle, (5) O ring.

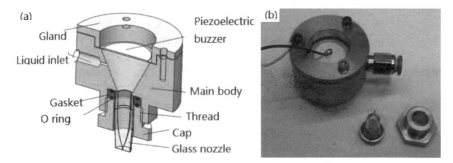

Fig.1 The printhead: (a) Schematic showing of the design. (b) A self-fabricated printhead following the design.

The selected material is brass, which has good conductivity. A wire is separately welded to the the piezoceramic layer, while the other wire is sticked to the main body by a screw. The wires connect the printhead to the piezo driver, as shown in Fig.1b.

Interchangeable nozzle design. The nozzle is fabricated by heating and pulling a glass tube. The Microsystems Laboratory in Nanjing University of Science and Technology helps us manufacture the glass nozzle, as shown in Fig.2a. The major advantages of this nozzle fabrication method are ease of manufacture and low cost. However, it is difficult to precisely control the taper angle of the nozzle and the size of the orifice.

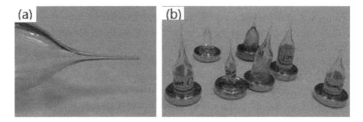

Fig.2 The nozzle: (a) Glass nozzle. (b) Interchangeable nozzle.

The nozzle is fixed to a brass cylinder (gasket in Fig. 1) by araldite epoxy adhesive, and Fig.2b shows how the interchangeable nozzle design is implemented. Then this interchangeable part is placed inside main body that has an inside thread which is tightly fitted to the outside thread of the cap. An O-ring must be used here to prevent cracking of the nozzle from overtightening of the threads and divulging of the fluid.

Experimental testing of the printhead

Experimental setup. The experimental setup is comprised of a computer an air compressor, a pressure regulator, a liquid reservoir, a piezoelectric actuated printhead, a piezo driver, a light, a CCD camera and so on, as shown in Fig.3.

The fluid to be dispensed is filled into a 300 ml plastic reservoir which is mounted on a Z motion stage. The reservoir is lower than the printhead, so that combination of the air compressor and the pressure regulator provides a positive pressure in the reservoir to hold up and prevent the liquid from leaking out of the orifice of the printhead. Electric signals are sent by a JetDrive™III (from Microfab

Technologies Inc.) to the piezoelectric buzzer, causing alternating expansion and contraction of the buzzer as well as the printhead chamber, ultimately, the liquid inside the chamber is pushed into the chamber and droplets are ejected out of the orifice.

The stroboscopic technique has been applied by mounting a LED light and a CCD camera. The shortest exposure time of the CCD camera is 10 μs, which makes the camera be capable of freezing an image of the high-speed droplet under   irradiating of the LED light. Another signal with the same frequency is also sent by the jet driver to the CCD camera. By varying the time delay between the signal for the camera and signal for the piezoelectric buzzer, sequential images of the droplet during its motion are captured with known time differences.

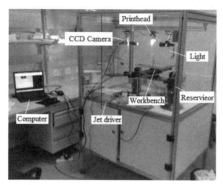

Fig.3 The drop-on-demand inkjet printing system used in the experiment

Experimental conditions. The jet driver made by Microfab Technologies Inc. is able to send out voltage pulses with designed profiles. Up to 12 points can be set to form the signal waveform. Commonly used signals are of uni-polar, bi-polar or sinusoidal shape. The maximum allowable amplitude and frequency for the pulse is ±140 V and 30 kHz, respectively.

Fig.4 shows a uni-polar pulse employed in our experimental study. The zero line represents the equilibrium state of the piezoceramic buzzer, without any external voltage. During the time of t_rise, the piezoceramic buzzer contracts inward to its minimum inner volume and holds that state for a time of t_dwell. During the time of t_fall, the piezoceramic buzzer expands outward, to its equilibrium state. The expansion and contraction of the piezoceramic buzzer leads to droplet ejection.

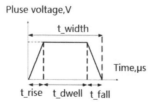

Fig.4 The uni-polar pulse waveform

During all the experiments, t_rise and t_fall were kept at 20 μs, with the purpose of introducing instant action of the piezoceramic buzzer. Static pressure needs to be applied to the reservoir, so that the liquid will not flow out of the nozzle under the hydrostatic pressure.

Testing results. Testing experiments are carried out by using the self-developed bend mode piezoelectric inkjet printing system.Fig.5 shows the normal injection of piezoelectric printhead, when the pluse amplitude is 70V, t_dwell is 550μs and the frequency is 330Hz.

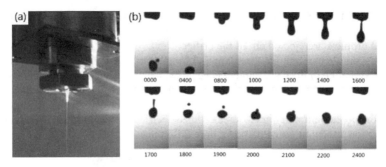

Fig.5 The normal inkjection : (a)Photo.(b)Jet formation.

The white line in Fig.5a stand for continuous droplets ejected. The jet formation observed by the stroboscopic technique is shown in Fig.5b.The number under the picture represents the delay time delay between the signal for the camera and signal for the piezoelectric buzzer, and the unit is μs.

Droplet ejection is usually to form a long liquid band in the beginning, then the long liquid band fracture, several micro-droplets are formed under the action of the liquid surface tension, the speed of the individual droplets are different, so merger between the micro-droplets may occur, which has been verified in Fig.5b.The largest micro-droplet is usually called the main droplet, and the rest are known as satellite droplets.

Experimental results

Satellite droplets. Fig.6 shows the formation of the satellite droplets. The formation of satellite droplet depends on the following factors: 1) unconstrained long liquid band; 2) the shrink velocity of the long liquid band; 3) the fracture time of the long liquid band. If the second shear doesn't appear before the liquid band shrinks into a spherical shape[10].

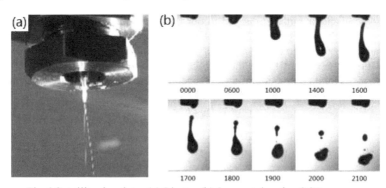

Fig.6 Satellite droplets: (a) Photo. (b) Image taken by CCD camera

Effect of the drive signal. Fig.7 visualizes the definition of maximum length of the long liquid band. For printhead, the effects of pulse width on maximum length and fracture time are investigated by keeping the pulse amplitude constant at 100 V. Note that here the pulse width represents the duration of t_dwell in Fig.4. The jetting frequency is kept constant at 30 Hz. Fig.8a shows that with the growth of the pulse width, the maximum length increases smoothly, but the fracture time Increase at first and then decreases.

Fig.7 Definition of maximum length

The effect of voltage pulse amplitude is shown in Fig.8b, when the applied pulse width is 1500μs and the jetting frequency is 30Hz. With the growth of the pulse amplitude, the maximum length increases smoothly, and the fracture time doesn't show a significant change law.

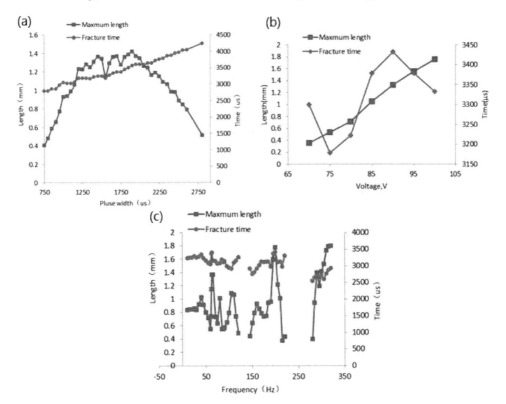

Fig.8 The effect of drive signal: (a) pulse width. (b) voltage pulse amplitude. (c) frequency

The effect of voltage pulse amplitude is shown in Fig.8c, when applied pulse width is 1500μs, and the voltage pulse amplitude is 80V. With the growth of the jetting frequency, the fracture time remains relatively stable, but the maximum length changes widely. The blank among the curve in Fig.8c means the failure of the dispensing.

Conclusions

In this paper, a piezoelectric bend mode drop-on-demand inkjet printhead with interchangeable nozzle is designed, fabricated and tested. All the parts of this droplet generator are low cost and easy to manufacture. The interchangeable nozzle design enables the same printhead to be fitted with nozzles of different orifice size, thus a clogged nozzle can be easily removed for cleaning or replacement. An experimental platform for micro-droplet jetting is built in this paper.

To get a better understanding of the behavior of droplet injection, the water is used to experiment in several conditions when testing self-developed inkjet printhead. The injection parameters, such as frequency, pulse amplitude and pulse width, are studied in the experiment. The frequency has a significant impact on the injection of the printhead. It may be caused by the frequency response characteristics of the piezoelectricceramic. In order to obtain more precise experimental conditions, multilayer piezoelectric actuator can be used to replace the piezoelectricceramic.

Acknowledgements

This paper is sponsored by Science Fund for Creative Research Groups of National Natural Science Foundation of China (No.: 51221004),) the Program for Zhejiang Leading Team of S&T Innovation (No. 2009R50008).

References

[1]   Feng, W., J.Y.H. Fuh, and Y.S. Wong. Development of a Drop-On-Demand Micro Dispensing System[J]. Mater. Sci. Forum (Switzerland), 2006.

[2]   Sirringhaus, H., T. Kawase, R.H. Friend, T. Shimoda, M. Inbasekaran, W. Wu, and E.P. Woo. High-Resolution Inkjet Printing of All-Polymer Transistor Circuits[J]. Science, 2000. 290(5499): 2123-2126.

[3]   Liu, Z., Y. Su, and K. Varahramyan. Inkjet-printed silver conductors using silver nitrate ink and their electrical contacts with conducting polymers[J]. Thin Solid Films, 2005. 478(1–2): 275-279.

[4]   Street, R.A., W.S. Wong, S.E. Ready, M.L. Chabinyc, A.C. Arias, S. Limb, A. Salleo, and R. Lujan. Jet printing flexible displays[J]. Materials Today, 2006. 9(4): 32-37.

[5]   Saunders, R.E., J.E. Gough, and B. Derby. Delivery of human fibroblast cells by piezoelectric drop-on-demand inkjet printing[J]. Biomaterials, 2008. 29(2): 193-203.

[6]   Wu, B.M., S.W. Borland, R.A. Giordano, L.G. Cima, E.M. Sachs, and M.J. Cima. Solid free-form fabrication of drug delivery devices[J]. Journal of Controlled Release, 1996. 40(1–2): 77-87.

[7]   Li, E.Q., Q. Xu, J. Sun, J.Y.H. Fuh, Y.S. Wong, and S.T. Thoroddsen. Design and fabrication of a PET/PTFE-based piezoelectric squeeze mode drop-on-demand inkjet printhead with interchangeable nozzle[J]. Sensors and Actuators A: Physical, 2010. 163(1): 315-322.

[8]   Fan, K.-C., J.-Y. Chen, C.-H. Wang, and W.-C. Pan. Development of a drop-on-demand droplet generator for one-drop-fill technology[J]. Sensors and Actuators A: Physical, 2008. 147(2): 649-655.

[9]   Lee, E.R. Microdrop Generation[M]. Boca Raton: CRC. 2003

[10] HongMing, D.Drop-on-Demand ink-jet:Drop formation and deposition[D].Georgia Institute of Technology:2006

Advanced Materials Research Vol. 819 (2013) pp 317-321
© (2013) Trans Tech Publications, Switzerland
doi:10.4028/www.scientific.net/AMR.819.317

Design and Implement of a Low Cost Drop-On-Demand Inkjet Printing System

Ziyuan Wei[a], Senyang Wu, Yong He and Jianzhong Fu

Department of Mechanical Engineering, Zhejiang University, Hangzhou 310000, China

[a]wzy@zju.edu.cn

Key words: DOD inkjet printing; back pressure control; droplet observation

Abstract: Nowadays, drop-on–demand inkjet printing technology has shown its potential in many fields. This paper presents designing and implementing of a low cost drop-on–demand inkjet printing system. Inkjet printing system can be divided into four parts, and back pressure control system and droplet observation system are two key components, both of which are described in details. In order to have a further control of accuracy and stability of the system, this paper discusses the parameters designed in the system. In pressure control system, the designed system has higher accuracy compared with the Microfab's product. Images can be taken through CCD camera instead of ultra-high-speed camera by using droplet observation system. Some experiments also have been done to verify the feasibility of the device. Accuracy of pressure control, stability of the system and reproducibility of inkjet process can be verified from the experiment.

Introduction

Inkjet printing has played an important role in 3D printing of rapid prototyping since the last two decades. Due to its advantages of non-contract, environmental friendly, low cost and wide application of materials, inkjet printing technology has drawn more and more attention in novel manufacturing method. Various of fields have applied drop-on-demand (DOD) inkjet printing technology such as cell printing[1-2], drug delivery[3], printed circuit board (PCB)[4] and electrode of liquid crystal display (LCD)[5].

DOD inkjet printing technology has many advantages over the others because of its dropping size (range from micron to millimeter), precise deposition and high rate production. However, the device for inkjet printing that the company produces is far too expensive, low cost drop-on-demand inkjet printing system has shown its necessity on the way to commercial use.

In this paper, a low cost DOD inkjet printing system design will be introduced. The stability of the system and the reproducibility of the jetting process will also be taken into consideration. Meanwhile some improvements and modification are employed in the control of back pressure and image taken. Experiments have been done to verify the accuracy and stability of jetting process and the feasibility of the device. Conclusions are drawn at the end of the paper.

DOD inkjet printing system architecture

DOD inkjet printing system consists of 4 major components: (1) back pressure control system, (2) Droplet observation system, (3) droplet delivery system, (4) motion system. The design of the system is illustrated in Fig.1. Through controlling back pressure, liquid in ink reservoir will be pushed into print head. Then the liquid restored in the print head will be jetted onto a glass dish which is mounted on a 3D motion platform. And the print head will start jetting when signal is given from the driver. During the process, CCD camera with magnifier will catch the image which will be shown onto a screen. To achieve synchronization, a trigger signal applied to CCD camera and a driving pulse applied to the print head is sent from the same driver which can control the delay time more precisely.

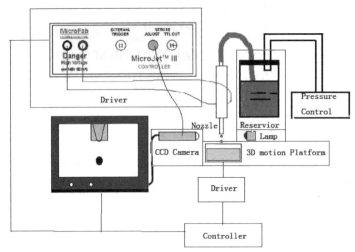

Fig.1 The components of DOD inkjet printing system

Back pressure control system. During the printing progress, four main problems need to be solved: firstly, how to press the liquid into print head while jetting; secondly, how to hold up and prevent the liquid from leaking out of the print head; thirdly, how to make the clean fluid moving up and down in print head to clean the deposits which generate a huge impact on the jetting progress; ultimately, the most critical factor is that how to maintain the proper meniscus for jetting. In order to solve these problems, back pressure control system is introduced.

As shown in Fig.2, the major pressure regulator is a pressure relief valve (RP-0.5-2, from Fujikura Ltd.) of which the range is 0-50kPa, the sensitivity is 0.2%F.S. In other words, the maximum precision of this valve (range multiply sensitivity) can be as low as 100pa that has higher accuracy compared with the product CTPT controller (from Microfab Technologies Inc.) of which precision is 200pa. Meanwhile in order to have further control of meniscus's position, a minor pressure regulator is introduced. The minor pressure regulator is a lifting platform that each step raises 1mm. Precision of this device's pressure control can be calculated as follows:

$$P = \rho g h \tag{1}$$

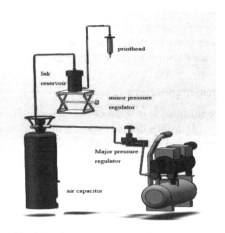

Fig.2 Back pressure control system

Where P represents pressure, ρ represents liquid density, h represents the height of liquid level between ink reservoir and print head.

If turn the liquid into water, each step the lifting platform raises, and the pressure will enhance 10pa, which is more precise compared with the product that other companies produce.

With the purpose of enhancing stability of the system during printing, an air capacitor is introduced. On one hand, the air pressure is transient instable coming out of the pressure relief valve. The jetting process will be instable if the air is directly into ink reservoir. Adding air capacitor into the system between ink reservoir and pressure relief valve can solve this problem. On the other hand, after printing several minutes, the height of the liquid level will drop, which will affect the pressure in return, finally have an effect on the stability of the printing system. If having an air capacitor, pressure decreasing caused by space increasing can be ignored since the space of air capacity is far larger than the space changing caused by jetting process. In the present experiments, the duration of printing is much longer if the system equipped with an air capacitor.

Droplet observation system. It is difficult to observe the droplets with naked eye because they are fast in speed and small in size. Nowadays, one solution is to use ultra-high-speed camera that can take more than 104 frames per second (fps). However it is far too expensive in this situation. Therefore, a special observation system is developed.

For DOD inkjet printing process, a droplet's velocity ranges from 0 to 5m/s, and a droplet cycle has a period of about 1 millisecond which means camera takes more than 10^3 fps can satisfy this situation. Generally, a CCD camera can take about 100 fps less than required 10^3 fps to illustrate jetting process. With the consideration of jetting process is stable, and the result of jetting is repeatable from one droplet to another, the process can be illustrated through different cycles with various exposure delay time. As shown in Fig.3, a LED light is used to illuminate the droplet required by a CCD camera which runs under computer's control.

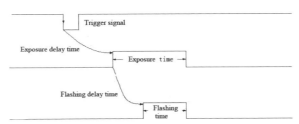

Fig.3   Timing diagram of observation system

A CCD camera can obtain a sharp image only when a light pulse with required short duration and intensity is provided. And a CCD camera is working only when it is in the exposure time and trigger signal is sent, so different period of process will be obtained through changing the known exposure delay time. In our system, the exposure delay time is controlled by JetDrive[TM]III (from Microfab Technologies Inc.) can be accurate to 5 microseconds. .

Another key factor of the system is exposure time. If exposure time takes too long, the image will be blurred, however, if exposure time takes too short, CCD camera won't work, so exposure time is limited. In order to obtain a clear image of droplets, exposure time must be calculated. The approximate equation of exposure time is [6]:

$$T = \frac{2 \times P_i}{v \times M} \tag{2}$$

Where T, P_i, v, M are exposure time, pixel size, velocity of the droplet and magnification of camera lens. In our experiment, P_i is 7 μm, M is 2, when v is about 1m/s, T turns out to be 10 microseconds, which our device can match with.

In order to have a convenient way helping capture the image of print head in CCD camera, the camera and print head are positioned onto a movable platform with resolution of 10μm in each direction.

Experiment

One of the crucial concerns in DOD inkjet printing system is accuracy. Since precise deposition is required, droplet from printhead leaking out must be forbidden, which means back pressure control must be precise and last long. In order to verify the accuracy of the system, some experiments have been done. Fig.4 shows the process of adjust back pressure in nozzle. Adjusting major pressure regulator to a proper pressure, the liquid in printhead can be held in state A. Then carefully changing minor pressure regulator, the liquid level in state B will be seen. In order to enhance the stability of jetting process, pressure adjustment must continue. When in state C, the pressure adjustment is completed. In the present experiments, the liquid level in state C can last for hours.

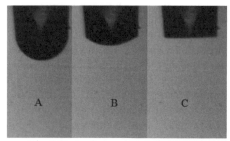

Fig.4 States in pressure adjustment

Imaging system is the way to describe jetting process. Fig.5 shows the process of droplet jetting. The interval between each image is 20 microseconds. From images every step of jetting process can be recognized, and separation of droplets also can be seen from series of images. Theoretically the influence of gravity can be ignored at the scale of inkjet droplets[7]. From series of images this theory can be verified, at the same time, velocity of the droplet can be calculated from these images. Meanwhile the stability of the system is verified as well.

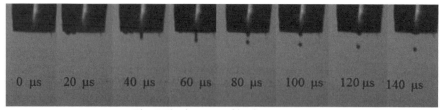

Fig.5 The images of jetting process

Since it is assumed that images taken from different cycle of droplets in same delay exposure time are the same, it is important to verify the reproducibility of the process. As shown in Fig.6, 5 images are taken at the same exposure time to describe the separation of droplet. Compared with each image, a conclusion can be drawn that the image taking technology applied in inkjet system is available.

Fig.6 The reproducibility of inkjet process

Conclusion

This paper presents a design of low cost DOD inkjet printing system, containing back pressure control system, droplet observation system, droplet delivery system, and motion system. Besides its low cost, this inkjet printing system has higher accuracy in pressure control than commercial pressure control device, and droplet observation system has similar accuracy with ultra-high-speed camera in image taking. Several experiments to verify the stability and reproducibility of the system have been done. A conclusion can be drawn that this system is available for inkjet printing.

Acknowledgements

This paper is sponsored by the National Natural Science Foundation of China (51175461), Group of scientific and technical innovation of Zhejiang Province (2009R50008), Zhejiang Provincial Natural Science Foundation of China (No. Y1100281).

Reference

[1] Paul Calvert. Printing cells [J]. Science, 2007, 318: 208-209.

[2] Xiaofeng Cui, Thomas Boland. Human microvasculature fabrication using thermal inkjet printing technology [J]. Biomaterials, 2009, 30: 6221-6227.

[3] Byung Kook Lee, Yeon Hee Yun, Ji Suk Choi, et al. Fabrication of drug-loaded polymer microparticles with arbitrary geometries using a piezoelectric inkjet printing system. [J] International Journal of Pharmaceutics, 2012, 427: 305-310.

[4] Hsien-Hsueh Lee, Kan-Sen Chou1 and Kuo-Cheng Huang. Inkjet printing of nanosized silver colloids [J]. Nanotechnology 2005, 16: 2436-2441.

[5] H.S. Koo, M. Chen, P.C. Pan, et al. Fabrication and chromatic characteristics of the greenish LCD colour-filter layer with nano-particle ink using inkjet printing technique [J]. Displays, 2006, 27(3): 124-129.

[6] H.M Dong. Drop-on-Demand inkjet Drop Formation and Deposition [D]. Doctor Thesis. Georgia Institute of Technology, 2006.

[7] H Wijshoff. The dynamics of the piezo inkjet printhead operation [J]. Physics reports, 2010, 491: 77-177.

Advanced Materials Research Vol. 819 (2013) pp 322-327
© *(2013) Trans Tech Publications, Switzerland*
doi:10.4028/www.scientific.net/AMR.819.322

Design and Implementation of an Interpolation Processor for CNC Machining

Jingchuan Dong[1, a], Taiyong Wang[2,b], Bo Li[3,c], Xian Wang[2,d] and Zhe Liu[2,e]

[1]School of Electrical Engineering and Automation, Tianjin University, Tianjin, 300072, China

[2]School of Mechanical Engineering, Tianjin University, Tianjin, 300072, China

[3]Tianjin Key Laboratory of Equipment Design and Manufacturing Technology, Tianjin University, Tianjin 300072, China

[a]new_lightning@sohu.com, [b]tywang@139.com(corresponding author), [c]specterwaxz1314@163.com, [d]wangxian320@163.com, [e]tyssqlz@sina.com

Keywords: computerized numerical control (CNC), interpolation, parallel computing, field programmable gate array (FPGA)

Abstract. As the demand for high speed and high precision machining increases, the fast and accurate real-time interpolation is necessary in modern computerized numerical control (CNC) systems. However, the complexity of the interpolation algorithm is an obstacle for the embedded processor to achieve high performance control. In this paper, a novel interpolation processor is designed to accelerate the real-time interpolation algorithm. The processor features an advanced parallel architecture, including a 3-stage instruction pipeline, very long instruction word (VLIW) support, and asynchronous instruction execution mechanism. The architecture is aimed for accelerating the computing-intensive tasks in CNC systems. A prototype platform was built using a low-cost field programmable gate array (FPGA) chip to implementation the processor. Experimental result has verified the design and showed the good computing performance of the proposed architecture.

Introduction

Nowadays, high speed and high precision computerized numerical control (CNC) machining is becoming more and more widely applied in industry. In the high speed and high precision CNC systems, the interpolation period and accuracy are vital to achieve the desired control performance. Besides, to keep the smoothness of motion, the motion profile should be carefully planned to meet the dynamic constrains of the machine tools. Complicated algorithms, such as S-curve acceleration and look-ahead, are introduced to solve the problem. However, due to the cost and size limit, the commercial embedded processors are difficult to satisfy such requirements.

One possible method to enhance the processing capacity is employing hardware accelerators in the CNC controller. In the hardware accelerators, computations can be executed in parallel to achieve high performance. Recently, hardware accelerator schemes based on field programmable gate array (FPGA) chips are becoming popular. A FPGA device contains abundant basic logical cells and routing resources, which can be configured many times to implement different circuits. FPGA devices give freedom to the researchers to realize specific functions for the CNC controllers in a single low-cost chip. Jeon et al. [1] developed a FPGA based acceleration and deceleration circuit for CNC machine tools and robots. Jimeno et al. [2] applied FPGA to compute tool path for shoe last machining on a CNC lathe. Yau et al. [3] implemented fast Cox-de Boor algorithm on FPGA to perform real-time NURBS interpolation. Osornio-Rios et al. [4] implemented a polynomial acceleration profiles generator on FPGA to control a servo motor driving CNC milling machine axis. Morales-Velazquez et al. [5] proposed a FPGA based open-architecture reconfigurable multi-agent platform for CNC machines.

Previous studies have demonstrated the benefits for efficiency by using hardware accelerators. However, there are still unsolved problems that prevent hardware accelerators being widely adopted. First, the trajectory plan and interpolation are the most time-consuming tasks in high speed machining. However, to the best of our knowledge, trajectory plan algorithms are not realized by hardware accelerators. The complexity of high speed trajectory plan algorithms makes it difficult and inefficient to be directly mapped to hardware circuits. Second, almost all of prior studies were using fixed-point algorithms on hardware accelerators. Fixed-point data is simple to implement and can save hardware resource. However, compared to the float-point data, fixed-point data is more likely to overflow due to a large resultant value, especially in non-linear operations. On the other hand, the accuracy of fixed-point data will also be decreased when the value is very small. Third, once the FPGA is configured, the algorithm will be fixed in the hardware and cannot be modified in the application filed.

In this paper, a novel interpolation processor for CNC controller is introduced. The processor takes advantages of parallel computing in hardware to improve real-time performance of the controller. The processor avoids the problems arising from conventional approaches by architecture design. The firmware running on the processor realizes float-point look-ahead trajectory plan and interpolation algorithms for high speed machining. Experimental results show the feasibility of the proposed design.

Interpolation Algorithm

The data process flow in a CNC controller is illustrated in Fig. 1. The NC program is interpreted by the controller to extract machining process information, including tool path, feed rate, spindle speed, etc. To guarantee the smoothness of the feed motion along tool path, the motion profiles must be planned under kinetic and dynamic constraints. The interpolator samples the tool path according to the planned motion profiles to generate position commands for each feed axis of the machine tools. The reference position commands for individual axis will be sent to the axis control modules to drive the servo motors for feeding.

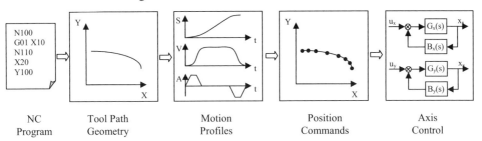

NC	Tool Path	Motion	Position	Axis
Program	Geometry	Profiles	Commands	Control

Fig. 1 CNC data process flow

If the length of tool path is S_T, the tool path curve can be described as

$$\mathbf{X} = \mathbf{X}(S) \quad (0 < S < S_T) \tag{1}$$

where S is the displacement along the tool path curve. After trajectory plan, the motion profile can be determined by the function of interpolation time t and displacement S

$$S = S(t) \quad (t \geq 0) \tag{2}$$

The essence of interpolation is sampling the curve at the interpolation period. Suppose the interpolation period is Δt, the interpolated points $\mathbf{X}_i$ $(i = 0,1,2...)$ is given by

$$\mathbf{X}_i = \mathbf{X}(S(i \cdot \Delta t)) \quad (i = 0,1,2...) \tag{3}$$

If the planned feed speed is v, the length between 2 interpolation points, Δs, will be

$$\Delta s = \|\mathbf{X}_{i+1} - \mathbf{X}_i\| = v \cdot \Delta t \tag{4}$$

As Eq.(4) suggests, Δs is proportional to v and Δt. To ensure high machining accuracy, the adjacent interpolation points should be close enough and Δs should be small. Since v is determined by the NC Program and the trajectory plan algorithm, the only method to reduce Δs is to reduce Δt, i.e. increase the interpolation rate. Therefore, in high speed and high precision machining, the interpolation task requires huge real-time computing power. For instance, if the v is 60m/min and Δt

is 0.1ms, Δs will be 0.1 mm and the interpolation algorithm will be performed 10,000 times per second.

The smoothness of feeding motion affects the impact to the servo system, the structure vibration and the surface quality of the finished parts. In high speed machining, the velocity profile must be carefully planned under the constraints of the machine tools. The jerk limited S-curve trajectory plan algorithm is widely applied in the CNC controller. A complete S-curve acceleration profile contains 7 polynomial curve segments. In the accelerating and decelerating phase, the jerk is limited to a constant value and the acceleration changes continuously.

If the jerk limit is J_{max}, the maximum acceleration in accelerating and decelerating phase is a_{max} and d_{max}, the motion time and velocity of each segments is T_i and v_i, the relationship of the displacement, S, and time, t, in the jerk limited S-curve acceleration can be written as

$$S(t) = \begin{cases} v_s\tau_1 + \dfrac{1}{6}J_{max}\tau_1^3 & t_0 \le t \le t_1 \\[2mm] S_1 + v_1\tau_2 + \dfrac{1}{2}a_{max}\tau_2^2 & t_1 \le t \le t_2 \quad S_1 = v_sT_j + \dfrac{1}{6}J_{max}T_1^3 \\[2mm] S_2 + v_2\tau_3 + \dfrac{1}{2}a_{max}\tau_3^2 - \dfrac{1}{6}J_{max}\tau_3^3 & t_2 \le t \le t_3 \quad S_2 = S_1 + v_1T_2 + \dfrac{1}{2}a_{max}T_2^2 \\[2mm] S_3 + v_3\tau_4 & t_3 \le t \le t_4 \quad S_3 = S_2 + v_2T_j + \dfrac{1}{3}J_{max}T_1^3 \\[2mm] S_4 + v_4\tau_5 - \dfrac{1}{6}J_{max}\tau_5^3 & t_4 \le t \le t_5 \quad S_4 = S_3 + v_3T_4 \\[2mm] S_5 + v_5\tau_6 - \dfrac{1}{2}d_{max}\tau_6^2 & t_5 \le t \le t_6 \quad S_5 = S_4 + v_4T_j - \dfrac{1}{6}J_{max}T_5^3 \\[2mm] S_6 + v_6\tau_7 - \dfrac{1}{2}d_{max}\tau_7^2 + \dfrac{1}{6}J_{max}\tau_7^3 & t_6 \le t \le t_7 \quad S_6 = S_5 + v_5T_6 - \dfrac{1}{2}d_{max}T_6^2 \end{cases} \tag{5}$$

and velocity is
$$v(t) = S'(t) \tag{6}$$

where $\tau_i = t - t_{i-1}$, $t_0 = 0$, $t_{i+1} = t_i + T_i$, $v_i = v(t_i)$, $JT_1 = a_{max}$ and $JT_5 = d_{max}$. Given the starting velocity v_s, the target velocity v_t and the expect end velocity v_{et}, the following constraints should also be satisfied

$$\begin{cases} v(0) = v_S \\ v(t_T) \le v_{et} \\ 0 \le v(t) \le v_t \end{cases} \qquad \begin{cases} S(0) = 0 \\ S(t_7) = S_T \end{cases} \qquad \begin{cases} a_{max} \le a_{lim} \\ d_{max} \le a_{lim} \end{cases} \tag{7}$$

where $t_T = \sum_{i=1}^{7} T_i$. To improve efficiency, the machining time, t_T, for each piece of tool path should be minimized. Therefore, the trajectory plan becomes an optimization problem to determine the motion time of each segment. In the machining process, the look-ahead algorithm read multiple machining blocks in advance to increase the allowable end velocity, v_{et}. When the number of pre-read blocks increases, the total machining time will approximate to the globe optimum. In high speed machining, a large number of program blocks are used in the real-time look-ahead, and the algorithm must analyze the pre-read tool-path in a deterministic short time.

After optimization, some of the segments may be removed ($T_i = 0$) and the total segments of the profile will be less than 7. In the optimization algorithm, high order polynomial equation may be encountered, such as

$$Ax^4 + Bx^3 + Cx^2 + Dx + E = 0 \tag{8}$$

In Eq. (8), A, B, C, D and E are coefficients. The solution is obtained by Newton-Raphson's iteration method [6]. To guaranty the accuracy, the iteration algorithm must be executed multiple times in the real-time trajectory plan, which is a challenge to the CNC processor.

To accelerate the interpolation algorithm, we designed a dedicate processor to assist the main processor in the CNC controller.

Processor Architecture

The interpolation processor is based on a classical reduced instruction set computer (RISC) model. Some of the integer instructions and float-point instructions are listed in Table 1. Different from other architectures that take multi-cycles to wait for execution of float-point instructions (Fig. 2), the proposed processor executes the pipelined functions asynchronously (Fig. 3). As Fig. 3 shows, after sending the operands to the inputs of the corresponding float-point pipeline by the "send" instruction, the processor utilizes the wait cycles to perform other instructions. When the execution of pipelined float-point instruction completed, another "get" instruction will write the result to the designated register. As a result, the asynchronous execution mode eliminates the wait states in float-point instructions to achieve high performance.

Table 1 Integer instruction examples

syntax	Description	Function	Cycles
AND R_d, R_x, R_y	and	$R_d = R_x$ & R_y	1
ADD R_d, R_x, R_y	add	$R_d = R_x + R_y$	1
SUBI R_d, R_x, C	subtract immediate	$R_d = R_x - C$	1
JMP lable	absolute jump	PC = lable	2
JZ.sel lable	jump if result is zero	if true goto lable	1/2
CALL lable	subroutine call	R15 = PC + 4; PC = lable	2
ST [R_x], R_y	store	memory[R_x] = R_y	1
LD R_d, [R_y]	load	R_d = memory[R_y]	1
PUSH R_x	push to stack	memory[R14] = R_x ; R14 = R14 - 4	1
FMOV FR_d, FR_x	move	$FR_d = FR_x$	1
FADD.SEND FR_x, FR_y	start binary addition	add/sub pipeline input 1 = FR_x; add/sub pipeline input 2 = FR_y;	1
FADD.GET FR_d	get addition result	FR_d = operator pipeline output	1
FABS FR_d, FR_x	get absolute value	$FR_d =	FR_x |$	1
FSTK.SEND.L FR_x	push lower word	memory[R14] = FR_x[31:0]; R14 = R14 - 4	1

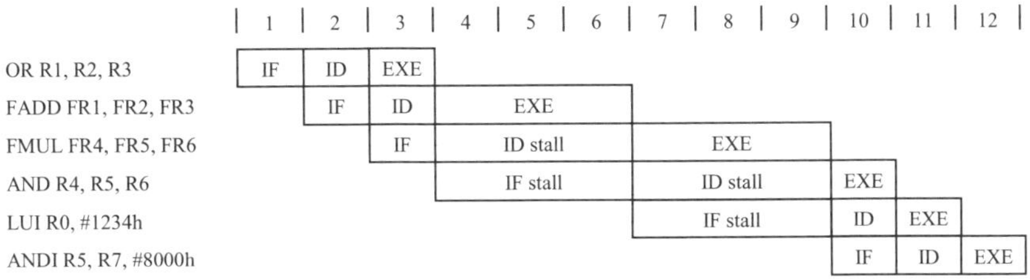

Fig. 2 Float-point instructions delayed by multi-cycle execution

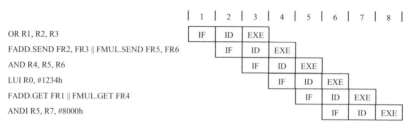

Fig. 3 Parallel and asynchronous float-point instruction execution

The hardware architecture of the proposed processor is shown in Fig. 4. The architecture exploits the parallel nature of hardware logic. The processor employs a 3-stage instruction pipeline, namely instruction fetch (IF) stage, instruction decode (ID) stage and execution (EXE) stage. The 3-stage pipelined design allows 3 instructions to be processed at a same time, which improve the hardware resource usage. The float-point instructions in the processor is a very long instruction word (VLIW) design that enables two sub-instruction to be executed in one cycle. The ID stage includes two float-point control units to dispatch the sub-instructions to different float-point function areas. In the EXE stage, 5 float-point functions are realized in hardware circuits, and 4 of them are pipeline designed to improve the throughput.

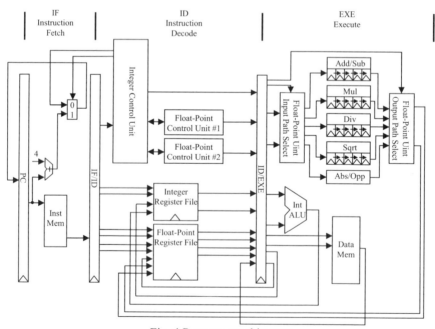

Fig. 4 Processor architecture

Experimental Verification

The proposed processor was realized on a reconfigurable CNC controller based on FPGA. The hardware logic of the processor was written in Verilog hardware description language and verified using behavior simulation software. The processor was then implemented on the Spartan-3E 500E FPGA chip in the controller. The firmware realized the jerk limited S-curve acceleration trajectory plan algorithm with look-ahead ability, and the 3-axis interpolation algorithm was also included. The main processor of the CNC communicated with the interpolation processor through shared memory. A 3D milling program was used to test the performance of the processor. The interpolation processor was running at 12.5MHz, the interpolation period was set to 2ms and the feed speed was set to 6000mm/min. In the test, when the interpolation processor was inactive, the usage of main processor was approximate to 60%. The usage was significantly reduced to less than 15% after activating the interpolation processor. The test has successfully verified the proposed processor architecture and demonstrated the power of parallel computational capacity of the new controller.

Conclusion

The high speed and high precision CNC machining requires fast and accuracy real-time execution of complex calculations. A novel processor architecture has been proposed to accelerate trajectory plan and interpolation algorithms in CNC systems. The processor utilizes parallel computing techniques, including the instruction pipeline, VLIW structure and asynchronous execution, to achieve high performance. Moreover, the processor supports hardware float-point functions that not only facilitate firmware design but also enables high accuracy and high throughput real-time computing. Experiments have successfully verified the proposed interpolation processor architecture. Further researches will focus on expanding the application fields and improving the performance of the new architecture.

Acknowledgment

This research is sponsored by the National Science and Technology Support Program, China (No. 2013BAF06B00), the Research Fund for the Doctoral Program of Higher Education, China (20100032110006) and the Innovation Foundation of Tianjin University, China (No. 60302046).

References

[1]　J.W. Jeon, Y.K. Kim, FPGA based acceleration and deceleration circuit for industrial robots and CNC machine tools, Mechatronics 12 (2002) 635–642.

[2]　A. Jimeno, J.L. Sanchez, H. Mora, J. Mora, J.M. Garcia-Chamizo, FPGA-based tool path computation - an application for shoe last machining on CNC lathes, Comput. Ind. 57 (2006) 103-111.

[3]　H.T. Yau, M.T. Lin, M.S. Tsai, Real-time NURBS interpolation using FPGA for high speed motion control, Comput. Aided Design 38 (2006) 1123-1133.

[4]　R.A. Osornio-Rios, R.J. Romero-Troncoso, G. Herrera-Ruiz, R. Castaneda-Miranda, FPGA implementation of higher degree polynomial acceleration profiles for peak jerk reduction in servomotors, Robot. CIM-Int. Manuf. 25 (2009) 379-392.

[5]　L. Morales-Velazquez, R.J. Romero-Troncoso, R.A. Osornio-Rios, G. Herrera-Ruiz, E. Cabal-Yepez, Open-architecture system based on a reconfigurable hardware–software multi-agent platform for CNC machines, J. Syst. Architect. 56 (2010) 407-418.

[6]　K. Erkorkmaz, Y. Altintas, High speed CNC system design. Part I: jerk limited trajectory generation and quintic spline interpolation, Int. J. Mach. Tool. Manu. 41(2001) 1323–1345.

Advanced Materials Research Vol. 819 (2013) pp 328-333
© *(2013) Trans Tech Publications, Switzerland*
doi:10.4028/www.scientific.net/AMR.819.328

Dynamics of Deployment for Mooring Buoy System based on ADAMS Environment

Zhongqiang ZHENG [1,a], Yuan DAI [1,b], Daxiao GAO [1,c], Xinlei ZHAO [2,d], Zongyu CHANG [1,e]

[1]Engineering College, Ocean University of China, Qingdao 266100, China

[2]Offshore Oil Engineering (Qingdao) Co.,LTD, 492 Lianjiang Road, Qingdao, 266100, China

[a]zhongqiangzheng@qq.com, [b]dyrain@yahoo.cn, [c]zxl@mail.cooec.com.cn,
[d]daxiaogao@yahoo.com.cn, [e]zongyuchang@ouc.edu.cn(corresponding author)

Keywords: Deployment; Mooring Buoy System; Multi-body Dynamics; ADAMS

Abstract. Buoy is widely applied in oceanographic research, ocean engineering and so on. The high cable tension and location derivation of its anchor can greatly affected the working condition of buoy. Multi-body software ADAMS is used to analyze the dynamics of deployment for buoy. Firstly, model of mooring buoy is developed. The forces on components of buoy are given such as drag forces, inertia force and impact force between bottom and anchor. The program is developed by macro commands in ADAMS environment. The trajectory and tension force of different nodes on buoy are calculated. This method can simulate the dynamics of deployment process for buoy system. And it is also helpful for guidance for buoy design and deployment process planning.

Introduction

Buoy is widely used in oceanographic research, ocean engineering and so on. It is one of the important equipments of multi-parameters ocean observation on fixed location for long period. Buoys suffer serious sea load like wind, wave, and current and so on, especially during the deployment and recovery period. Large dynamic tension load on cable may lead to damage and lost of buoy. So it is essential to study the dynamic of buoy in deployment process.

Some researchers studied the dynamics of buoy system. Dewey[1], Grobat and Grosenbaugh[2] developed the software package for the oscillation of buoy in the sea. Augusto and Andrade [3] development a software of optimization and automation design for deep-sea anchor deployment planning in offshore engineering. Baddour and Raman-Nair [4] studied dynamics of deployment and recovery process on oscillation platform by applying multi-boy approach and Kane's equations; Umar and Datta [5] studied the nonlinear behaviors of multi-mooring lacking buoy under the 1st and 2nd wave force. Chang et al. [6] modeled the buoy system by multi-body dynamic presented by absolute coordinate system and calculated the motion of buoy system during deployment process. Skeen et al. [7] studied that ADAMS dynamics simulation software has been used to model and test various buoy and cable deployments systems for Lucent Technologies.

With the development of marine science research, more and more surface/subsurface buoys are applied for ocean environment observation. Lots of research has been done. In this paper, multi-body software ADAMS is used to simulate the process of deployment for buoy system. Buoy system is modeled by lumped mass method firstly. Then a dynamic model is developed by macro commands in ADAMS environment. The trajectory and tension force of different nodes on buoy can be obtained by using the program.

Motion Equations for Mooring Buoy System

Mooring buoy system can be viewed as a multi-body system of nodes connected with massless spring. In the process of deployment, buoy, cable with components and anchor suffer with the forces like weight $\vec{W}_i$, buoyancy $\vec{B}_i$, drag force $\vec{F}_{Di}$, inertia force $\vec{F}_{Ii}$ and tension force $\vec{T}$ and so on as shown in Fig. 1.

According to Morison Equation, the hydrodynamic drag force on lumped mass point is described as

$$\vec{F}_D = \frac{1}{2}\rho_w C_D A \left| \vec{U}_w - \vec{U}_i \right| (\vec{U}_w - \vec{U}_i) \tag{1}$$

Where C_D is the drag coefficient of hydrodynamics; A is the immersed area; ρ_w is the density of sea water; $\vec{U}_w$ is the velocity of water; $\vec{U}_i$ is the velocity of mass point.

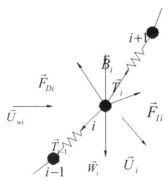

Fig. 1   Scheme of forces on classic nodes i

The Inertia force is calculated by

$$\vec{F}_{Ii} = m_{ai}\left(\frac{d\vec{U}_w}{dt} - \frac{d\vec{U}_i}{dt}\right) + \rho_w V_i \frac{d\vec{U}_w}{dt} = m_{vi}\frac{d\vec{U}_w}{dt} - m_{ai}\frac{d\vec{U}_i}{dt} \tag{2}$$

Where m_{ai} is the added mass ($m_{ai} = C_m\rho_w V_i$); m_{vi} is the virtual mass ($m_{vi} = (1+C_m)\rho_w V_i$); C_m is the coefficient of added mass.

As to weight $\vec{W}_i$, it is supposed to be constant. But for the buoyancy, in the process of deployment, Buoyancy of some bodies varied with the relatively displacement between buoy and water. It depends on the volume submerged in water as $B_i = \rho_w V_i g$, which is the function of displacement of mass underwater and the shape of body.

Tension force is an important factor to affect the motion of mooring system in deployment. According to the space discrete model, the tension force between node i and node $i+1$ can be determined by the relative distance between adjacent nodes and stiffness of cable. In this spring-mass model, the tension force could be expressed as:

$$\left| \vec{T} \right| = k\left(\sqrt{(x_{i+1} - x_i)^2 + (y_{i+1} - y_i)^2} - l_{i,i+1}\right) \tag{3}$$

Where $l_{i,i+1}$ is the free length of cable between node i and node $i+1$; k is the stretching stiffness of cable, and it is piecewise function and depends on the compliance between two nodes

$$k = \begin{cases} k_0 & \sqrt{(x_{i+1}-x_i)^2 + (y_{i+1}-y_i)^2} > l_{i,i+1} \\[2mm] 0 & \sqrt{(x_{i+1}-x_i)^2 + (y_{i+1}-y_i)^2} \le l_{i,i+1} \end{cases} \tag{4}$$

The tension force of each node can be divided into two components of forces in x, y two directions as

$$T_{ix} = k(\sqrt{(x_{i+1} - x_i)^2 + (y_{i+1} - y_i)^2} - l_{i,i+1}) \frac{(x_{i+1} - x_i)}{\sqrt{(x_{i+1} - x_i)^2 + (y_{i+1} - y_i)^2}}$$

$$+ k(\sqrt{(x_{i-1} - x_i)^2 + (y_{i-1} - y_i)^2} - l_{i,i+1}) \frac{(x_{i-1} - x_i)}{\sqrt{(x_{i-1} - x_i)^2 + (y_{i-1} - y_i)^2}}$$

(5)

$$T_{iy} = k(\sqrt{(x_{i+1} - x_i)^2 + (y_{i+1} - y_i)^2} - l_{i,i+1}) \frac{(y_{i+1} - y_i)}{\sqrt{(x_{i+1} - x_i)^2 + (y_{i+1} - y_i)^2}}$$

$$+ k(\sqrt{(x_{i-1} - x_i)^2 + (y_{i-1} - y_i)^2} - l_{i,i+1}) \frac{(y_{i-1} - y_i)}{\sqrt{(x_{i-1} - x_i)^2 + (y_{i-1} - y_i)^2}}$$

(6)

Impact forces between components and bottom can follow the approach in Gobat [2]. The unilateral boundary condition at the sea floor is modeled as an elastic foundation with linear stiffness and damping properties. The interaction force f_{ciy} in vertical coordinate varies with vertical position and velocity as follow:

$$f_{ciy} = \begin{cases} 0 & -h < y_i \leq 0 \\ k'(-h - y_i) - c'\dot{y}_i & y_i \leq -h \end{cases}$$

(7)

Where h is the depth of bottom; k' is stiffness of seabed; c' is damping coefficient. The impact force is defined as the displacement, velocity between anchor and seafloor, and the properties of sea soil. In the horizontal direction, the interaction force f_{cix} in anchor is combined with damping force and Coulomb friction force as follow:

$$f_{cix} = \begin{cases} 0 & -h < y_i \leq 0 \\ -c'\dot{x}_i - \mu f_{cy} sign(\dot{x}_i) & y_i \leq -h \end{cases}$$

(8)

Where μ is the Coulomb friction coefficient; $sign(\dot{x}_i)$ means the sign of horizontal velocity. For most problems, the gross response of the cable tension during deployment is largely insensitive to the choice of k' and c'. This has been proved by the following numerical simulation in which the above parameters could be changed.

According to the aforementioned and Ref. [7], the motion dynamics of the model can be obtained as follows:

$$(m_i + m_{ai}) \frac{d\vec{U}_i}{dt} = m_{vi} \frac{d\vec{U}_{wi}}{dt} + \vec{T}_i + \vec{F}_{Di} + \vec{f}_{ci} + \vec{B}_i + \vec{W}_i$$

(9)

Virtual Prototype for Mooring System based on ADAMS

In this section, an example of taut single-point surface buoy system of Berteaux [9] is given including surface buoy, chain, wire cable, recording current meter, floating balls, acoustic releasers and gravity anchor as shown in Fig. 2. According to previous method, the mooring buoy system can be viewed as a multi-body system of nodes connected with massless spring just like Fig. 1.

Then the whole system is simplified to 10 discrete nodes as shown in Fig. 3. Among these nodes, this paper emphasizes the nodes of anchor, floater group with 13 floating balls, RCM 2, RCM3 and surface buoy, which are respectively represented by nodes 1,3,5,7, and 10. The main physical and hydrodynamics parameters are shown in Table 1 and Table 2, respective. The coefficient of the added mass C_m and the drag coefficient of hydrodynamics C_D refer to Berteaux [9].

Table 1 Physical parameters of the buoy system

Node i	1	2	3	4	5	6	7	8	9	10
Mass of node m_i [Kg]	1817	72	195	26.5	26.5	63.5	26.5	31.75	26.5	300
Buoyancy B_i [N]	2018	400	5291	85	85	80	85	40	85	19600
Initial coordinate values (x,y) [m]	0,0	19.5,0	60.25,0	100.75,0	601.25,0	1101.75,0	1602.25,0	1852.75,0	2103.25,0	2115,0
Immersed area A_x [m^2]	0.2	0.2	1	0.1	0.1	4	0.1	2	0.1	1.5
Immersed area A_y [m^2]	1	0.2	0.5	0.1	0.1	4	0.1	2	0.1	3

The virtual prototype model can be developed conveniently in ADAMS as a multi-body system. Each element with parameters such as density, location, geometry and mass and so on can be given according to actual physical prototype. If the magnitude of nodes is huge, cycle operation can be processed with macros of ADAMS/View and the whole model can be automatically generated [8].

Table 2 Hydrodynamics parameters of the mooring system

The coefficient of added mass C_m	0.5
The drag coefficient of Hydrodynamics C_D	1.2
The stretching stiffness of cable k [N/m]	1×10^5

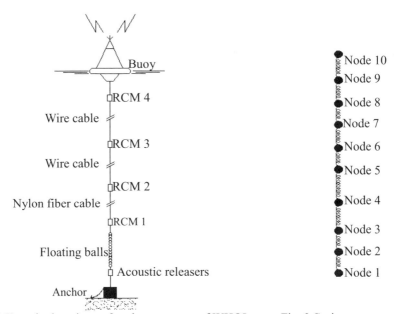

Fig. 2 Taut single point surface buoy system of WHOI   Fig. 3 Spring-mass model

According to Eq.1-3, the forces on each node should be given by the sforce function in the function library of ADAMS/View. Most forces are relaying on the kinematics parameters like displacement, velocity and acceleration, which can be acquired by the Design-time Functions and Run-time Functions. For example, we give the process of a drag force built in ADAMS. According to Eq.1, C_D, ρ_w and A can be considered as the constant coefficient that can be obtained from data and the physical truth such as Table 1. As to $\vec{U}_i$, it always change in different time and different location and it can be expressed by VX (To Marker, From Marker, Along Marker,

Reference Frame) In ADAMS. As to $\left| \vec{U}_w - \vec{U}_i \right|$, it can be expressed by ABS (X) in ADAMS. In the model the drag force is divided into two components of forces in x, y two directions .So the drag force can be written as Fig. 4 in the x direction.

As to example of impact forces between components and bottom, the magnitude of force is written as

$$F = \begin{cases} k' \delta^e + C(\delta)\dot{\delta} & \delta \geq 0 \\ 0 & \delta < 0 \end{cases} \tag{10}$$

Where, δ is deformation, k' is material stiffness, e is exponent, $C(\delta)$ is expressed by *step* function in ADAMS, $C(\delta) = step(\delta, 0, 0, d, c)$, c is the maximum of damping coefficient, d is the penetration depth when the damping attain the maximum.

It considered the nonlinear relationship between forces and deformation and is improved form as to Eq.7 and Eq.8. The force is assigned in ADAMS/View in contact force function. In the impact force model the coefficients are shown as Fig. 5.

Fig. 4 The drag force in the x direction   Fig.5 The impact force

Given the initial values of mooring system and situation of environment, the simulation can be obtained in working situation or deployment operation.

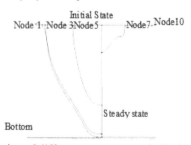

Fig. 6 Trajectories of different components in deployment process

The trajectories of different components in deployment process can be obtained in Fig. 6. During the deployment, the surface buoy moved forward while the anchor dropped towards the direction of the surface buoy. The horizontal derivation distance of anchor is about 1151m. It is related to the configuration and physical parameters of the buoy system. According to the simulation, the dynamic tension forces can be obtained. The tension forces of different components over time are shown in Fig. 7. The maximum tension force appeared as the anchor reached to the seabed. The tension force above anchor is lager than other tension forces during deployment.

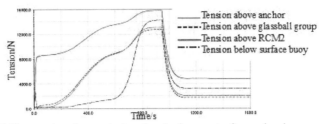

Fig.7 The tension forces during the deployment of mooring buoy system

Conclusions

The paper analyses the dynamic response of mooring buoy system during deployment process. The mooring buoy system is considered as spring-mass model and built using the macros in ADAMS software. According to the core of ADAMS software, the dynamic response can be easily obtained. The trajectories and the tension force of different components have been analyzed. The method is helpful for guidance for buoy design and deployment process planning.

Acknowledgments

The authors appreciated the support of NSFC (No. 51175484) and Science Foundation of Shandong province (No. ZR2010EM052). The authors are also grateful for the help from Key Lab of Ocean Engineering of Shandong province.

References

[1]   R K Dewey, Mooring Design & Dynamics - a Matlab (R) package for designing and analyzing oceanographic moorings, Marine Models. 1 (1999) 103-157.

[2]   J I Gobat, M A Grosenbaugh, Time-domain numerical simulation of ocean cable structures, Ocean Engineering. 33 (2006) 1373-1400.

[3]   O B Augusto, B L Andrade, Anchor deployment for deep water floating offshore equipments, Ocean Engineering. 30 (2003) 611-624.

[4]   R. E. Baddour, W. Raman-Nair, Marine tether dynamics: retrieval and deployment from a heaving platform, Ocean Engineering. 29 (2002) 1633-1661.

[5]   A Umar, T K Datta, Nonlinear response of a moored buoy, Ocean Engineering. 30 (2003) 1625-1646.

[6]   Z Y Chang, Y G Tang, H J Li, J M Yang, L Wang, Analysis for the deployment of single-point mooring buoy system based on multi-body dynamics method, China Ocean Eng.. 26(3) (2012) 495-506.

[7]   T Skeen, S. McDonald, T DePauw, Shipboard Buoy Cable Deployment System In ADAMS, Mechanical Dynamics Inc..

[8]   Using ADAMS/View, Mechanical Dynamics Inc..

[9]   H O Berteaux, Buoy Engineering, John Wiley and Sons, New York, 1976.

Advanced Materials Research Vol. 819 (2013) pp 334-338
© *(2013) Trans Tech Publications, Switzerland*
doi:10.4028/www.scientific.net/AMR.819.334

Design of Machine Operating Panel based on CAN Bus for the PC-based CNC System

Kaifa Wu[1, a], Taiyong Wang[1,b], Jingchuan Dong[1,c], Qingjian Liu[1,d], Fuxun Lin[1,e] and Ruoyu Liang[1,f]

[1]Key Laboratory of Mechanism Theory and Equipment Design of Ministry of Education, Tianjin University, Tianjin300072, China

[a]sdwkf.good@163.com, [b]tywang@189.cn, [c]new_lightning@sohu.com (corresponding author), [d]lklqj-5759@163.com, [e]lfx3107@163.com, [f]lryasa@yahoo.cn

Keywords: CAN, Remote IO, Operating panel, CNC

Abstract. Machine operating panel is an important component of controlling machine and the human-computer interaction. As CAN bus has the remote and stable transmission characteristic, the remote IO board is designed. A design of machine operating panel based on CAN bus for the PC-based CNC system is introduced, including the overall architecture design and software design method. This paper focuses on the remote IO board, as well as the practical application in the CNC system of TDNC-H8.

Introduction

Machine operating panel is the man-machine interface for controlling machine. And whether the information delivery and response of the CNC system is stable and timely directly affects the performance and user experience of CNC machine. Therefore, doing research in operation panel response mechanism with the characteristics of transmission stability and response timely is of great significance. Pc-based CNC system has become the mainstream of development for its multi-tasking nature, extensible and man-machine interface friendly [1]. Presently field bus is one of the hot spots of the automation technology development. And it is used in the production locale and bidirectional serial multi-node digital communication between computerized measurement and control equipment. It is also known as the open, digitized, multi-point communication control network. In many types of field bus, CAN bus adopt a number of new technologies and unique design compared with the general communication bus [2]. Its data communication has outstanding reliability, real time and flexibility. These provide a good foundation and conditions to designing stable and timely response operation panel communication system.

The overall architecture design

This system's overall architecture is shown as Fig.1. Machine operating panel adopts standard panel design, and has 48 input points and 36 output signal lights. In these points, machine mode selecting knob, feed scale selecting knob and tool number use the way of binary encoding to achieve switching and distinction. The rest of the buttons use the way of one to one to achieve its function.

The pressing button information from machine operating panel and the feedback information from CNC are processed by a self-designed remote IO board. This board uses CANBUS protocol and it has double CAN ports. It can realize transferring data stably and long-distance with high-speed.

The interaction between the CAN port and PC is executed by PCI-1680U. PCI-1680U is communication card dedicated connecting between CAN network and PC. PCI-1680U realizes bus arbitration and error detection function by automatic retransmission feature. It can greatly reduce the chance of data loss and ensure system reliability. Onboard CAN controllers are located in different locations of the memory. Then two CAN controllers can run simultaneously and independently [3]. It has a universal PCI interface, high-speed transfer rate of 1Mbps, 16 MHz CAN controller frequency as well as 1000 VDC optical isolation protection to enhance system reliability. The CAN port of

PCI-1680U is a nine-pin DB head and only pin 2, pin 3 and pin 7 are effective. Pin 2 is CANL, pin3 is CANH and pin 7 is CANGND.

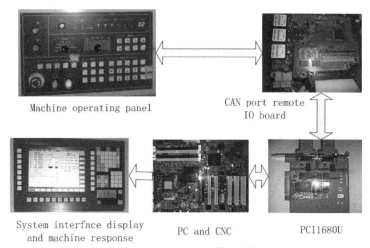

Fig. 1 The system's overall architecture.

Remote IO board with CAN port design

CAN (Controller Area Network) is called control LAN, in line with the international standard ISO 11898. It initially is a bus communication network used for car monitoring, controlling system designed by Germany BOSCH [4]. As a result of the many new technologies and unique design used, compared with normal bus, CAN bus's data communication has outstanding reliability, real-time and flexibility.

Hardware design. The board adopts the microcontroller called STM32F103C8T6 as Embedded - micro-controller (MCU). And it has 64kB internal flash memory and stable performance. External input power supply is 24 volts, and is converted to 5V DC by three power conversion modules named NR24S5/100A. Its IO module uses Xilinx's XC95144XL. Its I / O points are distributed as shown in Fig.2. We can get the result that its IO point number is more than enough, and can meet the requirement of the operator panel.

The board uses PCA82C250 as CAN transceiver. It is the interface between the CAN protocol controller and the physical bus and provides differential transmit capability to the bus and differential receive capability to CAN controller. PCA82C250 is the world's most widely used CAN transceiver. CAN circuit design is shown in Fig. 3. The CAN port consists three wiring named as CANH, CANL and CANGND.

CAN program design. In order to realize the synchronous communication, we must do initialization and internal parameter setting to STM32F103C8T6 internal CAN controller. This is achieved by CANConfig (void) function, as shown in Fig.4.

In order to test remote IO board, self-test program is written specifically. We did communication test connecting two boards with standard CAN lines. It confirms that this design is stable and viable. The core part of the program is shown in Fig. 5.

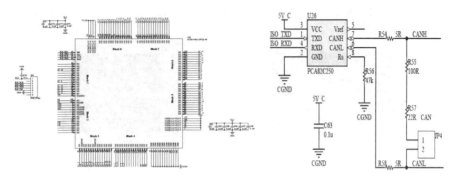

Fig. 2 The IO module, XC95144XL   Fig. 3 CAN circuit design

```
/* transmit */
TxMessage.StdId = 0x321;
TxMessage.ExtId = 0x01;
TxMessage.RTR = CAN_RTR_DATA;
TxMessage.IDE = CAN_ID_STD;
TxMessage.DLC = 6;
```

Fig. 4 Initialization and internal parameter setting

```
CAN_Receive(CAN1, CAN_FIFO0, &RxMessage);
if ((RxMessage.StdId == 0x321)&&(RxMessage.IDE == CAN_ID_STD)&&(RxMessage.DLC == 4))
{
    IO_Write(1, 0, RxMessage.Data[0]);
    IO_Write(1, 1, RxMessage.Data[1]);
    IO_Write(2, 0, RxMessage.Data[2]);
    IO_Write(2, 1, RxMessage.Data[3]);
    TxMessage.Data[0] = IO_Read(1, 2);
    TxMessage.Data[1] = IO_Read(1, 3);
    TxMessage.Data[2] = IO_Read(1, 4);
    TxMessage.Data[3] = IO_Read(2, 2);
    TxMessage.Data[4] = IO_Read(2, 3);
    TxMessage.Data[5] = IO_Read(2, 4);

    CAN_Transmit(CAN1, &TxMessage);
}
```

Fig. 5 Remote IO board self-test program

Software realization way

The operating panel design is applied to TDNC-H8 CNC system, which is a high-end CNC system based on the PC architecture. We use VC programming. Specific ways are shown in Fig.6.

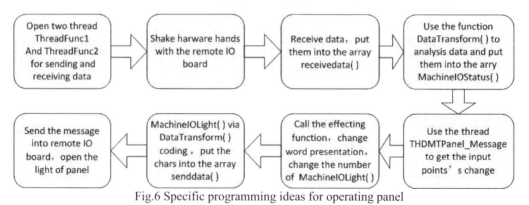

Fig.6 Specific programming ideas for operating panel

According to the number of IO points, we set two bool type arrays named MachineIOStatus [72] and MachineIOLight [48], respectively, to record the input and output of the operation panel. If one value of MachineIOStatus is 1, it means the button is pressed. If one value of MachineIOLight is 1, it means the indicator light on the button will be lit.

We newly built a class CCANPCI and in this class open the two threads ThreadFunc1 and ThreadFunc2, respectively, to the two CAN ports for receiving and sending data. We use hDevice1 = acCanOpen (szCurrentPort, FALSE) to open the CAN port, use nRet = acSetBaud (hDevice1, CAN_TIMING_250K) to set the baud rate, use nRet = acSetTimeOut (hDevice1, 3000, 3000) to set the delay and use msgWrite.id = 801 to set the IO board ID. So we shake hardware hands with the remote IO board. The data transferred from the remote IO board is 9 char type numbers and is put into the arry named as receivedata ( ). This array will be translated into 1 or 0 by function DataTransform (), and stored in the corresponding arry named as MachineIOStatus[72].

We open another thread THDMTPanelMessage to cyclically scan MachineIOStatus changes, then through the VC message response mechanism call corresponding effect functions and do text presentation in the system interface and set necessary MachineIOLight [i] to 1for lighting the indicator light on the control panel. The following is function DataTransform().

```
    void CTDNC_H8Dlg::DataTransform()
{               int mm=0;
                for (int i=0;i<9;i++)
                { int inNum=receivedata[i];   char temp[8]; int resiNum, bm =0;
                  for (bm=0;bm<8;bm++)
                {       resiNum = inNum % 2;
                        If (resiNum!=0)
                        {       temp[bm]='1';
                                inNum=(inNum-1)/2; }
                        else
                        {  temp[bm] = '0';
                                inNum /= 2; }
                MachineIOStatus[mm]=temp[bm];
                        mm++;}}}
```

Conclusions

This Machine operating panel communication system has been used in self-developed numerical control system called TDNC-H8. With practical testing and using, this system's delay time is not exceeding 500ms, and the functions response in time and stably. It is proved that the system can meet practical using need perfectly.

Acknowledgements

This work was financially supported by National Key Technology Research and Development Program (2013BAF06B00), Tianjin application foundation and frontier technology research program (12JCQNJC02500), and the Research Fund for Doctoral Program of Higher Education, China (20100032110006).

References

[1] Xiongbo Ma, Zhenyu Han: Development of a PC-based Open Architecture Software-CNC System, Chinese Journal of Aeronautics , 2007, 20(3): p. 272.

[2] Lei Lu, Deying Gu: The Data Communication Based Oil Canbus, Chinese Journal of Scientific Instrument, Vol.26 No.8, Aug. 2005, p. 432.

[3] http://www.advantech.com/pci-1680u

[4] Chen Peiyou ,Tong Weiming: Research on CAN bus layered structure and MAC mechanism, Chinese Journal of Scientific Instrument, 2006, 27(z3), p2422.

Advanced Materials Research Vol. 819 (2013) pp 339-343
© (2013) Trans Tech Publications, Switzerland
doi:10.4028/www.scientific.net/AMR.819.339

Research on Product Virtual Prototyping Technology and Design Method for Concurrent Engineering

Li Xiaona[1, a], Liu Xin[2,b]

[1] School of Mechanical Engineering Tianjin Commerce University, Tianjin, China

[2] Department of Design Center of Tianjin Optical Electrical Group Co., Ltd. , Tianjin, China

[a] xnsmile@163.com, [b] catlxsukur@163.com

Keywords: virtual prototyping; simulation; concurrent engineering; PLM; bicycle

Abstract: Concurrent engineering claims all influence factors within product lifecycle should be took into account during the stage of product design. After analyzing the relationship between concurrent engineering and virtual prototyping, the human-machine system simulation process and method based on virtual prototyping technology were discussed and illustrated with bicycle design, and the product design method based on virtual prototyping was proposed in this paper.

Introduction

With the technology development and consumer's updated demands, manufacture industry is facing more and more challenges. Research has shown that the cost in the stage of product development, including concept design, preliminary design and detailed design, accounts for only 18% of the total cost, but it will determine the product's value of 80% - 90%[1]. The product design implementation directly influences the development cycle, product quality, product cost, customer satisfaction, etc. That is the important thoughts of concurrent engineering. Therefore, the product design has the critical relationship with the concurrent engineering development. In this stage, the factors of product lifecycle must be considered adequately. Virtual prototype technology provides a quick and efficient way for product design. Through the man-machine system simulation and analysis, the function of products can be evaluated in the design stage and the optimization design can be done. This paper tried to put forward a kind of product design method based on the thought of concurrent engineering and the virtual prototype technology.

Simultaneous Engineering and virtual prototyping

Concurrent engineering, also called Simultaneous Engineering, is a integrated and synchronized product design method. It aims to short the product development cycle and time to market, improve product quality and reduce the product cost. Concurrent engineering was proposed by the Institute of Defense Analyze in 1988. This method emphasizes that the work teams with various function should be organized in the early stage of product development, so as to make the relevant personnel get more information from the product concept design stage, and carry out the work of their respective departments as soon as possible. Also, this method can strengthen the cooperative work between different departments, so that many problems in the early stage of development can get resolved, so as to avoid a lot of rework waste, and ensure the quality of product design [2]. This model enables developers to consider the product lifecycle from the start.

According to the characteristics of the product design, based on the concurrent engineering thought of "start work as soon as possible" and "multidisciplinary joint participation", only effective simulation analysis and optimization work were carried out in the stage of product design, cooperation within different departments was strengthened, the overall optimization target was emphasized, and the concurrent engineering of product development was carried out efficiently, the goals of shorten development cycle, quality improvement and cost reduction could be achieved.

With the development of information technology and network technology, modern design technology is progressing with the features of integration, agility and virtualization. Kinds of production mode came out, such as Lean production, agile manufacturing, virtual manufacturing, and mass customization and so on, which all based on concurrent engineering and concurrent design. The new technology of virtual prototyping arises at the right time and conditions.

The virtual prototype is some kind of digital model built with software before the first physical prototype[3]. It can reflect the actual product characteristics, including its spatial relations, as well as the characteristics of kinematics and dynamics. With the simulation and analysis based on the entity visualization, virtual prototype can reveal the kinematic and dynamic characteristics of the system under the real working environmental conditions, thus revised design program can be done repeatedly, and the optimal design scheme can be achieved ultimately.

Man-machine system simulation based on virtual prototyping

Based on the above analysis, this paper discussed product design method based on virtual prototyping technology, and illustrated with bicycle's man-machine system design through the process of the establishment of man-machine digital model, man-machine interactive simulation, simulation results analysis and optimization design. These were all described as follow.

Establishment of virtual prototyping. The virtual man-machine system is composed with bicycle digital model and human body digital model. Bicycle digital model was built using the three-dimensional modeling software of Pro-Engineer. Digital model of the human body was built using the biomechanics modeling software of BRG. LifeMOD. Both of them were jointed using the software of ADAMS, and completed the dynamics and kinematics simulation.

ADAMS, which is the abbreviation of Automatic Dynamic Analysis of Mechanical Systems, is the virtual prototype analysis software developed by MDI Company. Biomechanics modeling software of BRG. LifeMOD, developed by Biomechanics Research Group, is the most advanced and integrated human body modeling software, which can be used to establish any biological system based on the real physical biomechanical model. LifeMOD is the common plug-in software of ADAMS. It can generate passive and dynamics biological model accurately which can interact with implements and equipments with Integrated modeling environment. LifeMOD is a complete, state-of-the-art virtual human modeling and simulation software solution. Its advanced capabilities and intuitive graphical interface, developed and refined over two decades, enable engineers, designers, and others interested in biomechanics to create human models of any order of fidelity, report true engineering data, and enable rapid and repetitive testing of designs, all while slashing time, cost, and risk from new product development[4].

ADAMS - LifeMOD simulates human motions by taking human limb segment as a multi-link system, and reveals the real interaction between human and machine. The simulation output can be used for the real reflection of human body, then can help designer to analyze the problems of the design and make corresponding improvement.

Virtual prototyping system simulation. Based on the setting of the parameters, the man- bicycle virtual prototype system was built up, and then the reverse dynamic and positive dynamics simulation can be done. The simulation results, including lower limb muscle tension analysis, the main muscle tension contrast analysis with long and short sitting tube, can be obtained. Through the analysis comfort evaluation on bicycle design guidance can be found, and then the further optimization design. Also, through the joint torque the joint comfort and fatigue can be simply analyzed (not discussed in this paper).

Fig.1. Simulation of the man-bicycle system

The simulation takes crank's 360 degrees as a cycle. In one riding cycle, when the pedal with counterclockwise operation is at the front (horizontal), the angle is 0 degrees, in the end is 180 degrees, and at its lowest point is 90 degrees, the upper point is 270 degrees, and again to the front end for 360 degrees is a complete cycle.

Taking the left leg as an example, this paper analyzes the leg muscle tension changes in a cycle. And leg muscles' roles and conditions at any time of the cycle can be summarized. For example, musculus glutaeus maximus tension is shown as follow.

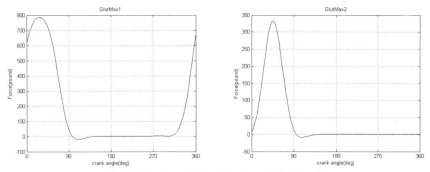

Fig.2. Tension of GlutMax1 and GlutMax2

In order to observing each muscle's tension in the riding process, and the influence to human body when changing the parameters of the bicycle frame, the bicycle saddle was risen, which means the length of seat post increased. We can see the muscle tension contrast in both cases. The contrast of the simulation muscle force with long seat post and short seat post can be summarized.

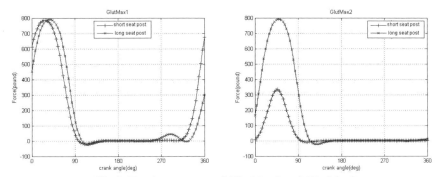

Fig.3. Tension contrast of GlutMax1 and GlutMax2

Simulation analysis. Reducing the user's feeling of fatigue is one of the main performance indicators of the bicycle product. Through the above simulation of the virtual prototyping system, the results can be used with the muscle fatigue analysis methods to evaluate the product, so as to provide a basis for the optimization design. Crownishield R.D. puts forward muscle stress's sum of squares minimum standard to measure muscle fatigue [5] [6]. This is the most common evaluation standard.

$$J = \min \sum \sigma^2 = \min \sum_i (F_i / S_i)^2 \tag{1}$$

In the equation, F_i is the Physiological Sectional Area of the i th muscal. S_i is the tensile force of the i th muscal. So the changes of muscle tension in the movement can be used to help judge muscle fatigue situation.

Through the equation(1), the curve showed the main changes of the lower limb muscle stress with time by using MATLAB software programming calculation.

As can be seen, in a cycle, when the pedal was at the position of about 45 degrees, 200 degrees and 290 degrees, muscle stress squares were very large. That means when the left leg was under the hard push at the three phases, the muscle was at the most nervous situation, and each part of the muscle tension were bigger, so the leg would be easily fatigue. When at the lowest, it was the right leg which takes the role, so the left leg became more relaxed.   After the peak, about 290 degrees or so, muscle stress reached a maximum and then decreased gradually until the minimum, and then a new cycle started.

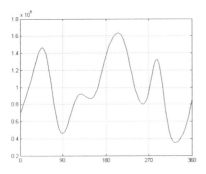

Fig. 4. The quadratic sum of left leg muscle stress in a cycle

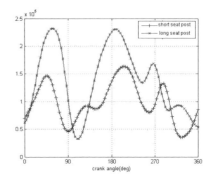

Fig.5. The contrast of quadratic sum of left leg muscle stress with short and long seat post

After lengthening the seat post, the contrast of quadratic sum of left leg muscle stress changed as shown in figure 5. It can be seen, after raising saddle, the muscle stress change rules were similar, but the size of the value increased significantly, indicating that the human body lower limb muscle more easily fatigue in this kind of circumstance. That is, lower the saddle's height will be more conducive to the user's riding for this bike.

Product design method

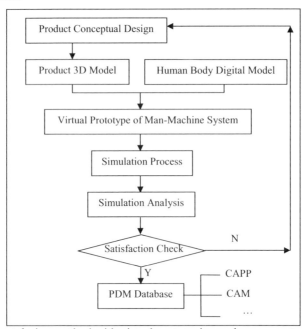

Fig.6. Product design method with virtual prototyping and concurrent engineering

Through the above it can be seen that the main basis of bike optimization design can be gained through virtual prototype simulation analysis, this method can be so on to other products. Man-machine system simulation, through the real people in the process of interaction with the product, can reveal various reflections of the human body, and then to judge the rationality and accuracy of the design according to the result of simulation, and therefore can predict the user-product condition in the process of design. Through feedback optimization design can be done until the best design effect achieved.

Method of product design based on virtual prototype technology can be summed up. 3D virtual prototype of the initial product design, simulation and analysis of man-machine system, and the processing of final physical prototype, can be unified harmoniously, in order to achieve the collaboration of the product design, analysis and management, so as to provide the basis for the integration of concurrent engineering.

The specific implementation process is as shown in figure 6.

Summary

Product design is an important part of the whole product life cycle. On the basis of concurrent engineering thought, the man-machine system simulation based on virtual prototype technology for product design stage provides an effective method. It is crucial for the optimization of product design, as well as product functionality and quality. The method has strong advantages and development potential, and will become the future mainstream of product research and development.

References

[1] Edlin N. Management of engineering/design phase. Journal of Construction Engineering and Management ASCE, 1991, 12(117):63- 75

[2] BAIDU BAIKE. Concurrent Engineering [EB/OL]. http://baike.baidu.com/view/50733.htm, 2013-02-16

[3] BAIDU BAIKE. Virtual Prototyping [EB/OL]. http://baike.baidu.com/view/5883974.htm, 2013-02-25

[4] LifeModeler. lifeMOD[EB/OL]. http://www.lifemodeler.com/products/lifemod/, 2013-01-23

[5] Crownishield R D, Brand R A, A Physiologically based criterion of muscle force prediction in locomotion[J]. Journal of Biomechanics, 1981,14(11): 793-801

[6] XU Li, GUO Qiao, CHEN Haiying. Optimum Solution and Analysis of Redundant Muscular Force in Lower Extremity During Running [J].Transactions of Beijing Institute of Technology, 2004,10(24):869-873

Advanced Materials Research Vol. 819 (2013) pp 344-349
© *(2013) Trans Tech Publications, Switzerland*
doi:10.4028/www.scientific.net/AMR.819.344

The Optimal Design of the Plasma Discharge Structure of the Far Zone of the Hollow Cathode

WEN Yifang [1,2,a], RUI Yannian[2,b], CHEN Chuang[2,c], WANG Hongwei[3,d]

[1]Suzhou Institute of Industrial Technology, Suzhou 215104, China

[2] Soochow University, Suzhou 215021, China

[3]OPS Plasma Technology Co., Ltd, Suzhou 215000, China

[a]wenyf@siit.edu.cn(corresponding author), [b]ryn1951@sina.com, [c]452782381@qq.com, [d]szomega@126.com

Key words: Hollow cathode discharge; Hollow cathode nozzle; Optimal design

Abstract: The discharge of RF hollow cathode far zone plasma has advantages of high ion concentration, easy to implement the processing of a large area, given more and more attention. The characteristics of hollow cathode plasma flow have a great relationship with the hollow cathode nozzle structure. How to design the hollow cathode nozzle discharge structure accurately and conveniently is the key problem of the hollow cathode far plasma surface treatment. This article builds a hollow cathode plasma discharge self-consistent model; derive the relationship between the discharge current of the hollow cathode plasma and the hollow cathode nozzle structure. Optimize the design of the hollow cathode discharge structure using discharge particle simulation software, to achieve a fast and accurate design and the purpose of efficient plasma surface modification.

Introduction

The far zone of RF hollow cathode plasma discharge is the discharge system composed by the anode and cathode having a hollow structure, with the samples placed in outside the yin and yang level.

Comparison with ordinary flat direct plasma treatment, it has a high ion concentration, and the advantages of achieving a large area processing easily, increasingly attracted the attention of researchers, as a novel plasma surface treatment method.

The flow characteristics of the hollow cathode plasma have a great relationship with the hollow cathode nozzle structure.

The article build a hollow cathode plasma discharge self-consistent model, derived the relationship between the discharge current of the hollow cathode plasma and the hollow cathode nozzle structure, optimized the design of the hollow cathode discharge structure applying discharge particle simulation software, achieved a rapid and accurate design and reached efficient plasma surface modification purposes.

The self-consistent model hollow cathode plasma discharge

The hollow cathode discharge structure generally refers to a hollow cathode nozzle assembly comprising a hollow cathode, the anode and the insulating layer.

In the hollow cathode apparatus, the radio frequency hollow cathode is composed of a plurality of individual nozzle single hollow cathode. The cathode and the anode aperture pairs arranged axially, and as a nozzle. Each nozzle forms a low temperature plasma nozzle, and the formation of the hollow cavity. Specifically as shown in Figure 1.

The hollow cathode nozzle District discharge self-consistent model shown in Figure 2.The model includes a main plasma region and two of the sheath region. The diameter of the main plasma area is d, and the length of t_2.The two sheath region is divided into the cathode sheath and the anode sheath, the thickness of the sheath region is S_{ma} and S_{mc}, respectively. Absorbed power in the main plasma region from holmic heating, the sheath electric field generated by the change over time, the current flowing through the two sheaths is substantially the displacement current. Therefore, the power absorption in the region of the sheath random heating caused by the collision of electronic and sheath.

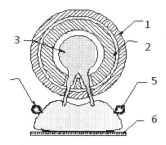

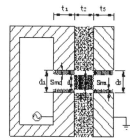

Fig.1 Schematic diagram of hollow cathode discharge

Fig.2 Hollow cathode nozzle area discharge self-consistent

1-Anode;2- Cathode;3- Hollow cathode plasma;4- Tracheal graft polymerization of monomers;5- Graft polymerization of monomers;6- Base mat- erial

Hollow cathode nozzle area plasma discharge current

Fig 3 is a schematic diagram of the main plasma region and RF plasma density of the sheath. n_s is the sheath at the boundary of the plasma density, n_o is at the center of the plasma density.

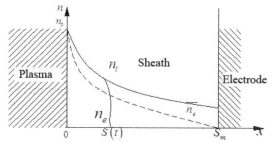

Fig 3 Cathode nozzle area plasma density schematic diagram

Ions across the sheath boundary at $x = 0$ into sheath and accelerated, Finally, punishable by a high energy collision electrode at $x = x_m$. Ion flux density is a conserved quantity, when the ions through the sheath, u_i increasing, n_i declining. According to literature [4] [5] and Childe law, Ion current $\overline{J_i}$ is obtained by plasma boundary at the Bohme flux.

$$\overline{J_i} = e n_s u_B = e \times 0.80 n_0 \left(4 + \frac{d_3 n_g \sigma_i}{4} \right)^{-1/2} (eT_e / m_i)^{1/2} = e \times 0.80 \times \frac{P_{abs}}{e(eT_e/m_i)^{1/2} A_{eff}(\varepsilon_c + \varepsilon_e + \varepsilon_e')} \left(4 + \frac{d_3 n_g \sigma_i}{4} \right)^{-1/2} (eT_e/m_i)^{1/2} \quad (1)$$

Meanwhile

$$\frac{n_s}{n_0} \approx 0.80 \left(4 + \frac{d_3 n_g \sigma_i}{4} \right)^{-1/2}, n_0 = \frac{P_{abs}}{e(eT_e/m_i)^{1/2} A_{eff}(\varepsilon_c + \varepsilon_e + \varepsilon_e')}, A_{eff} = \frac{\pi d^2 h_l}{4} + \pi d t_2 h_R, h_l = \frac{n_{sl}}{n_0} \approx 0.86 \left(3 + \frac{t_2}{2\lambda_i} \right)^{-1/2},$$

$$h_R = \frac{n_{sR}}{n_0} \approx 0.80 \left(4 + \frac{d}{4\lambda_i} \right)^{-1/2}.$$

In the above formula,

$\overline{J}_i$ - Plasma discharge current;

e - The charge of the electron;

P_{abs} - Plasma total absorbed power;

T_e - The temperature of the electron;

A_{eff} - The effective area of particle losses;

ε_c - Electron - ion collision energy caused by the loss;

ε_e - Every loss of a loss of the average kinetic energy of the electrons;

ε_e' - Ions caused by the loss of one electron loss of kinetic energy;

n_g - Discharge gas density;

d_1 - The cathode aperture of a hollow cathode;

d_3 - The anode aperture of a hollow cathode;

t_2 - The distance between the cathode and the anode

σ_i - The total ion momentum transfer cross section;

m_i - Positive ion mass;

λ_i - Ion - neutral particle collision mean free path

It Shows that the size of the plasma discharge current $\overline{J}_i$ is associated with hollow cathode aperture d_1, the anode aperture d_2, the distance t_2 between the cathode and the anode, the applied RF power supply voltage, the current and the angular frequency, a discharge power, etc.

The simulation Hollow cathode plasma far zone discharge particle

From the above, the hollow cathode nozzle structure parameter is the decision for the characteristic parameters of the plasma discharge current size. Its design can be optimized and the hollow cathode optimal nozzle structure parameters determined.

Particle simulation method (PIC) starts from the basic electromagnetic law of motion and particle motion mechanical laws directly, can truly reflect the actual physical process. The topics refer the same type of editing software Track program of Plasma Physics, Chinese Academy of Sciences in cooperation and the Japan Atomic Energy Research Institute exchange positive ion source for neutral beam extraction system, makes simulation design of hollow cathode nozzle structure.

According to the nozzle structural design requirements and simulation software function, in the case of the applied RF power, voltage, current, angular frequency values are fixed, Mainly on the cathode diameter, the anode diameter and other parameters will be simulated.

Simulated conditions: ① Discharge conditions: frequency of 13.56MHz, RF power of 70W, RF voltage of 230V; ② vacuum conditions: vacuum pressure 5Pa, discharge gas is hollow, the air flow rate of 2.5L/min; ③ the condition of hollow cathode structure: 2.5 mm gap between the cathode and the anode, the distance of the material to be treated cathode distance of 15mm; ④ discharge time: 1.5min.

When the hollow cathode of the cathode and the anode aperture are 4mm, we can obtain the simulation results shown in Figure 4.

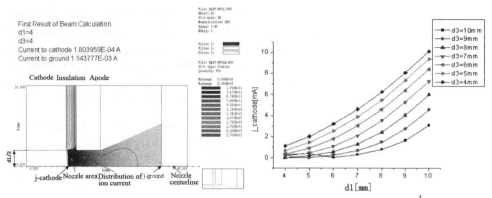

Fig 4 Simulation discharge figure   Fig 5 The impact of anode aperture change d_3 to j-cathode

The Ion current refers to the average ion current analog in the hollow cathode nozzle area and the processing area (substrate) obtained, and j-cathode [mA] and j-ground [mA] respectively said.

It can be seen from the simulation results, plasma discharge region, mainly in the area of the cathode and the anode nozzle, in the closer region away from the nozzle, the higher the value, the greater the ion current of the plasma density.

Diameter variation range of the cathode and the anode are the 4-10mm, the ion current is obtained by simulation at the cathode terminal and the male terminal ion current average, j-cathode [mA] and j-ground [mA] respectively said. j-cathode [mA] refers to the average value in the nozzle zone of the ion current of the hollow cathode. j-ground [mA] is the average ion current reaches the region of the material to be treated.

The impact of Changes in the structure of the hollow cathode to the ion current j-cathode.

The impact of the anode aperture change. The impacts of the anode aperture change to the ion current j-cathode seen in Fig 5. It shows the anode diameter, directly affects the magnitude of ion current, with the increase of the anode aperture, j-cathode value decreasing. So the anode aperture may be smaller. However, if the anode aperture is too small, active particles ejected from the anode plasma to orifice, directly affect the grafting polymerization effect.

The impact of the cathode aperture change. The impacts of the cathode aperture change to the ion current j-cathode seen in Fig 6.

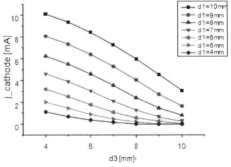

Fig.6 The Impact of anode aperture d_1 change to
j-cathode

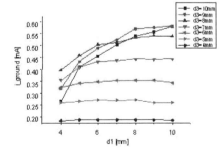

Fig.7 The Impact of changes of the anode
aperture d_3 to j-ground

It is shown in the Figure 6, when the anode aperture diameter d_3 is fixed, for example, $d_3 = 4mm$ the increase of the pore diameter d_1, with the cathode, the ion current density of the hollow cathode nozzle area increases. This shows that in the discharge power and the discharge gas flow rate to maintain under certain circumstances, as the cathode diameter increasing, the increase in gas density increases along smaller obvious effects of collisions between the gas, to produce ionization probability, the activation of the plasma state the particles increased, the plasma generated is also increased, i.e. cathode aperture and a large ion current.

Figure 5, 6 shows, the hollow cathode ion current generation is affected by the cathode and the anode aperture, if the cathode aperture big, the anode aperture will be small, and the larger nozzle area ion current. j-cathode only, however, that the size of the ion current near the nozzle, the nozzle area of the role of plasma surface modification of materials are limited. Really played a key role on the material surface modification is the ion current j-ground reach the surface of the material area.

The impact of Changes in the structure of the hollow cathode to the ion current j-ground.

The impact of the anode aperture change. The impact of the anode aperture changes to the ion current j-ground seen in Fig 7.

Seen from figure, when cathode diameter is fixed, the ion current is affected by the anode diameter change is very big. If cathode are 10 mm diameter, the anode is 4 mm in diameter, ion current density of 0.18601 mA, and when the anode is 10 mm in diameter, ion current density of 0.57845 mA, the growth of the amplitude is 0.39244 mA, the latter is the former three times. And with the increment of anode diameter, the ion current also will increase.

Therefore, the design of anode structure is directly related to the ion current densities, which is the number of active particles, and at the same time, also affect the uniformity of hollow cathode grafting polymerization.

When the anode fixed in 4 to 7 mm in diameter, as the change of cathode diameter, due to the effect of diffusion of the plasma, the material on the surface of the plasma is relatively uniform, ion current density basic remain unchanged. But, when the anode, 8 -- 10 mm in diameter, as the change of cathode diameter, the uniformity of plasma larger mutation happened, suggesting that the anode diameter directly affect the active particles distributed uniformly on the surface of the material, ultimately affect the uniformity of graft polymerization.

Comprehensive above, can get the best anode diameter of 8 mm.

The impact of the cathode aperture change. The impacts of the cathode aperture change to the ion current j-ground seen in Fig 8.

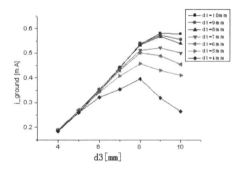

Fig 8 The Impact of anode aperture d_l change to j-ground

The figure 8 shows, when the fixed anode diameter d_3, cathode diameter d_1 is bigger in the process, overall ion flow is big, but big trend is more and more small, the gap is more and more small. When the anode is less than 8 mm diameter d_3, the cathode diameter d_1, j-ground impact is not big.

When the anode diameter $d_3 = 8mm$, j-ground began to appear a turning point, with the increase of cathode diameter d_1, j-ground there will be a first increases and then process, which indicates that the cathode can't unlimited increase the diameter, the best cathode diameter d_1 value is 8 mm.

Conclusions

Through the above research, the paper work and conclusions are as follows:

Build the hollow cathode plasma discharge self-consistent model, deduced the hollow cathode plasma discharge current and the relationship between the hollow cathode nozzle structure, the application of plasma particle simulation software, the optimization design was carried out on the structure of hollow cathode discharge.

Diction in the hollow cathode plasma discharge area, namely the cathode and anode nozzle area, the ion current by the influence of cathode and anode aperture, if the aperture large cathode, anode aperture is small, the nozzle area of the ion current is bigger.

Area away from hollow cathode nozzle, namely, material surface treatment area, with the increment of anode diameter, the ion current also will increase.

Hollow cathode and the nozzle structure of the ion current density is directly related to the design size, namely the number of active particles, at the same time, also affect the material's surface activity of uniform particle distribution, and ultimately affect the uniformity of graft polymerization. Best anode diameter is 8 mm; best cathode diameter value is 8 mm.

References

[1] D.Korzec, J.Engemann, Multi-jet hollow cathode discharge for remote polymer deposition. Surface and Coatings Technology. 93 (1997)128-133.

[2] L. Bardos, Radio frequency hollow cathodes for the plasma processing technology. Surface and Technology. 86-87 (1996)648-656.

[3] H.Barankova, L.bardos, Hollow cathode cold atmospheric plasma source with monoatomic and molecular gases. Surface and Coatings Technology. 163-164 (2003)649-653.

[4] A Michael, J. A. Inberg, Plasma discharge principle and material processing, Beijing, 2007.

[5] M. Naddaf, S. Saloum, B. Alkhaled, Atomic oxygen in remote plasma of radio-frequency hollow cathode discharge source. Vaccum, 8(2010)4-12.

[6] H. S. Zhen. Plasma processing technology, Tsinghua university press, 1990.

Advanced Materials Research Vol. 819 (2013) pp 350-355
© (2013) Trans Tech Publications, Switzerland
doi:10.4028/www.scientific.net/AMR.819.350

The topology optimization of the car sub-frame based on APDL language

LI Jian-min[1,a] YANG Zhang-cheng[1,b] SUN Chuan-yang[1,c] XIE Hai[1,c]

[1]Zhejiang Province's key laboratory of reliability technology for Mechanical and Electronic product, Zhejiang Sci-Tech University, Hangzhou 310018, China
[a]ljmzrz@163.com, [b]yzc1988628@163.com

Key words: the car sub-frame; parametric modeling; topology optimization

Abstract: Using APDL language, the car sub-frame finite element parametric model is created. It indicates that the emergency braking is the dangerous conditions of sub-frame. The sub-frame is topology optimized under the emergency braking, and gets the element pseudo density distribution map; By modifying the model according to the pseudo-density, the optimization results meet the design requirements. Compared to the original model, it gets a 9.05% reduction in volume, and achieves lightweight of the sub-frame.

Introduction

The cars sub-frame is important structural parts of the car chassis. Lighter chassis structure can help reduce a car's fuel consumption and save energy. It is one of the most important works of chassis design to develop the sub-frames, which should meet the strength and stiffness requirement and be with a lighter weight [1]. In this paper, we select sub-frame as the research object. We make analysis of sub-frame with multidisciplinary subjects such as parametric modeling with APDL language [2], finite element, multi-body mechanics structural and structural optimization and design. The optimized sub-frame gets a weight reduction of about 9.05% under the premise of keeping the maximum stress essentially the same.

Taking into account the sub-frame for the steel plate welded structure which withstands buckling deformation in the work, it will use shell elements in the Finite Element Analysis. Therefore, it does not need to complete the three-dimensional model. Instead, it uses a two-dimensional surface model. Plate thickness is given in the form of a real constant.

The geometric modeling of complex structures is important work of mechanical CAD, and also the basis of the CAE work such as structural analysis. The specific critical parameters are such as Table 1. It gets sub-frame parametric model when running the APDL program, as show in Fig.1.

Table 1 The main parameters of characteristics

Thickness	Centerline distance	The distance between the plane	Total length	Opening radius
2.5mm	612mm	32.6mm	804mm	29mm
Opening length	The radius of Ribs 1.	The length of ribs 1.	The Radius of rib 2.	The length of rib 2.
40mm	24mm	54mm	20mm	88mm

Finite Element Analysis of sub-frame

The sub-frame structure is steel stamping and welded. Its main deformation is bending. We use SHELL63 elements for finite element analysis. To mesh the parametric geometry model of sub-frame with the cell of the minimum size of 5mm. the sub-frame is divided into 26153 elements and 77218 nodes, as shown in Fig.2. The material of the sub-frame is automotive structural steel, and its elastic modulus is 206GPa, and the Poisson' ratio is 0.3, and with a density of $7.8 \times 10^{-9} T / mm^{-3}$ and the yield limit of 345MPa.

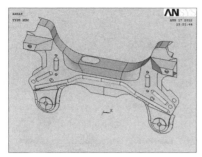
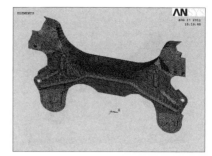

Fig.1 Parametric model   Fig.2 Finite element model of sub-frame

As the car body's stiffness is far greater than the sub-frame. Therefore, the four connection points with the sub-frame and the body are considered the full constraint [3].

When the cars work, the sub-frame needs to withstand the gravity of the engine compartment and the inertia force that generated by shifting movement acceleration, while the loads produced by body pass primarily through independent suspension. Therefore, the most dangerous condition is when the inertial force is the largest for the sub-frame. Compare the car's typical conditions such as starting condition, driving condition and braking condition:

Starting condition: the car takes time of 9.75s in the 100 kilometers acceleration. Assuming acceleration is constant, and acceleration is $2.85m/s^2$.

Driving condition: it is generally considered the driving acceleration is approximately zero with constant speed.

Braking condition: It will be stopped immediately to avoiding accident when the car is drove in a high speed. There is a great negative acceleration in an emergency braking process which will produce a great inertial force on the chassis, and therefore the sub-frame of the braking phase is always the dangerous working conditions [4-5]. The maximum speed on the freeway is generally 120km/h in China's traffic law. The prescribed safe driving distance is 100m. However, in reality, a lot of cars can't keep a safe distance, usually 50-60m. If stopping the car in the distance of 55m, the braking acceleration is $10.05m/s^2$.

Compare the above acceleration of three conditions, the braking condition is the most dangerous condition.

The simulation results based on ADMES/CAR in braking condition show that the force of the four suspensions point are up to 980N. Applied the load of 980N on these suspension, then analysis sub-frame using FEM. The result is shown in Fig.3 and Fig.4.

The maximum of sub-frame' deformation which can be seen from the analysis results is 0.382mm, which occurs at connection points of the left and right rear of engine and sub-frame suspension. The maximum stress is 307.073MPa which occurs at the right rare of suspension. In order to fully reveal the stress and deformation of the sub-frame, the local area of sub-frame is analyzed.

Path 1 is taken to outside at the node of maximum stress of sub-frame. The equivalent stress along the path is shown in Fig.5. From the figure, the stress is concentrated in the above of 190MPa within 5mm of the connection point, which fully reflects the former of the stress concentration of the point due to the connection. This area is the connection region of the engine and sub-frame. The results show that the engine load is the most important factor of high stress of the sub-frame. The stress continued to decline from 190MPa to about 80MPa with radius decreases away from range of the 6mm-17mm of the connection point; the stress is gently reduced to 68MPa with radius decreases away from range of the 17mm-35mm. This area is the danger region of sub-frame, and the maximum stress is less than the yield limit of material, which indicating that the sub-frame has the space of optimization.

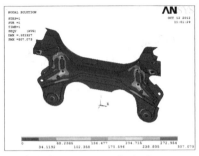

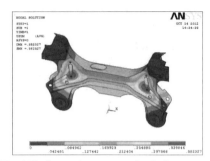

Fig.3 Stress maps of sub-frame   Fig.4 Deformation maps of sub-frame

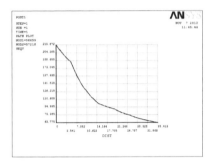
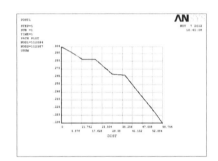

Fig.5 Stress maps of path 1   Fig.6 Deformation maps of path 2

Path 2 is taken to outside at the node of maximum deformation of sub-frame. The displacement and deformation along the path is shown in Fig.6. From the figure, the value of the displacement and deformation is gradually reduced along the path from 0.304mm to 0.205mm. The overall deformation is very small, which indicating the sub-frame has good stiffness.

The topology optimization of sub-frame

The topology optimization of static structure is to select an optimized subset in the design area, which meets the constraint. And the objective function reaches its minimum. The most commonly used method to solve topology optimization problems is the homogenization method or relative density method [6].

In the finite element analysis, the equilibrium equations of static structural [7].

$$[K_0]\{q_0\} = \{F\} \tag{1}$$

The total strain energy is given by

$$W_{E_0} = \frac{1}{2}\{F\}^T\{q_0\} = \frac{1}{2}\{q_0\}^T(\sum_e [K_e])\{q_0\} \tag{2}$$

It shows that the maximum of structural stiffness equivalent to the minimization of total structural strain energy in Eq (2). The structural stiffness changes in the topology optimization process when delete element "e".

$$[\Delta K] = [K] - [K_0] = -[K^e] \tag{3}$$

$[K]$ is the stiffness matrix after the change and $[K_0]$ is the original stiffness matrix and $[K]$ is the stiffness matrix of the deleted element.

The changed model is then given by

$$[K]\{q\} = ([K_0]+[\Delta K])(\{q_0\}+\{\Delta q\}) = \{F\} \tag{4}$$

Expand this formula

$$\{\Delta q\} = -[K_0]^{-1}[\Delta K]\{q_0\} = -[K_0]^{-1}[K^e]\{q_0\} \tag{5}$$

Where $\{\Delta q\}$ is displacement increment when delete e-element.

The increment of the total strain energy is

$$\Delta W_E^e = W_E - W_{E0} = -\frac{1}{2}\{q_0^e\}^T[K^e]\{q_0^e\} \tag{6}$$

In order to eliminate the impact of cell size on ΔW_E^e, it defines density of strain energy of element as a structural topology modified stiffness sensitivity

$$S_{V_E}^e = \frac{\Delta W_E}{V_E} \tag{7}$$

V_E is the volume of the e-element in Eq. (7).

Based on the Equations can calculate the topology optimization of stiffness sensitivity of each element S_j^e (j=1,2...n), and calculating the dimensionless factor

$$S_j = \frac{S_j^e}{\max(S_1^e,S_2^e,\ldots,X_n^e)} \tag{8}$$

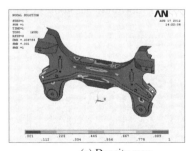 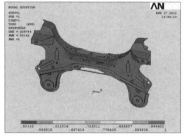

(a) Density map   (b) The saved elements

Fig.7 Density cloud map of topology optimization

Where n is the number of elements, and $\max(S_1^e,S_2^e,\ldots,X_n^e)$ is the maximum stiffness sensitivity of the all elements. Use the following formula as the criteria of deletion element

$$S_j < S_c \tag{9}$$

S_c is the threshold value of deletion element given in advance. If the j-element's stiffness sensitivity meets the criteria, the element can be deleted. On the contrary be saved.

It gets pseudo-density maps by the topology optimization in ANSYS as show in Fig.7. The blue areas show relative density of smaller value for topology optimization, which will be deleted. The red areas show relative density of the largest value, which will be saved. The yellow areas show relative density of larger value (0.5-0.9), which also will be saved.

Model modifications and strength analysis

We modify the sub-frame model based the APDL language. The main change is to remove the rear overhang fusion part of sub-frame and to increase the opening of the front sub-frame. The topology optimization CAD model is shown in Fig.8. In order to assess the strength of the modified sub-frame, we make the finite element analysis on it. The displacement distributing and the stress distributing of modified displacement sub-frame are shown in Fig.9.

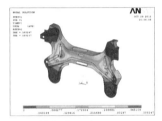 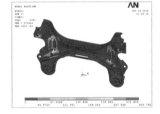 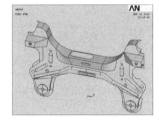

(a) Deformation maps   (b) Stress maps

Fig.9 Stress and deformation maps of modified sub-frame   Fig.8 Optimized sub-frame

The local area of modified sub-frame is analyzed. Path 2 is taken to outside at the node of maximum stress of sub-frame. The equivalent stress along the path is shown in Fig.10 and Fig.11. The stress is concentrated in the above of 240MPa within 5mm of the connection point. The stress continued to decline to about 40MPa with radius decreases away from range of the 5mm-15mm of the connection point. This area is the danger region of sub-frame, and the maximum stress is less than the yield limit of material, which indicating that the modified sub-frame meets the strength requirements. Path 1 is taken to outside at the node of maximum deformation of sub-frame. From the figure, the value of deformation is gradually reduced along the path from 0.349mm to 0.179mm. The overall deformation is very small, which indicating the modified sub-frame has good stiffness.

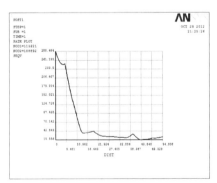 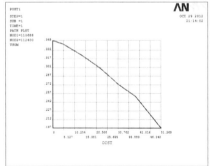

Fig.10 Stress maps of path 1   Fig.11 Deformation maps of path 2

Compare the original sub-frame and the modified sub-frame, the result is shown in Table2. The results show that modified sub-frame is still being designed within the requirement of the stiffness and strength. The volume of the modified model decreases by 9.05%.

Table 2 The results of the comparison

	Maximum stress of local area/MPa	Maximum deformation of local area/mm	Volume/m3
Original sub-frame	307	0.368	0.00159898
Optimized sub-frame	306	0.388	0.00145424
Results of the comparison	-0.3%	5.5%	-9.05%

Conclusions

Using APDL language, finite element parametric model is created by APDL which is a secondary development tool based on ANSYS.

It indicates that the emergency braking is the dangerous conditions of sub-frame by car condition analysis. The sub-frame is topology optimized under the emergency braking condition, and gets the element pseudo density distribution map;

By modifying the model according to the pseudo-density and making finite element analysis on modified sub-frame. The optimization results meet the design requirements. Compared to the original model, it gets a 9.05% reduction in volume, and achieves lightweight of the sub-frame.

References

[1] CHEN Jia-rui Automobile structure [M] BeiJing: China Machine Press 2008.1

[2] Ding Feng The finite element analysis manual of ANSYS 12.0 [M] Publishing House of Electronics industry 2011.1

[3] GAO Jing; SONG Jian, Prediction on fatigue life of vehicle's driving axle house based on MSC.Fatigue

[4] CHEN Meng, Design of front sub-frame of Roewe car [D] Hefei University of Technology 2010.5

[5] ZHANG Hai-yan, Finite Element Analysis of a Car's Front Sub-frame

[6] FENG Zhen, YU Tao, Application and attempt of topology optimization in design [J] Machinery Design & Manufacture 2007.3

[7] GUO Li-qun, Research on Topological Optimization Light-weight Design Method of Truck Frame [D] Jilin University 2011.6

Advanced Materials Research Vol. 819 (2013) pp 356-361
© *(2013) Trans Tech Publications, Switzerland*
doi:10.4028/www.scientific.net/AMR.819.356

Topology Optimization of Suspension of the Hard Disk Drive based on SIMP Method

Yang Shuyi[1,a], Li Hong[1,b], Ou Yangbin[1,c]

[1] Hunan University of Science and Technology, Xiangtan 411201, China

[a]ysy822@126.com   [b]15973231761@163.com   [c]329364863@qq.com

Keywords: Suspension; Topology optimization; SIMP; Multi-objective; Frequency.

Abstract. In order to guarantee the reliability of the hard disk drive data read/write and slider positioning accuracy, the suspension first order bending frequency, first order torsional frequency, first order sway frequency single objective topology optimization model which were based on solid isotropic elastic material penalty (SIMP) method were put forward, and the suspension multi-objective topology optimization model was defined by using the weighted method. Through the topology optimization design the hard disk drive suspension new topological structure was obtained. The results show that hard disk drive suspension single objective and multi-objective topology optimization design objective frequency are large promote than that of the initial design.

Introduction

With the rapid development of information technology, the hard disk drive has been widely used in personal computer, portable equipment and electronic products and other fields, and it has become one of the important program and data carrier. When the hard disk drive in operating state, slider is driven by the suspension to read the data in the disk surface movement and positioning in the objective magnetic track. The suspension belongs to the thin shell structure, and it plays an important role in track seeking servo and magnetic track following. The suspension is very sensitive to the disturbance of air bearing force of head disk system, while the frequent start- stop of the hard disk drive is easy to generate excitation and produce resonance to the modal of the suspension, and thereby the positioning accuracy of the slider is reduced. In order to guarantee the reliability of the hard disk drive data read/write and slider positioning accuracy, it is necessary to suspension topological optimization design, and make its dynamic response minimum.

At present, the related study of suspension optimization has been performed by domestic and foreign scholars. Kilian[1] developed the optimal object for the topology optimization and shape optimization and topology morphology combinatorial optimization design by regarding maximize suspension first order torsional frequency and first order sway frequency. Yu and Liu[2-3], based on material distribution method, the natural frequency and supple of the maximum suspension are regarded as the optimization objective for multi-objective topology optimization. Sequential quadratic programming method and the optimal criterion method of two kinds of optimization method are applied by Pan[4] to solve the maximum suspension first order torsional frequency and first order sway frequency problem. Topology optimization of suspension component is performed by Lau[5] based on the modal tracking method of vibration mode weighting coefficient. Zhu Denglin[6-7] is based on dynamic compliance for the topology optimization design. Static and free vibration characteristics of micro hard disk drive suspension is treated by Wu Han[8] as optimization objective to study the topology optimization problem of the hard disk drive arm.

In this paper, multi-objective topology optimization model is established based on solid isotropic elastic material penalty (SIMP) method by comprehensive consideration of the hard disk drive suspension first order bending frequency, first order torsional frequency and first order sway frequency; Topology optimization results of the hard disk drive suspension are obtained by using Altair Opristruct finite element software and applying an iterative convergence of optimal criteria method.

The finite-element analysis of suspension

Initial design of suspension. Using the Hypermesh software to establish the finite element model of the initial suspension design is as shown in Fig1, the model parameters are shown in Table1. In Fig1, the left region between inner circle and outer square is the connection part of suspension and actuator. The stiffness of the actuator is much more than suspension. Thereby, this part can be considered as a rigid body. In the topology optimization, the freedom degree of this part is completely limited.

Table 1 The finite element model parameters of suspension

Structure	Thickness (mm)	Elastic modulus(GPa)	Density (g/mm)	Type	Number of element	Number of node	Number of DOF
Suspension	0.1	73	2.7×10^{-3}	Shell	4818	5215	23640

Fig.1 The finite element model of the initial design of suspension

Fig.2 Suspension topology optimization boundary conditions

When hard disk drive is operating, suspension and slider are flying by the action of air bearing force in a certain height above the disk. In the model, air bearing force is simplified as the concentrated force acting on the four vertices of the slider in contact with the suspension, the size of 0.001N, shown in Fig2.

Modal analysis of suspension. Modal analysis of the suspension is performed by using Block Lanczos method in Optistruct to get the frequency values of first to ten order vibration mode, such as shown in Table 2.

Table 2 The modal analysis results of suspension

Order	Frequency (Hz)	Vibration mode	Order	Frequency (Hz)	Vibration mode
1st	421.96	1st bending	6th	11735.48	4th bending
2nd	2441.93	2nd bending	7th	17166.94	3rd torsional
3rd	3693.27	1st torsional	8th	20169.78	1st sway
4th	6134.48	3rd bending	9th	20275.56	5th bending
5th	10028.33	2nd torsional	10th	25319.90	4th torsional

From Table 2, it is known that the suspension topology optimization is mainly concerned with the first order bending, first order torsional and first order sway of natural frequencies and the value are respectively 421.96Hz, 3693.27 Hz, 20169.78 Hz. Fig3, 4, and 5 are shown as suspension first order bending, first order torsional and first order sway modal shapes.

The Topology optimization model

The SIMP method is the variable density method with penalty factor, and the corresponding relations between the relative density of the element and elastic modulus is explicitly expressed by a continuous variable density function form[10] .

Fig.3 The first bending modal shape of suspension

Fig.4 The first torsional modal shape of suspension

Fig.5 The first sway modal shape of suspension

The single objective topology optimization of suspension. Fig 6, 7, 8 are respectively the structure diagram of the topology optimization of first order bending frequency, first order torsional frequency and first order sway frequency.

Fig.6 The 1st bending frequency topology optimization of structure

Fig.7 The 1st torsional frequency topology optimization of structure

Fig. 8 The 1st sway frequency topology optimization of structure

Comparing Fig 1 and Fig 6,7,8 show that the variation of topology optimization structure of a single objective in suspension structure is mainly in the front end, and the inner of the front end material produces the butterfly empty material in the maximum first order bending frequency, on the contrary, the empty material is produced in the outer of the front end material in the maximum first order torsional frequency, in the maximum first order sway frequency, the empty material is produced in the inner and outer of structure material.

Fig 9, 10, 11 are respectively the objective function iterative curve of first order bending frequency, first order torsional frequency and first order sway frequency. From Figure 9, 10, 11 show that frequency convergence is respectively produced in after 47,17 ,10 iterations.

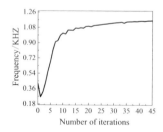

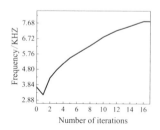

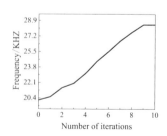

Fig.9 The 1st bending frequency Fig.10 The 1st torsional frequency Fig. 11 The 1st sway frequency

Table.3 Suspension single objective optimization structure frequency comparison in topology optimization

Vibration mode	Original Frequency (Hz)	The optimized frequency of 1st bending/ Increasing rate	The optimized frequency of 1st torsional/ Increasing rate	The optimized frequency of 1st sway/ Increasing rate
1st bending	421.96	1159.88/+174.88%	694.53/+64.60%	424.59/+0.62%
1st torsional	3693.27	5139.29/+39.15%	7808.23/+111.42%	715.70/+0.61%
1st sway	20169.78	17490.91/-13.28%	21646.13/+7.32%	28451.34/+41.16%

We can see from Table 3 that, after the single objective optimization, compared each order frequency with the initial design has greatly improved. Among them, the first bending frequency is increased from 421.96 Hz to 1159.98 Hz, an increase of 174.88%; the first torsional frequency is ranged from 3693.27 Hz to 7808.23Hz, increased by 111.42%; the first sway frequency is enhanced from 20169.78Hz to 28451.34 Hz, an increase of 41.16%. In a single objective topology optimization, except the objective function is the maximum first order bending, and the first order sway frequency is dropped by 13.28, the frequency of the non-objective function than the original frequency can be increased in the other topology optimization.

The multi-objective topology optimization of suspension. Hard disk drive is a precision micro electromechanical system, and suspension is often affected by many uncertain excitations in the operating state, considering only the influence of a single frequency, it is difficult to improve the dynamic characteristics of the suspension. Therefore, the topological optimization model of suspension multi-objective is established by using weighted method in the first order bending frequency, first order torsional frequency and first order sway frequency, as shown in equation (1).

$$Max : w_1 f(x_1) + w_2 f(x_2) + w_3 f(x_3)$$

$$Subject\ to : \frac{V}{V_0} \leq V^*$$

$$\left| K - \omega_i^2 M \right| = 0$$

$$0 < x_{min} \leq x_i \leq 1$$

(1)

Where: w_1 , w_2, w_3 is respectively the weighted coefficient of first order bending frequency, first order torsional frequency and first order sway frequency. $f(x_1)$, $f(x_2)$, $f(x_3)$ is the response frequency of first order bending, first order torsional, and first order sway. V 、 V_0 are respectively the volume of the material and the volume of the initial design. ω_i is frequency, and M is mass matrix. x_i is a design variable that represents the relative density of the i-th unit in the optimization. x_{min} is the value of the minimum density of the material in the blank material , and the value is 0.001.

In the suspension multi-objective topology optimization, the function of first order bending, first order torsional, and first order sway is measured by three frequency value. Assume the above weighting factor of three frequencies is respectively 0.5, 0.3, 0.2. Fig12, 13 is respectively the multi-objective topology optimization structure of suspension and the iterative curve of multi-objective function.

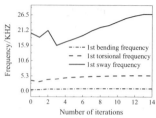

Fig.12 The multi-objective topology optimization structure Fig.13 The iterative curve of multi-objective function

According to Fig.13 shows that the objectives function of suspension multi-objective topology optimization produces convergence after 14 iterations. Compared the suspension structure, it can be found that the first torsional frequency and first sway frequency have greatly effect on the multi-objective optimization design of structure.

Table 4 Multi-objective topology optimization structure frequency comparison

Vibration mode	Original frequency(Hz)	The optimized frequency (Hz)	Increasing rate
1st bending	421.96	818.75	94.03%
1st torsional	3693.27	5409.35	46.47%
1st sway	20169.78	26766.61	32.53%

It is seen from Table 4 that, compared the frequency of each order objective function with the frequency of initial design, the first order frequency of bending, torsional, and sway of three mode are increased. Among them, the first order bending frequency is ranged from 421.96Hz to 818.75Hz, an increase of 94.03%; the first torsional frequency is increased from 3693.27Hz to 5409.35Hz, increased by 46.47%; and the first sway frequency is enhanced from 20169.78Hz to 26766.61 Hz, increased by 32.53%.

Conclusion

A single objective optimization model is established through the analysis of the suspension motion characteristics of the hard disk drive based on the hard disk drive suspension first order bending frequency, first order torsional frequency and first order sway frequency of SIMP method. Considering the influence of various modes, the multi-objective optimization model of suspension is established by using the weighted method, and the topology optimization structure of suspension is obtained by Optistruct finite element software to solving, the first order frequency value of various modes is more than the initial frequency value. The hard disk drive suspension topology optimization based on SIMP method provides a theoretical basis for structure design of suspension.

Acknowledgements

This research is sponsored by the Natural Science Foundation of China (Grant No. 51075139) and the Key Project of Chinese Ministry of Education (Grant No.211120) and Scientific Research Fund of Hunan Provincial Education Department (Grant No. 10B030).

References:

[1] S.Kilian, U.Zander, F.E.Talke, et al. Suspension modeling and optimization using finite element analysis[J]. Tribology International,2003,36(1):317-324.

[2] Yu.S.K, Liu. Optimal suspension design for femto sliders.IEEE Trans Magn, 2003, 39(5): 2423-2425

[3] Yu.S.K, Liu. Topology optimization for femto suspension design.Microsyst Technol,2005,11:851-856

[4] Pan.L, Lau.G.K, Du,H, Ling.S.F.On optimal design of HDD suspension using topology optimization [J]. Micro-system Technologies,2002,9:137-146

[5] Gih.Keong.Lau, Hejun.Du.Topology optimization of head suspension assmblies using modal participation factor for mode tracking.Microsyst Technol,2005,11:1243-1251

[6] Zhu.D.L, Wang.A.L, Jiang T.Topology design to improve HDD suspension dynamic characteristics. Struct   Multidisc Optimization,2006,31:497-503

[7] Zhu.D.L, Wang.A.L. Topological design of HDD(Hard Disk Drive) suspension based on dynamical compliance.Journal of Mechanical Strength, 2005,29(2):269-273.

[8] Wu Han. Topology Optimization Design of Micro Hard Disk Drive Actuating Arm. Harbin. Harbin Engineering University.2011

[9] Fan Wenjie,Fan Zijie,Su Ruiyi. Research on Multi-objective Topology Optimization on Bus Chassis Fram. China Mechanical Engineering, 2008,19(12):1505-1508.

[10]Su Shengwei. Application and Research of Topology Optimization with Optistrcut. Harbin.Harbin Engineering University.2008.

Advanced Materials Research Vol. 819 (2013) pp 362-367
© (2013) Trans Tech Publications, Switzerland
doi:10.4028/www.scientific.net/AMR.819.362

Unified Modeling for Input Forces and Input Vectors for Two Scissor Lifting Mechanisms

ZHANG Wei [1,2,3 a], ZHANG Xuefei[4,b] and YAN Chao[1,4,c]

[1]Airport College, Civil Aviation University of China, Tianjin, 300300

[2]Sino-European Institute of Aviation Engineering, Tianjin, 300300

[3]Aviation Ground Special Equipment Research Base, Tianjin, 300300

[4]Aeronautical Automation College, Civil Aviation University of China, Tianjin, 300300

[a]drwadecheung@gmail.com, [b]wuyulunbi2047@126.com, [c]yc123926@163.com

Keywords: Scissor lifting mechanism; Kinematics; Unified expression.

Abstract. Scissor lifting mechanisms are widely equipped on airport ground support facilities, in this paper, we propose a unified modeling process for two special SLMs by establishing the expression of input motion vectors and input force vectors and test the results by three proposed cases. A characteristic triangle and corresponding parameters are introduced for any configuration. The presented method provides a reference mode for studying scissor lifting mechanism.

Introduction

Airport ground support facilities are often mounted on scissor lifting mechanisms (SLMs), such as catering trucks[1] , platform trucks[2-3], mobile boarding bridges[4] and mobile towers. These SLMs use scissor lifting units (SLUs); which are often driven by a ball screw, chain drive or hydraulic actuator. Generally, input vector positions vary from different drive modes. Previous research on SLMs can only deal with one certain input configuration [4-14]. Unified expressions are significant for evaluating different input configurations, such as external supported and symmetric articulated SLM in this paper, for a further research of design, analysis and optimization.

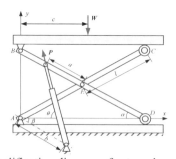

Fig.1 Simplification diagram of external support SLM

Unified Modeling for Input Forces and Input Vectors for External Supported SLM

Model Simplification. As shown in Fig. 1, based on its symmetry, a SLU can be simplified in 2-dimensional space. The input component can be expressed as a scalable linear beam. The length of the scissor arm is $2l$, the height of top platform is H, the length of input beam is I, the lifting angle is α ($\alpha \in (0, \frac{\pi}{2})$) and the load is a concentrated vertical force. β , a, b, and c are known parameters. A SLU is a rod system, the load itself is ignored. Theoretically there are 3 configurations which are showed in Fig. 2.

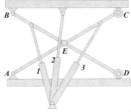

Fig. 2 Parameters for all input configurations

Output Vector Expression. Because the top platform can only move upward with respect to the lower platform, the output vector is:

$$H = \{0, 2l\sin\alpha\}^{T} \tag{1}$$

Calculation of Input Vector

A Characteristic Triangle. According to the geometric properties of SLUs, a characteristic triangle can be proposed. In each type of input configuration, the output side, the right-angle side and the scissor arm as the hypotenuse make a right triangle (as shown in Fig. 3). Inside this right triangle, the input side is a segment from the right-angle side to the hypotenuse. Thus, this paper proposes a characteristic triangle to model and solve of input vectors.

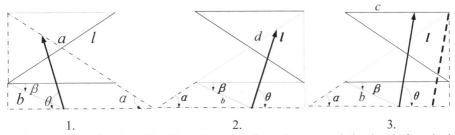

1.   2.   3.

Fig. 3 Characteristic triangles of 3 different input configurationsaracteristic triangle for calculating the input vector

Each characteristic triangle described in Fig. 4 has a right-angle side BE whose length is H. The length of BF is $2l$. The hypotenuse BF is a SLU arm with length $2l$, The side BC is an expanded SLU arm with length L. The length of the input vector is expressed by segment $B'B''$.

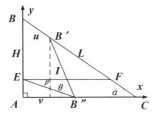

Fig. 4 Characteristic triangle for calculating the input vector

Length of the Input Vector. Applying the law of cosines to the triangle in Fig. 4, B'B'' is the length of input vector, namely

$$I = \sqrt{(N\sin\alpha + b\sin\beta)^{2} + (M\cos\alpha - J)^{2}} \tag{2}$$

The values of the parameters are given in Table 1.

Table 1. Values of K, J, and M under different input configurations

Configuration	1	2	3
K	0	$2l$	0
J	$-b\cos\beta$	$b\cos\beta$	$b\cos\beta - c$
M	$a-l$	$l+d$	0
N	$l+a$	$l+d$	$2l$

Expression of the Input Length Vector. According to Fig. 4, the input angle can be defined as: $\theta = \angle AB'B''$, the trigonometric values of the input angle are:

$$
\begin{cases}
\sin\theta = \dfrac{N}{I}\sin\alpha \\[2mm]
\cos\theta = \dfrac{M\cos\alpha - J}{I}
\end{cases}
\tag{3}
$$

So, the expression of the input vector in reference system $\{xy\}$ is:

$$
\bar{I} = \{J - M\cos\alpha \quad N\sin\alpha + b\sin\beta\}^T
\tag{4}
$$

The expression of the input vectors for a fixed reference system such as that described in Fig. 1 can be achieved through rotating or mirroring the reference system of characteristic triangles.

Unified Expression of Input Force. Ignoring mechanism quality, research results under equilibrium can be extended to the moving situation. Consider a mechanism in a stationary state as shown in Fig. 1, where the load $\overline{W}$ is paralleled to the output vector $\overline{H}$. Meanwhile, the input component is a two-force rod. The input vector $\overline{I}$ is paralleled to input force $\overline{P}$. If the SLU is an ideal constraint system, analysis of the input force can also be performed using a characteristic triangle. As shown in Fig. 5:

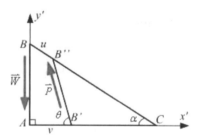

Fig. 5 Characteristic triangle of the input force

$$
\begin{cases}
\overline{W} = \{0 \quad -W\}^T \\[2mm]
\overline{P} = \{-P\cos\theta \quad P\sin\theta\}^T
\end{cases}
\tag{5}
$$

According to the principle of virtual displacement:

$$
\overline{W} \cdot \delta\overline{H} + \overline{P} \cdot \delta\overline{I} = 0
\tag{6}
$$

Considering equation (1) ,

$$\delta \overline{H} = \{0 \quad 2l\delta\alpha\cos\alpha\}^T \tag{7}$$

Using equation (4),

$$\delta \overline{I} = \delta\alpha \cdot \{M\cos\alpha \quad N\sin\alpha\}^T \tag{8}$$

According to the principle of virtual displacement and the results above, there is:

$$P = \frac{2lW\cos\alpha}{N\sin\theta\cos\alpha - M\cos\theta\sin\alpha} \tag{9}$$

Unified Modeling for Input Forces and Input Vectors for Symmetric Articulated SLM

Model Simplification and Characteristic Triangle. To symmetric SLM in Figure 6, both sides have the same properties, similar to Figure 4 and equation (4),

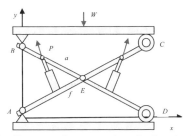

Fig. 6 Principal schematic diagram of symmetric articulated scissor lifting mechanism

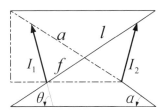

Fig. 7 Characteristic triangles

$$\left.\begin{matrix}\overline{I_1}\\\overline{I_2}\end{matrix}\right\} = \{J \mp M\cos\alpha \quad N\sin\alpha\}^T, \begin{cases} M = a - f \\ N = a + f \\ J = 0 \end{cases} \tag{10}$$

Unified Expression of Input Force. Similarly to the analysis of external support scissor lifting mechanism,

$$\overline{W} \cdot \delta\overline{H} + \overline{P_1} \cdot \delta\overline{I_1} + \overline{P_2} \cdot \delta\overline{I_2} = 0 \tag{11}$$

Namely:

$$-2lW\cos\alpha + P_1\cos\theta M\sin\alpha + P_1\sin\theta N\cos\alpha + P_2\cos\theta M\sin\alpha + P_2\sin\theta N\cos\alpha = 0 \tag{12}$$

Because of symmetic input, $\left|\overrightarrow{P_1}\right| = \left|\overrightarrow{P_2}\right|$. It is easily concluded that input force has a half value as equation (10).

According to equation(11), if there are n $(n \geq 2)$ input vectors, the equation can be expressed as:

$$\overline{W} \cdot \delta\overline{H} + \sum_{i=1}^{n} \overline{P_i} \cdot \delta\overline{I_i} = 0 \tag{13}$$

Examples Illustration

Many articles have presented the mechanical properties of a single input SLM, and these reported examples are used to verify the theory of this paper. There are 3 examples provided.

Example 1: LI EMin [8] investigated the kinetic analysis of an SLM which is an application of external supported SLM.(Type1).

Example 2: LI Emin [14] formulated the equation for the input force value for a symmetric articulated scissor lifting mechanism and expressed input force value as a function of output height .

Example 3: WEI Zhang [6] proposed the optimization of a support arm of one certain SLM, which belongs to the external support SLM (Type1).

Table 2. Results from various SLM studies

Example	Original data	Configuration and parameters	Results
1	$P = \dfrac{2l\cos\alpha}{a\sin(\theta+\alpha)+l\sin(\theta-\alpha)} \cdot W$	$M = a - l$ $N = a + l$ $\theta = \pi - \theta$	$P = \dfrac{2lW\cos\alpha}{N\sin\theta\cos\alpha - M\cos\theta\sin\alpha}$
2	$P = \dfrac{2lW\cos\alpha}{a\sin(\theta+\alpha)+k\sin(\theta-\alpha)}$	$k = b$ $M = a - k$ $N = a + k$ $\theta_a = \pi - \theta$	$P_a = \dfrac{2lW}{N\sin\theta_a - M\cos\theta_a\tan\alpha}$
3	$P = \dfrac{2l\cos\alpha}{l\sin(\theta+\alpha)+a\sin(\theta-\alpha)} \cdot W$	$M = a - l$ $N = a + l$	$P = \dfrac{2lW\cos\alpha}{N\sin\theta\cos\alpha - M\cos\theta\sin\alpha}$

Conclusion

The results are verified in all 3 examples. As mentioned before, the presented equations are suitable for not only single input mechanisms but also symmetric input SLM. In practice, 2, 4 or even more hydraulic drives are mounted with a SLU for many reasons. It can be easily implied that more complex SLMs can also be properly handled in with proposed method. The proposed unified method of input vectors and input forces provide a reference mode for analyzing SLM. Next work will apply the proposed method into more input configurations, like multi-group SLM, or parallel SLM.

Acknowledgement

This work is supported by the National Natural Science Foundation of China and the Civil Aviation Administration of China as a jointly funded project (Grant # U1233106), the Fundamental Research Funds for the Central Universities funded project (Grant # ZXH2012H007) and the university scientific research project of the Civil Aviation University of China (Grant # 2012KYE05). The corresponding author is grateful to all who have provided help with this research.

References

[1] X.L. Liu, J.Y. Tian, B. Han, Stress analysis of the catering truck scissors girder based on ansys, J. Shandong Jiao tong Univ.17 (2009) 74-76.

[2] B.M. Xie, M.H. Jiang, Experimental stress analysis of scissors mechanism of cylinder loader, J. Civ. Aviat. Univ. China 25 (2007) 39-41.

[3] H.L. Guo, The main framework virtual design of scissor assembly hydraulic cylinder loader, Civil Aviation University of china, Tianjin, 2008.

[4] L.P. Zhou, X.L. Zhang, Z.W. Xing, Design of lifting machine of airport lift bus, J. Civ. Aviat. Univ. China 27 (2009) 23-25.

[5] W. Zhang, X.X. Wang, L.W. Wang, Strength optimization design of box-shape arms of a scissor lift mechanism with single hydraulic cylinder, Appl. Mech. Mater. 141 (2012) 513-518.

[6] S.T.Yan, J.Ye, Z.L.Tang, Control system based on PROFIBUS2DP of orchestra lift [J], Mech. Res. Application, 31 (2005) 47-50.

[7] W.Y.Gao, M.B.Jin, J.Chu, Z.M.Zhu, PID control of scissors lift platform [J], Mech. Res. Application, 33 (2007)43-46.

[8] E.M. Li, Kinematic and dynamic analysis of hydraulic cylinders driven scissors mechanism [J],Gansu Univ. Technol. 20 (1994) 34-37.

[9] Y.J. Song, Z. Liu, Kinematics and kinetics analysis of hydraulic cylinder powered scissors mechanism, Hoisting Conveying Mach. 2 (2004) 41-43.

[10] F.K.Wei, Dynamic property research of orchestra lift, Mech. Res. Application, 23 (1997) 48-50.

[11] E.M. Li, J.T. Li, Analysis and comparison of two disposition ways of hydraulic cylinder in scissors mechanism, J. Gansu Univ. Technol. 26 (2000) 54-57.

[12] D.X. Xiang, F.K. Wei, A structure design of orchestra lift, Mech. Res. Application, 12 (2007) 86-87.

[13] X.Z. Hu, J.K. Hu, G.H. He, The modeling of scissors elevation mechanism and research of key parameters, Mech. Res. Application, 19 (2006) 84-85.

[14] E.M. Li, W.H. Qi, B. Wang, Y. Yang, C. Zhang, The kinematic and dynamics analysis of scissors mechanism driven by hydraulic cylinder, Mach. Tool Hydraulics 39 (2011) 71-72.

Advanced Materials Research Vol. 819 (2013) pp 368-372
© *(2013) Trans Tech Publications, Switzerland*
doi:10.4028/www.scientific.net/AMR.819.368

User-friendly Design of Modern Forest Fire Helmet

HongzeYang[1,a], Bo Li [1,b], Yanan Wu[1,b],Bin Li[1,b]

[1]College of Mechanical and Electrical Engineering, Northeast Forestry University, Harbin 150040, China

[a]9316762@qq.com, [b]id-workshop@sohu.com

Keywords: Forest fire helmet industrial design user-friendly design

Abstract: This article researches and discusses the current situation of the modern forest fire helmet. It focuses on the problems and difficulties in the modern forest fire helmet design and analyses them. With the concept of user-friendly design, this article advances the user-friendly design principles in modern forest fire helmet design.

Introduction

Fire helmet is important protective equipment which helps fire fighters protect their heads when they are putting out the forest fire. However, due to high incidence and the danger of the forest fire, thousands of fire fighters have hurt or even dead in the world annually during the fire extinguishing. As the development of the technology, the flexibility and protecting ability of outdated forest fire helmet cannot meet the requirement of modern forest fire fighters, modern forest fire helmet comes on the stage. In aspect of the function, this helmet integrate the advanced technology such as electromechanical integration, intellectualization and networking; in the aspect of structure, every function module have been reasonably integrated to the internal helmet body; in the aspect of modeling, sense of aesthetic has been considered in the design. So during the process of design, not only should put every function module to the forest fire helmet to achieve the protecting function, but also begin with the user's operating process and fully consider the process of wearing and operating. To achieve this, humanized design method should be used in the design of the forest fire helmet.

Current situation of forest fire helmet at home and abroad

With the development of the technology, the protecting ability and the using ability of forest fire helmet have been greatly improved. The comparison of the forest fire helmet at home and abroad is below.

Material. The development of technology and materials at home and abroad is becoming increasingly mature and most shell material is PEI and PC plastic. These materials have a smooth surface and a strong ability of high temperature resistant which combines both strong ability of resistance and comfortable wearing. So in the aspect of material choosing, the development of material of forest fire helmet at home and abroad has been advanced.

Appearance. Overall, the appearance of fire helmet abroad is designed by means of streamline which make the helmet artistic while the appearance of fire helmet abroad is cumbersome. Because the forest fire helmet should be integrated multiple functions, designers must consider the appearance as well as put the multiple functions together. In this aspect, comparing to forest fire helmet in Europe and America, products in China is in a disadvantage state. Taking wireless transmission system in Taizhou Fire Company, many components are bare outside the helmet body, making it not artistic and unsafe. In contrast France Leijia F1 helmet assembled many advanced technological functions and the designers integrate the functions into the inner helmet so the sententiousness and aesthetic aren't be influenced.

Functions. Based on the achievement of the elementary head protecting functions, more advanced technological functions such as communication, intercom, help calling, lighting and camera shooting have been successively applied to the design of forest fire helmet. As the development of the technology has been advanced, advanced technology has been applied in some forest fire helmets in Europe and America such as France JiaLei F1 helmet and British MSA F2 forest helmet. By contrast, in China, the materials of helmet body has been considered into the domestic forest helmet and the advanced materials and processing technology harder the impact resistance which prolongs the working life of the helmet and improves the shock absorption and comfort level of the forest fire helmet, but there is a long way to go to improve the fire helmets in China. So applying the advanced technology to the fire helmet and integrating every function module has become one of the important tendencies in the further design of forest fire helmet.

The problems and difficulties in design

To achieve every function of modern forest fire helmet, some advanced technology should be used in the helmet. These technologies contain wireless communication, infrared thermal imaging, digital image processing, GPS positioning, human health detection and sensing and oxygen making. As the fire helmet is a system integrated multiple functions, when we conduct the project, but are not able to consider comprehensively. One function is used into the helmet with the restriction of other functions. So when we integrate these technologies to one product, we also engage in each module structure design reasonably, and it has been a difficult point of this design.

Humanized design method in modern forest fire helmet

According to the human's behavior custom, physiological structure, psychological situation and cognitive process, humanized design is a design method which is accorded to the basic function and performance of original design and optimizes the function to make the user conduct a more convenient and comfortable use. And during the process of humanized design, human's physiological need and pursuit of spirit must be respected and satisfied, so humanized design is a humanistic concern and respect to human. Overall, humanistic concern is a combination of science and art, technology and humanity. Technology brings solid structure and good function to the design, and art gives design a sense of beauty with appeal and vigor.

If we want to put the humanized design method into the design of modern forest helmet, we should consider human first, analyze psychological characteristics of firemen and apply relative discipline such as product semantics, modeling aesthetic and color science etc to meet firemen's psychological requirement in appearance of the product. And then we should analyze the operating process in the fire extinguishing and apply some relative knowledge of ergonomics to arrange the position go every function module in the helmet. We also should design the operation buttons and display structure in the helmet reasonably which makes the operating process convenient and meet the firemen's needs.

The application of design psychology in modern forest helmet. Fire fighters have donated themselves to the protection of people and their property. Due to the extreme special environment of living and working, they are under big pressure which leads to some psychological problems in the fire fighters. Most firemen have the reaction of negative mood and the performances of these moods are obsession, anxiety, helplessness, passiveness, inferiority. These negative moods can cause behavior disorders and the serious patients even commit suicide. These negative moods are formed from the situation of danger and intensive fire extinguishing. Under the stimulation of the big fire crisis, the firemen also have mood changes and these mood changes are the consequence of frustration. During dealing of the stimulation, if his physical ability is not able to deal with the change of the environment and fail to control the stimulation, the fireman can have series of emotional response of frustration. The reasons of negative moods which are caused in the process of fire extinguishing are classified into two points.

External environment stimulate: these stimulation mainly comes from the sensible stimulation of vision, hearing, smell, touch coming from the external environment. The dazzling light caused by the burning fire, the noise caused by the falling of burning trees and branched firing and high temperature can cause the fidget of the fire fighters and psychological pressure which makes the fire fighters fail to take action to the fire.

Personal psychological fear: this psychological is cause by a sense of failure due to mistakes in the operation, which makes the fire fighters lose a sense of security. The reasons for the psychological fear can be classified into: the shortage of fire extinguishing equipment, fire fighters not having confidence in self-defense, the panic caused by the lack of contact with his companion, the shortage of acknowledge of the station of the fire. All these make fire fighters lose confidence and in the process of fire extinguishing, the firemen are not able to make a correct judgment and are not able to understand the dangerous atmosphere, and then they make the wrong firefighting measures. So reasonably solve the shortcomings of fire equipment is the key point of solve the problem of firemen's psychological fear.

If we want to apply design psychology into the design of forest fire helmet, firstly we should analyze the psychological problems of firemen caused in the fire extinguishing. After summarizing the causes of psychological problems, we should put forward corresponding design scheme for each question.

Table 1. Design proposals to the psychological problems

Cause	Detail	Design proposals
Stimulation of external environment	Strong light	Goggles are brunet which have good resistance of light.
	Noise	Cap shell has good sound insulation ability.
	Smell	Mask has the isolation function and be provided with breather.
	High temperature	Cap shell choose high temperature resistant materials
Personal psychological fear	Lack of contact with companions	Apply modern communication technology and install the devices of voice receiving and sending.
	Lack of correct judgment to the fire station	Apply Infrared Thermal Mapping Technique and GPS positioning system to judge fire correctly.
	Lack of reasonable strategy	Apply digital image processing technology to receive and display the orders from commanding center.
	Lack of security	Apply human body health testing technology to know the health condition.

Dealing with the problems of firemen's psychological problems has been a topic in every fire company in China. Besides the guidance in the center of fire companies, more successful extinguishing during the fire fighting is the basic method. Design psychology comes from the users' psychological requirement, puts forward a set of solutions to each cause of the psychological problems. This design method promote the fire equipment's function appropriately and enhances the equipment's design actually, which makes fire fighter acquire a sense of security and confidence in the fire extinguishing and hold the belief that the plan is correct and they are able to escape the dangerous in time. This has solved the problems of the firemen's psychological problems and enables them to work in the operation of fire extinguishing. This requires that the fire helmet can protect fire fighter's head better and the function of helmet can satisfied the psychological requirement of fire fighters.

The application of ergonomics in modern forest fire helmet. Ergonomics is a discipline which studies the element of human-machine-environment, and through the study of these three elements, the designers want to make sure the optimization of total function. The forest helmet is wore on the head of fire fighters, which makes many position of the helmet contact with the users' head, so the function of modeling have an direct effect on the comfort and user's health. Moreover, because the forest fire helmet integrates many advanced technological functions and the fire fighters have to operate it, the ergonomics method should be applied in the modern forest fire helmet to reasonably arrange every operating module and reasonably design the display devices to make the helmet accord with the humanized design principle of forest fire helmet. The application of ergonomics in the helmet mainly reflects as the aspects below:

Human dimension. The users of the forest fire helmet are male mostly. As a result, the dimension design should be mainly considered the group of grown-up males and the helmet dimension should be taken the percent of 90%, which means that the helmet should adjust to over 90% the males and the 90% males' data should be chosen.

Human dimension is one of the basic parameter of forest fire helmet which directly determines whole helmet's dimension and the arrangement and position of correlative component s. To meet this, the knowledge of human dimension should be uses in the design. So the designer should choose the appropriate data in static anthropometry and dynamic human body measure to determine the helmet dimension. In static state, the designer should make the helmet adjust the shape and the size of user's head and in dynamic state, correspond the firemen's active regulation to ensure the safety and effective wear of the fire fighters.

Display equipment. Display equipment is the design of man-machine interface is the display and operation interface which transmit the information between the helmet and fire fighters. During the design of man-machine interface the designers should base on the general design principle of and consider the physical features and active custom in the fire extinguishing. Moreover, they should pay attention to the analyze of psychological requirement of fire fighters to achieve the information interaction between firemen and fire helmets to achieve the free, convenient, comfortable, safety control of the helmet. The main design content of forest fire helmet in the study of man-machine interface contains two points: One is the display of goggles and the other is the display of wrist controller. These designs of display equipment must reasonable and cleat to convenient the fire fighters' understanding.

Arrangement of operation. The more functions the production owns, the bigger possibility of error operation emerges. As modern forest fire helmet integrates multiple functions, the design the arrangement of operating components is essential. One basic operating buttons should be settled into the helmet's shell, so the designers should reasonably arrange the buttons according to the using habit; the other operating buttons of realizing advanced technology should be integrated in the wrist controller, which unites the operating process of fire fighters working and convenient the fire fighters to operate and observe.

Application of modeling aesthetic in modern forest fire helmet. The modeling design of products is one aspect to reveal the product humanized design. The model is reflected through the product appearance. The shape of the product is constituted of different form, face, line, which reflects in the entirety to form the product appearance. Foreign fire helmets mostly are used streamlining design which fix reinforced muscle and makes total appearance clear and fluent. By contrast, domestic forest fire helmets' form is simple and are lack of aesthetic feeling which are looked mass. This appearance can create a psychological suggestion of affording weight when the fire fighters are wearing them. So, in the fire helmet design, the design method of modeling aesthetic should be applied and reliving the weight of the helmet is the key point of model design.

Conclusion

The research and the manufacturing of modern forest fire helmet have a significant meaning to the safety of forest fire fighters and protection of forest fire. This paper analyzes the design difficulties of the forest fire helmet and based on the principle of humanizing, puts forward optimum design scheme according to the problems in the fire extinguishing, which provides a referential technical guidance to further design of forest fire helmet.

Acknowledgements

The project was supported by the Fundamental Research Funds for the Central Universities (Grant No.DL11BB39) and Supported by Foundation of Heilongjiang Educational Committee (Grant No.12523015).

References

[1] Bo Li,Shuyang Wang and Hongze Yang, "The development of modern forest fire helmet," Forestry Labour Safety, edited by China,Vol.21 No.2 (2008), p.14

[2] Jiaojiao Ma, "Development of a new fire helmet," Master's thesis of Shanghai Jiao Tong University. China,(2007) 6~11 13~22 27~34 38~39

[3] Bo Zhou, "A Short History of the evolution of the fire helmet." Anhui Fire, edited by China, NO.8(1996), p33

[4] Jiaojiao Ma, Kang Sun, Chuanping Hu, "The fire helmet woodpecker head impact protection technology research based on the characteristics." Fire Science and Technology, edited by China, NO.1(2007), p69

[5] Wenbin Li. "Assessment on the performance of hand pumps for forest fire suppression based on operator s heart rate and water spraying efficiency"[J].Journal of Beijing Forestry University, 2009, 1: 134-138

Advanced Materials Research Vol. 819 (2013) pp 373-378
© (2013) Trans Tech Publications, Switzerland
doi:10.4028/www.scientific.net/AMR.819.373

Gearbox Fault Simulation and Analysis Based on Virtual Prototype Technology

Chen Xue[a], Cui Lingli

Key Laboratory of Advanced Manufacturing Technology, Beijing University of Technology, Beijing, 100124, China

[a]xuming.1005@163.com

Keywords: gearbox; virtual prototype; ADAMS; fault diagnosis

Abstract. Simulation study of the failure of gearbox system using virtual prototype technology and analysis of its vibration signal is a foreword topic in recent years. This paper established a rigid-flexible coupling model of gearbox system using SolidWorks, ANSYS and ADAMS. Then, make simulation under the following three conditions: normal, gear fault and bearing outer ring fault. Next, make time domain and frequency domain analysis of the simulated fault signal using MATLAB. Simulation results are consistent with gear vibration mechanism and signal modulation theory, which verified correctness and feasibility of virtual prototype technology in gearbox fault simulation. Lay foundation for gearbox fault diagnosis.

Introduction

Gearbox as kind of necessary mechanical equipment, which can make connection and transfer power, has been widely used in metal-cutting machine tools, aviation, electric power systems, agricultural machinery, transportation machinery and other industrial equipment. But, gearbox is prone to be damaged due to the intensity of the work, poor working conditions and its complex structure and other reasons. Now mechanical equipment is becoming larger, complicated, automated and continuous. When the gearbox system appears fault and failure, it will cause big loss to the production and society. So making status monitoring and fault diagnosis for gearbox system to prevent the occurrence of the fault has an important value and significance.

Vibration signal acquisition, analysis, processing and accurate extraction of fault characteristics are the key points in gearbox fault diagnosis. Traditional fault feature extraction is mainly through large number of experiments and theoretical analysis. Laboratory personnel set all kinds of faults through making damage on physical prototypes artificially, then doing experiments to get vibration signal data. Feature extraction is Completed using the gear vibration mechanism theory. We can also simulate various failures of the mechanical system with virtual prototype technology[1,2,3] . We get various characteristics of the fault system through simulation, then compare with the characteristics of the normal system, which can extract fault features and provide basis for mechanical system. Compared with the traditional fault feature extraction method, virtual prototype technology can not only save costs, shorten test time but also simulate all kinds of possible failure of the mechanical systems.

The fault type of large gear transmission equipment can be broadly divided into the shaft fault, the bearing fault, the gear fault and the box fault, among which the gear fault accounts for 60%, the bearing fault accounts for 30%, this two kind of faults account for the most part of the gearbox fault. So, this paper established the rigid-flexible coupling model of gearbox system. Making simulation and analysis of fault happens mainly on gear and bearing. At last this paper verified the correctness and feasibility of virtual prototype technology in gearbox fault simulation.

Establish simulation model

The emphasis and difficulty of gearbox system simulation is to establish a correct virtual prototype model. We establish the rigid-flexible coupling model of gearbox system, which refers to the Qianpeng test bench of our laboratory, considering the bearing stiffness, gear contact force relations and gearbox system dynamics [4].

Establish rigid-flexible coupling model. First, the multi-rigid-body model of gearbox was established with the three-dimensional entity modeling software SolidWorks using bottom-up modeling method [5]. The size data of the model refers to the Qianpeng test bench. The multi-rigid-body model is shown in Fig.1. The upper box has been hidden. Bolt, screw hole and chamfering which have little effect on the simulation results have been omitted. Bearing model is 6206. Gear parameters are shown in table 1.

Fig.1. Multi-rigid-body model of gearbox

Table 1.Gear parameter

Gear	Module [mm]	Teeth number	Press angle [∘]	Tooth width [mm]
Drive wheel	2	55	0	30
Driven wheel	2	75	0	30

Gearbox has a larger deformation and vibrates in the actual operation. The simulation result will be closer to the real situation when we replace rigid body with a flexible body [6]. The input and output shaft are easily deformed, and upper and lower box are easy to vibrate. So we transmit the shafts and boxes into flexible bodies using the finite element analysis software ANSYS [7]. Specific steps are shown in Fig.2. Lead the MNF file into ADAMS [8]. The rigid-flexible coupling model of gearbox is established through the rigid to flex model in ADAMS, which is shown in Fig.3. The upper box has been hidden.

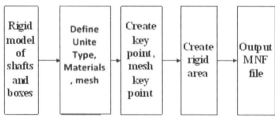

Fig.2. Steps of transmit rigid body to flexible body

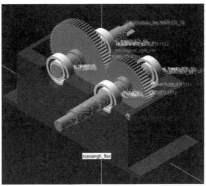

Fig.3. Rigid-flexible coupling model of gearbox

Solve the contact parameters of gear and bearing. The gear transmission in the model is achieved by the contact of two gear teeth. The contact parameters are gotten based on Hertz contact theory and empirical data. By Hertz contact theory, the contact stiffness coefficient between gears is defined as[9]:

$$k = \frac{4E^*}{3} \left[\frac{R_1 R_2}{R_1 + R_2} \right]^{1/2} \tag{1}$$

$$\frac{1}{E^*} = \frac{(1 - \mu_1^2)}{E_1} + \frac{(1 - \mu_2^2)}{E_2} \tag{2}$$

Where, μ_1, μ_2 represent the Poisson's ration of the two contacting objects; E_1, E_2 represent the Modulus of elasticity of the two contacting objects; R_1, R_1 represent the Equivalent of curvature of the two contacting objects.

The comprehensive stiffness of bearing contains contact stiffness and oil film stiffness. The equivalent comprehensive stiffness within bearing ball and rings is defined as[10]:

$$\begin{cases} K'_{i\phi} = \dfrac{K_{yi\phi} K_{i\phi}}{K_{yi\phi} + K_{i\phi}} \\ K'_{e\phi} = \dfrac{K_{ye\phi} K_{e\phi}}{K_{ye\phi} + K_{e\phi}} \end{cases} \tag{3}$$

Determination of drivers and constraints. By the kinematic characteristics of the gearbox, the applied constraint and driver are determined as follows:

Applied fix joint between lower box and ground;

Applied fix joint between lower box and upper box;

Applied fix joint on the driving wheel and the input shaft, the driven wheel and the output wheel respectively;

Applied fix joint between the four bearing inner rings and the two shafts respectively;

Applied contact force between the two gears;

Applied revolute pair between the inner rings and the outer rings;

Applied contact between one bearing's balls and rings, compressed inner rings and balls of the other bearings;

Applied driven on input shaft, load torque on output shaft;

Model correctness preliminary verify.

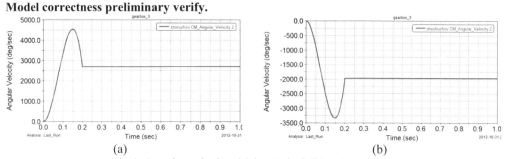

(a)   (b)

Fig.4. Angular velocity: (a) input shaft (b) output shaft

We set the input shaft speed for 2700°/s (450r/min), load torque for 50000 Nmm. The angular velocities of the two shafts obtained by simulation are shown in Fig.4. From the figure we see that the input shaft angular velocity is 2700°/s, the output shaft angular velocity is -1978°/s, the negative sign indicates the direction is opposite. The theoretical value of the output shaft angular velocity calculated by the transmission ration is 1980°/s. The simulation result is in good agreement; the relative error is 0.1%, which proved that the constraint applied on the model is correct.

Fault simulation and analysis

Gear fault simulation and analysis. One tooth of the gear is cut using SolidWorks software to make broken teeth fault artificially. The broken tooth gear is shown in Fig.5 (a).

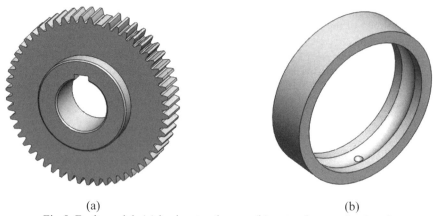

(a)   (b)

Fig.5. Fault model: (a) broken tooth gear; (b) outer ring erosion bearing

Simulation begins after the normal gear is replaced by the fault gear. Set rotation speed 450r/min, step size 0.0001s. The vibration acceleration signal of the upper box is collected in the post-processing module of ADAMS. The time domain and frequency domain diagram of the signal are shown in Fig.6 (a). In Fig.6 we see that side frequency exists. Use Hilbert demodulation to the signal. The demodulation spectrum is shown in Fig.6 (b). The gear meshing frequency is defined as:

$$f = {zn}/{60} \qquad\qquad\qquad\qquad\qquad\qquad\qquad\qquad (4)$$

The theoretical value of the meshing frequency is 412.5Hz. In Fig.6 (b) we can see 412.2Hz and its double frequency. The analysis result is coinciding with the theoretical value; the relative error is 0.073%.

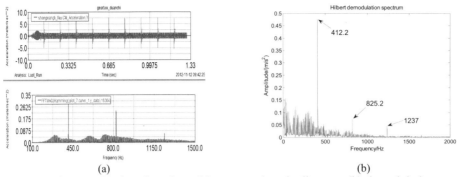

Fig.6. Signal diagram: (a) time domain and frequency domain diagram; (b) demodulation spectrum

Bearing outer ring fault simulation and analysis. The point corrosion of bearing outer ring is shown in Fig.5 (b). Simulation begins after the normal bearing is replaced by the fault bearing. Set rotation speed 1280r/min, step size 0.00001s. The vibration acceleration signal of the upper box collected from the post-processing module of ADAMS is shown in Fig.7 (a). The spectrum and demodulation spectrum are shown in Fig.7 (b).

Bearing outer ring fault frequency is defined as:

$$f_o = \frac{z_{bear}}{2}\left(1 - \frac{d}{D}\cos\alpha\right)f_r \tag{5}$$

6206 bearing parameters are substituted into formula (5), then get the outer ring fault frequency is 76.55Hz. In the demodulation spectrum, we can see the frequency close to the outer ring fault frequency and its harmonics. This proves that the simulation signal is correct.

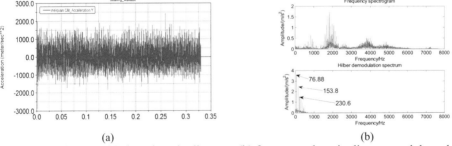

Fig.7. Signal diagram: (a) time domain diagram; (b) frequency domain diagram and demodulation spectrum

Conclusions

(1)This paper firstly establishes a multi-rigid-body model of gearbox using three-dimensional modeling software SolidWorks, then translates the shafts and boxes into flexible body, completes the rigid-flexible coupling virtual prototype model. Finally, this paper verifies the accuracy of the model through rotate speed.

(2) Use ADAMS to make simulation under the following three conditions: normal, gear fault and bearing outer ring fault. Make signal process to the simulation signal. There is side frequency in spectrum diagram, and we can see the fault characteristic frequency and its harmonics. Simulation results are consistent with gear vibration mechanism and signal modulation theory, which verifies correctness and feasibility of virtual prototype technology in gearbox fault simulation. And this paper provides a new idea for the other fault simulation of gearbox and other mechanical fault simulation.

Acknowledgements

This work is supported by National Natural Science Foundation of China (51175007).

References

[1] Lassaad Walha, Tahar Fakhfakh, Mohamed Haddar. Nonlinear dynamics of a two-stage gear system with mesh stiffness fluctuation, bearing flexibility and backlash [J]. Mechanism and Machine Theory, 2009, 44:1058-1069.

[2] Li Yaqiang, Chen Dingfang, Zu Qiaohong. Kinetic simulate of reducer based on ADAMS [J]. Journal of hubei university of technology, 2008, 4:41-43.

[3] Haastrup, Morten; Hansen, Micheal R.; Ebbensen et al. Modeling and parameter identification of deflections in planetary stage of wind turbine gearbox. Modeling identification and control, 2012,33(1):1-11.

[4] Li runfang, Wang jianjun. Gear system dynamics [M]. Beijing: science press, 1997.

[5] Zhao chunxiao, Zhang baoxia, Cun ligang et al. virtual design of gearbox based on solidworks[J]. Coal mine machinery, 2011, 32(1):224-226.

[6] Wang yan, Ma jisheng, Meng gang et al. the modeling and simulation research on rigid-flexible coupled gear system [J]. Journal of mechanical transmission, 2009, 4:32-35.

[7] Wang jinlong. ANSYS finite element analysis and example analysis [M]. Beijing: machinery industry press, 2010.

[8] Li zenggang. A detailed introduction to ADAMS and examples [M]. Being: national defense industry press, 2006.

[9] Hua shungang, Yu guoquan, Su tieming. Modeling and dynamic simulation of reducer virtual prototype based on ADAMS [J]. Machine design and research, 2006, 22(6):47-52.

[10] Zhu lijun, Tan jing, Huang dishan et al. simulation analysis of deep groove ball bearings based on ADAMS [J]. Bearing, 2011, 2:3-6.

Advanced Materials Research Vol. 819 (2013) pp 379-383
© (2013) Trans Tech Publications, Switzerland
doi:10.4028/www.scientific.net/AMR.819.379

Research on Coverage Path Planning of Mobile Robot Based on Genetic Algorithm

Sanpeng Deng[1,a], Zhongmin Wang[2,a], Peng Zhou[3,a], Hongbing Wu[4,a]

[1]Tianjin Key Lab of High Speed Cutting and Precision Maching,Tianjin University of Technology and Education Tianjin 300222, China

[a]sanpeng@yeah.net, [b]wangzhongmin@163.com, [c]zhoupeng@yeah.net, [d]wuhongbing@yeah.net,

Key words: Mobile robot; complete coverage path planning; genetic algorithm

Abstract: This paper presents a complete coverage path planning method, which combines local space coverage with global motion planning. It is realized by modeling mobile robot environment based on Boustrophedon cell decomposition method; and according to the characteristics of regional environment model, the connectivity of the traversing space is represented by a complete weighted connected matrix. Then Genetic algorithm (GA) is used to optimize the subspace traversal distance to obtain the shortest global traversal sequence of mobile robot.

Introduction

Complete coverage path planning of mobile robot is that a mobile robot in certain index completely covers the target environment in a reachable region[1]. It is widely used in the fields of cleaning robot, robot mower, EOD robot and so on [2].Usually, complete coverage path planning algorithm includes template model method and cell decomposition method[3]. Also, model is lack of overall planning for the whole environment. With low efficiency, mobile robots often run into dead circulation state, unable to handle environmental changing[4]. The cell decomposition method, based on the distribution of obstacles, divide environmental space into a series of limited regions with no coincidences and no obstacles to realize each traversal[5].

Genetic algorithm[6] (GA) is put forward by American professor Holland in University of Michigan. The basic idea of GA lies in the theory of biological evolution of species and genetic variation. Firstly, function value chooses the individual, and the parent crossover to produce progeny. Then, the offspring operates according to mutation probability, to conduct screening operation for the individual in accordance with the new screening method. Through screening strategies of individuals, it calculates the fitness of offspring. Finally, new generation joins into form a new generation of (offspring), after repeated iteration, to get the optimal solution.

Environment map modeling based on Boustrophedon cell decomposition method

In view of the environment map for mobile robot as shown in Figure 1, the hazard or coverage zone information is represented by environment map, with property of each space and coordinate information of regional peaks being contained in each of the divided region. Boustrophedon cell decomposition method applies a virtual scanning line to scan sequentially from left to right map, whose connectivity changes generate coverage zones. For coverage zone must have two parallel and adjacent edges, so if the mobile robot traverses to the public edge, it also goes to public region. When the region traversal sequence determined, repeated single region traversal method of mobile robots can achieve complete coverage of the environment.

Figure 2 is a graph of Figure 1, which satisfies the sufficient condition of Hamilton channel definition. So there must be a path, which passes through each sub region once and only once. However, in accordance with the connection Figure 2, it is very difficult to implement traversal algorithm, for a region can only be visited once, and their names must be added to a taboo list, to make complete coverage more difficult. In view of this situation, this paper defined connectivity, distance, obstacles between regions in Figure 1, to add further connections between the existing zones, to form a fully connected graph, as shown in Figure 2.

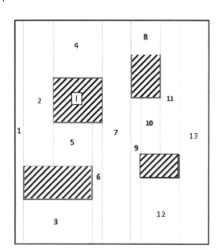

Fig.1 Structure chart based on Boustrophedon unit decomposition

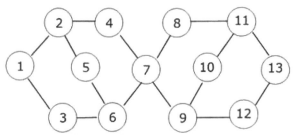

Fig.2 Hamilton connective path

According to connectivity in Figure 2, it can be referred that arbitrary connection within two regions can be reached by the six connections, as shown in Equation 2. Based on multiplication of elements in the same position of the obstacle distance matrix, connectivity matrix, and the multiplication of connected region distance of more than once and a coefficient, to finally get a redefinition of the integrated distance matrix as shown in Equation (2) [7].

$$
D' = \begin{bmatrix}
0 & a & a & 30b & 152.4b & 70b & 120b & 220b & 220b & 695b & 700b & 400b & 360b \\
a & 0 & 30b & a & a & 56.6b & 50b & 120b & 150b & 135.9b & 220b & 260b & 453b \\
a & 30b & 0 & 165.6b & 30b & a & 10b & 204b & 60b & 128b & 322.4b & 180b & 125b \\
30b & a & 165.6b & 0 & 40b & 40b & a & 30b & 36b & 134.4b & 90b & 120b & 196.8b \\
152.4b & a & 30b & 40b & 0 & a & 60.8b & 216.6b & 216.6b & 102b & 283.2b & 228b & 127.5b \\
70b & 56.6b & a & 40b & a & 0 & a & 67b & 30b & 67.2b & 100.5b & 120b & 80b \\
120b & 50b & 10b & a & 60.8b & a & 0 & a & a & 70.8b & 30b & 50b & 15b \\
220b & 120b & 204b & 30b & 216.6b & 67b & a & 0 & 40b & 40b & a & 20b & 331.2b \\
220b & 150b & 60b & 36b & 216.6b & 30b & a & 40b & 0 & a & 53.8b & 20b & a \\
695b & 135.9b & 128b & 134.4b & 102b & 67.2b & 70.8b & 40b & a & 0 & a & 72.8b & 30b \\
700b & 220b & 322.4b & 90b & 283.2b & 100.5b & 30b & a & 53.8b & a & 0 & a & 20b \\
400b & 260b & 180b & 120b & 228b & 120b & 50b & 20b & 20b & 72.8b & a & 0 & a \\
360b & 453b & 125b & 196.8b & 127.5b & 80b & 15b & 331.2b & a & 30b & 20b & a & 0
\end{bmatrix}
\tag{1}
$$

GA design

Targeting the path planning environment map of mobile robots shown in Figure 1 and the Hamilton connective path shown in Figure 2, procedures of applying GA to solve complete traversal are shown in the following:

(1) Set up encoding scheme and randomly initialize population. The coverage region decimal coding scheme is applied. The code is defined as reflecting a divisional traversal sequence of feasible solution from the solution space to the solution that GA can handle the search space. However, the decoding is converted by the solution space to the problem space.

(2) Set the fitness function, and calculate the fit value of each individual. The paper uses a fitness function $f(x)$:

$$f(x) = \sum_{i=1}^{n-1} d(i,i+1) + d(1,n) \tag{2}$$

Where $d(i,j)$ shows the distance between sub region i and sub region j. With fitness value being smaller and the path being shorter, it is more in line with the requirements of the individual.

(3) Judge whether the GA achieves convergence condition. If the algorithm is complete, output of search results is given. Otherwise, the following steps will be performed. Since the shortest path length is unknown in the beginning, iteration number is used as the termination condition.

(4) Determine the fitness function value. Then, related operations of copying and the like are implemented according to the fitness function value. Meanwhile, sort calculated fitness value, namely the path length from small to large order according to the fitness function value.

(5) Design crossover operator. Select a fragment from a parent and save it as the relative order of the region to construct offspring.

(6) Design mutation operator. The inversion mutation is applied to select two points randomly in chromosome, reverse the substrings between the two points.

(7) Returns to step (3).

Using GA algorithm to conduct mobile robot path planning algorithm is shown in Figure 3.

The simulation results

In this paper, C language is used to realize GA, and in computers with memory of AMD X2 240 2.81GHZ, 2.00GB. GA algorithm is run to obtain mobile robot path planning environment map shown in Figure 1 with the traversal sequence as $8 \rightarrow 4 \rightarrow 2 \rightarrow 1 \rightarrow 3 \rightarrow 5 \rightarrow 6 \rightarrow 7 \rightarrow 13 \rightarrow 12 \rightarrow 9 \rightarrow 10 \rightarrow 11$, whose optimal value is 77.9.

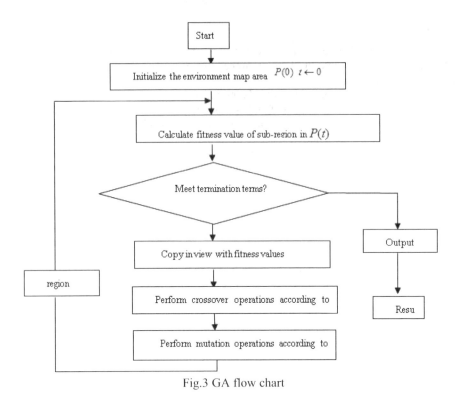

Fig.3 GA flow chart

Conclusion

This paper studies the coverage path planning problem of mobile robots based on GA. Firstly, Boustrophedon cell decomposition method is used to model environment map of mobile robots. Secondly, connectivity of the traversing space is represented by a complete weighted connected matrix. Then, qualities of GA, such as rapid random search, global convergence and easiness to obtain a better solution in evolutionary iteration search procedure, and so on, it make it possible to realize the complete coverage of mobile robots in environment map.

Acknowledgements

This work is supported by the Tianjin Science and Technology Key Research Program (10ZCKFSF01500), the Research and Development Projects of Tianjin University of Technology and Education (KJY11-06, KZ201240).

References

[1] D. Q. Zhu, M. Z.Yan. Survey on technology of mobile robot path planning[J]. Control and Decision，2010, 25(7):961-965.

[2] Z. M.Wang. Path planning and trajectory tracking of mobile robot[M]. Ordnance Industry Press, 2008: 9-16

[3] X. F.Yang. Mobile Robot Path Planning Based on Grid Algorithm and CGA [J]. Computer Simulation, 2012，29(7):223-226.

[4] R. N. Carvalho, H. A Vidal, P. Vieira. Complete coverage path planning and guidance for cleaning robots[C]. Proc. Of IEEE International Symposium on Industrial Electronics, Guimaraes, Portugal, 1997: 677-682.

[5] K. Liu. Research on Mobile Robot complete coverage systems [D].South University master thesis,2006.

[6] H. J. Liu. Research on Mobile Robots motion planning: A Survey[J]. China Mechanical Engineering, 2006，8(1):85-92.

[7] Z. Peng, Z.M. Wang, Z.N. Li, Y. Li. Complete Coverage Path Planning of Mobile Robot Based on Dynamic Programming Algorithm[C].In：2nd International Conference on Electronic and Mechanical Engineering and Information Teachnology. Shenyang:, 2012:1837-1841.

Advanced Materials Research Vol. 819 (2013) pp 384-388
© *(2013) Trans Tech Publications, Switzerland*
doi:10.4028/www.scientific.net/AMR.819.384

The Self-Synchronization of A Novel Vibrating Mechanism Excited by Two Unbalanced Rotors

Li He[1, a], Liu Dan[1,b] and Wen Bangchun[1,c]

[1]School of Mechanical Engineering & Automation, Northeastern University, Shenyang, China

[a]hli@mail.neu.edu.cn, [b]neuliudan@163.com, [c]bcwen1930@vip.sina.com

Keywords: self-synchronization, vibrating mechanism, stability

Abstract: This paper proposes an analytical approach to study self-synchronous motion and stabilizing conditions of a novel vibrating mechanism excited by two unbalanced rotors. This approach begins with utilizing Lagrange equation to establish differential motion equations of the system, and then obtains the dimensionless coupled equation of the unbalanced rotors with a modified average small parameter method. The zero solutions of dimensionless coupled equations are used to achieve the condition to implement self-synchronization of the vibrating system, and finally the Routh-Hurwitz criterion is used to derive the conditions of self-synchronous motion stability.

Introduction

The self-synchronization of vibrating system is based on the discovery of the self-synchronization. Self-synchronization theory was firstly studied by Huygnens[1], who discovered the phenomenon that two clock pendulums suspended from stiff wooden beams could run steady and move oppositely to each other. Synchronization (also known as "frequency capture") has been also investigated in nonlinear circuits by many researchers, such as Rayleigh[2], Vincent, Moler etc. Since 1890s. Blekhman[3-8] proposed the synchronization theory of mechanical exciters that two inertia rotors, driven by two induction motors installed in a vibrating system, could run synchronously without the force from gears. This would enable machines to achieve straight-line motion, circular motion and other motions required in engineering applications.

This paper focuses on the self-synchronization theory of a novel vibrating mechanism excited by two unbalanced rotors, which comprises an inner plastidium and an outer plastidium. The rotational centers of two unbalanced rotors are in line with the inner plastidium's center of mass along a vertical axis and their rotational planes are symmetric about the symmetric horizontal plane of the inner plastidium. This paper is structured in the following manner. Section 2 is devoted to establishing the equations of motion of the vibrating mechanism. The condition of achieving self-synchronization and motion stability are deduced in Section 3. Conclusions are demonstrated in Section 4.

Equations of motion of a vibrating mechanism

The dynamic model of a novel vibrating system is illustrated in Fig. 1, which consists of an inner plastidium m_1 and an outer plastidium m_2, and is driven by two unbalanced rotors m_{01} and m_{02} The inner plastidium m_1 is connected with the outer plastidium m_2 by springs in the x-axis and y-axis direction. The outer plastidium m_2 is supported by the elastic foundation k_z, and is fixed along the x-axis and y-axis direction. o_1 and o_2 are rotational centers of m_{01} and m_{02}, respectively. m_{01} and m_{02} are driven by two induction motors, which are installed symmetrically about the horizontal plane xoy of the centroid o of m_1. The rotational plane is δ against xoy. The rotational centers of two unbalanced rotors and the inner plastidium's center of mass are along the same vertical axis. The two motors rotate in the same direction from the top view.

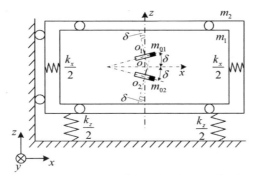

Fig. 1 Dynamic model of the vibrating mechanism

Applying Lagrange equation, we can obtain the differential equations of motion of the system as follows:

$$M_1\ddot{x} + f_x\dot{x} + k_x x = r\cos\delta \sum_{i=1}^{2} m_{0i}(\dot{\varphi}_i^2 \cos\varphi_i + \ddot{\varphi}_i \sin\varphi_i)$$

$$M_1\ddot{y} + f_y\dot{y} + k_y y = r \sum_{i=1}^{2} m_{0i}(\dot{\varphi}_i^2 \sin\varphi_i - \ddot{\varphi}_i \cos\varphi_i)$$

$$M_2\ddot{z} + f_z\dot{z} + k_z z = r\sin\delta \sum_{i=1}^{2} (-1)^{i-1} m_{0i}(\dot{\varphi}_i^2 \cos\varphi_i + \ddot{\varphi}_i \sin\varphi_i) \qquad (1)$$

$$(J_{01} + m_{01}r^2)\ddot{\varphi}_1 + f_1\dot{\varphi}_1 = T_{e1} + m_{01}r(\ddot{x}\cos\delta\sin\varphi_1 - \ddot{y}\cos\varphi_1 + \ddot{z}\sin\delta\sin\varphi_1)$$

$$(J_{02} + m_{02}r^2)\ddot{\varphi}_2 + f_2\dot{\varphi}_2 = T_{e2} + m_{02}r(\ddot{x}\cos\delta\sin\varphi_2 - \ddot{y}\cos\varphi_2 - \ddot{z}\sin\delta\sin\varphi_2)$$

where M_1 is the mass of the vibrating system in x-axis and y-axis directions, $M_1 = m_1 + m_{01} + m_{02}$; M_2 is the mass of the vibrating system in the z-axis direction, $M_2 = m_1 + m_2 + m_{01} + m_{02}$; k_x, k_y and k_z are the stiffness coefficients of the springs in x-axis, y-axis and z-axis directions; f_x, f_y and f_z are the damping coefficients in x-axis, y-axis and z-axis directions,; f_1, f_2 are the damping coefficients of the two motors.

The conditions of implementing self-synchronization and motion stability

As illustrated in Fig. 1, assuming the average phase of the two rotors and their phase difference to be φ and 2α, respectively, then we obtain

$$\varphi_1 = \varphi + \alpha$$
$$\varphi_2 = \varphi - \alpha \qquad (2)$$

The angular velocity of the two motors is ω_{m0}, when the system operates at the steady-state and the coefficients of instantaneous change of average angular velocity and the phase difference between the two exciters are ε_1 and ε_2, respectively, which are the functions of t, then we have Eq. 3

$$\dot{\varphi}_1 = (1 + \varepsilon_1 + \varepsilon_2)\omega_{m0}, \dot{\varphi}_2 = (1 + \varepsilon_1 - \varepsilon_2)\omega_{m0} \qquad (3)$$

$$\ddot{\varphi}_1 = (\dot{\varepsilon}_1 + \dot{\varepsilon}_2)\omega_{m0}, \ddot{\varphi}_2 = (\dot{\varepsilon}_1 - \dot{\varepsilon}_2)\omega_{m0}$$

By using Eq. 3 , we can obtain

$$(J_{01} + m_{01}r^2)\omega_{m0}(\dot{\bar{\varepsilon}}_1 + \dot{\bar{\varepsilon}}_2) + f_1\omega_{m0}(1 + \bar{\varepsilon}_1 + \bar{\varepsilon}_2) = \bar{T}_{e1} - \bar{T}_{L1}$$

$$(J_{02} + m_{02}r^2)\omega_{m0}(\dot{\bar{\varepsilon}}_1 - \dot{\bar{\varepsilon}}_2) + f_2\omega_{m0}(1 + \bar{\varepsilon}_1 - \bar{\varepsilon}_2) = \bar{T}_{e2} - \bar{T}_{L2} \qquad (4)$$

with

$$\overline{T}_{L1} = \chi'_{11}\dot{\overline{\varepsilon}}_1 + \chi'_{12}\dot{\overline{\varepsilon}}_2 + \chi_{11}\overline{\varepsilon}_1 + \chi_{12}\overline{\varepsilon}_2 + \chi_{f1} + \chi_a$$
$$\overline{T}_{L2} = \chi'_{21}\dot{\overline{\varepsilon}}_1 + \chi'_{22}\dot{\overline{\varepsilon}}_2 + \chi_{21}\overline{\varepsilon}_1 + \chi_{22}\overline{\varepsilon}_2 + \chi_{f2} - \chi_a$$

(5)

It can be seen from the above formulas, the system exerts resisting moment χ_a on the motor at a higher speed to slow it. Meantime, the system exerts driving moment χ_a on the motor at a lower speed to speed it up. And ultimately two motors reach the same speed.

And then writing them into the matrix form in the following manner: adding two formulas to get the first row as well as subtracting the second Formula from the first one to obtain the second row, we have

$$A\dot{\overline{\varepsilon}} = B\overline{\varepsilon} + u$$

(6)

Where,

$$\overline{\varepsilon} = \{\overline{\varepsilon}_1, \ \overline{\varepsilon}_2\}^{\mathrm{T}}$$
$$u = \{u_1, \ u_2\}^{\mathrm{T}}$$

(7)

If the zero solutions of Eq. 6 exist and are stable, the vibrating system can implement the synchronization of the two unbalanced rotors.

The condition of implementing the self-synchronization. If the system can achieve the synchronization of the two unbalanced rotors, in a period $T = 2\pi / \omega_{m0}$ the coefficients of instantaneous change are zero, i.e., $\varepsilon = 0$. Substituting $\varepsilon = 0$ into Eq. 6, we obtain $u = 0$. Rearranging $u = 0$, we have

$$\sin 2\overline{\alpha} = \left|\frac{\Delta T_R}{T_S}\right|$$

(8)

where T_S is the torque of frequency capture; ΔT_R is the difference torque of the residual electromagnetic torques of motor 1 and 2.

Because $\sin 2\overline{\alpha} \leq 1$, so the condition that the vibrating system can implement the synchronization of the two unbalanced rotors is

$$T_S \geq \left|\Delta T_R\right|$$

(9)

Therefore, the condition of implementing synchronization of the two unbalanced rotors is that the torque of frequency capture is greater than or equal to the difference torque of the residual electromagnetic torques of the two motors.

When the parameters of this system can meet the condition of implementing the synchronization, we solve ω_{m0} and $\overline{\alpha}$ to obtain their numerical solution, which are denoted by ω^*_{m0} and $\overline{\alpha}_0$ [9], respectively.

The conditions of self-synchronous motion stability. By linearizing Eq. (6) around $\overline{\alpha} = \overline{\alpha}_0$, considering $u = 0$, and $\Delta\dot{\overline{\alpha}} = \omega^*_{m0}\overline{\varepsilon}_2$, and writing the equations as a system of the three first order differential equations, using the notation $z = \{\overline{\varepsilon}_1 \ \ \overline{\varepsilon}_2 \ \ \Delta\overline{\alpha}\}^{\mathrm{T}}$, we obtain

$$\dot{z} = Cz$$

(10)

Solving the determinant equation $\det(C - \lambda I) = 0$, we obtain the characteristic equation for the eigenvalue λ as the following:

$$\lambda^3 + c_1\lambda^2 + c_2\lambda + c_3 = 0 \tag{11}$$

where,

$$c_1 = \frac{4\omega_{m0}^* H_1}{H_0}, \quad c_2 = \frac{2\omega_{m0}^{*2} H_2}{H_0}, \quad c_3 = \frac{2\omega_{m0}^{*3} H_3}{H_0}$$

$$H_0 = 4\rho_1\rho_2 - W_{cc}^2 \cos^2 2\bar{\alpha}_0 + W_{cs}^2 \sin^2 2\bar{\alpha}_0$$

$$H_1 = \rho_1\kappa_2 + \rho_2\kappa_1 - W_{cc}W_{cs} \tag{12}$$

$$H_2 = 2\kappa_1\kappa_2 + (\rho_1 + \rho_2)W_{cc} \cos 2\bar{\alpha}_0 + (\rho_1 - \rho_2)W_{cs} \sin 2\bar{\alpha}_0 +$$
$$W_{cc}^2(1 + \sin^2 2\bar{\alpha}_0) - W_{cs}^2(1 + \cos^2 2\bar{\alpha}_0)$$

$$H_3 = (\kappa_1 + \kappa_2)W_{cc} \cos 2\bar{\alpha}_0 + (\kappa_1 - \kappa_2)W_{cs} \sin 2\bar{\alpha}_0 + 2W_{cc}W_{cs}$$

Using the Routh-Hurwitz criterion, we may show that when

$$c_1 > 0, \; c_3 > 0, \; c_1c_2 > c_3 \tag{13}$$

The solution $z = 0$ is stable. Eq. 13 can respectively be rewritten as Eq. 14.

$$H_0 > 0, \; H_1 > 0, \; H_3 > 0, \; 4H_1H_2 > H_0H_3$$
$$H_0 < 0, \; H_1 < 0, \; H_3 < 0, \; 4H_1H_2 > H_0H_3 \tag{14}$$

Because working frequency of the vibrating system is greater than four times of the natural frequency in the non-resonant direction, and damping ratio is small, i.e., $\xi \leq 0.07$, we can know the structure of the vibrating system meets $H_0 > 0$ [10]. So the condition of stability of the synchronization is

$$H_1 > 0, \; H_3 > 0, \; H > 0 \tag{15}$$

where, $H = 4H_1H_2 - H_0H_3$.

Conclusions

This paper proposes a novel vibrating mechanism, consisting an inner plastidium and an outer plastidium. The rotational centers of two unbalanced rotors and the inner plastidium's center of mass are along the same vertical axis. Utilizing Lagrange equation, we obtained the differential equations of motions for the vibrating mechanism. The dimensionless coupled equation of the unbalanced rotors is obtained by using a modified average small parameter method. Finally, we deduced the conditions of achieving self-synchronization and motion stability. We obtain following conclusions:

(1) If $T_S \geq |\Delta T_R|$ is satisfied, i.e. the torque of frequency capture is equal to or greater than the difference torque between the residual electromagnetic torques of two motors, the vibrating system can achieve self-synchronous motion. And we know that χ_a is the coupling torque of the system. It acts on the faster motor as resisting moment and acts on the slower motor as driving moment. Ultimately two motors reach the same speed.

(2) By applying the Routh-Hurwitz criterion, we obtain that the conditions of self-synchronous motion stability are $H_1 > 0, H_3 > 0, 4H_1H_2 - H_0H_3 > 0$.

Acknowledgements

This work is supported by National Science Foundation of China (No. 51175071), the New Century Excellent Talent Project of the Ministry of Education of China (No. NCET-10-0271), Science Foundation of Liaoning Province (No. 201102072).

References

[1] C. Huygens. Horologium Oscillatorium. Paris, Frence, 1673.

[2] J. Rayleigh, Theory of Sound. Dover, New York, 1945.

[3] Blekhman I.I. Synchronization in Science and Technology, ASME Press, New York, 255(1988), (Engl. transl. of the Russian issue: Moscow: Nauka, 1981).

[4] Belkhman I.I. Synchronization of Dynamical System. Nauka, Moscow, 1971 (in Russian).

[5] Blekhman I.I. Vibrational Mechanics. Fizmatlit. Moscow: Nauka, 1994, p. 394 (in Russian, Engl. transl.: Singapore et al., World Scientific Publishing Co., 2000, p. 510).

[6] Blekhman I.I. Selected Topics in Vibrational Mechanics. World Scientific, Singapore, 2004.

[7] Blekhman I.I, Fradkov A.L, Tomchina O.P, et al. Self-synchronization and controlled synchronization: general definition and example design. Mathematics and Computers in Simulation, 58(4), 367-384 (2002)

[8] Blekhman I.I, Yaroshevich N.P. Extension of the domain of applicability of the integral stability criterion (extremum property) in synchronization problems. Journal of Applied Mathematics and Mechanics, 68(6), 839-864 (2004)

[9] Zhao Chun-yu, Zhu Hong-tao Bai Tian-ju, et al. Synchronization of two non-identical coupled exciters in a non-resonant vibrating system of linear motion. Part II: numeric analysis. Shock and Vibration, 16(5), 517-528 (2009)

[10] Zhao Chun-yu, Zhu Hong-tao, Zhang Yi-min, et al. Synchronization of two coupled exciters in a vibrating system of spatial motion. Acta Mechanica Sinica, 26(3), 477-493 (2010)

Advanced Materials Research Vol. 819 (2013) pp 389-392
© (2013) Trans Tech Publications, Switzerland
doi:10.4028/www.scientific.net/AMR.819.389

Study of Dynamics Characteristics for Precision Motor Spindle System

WANG HongJun [1,2, a], Han Qiushi [2,b] and Zheng Jun [2,c]

[1] Key Laboratory of Modern Measurement & Control Technology, Ministry of Education (BISTU), Beijing 100192, China

[2] School of Mechanical and electronic Engineering, Beijing Information Science & Technology University (BISTU), Beijing 100192, China

[a]wanghj86@163.com(corresponding author), [b]hanqs@bistu.edu.cn, [c]zhengjun@bistu.edu.cn

Keywords: motor spindle system, finite element analysis model, dynamic characteristics analysis.

Abstract. The structure of a precision high speed spindle system is described. The shaft and bearing finite element models of the spindle system are constructed by using the Timoshenko beam element and the spring-damper unit. The static mechanical properties of spindle system and modal analysis are studied with different spring element arrangement forms. The results showed that the rotary speed designed was far lower than the critical rotary speed which matched with the low-order modal natural frequency of the spindle,so resonance vibration could be effectively avoided.

Introduction

With the development of science and technology, high speed and high precision processing technology have been widely used in high-level equipment manufacturing industries. High speed and high precision CNC machine tool has become the key to modern manufacturing equipment. It is becoming increasingly important, to enhance the competitiveness of enterprises by improving the running machining precision reliability, stability and maintainability of the high speed and high precision CNC machine tool [1]. High speed and high precision machine tool performance depends on the spindle system of machine tools. For example, 30-70 percent of precision turning roundness errors were resulted by spindle rotation error [2]. Its dynamic performances have great influence on cutting resistance vibration, machining precision and surface roughness of a machine tool.

Many scholars have done a lot of research on the spindle system of machine tool and achieved amount of research results. GAO studied the spindle system dynamic research [3]; Chi-Wei Lin used modal analysis for characteristics of high speed machine tool spindle [4]; KOSMATKA built Timoshenko beam model, laying a foundation for modeling and analysis of spindle [5]; T. l. Schmitz used finite element method to study the spindle system dynamic performance [6]; Rantatalo noted that softened the stiffness of the bearing as an important factor affecting the dynamics characteristic of spindle system [7]; Sun Wei compared the dynamics characteristic of spindle speed [8].

This paper presented a method to analysis a precision high speed spindle system dynamic performance. First the structure of the spindle system is introduced; then the dynamic model of the spindle system was constructed by using the finite element analysis method. Finally the static characteristics and dynamic analysis of modal were analyzed and discussed.

Modeling Method for Precision High Speed Spindle System

The simple structure and rigidity motorized spindles with speed of tens of thousands of rpm or even a hundred thousand of rpm are used widely in the high-speed machine tools currently. A rational and scientific dynamic model is a very important means for prediction and evaluation of the dynamics characteristic of spindle system. A precision high speed spindle speed is 12000r/min to 20000r/min. Its front bearing is a 4-column high speed ball bearing, rear bearing is a single row cylindrical roller bearings, oil-air lubrication; spindle positioning system using face double-location design, preload, ensure that the spindle rotation accuracy of a spindle with high stiffness and good.

Spindle Modeling Method. Construct the spindle model (see Fig.1) based on the finite element method .There are solid unit, beam unit and pipe unit when modeling. Solid unit can be used for solving a system's stiffness or bearing radial load exactly, but speed is slow. The finite model is established by using TIMOSHENKO beam element with simple structure and high accuracy.

High speed spindle system dynamic is similar to the rotor dynamic, the moving equation of the beam unit is as the following:

$$M^b \ddot{X} - \Omega G^b \dot{X} + (K^b + K_p^b - \Omega^2 M_C^b)X = F^b \tag{1}$$

where, M^b -- quality matrix ; M_C^b -- additional quality matrix considering the centrifugal force effect; G^b -- Anti-symmetric gyroscopic matrix; K^b --stiffness matrix; K_p^b --additional stiffness matrix caused by axial load; F^b --external forcevector. Upper mark " b "-beam unit, Ω -speed。

Fig.1 Spindle Model

Ball Bearing Modeling Method. Angular contact ball bearings with low friction characteristics, can withstand both radial and axial loads from metal cutting, also meet the requirements of high speed machining. It has easy maintenance and low cost. When building the bearing dynamic model, the support stiffness and damping dynamic characteristics of bearings is importing the system. Jones bearing dynamic model is complete model now. Due to combined effects by the bearing geometry, preload and external load, the spindle system can be stated as a nonlinear system with variable damping performance and variable stiffness.

Combin14 spring unit itself does not take into account the length, only considering the elastic modulus and damping. With axial tension or torsion performance, better simulations of bearing stiffness. Combin14 spring unit is chosen as the model. Each bearing is simulated as an elastic damping unit, which simplified into 4 evenly over each bearing spindle cylindrical, as shown in Fig.2. Add full constraints at the node outside of the bearings and axial constraint on the contact surfaces of the inner ring. Using spring-damper unit simulation of elastic bearing supports, set up two sets of springs, three sets and five sets springs (see Fig.3). A single bearing preload Radial stiffness calculation formula is as the follows:

$$k_r = 17.7236 \sqrt[3]{Z^2 D_b} \frac{\cos^2 \alpha}{\sqrt[3]{\sin \alpha}} \sqrt[3]{F_{a0}} \tag{2}$$

where, D_b -roller diameter; Z -number of bearing roller; α -contact angle; F_{a0} -axial preload

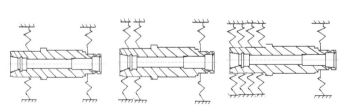

Fig.2 Spring Unit Layout   Fig.3 Spindle Spring Unit Sets Deposit Method

Modal analysis of the spindle system

The spindle material is defined as 20CrMnMoH.Here the bearings are SKF bearings. Tetrahedron method is used for generation mesh. According to the working condition, the spindle is fixed in axial direction while there is freedom in radial direction. The left end of the spindle is axially bound. The back support is free in axial direction. The front support is supported as the angular contact ball bearings. The inner nodes are constrained in axial and the back support inner nodes keep free state. The Young's Modulus is 2.06e+11 and the Poisson's ratio is 0.27.

The modal analysis of spindle of above is done based on the finite element model after building the spindle dynamics model. Modal analysis method is using the Block Lanczos method. Gives two sets of spring units, three spring units and 5 spring units' states spindle 6-order natural frequencies as shown in Table 1.

Table 1 Natural Frequencies on Different Spring Units (HZ)

Order	1	2	3	4	5	6
2 sets	1311.7	1312.6	2603.6	2606.9	3111.3	4582.7
3 sets	1357.6	1358.4	2656.8	2659.6	3192.6	4730.7
5 sets	1421.4	2715.3	2715.3	2717.8	3192.6	4730.7

From Table 1, we can see that three groups of spring-damper unit-order model was significantly greater than the use of two sets of spring-damper unit-order mode, and the five sets best. And the Natural frequencies on 5 and 6 order of 3 sets and 5 sets are same. In some case, if we just concerns that high order, the 3 sets can be used to instead 5 sets. It will make the analysis simple and easier. Based on the modal analysis results, the relationships by between critical rotary speed n and the natural frequency f can be calculated as the equation n=60 f. The critical rotary speed n corresponding to the related modal order can be seen in Table 2. It can be seen that the critical speed n is much greater than the experimental electro-spindle design maximum speed 12000 r/min, so the spindle system design is reasonable, can effectively avoid resonance area when the spindle is running. It can be seen that the critical speed n is much greater than the experimental electro-spindle design maximum speed 12000 r/min, so the spindle system design is reasonable, can effectively avoid resonance area when the spindle is running.

Table 2 Critical Rotary Speed n in Different Order (r/min)

Order	1	Max Working speed
2 sets	78707	12000
3 sets	81456	12000
5 sets	85260	12000

Summary

Spindle bearing uses a spring-damper unit simulation, finite element model on the construction of main shaft and bearing, under the analysis discusses the spring element arrangement static mechanical characteristics of spindle system, and modal analysis. This research indicated that the spindle design speed 12 000 r/min far less than lower-order modes 1311.7 Hz corresponds to the critical speed 78707 r/min, which can effectively avoid resonance. Therefore, the spindle/bearing unit design is reasonable.

Acknowledgements

This paper is sponsored by Natural Science Foundation of China Grant No. 51275052; Key Project of Natural Science Foundation of Beijing Municipalipality (KZ201211232039); Funding Project for Academic Human Resources Development in Institutions of Higher Learning under the Jurisdiction of Beijing Municipalipality (PHR201106132).

References

[1]   Xiong W L , Lv L , Yang X B, et al. High-order harmonic vibration of motorized spindle caused by high-frequency converting current and the suppressing methods. Journal of Vibration Engineering. 21(2008) 600-607.

[2]   CAO Hong-rui, LI Bing, HE Zheng-j ia. Dynamic modeling of high-speed spindles and analysis of high-speed effects. Journal of Vibration Engineering.25 (2012)103-108.

[3]   GAO Shanghan, LONG Xinhua,MENG Guang. Nonlinear response and nonsmooth bifurcations of an unbalanced machine-tool spindle-bearing system. Nonlinear Dyn. 54(2008)365-377.

[4]   Chi-Wei L, Jay F, Joe K. Integrated thermo mechanical dynamic model to characterize motorized machine tool spindles during very high speed rotation. International Journal of Machine Tool & Manufacture. 43 (2003) 1035-1050.

[5]   KOSMATKA J B. An improved two-node finite element for stability and natural frequencies of axial-loaded Timoshenko beams. Computers & Structures.57 (1995) 141-149

[6]   T. L. Schmitz, J. C. Ziegert, C. Stanislaus. A method for predicting chatter stability for systems with speed-dependent spindle dynamics. Transactions of the North American Manufacturing Research Institute of SME Conference, Charlotte, 32(2004) 17-23.

[7]   RantataloRantata lo M, A idanpaa J O, G or ansso n B, et al. Milling machine spindle analysis using FEM and non-contact spindle excitation and response measurement. International Journal of Machine Tools &Manufacture.47 (2007) 1034-1045

[8]   SUN Wei,WANG Bo, WEN Bangchun. Comparative Analysis of Dynamics Characteristics for Static and Operation State of High-speed Spindle System. JOURNAL OF MECHANICAL ENGINEERING. 48(2012) 146-152

Advanced Materials Research Vol. 819 (2013) pp 393-397
© (2013) Trans Tech Publications, Switzerland
doi:10.4028/www.scientific.net/AMR.819.393

The production system automatic layout based on simulation

Chao LV[a], Shuang Liu[b], Shiming Wang,Bei Cai

College of Engineering Science and Technology, Shanghai Ocean University, Shanghai, China

[a]clv@shou.edu.com, [b]s-liu@shou.edu.cn

Keywords: simulation; production system; automatic layout; simulation modeling

Abstract. The U-Shaped layout and function of production system is studied, the purpose is showing how the layouts and their main function influence each other, thus to provide enlightenments and references to improve the design of the production system layout. The simple automatic arrangement system is presented based on Plant Simulation software and focused on an equipment layout problem. The automatic layout system is an automatic layout system of workshop layout according to data; adopt the method of simulation modeling combined with optimization of the workshop layout. According to the different optimization objectives, the user can get a different optimal solution. Through simulation, we can get the default layout of the carrying amount, size and arrangement, combined with the enterprise's actual situation. The appropriate adjustments and production efficiency of production workshop are improved, and the operation time of manual modeling is shortened.

Introduction

In the competitive manufacturing industry, lowing the cost and raising the profit is crucial to a company's surviving and development. Among the influences, the establishment and improvement of the product distribution system usually play a vital role [1]. Meanwhile, support of mixed-flow variable product system is the main way to respond to the requirement of the market fleetly in the manufacturing industry. By adjusting system logical and physical layout, variable product system can be functional and variable. However, the system layout structural form is the key and base of the system physical and logical layout adjustment [2]. To satisfy the optimized design of changeable ability of the fleet-response-requirement better, it's important to research the characterized product system layout.

The Machine Layout of Production System

The plant layout forms are single-row layout and multi-row layout, single-row layout can be divided into line type, L-shaped layout, Z- shaped layout, U- shaped layout, S-shaped layout and semi-circle layout [3-7]. The most common layout form in the use of single-row layout is usually applied in unidirectional flow line (such as rigidity product line), group technology (GT) unit, flexible manufacturing system, FMS, workshops adopted the Just-in-time (JIT) system and manufacturing systems reconfiguration process, etc. [8-10]. Multiple-line layouts are basically evolved according to the basic layouts mentioned above. The basic layout form is shown in Fig.1.

The U-shaped layout in the single-row layout is studied in this paper. It's arranged based on the flexible principle. The arrangement is an agile application of technological principle and product principle arrangement, its market-oriented, workers and equipment can respond fleetly to the different requirement from customers and markets. It's a generic term of the product layout design to be accustomed to the individualized needs of marketing.

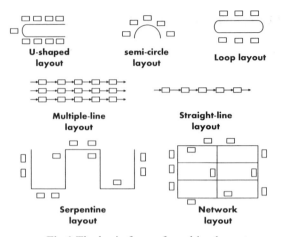

Fig.1 The basic form of machine layout

The System Framework and Simulation Modeling

System Framework

The system used the Plant Simulation software to design a control panel about automatic layout. The system is divided into three parts: basic technical data, automatic arrangement, and data information display, it is shown in Fig.2. The basic data chart is mainly supported by "TableFlie", automatic arrangement is to create and array machine automatically according to certain rules, data information display is mainly a visualization design of external sources alternation.

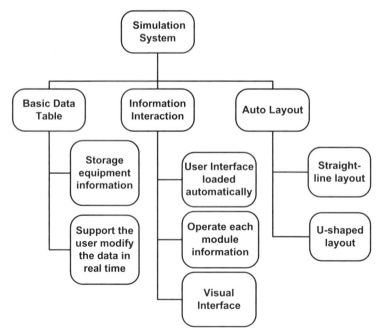

Fig.2 The system framework

The requirement is to design an automatic layout system. It can analyze the production bottleneck and find improve opportunities. The specific area of the workshop; only 10 machines can be placed; it is a mixed production line; the length of the equipment is unknown, but the user can determine the position of the first machine.

Simulation Modeling

According to the simulation targets which have been discussed above and combining the functional features of Tecnomatix Plant Simulation. The system is divided into three parts: basic technical data, automatic arrangement, and data information display. The basic data chart is mainly supported by "TableFlie", automatic arrangement is to create and array equipment automatically according to certain rules, data information display is mainly a visualization design of external sources alternation. The data stored in Table File contain a sequence number, machine number, X Position, Y Position, length, width, process time, the data to make automatic layout and bottleneck analysis come true.

Description: based on the given syntax, the object entity should be established in Frame automatically after statement execution. The Main program sentences are written as follows:

If. models. dialog. tablefile [4, I] =60

then

obji:=.models.dialog.singleproc..createobject(.models.dialog.tablefil[3.i],.models.dialog.tablefile[4,i]);

else

obji:=.models.dialog.singleproc1..createobject(.models.dialog.tablefil[3.i],.models.dialog.tablefil e[4,i]);

end;

obji.name:="M"+num_to_str(i);

for i:=1 to .model.dialog.tablefile.ydim-1 loop;

.materialflow.connector.connect(ste_to_obj(.models.dialog.tablefile[2,i]),(.models.dialog.tablefil e[2,i+1]));

. materialflow.connector.connect(ste_to_obj(.models.dialog.tablefile[2,i]),drain)

next;

Among these sentences, the two paragraphs in the brackets read the created x and y coordinates of "creatobject" individually, thus specific coordinate can be determined and be set up. The SingleProc is supposed to have samples made in Frame, that's how the sentence "creatobject" can create an entity. And singleproc1 which behind "else" is the entity of 180 degrees-rotation, therefore product stream is able to flow in a right direction.

The length of the equipment should be converted into the measurement in Tecnomatix Plant Simulation software. Then we need to calculate the length and width of the SingleProc. Here we can take advantage of object attribute. Secondly, select one of them and right click the mouse and choose "show the attributes and methods" or press the F8.Thirdly, find SingleProc's XPos in a pop-up window, then find SingleProc1's XPos in the same way. Fourthly, SingleProc1's XPos minus SingleProc's XPos is forty. After that, then set and if conditional statement, when the X coordinate of the device is larger than 600 (the right boundary).

Results and Discussion

Compared with the conventional layout situation, it implements the method that's based on the simulation and parameter-driven layout structure associated with the performance levels of analysis. Non-artificial arranged controls are realized, and interactions between emulation software and users based on Dialog interface are available. It accelerates the emulation towards the U-type and linear layout. The final layout result as shown in Fig.3.

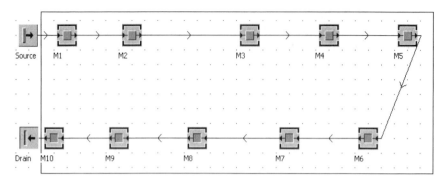

Fig.3 The example layout

The part of layout performance is shown in Fig.4.

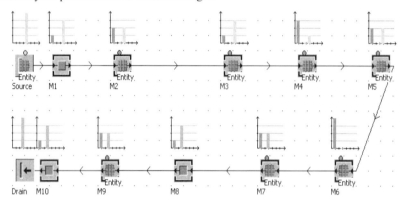

Fig.4 Bottleneck Performance Analysis

Conclusions

The product system layout forms nowadays, aims at the problems and troubles already existed in the process of predicting, combined with the use of simulation software, mainly set up the automatic layout plan design facing the U-shaped layout of mixed-flow product line. The plan basically overcomes part of the problem of resource-wasting during the process and improves the positive emulation of existing layout equipment form; meanwhile it evaluates the program based on the plan design according to the preinstalled targets, the program basically can run after continuous improvement and emulation to the software and interaction of the information. The system has the characters as follows:

1) The design of the system is mainly to solve the plan evaluation towards U- shape layout.
2) User-friendly operation by visual dialog interface.
3) Raise the systems' automatically by external information interaction.
4) Reduce the manual operation time, layout can be done automatically after sizes of equipment been input.

Acknowledgements

This work was financially supported by Shanghai ocean university key teaching reform project (B860610000113), Shanghai college students' innovative projects (B5106110073).

References

[1] Hassan M. International Journal Production Research , Vol. 32-11,1994, P.2559-2584.

[2] Solimanpur M, Vrat P, Shan Kar R. Computers&Operations Research, Vol.32-3, 2005, P.583-598.

[3] Valerie Maier-Speredelozzi, S.jack Hu. Selecting manufacturing system configurations based on performance using AHP. Society of Manufacturing Engineers, 2002,179:1-8..

[4] Peigen Li,Qiang Cheng,Xinyu Shao. A Novel Method of Adaptable Design for Production, The 16th CIRP international design seminar: design&innovation for a sustainable society, Kananaskis, Alberta, Canada: 2006,462-466.

[5] Corrpy P, Kozan E. Computational Optimization and Applications , Vol.28 -3, 2004 , P.287-310.

[6] O. Kulak, S. Cebi, C. Kahraman. Expert Systems with Applications, Vol.37-9 (2010), P. 6705-6717.

[7] Baykasoglu A, Dereli T, Sabuncu I. International Journal of Management Science, Vol.34-4 (2006), P. 385-396.

[8] Azab A, ElMaraghy H A. CIRP Annals-Manufacturing Technology, Vol.56-1 (2007), P. 467-472.

[9] BA ykasogky A, Derlei T, Sabuncu I. International Journal of Management Science, Vol.34 -4, 2006, P.385-396.

[10] Kwong K C,Bai H. Journal of Intelligent Manufacturing, Vol.13-5, 2002, P.367-377.

Advanced Materials Research Vol. 819 (2013) pp 398-403
© (2013) Trans Tech Publications, Switzerland
doi:10.4028/www.scientific.net/AMR.819.398

Transition Segments Processing Method Based On Curve Fitting Algorithm

Yue Zhang [1, 2,a], Qingjian Liu [*1,b], Lu Liu[3,c] and Taiyong Wang [1,d]

[1]Tianjin Key Laboratory of Equipment Design and Manufacturing Technology, Tianjin University, Tianjin 300072, China

[2]Department of Mechanical Engineering, Renai College of Tianjin University, Tianjin 300072, China

[3]Petrochina Pipeline R&D Center, Langfang City, Hebei 065000, China

[a]25325467@qq.com, [b]lklqj_5759@163.com, [c]lordman1982@aliyun.com, [d]tywang@189.cn

Keywords: Bézier curve, Fitting, Transition segments.

Abstract. In order to obtain better speed and high accuracy, real-time curve fitting and fitting processing at transition points are often necessary when machining complex contours. Based on the quadratic trigonometric Bézier curve, a new algorithm is presented to realize curve fitting only between transition segments. This method ensures the machining precision and increases the maximum allowable speed at the transition points. Simulation and experiment results show that the transition velocity changes smoothly and steadily as well as the processing efficiency is promoted.

Introduction

In NC machining process, the contours of mechanical parts are various and eventually comprise of segments (arc, straight line), so there must be transition points between adjacent segments. Considering computing accuracy and speed processing, the remaining of the trace segment is always unequal to the last interpolation cycle distance, thus transfer processing should be done with ensuring contour error [1-6]. If the turning corner is too large, velocity vectors change significantly and great impact comes up in NC machine, as a result the machining accuracy is affected, especially in the high-speed machining situation.

As the core competitiveness of modern manufacturing, tiny segment algorithms are confidential in abroad and there has less papers about these. Huang[7] and Peng[8] proposed real-time NURBS fitting algorithm, realized look-ahead control according to the information of parts' geometry features and machine tool's kinetic characteristics and obtained high profile precision and small speed vibration. But its computation is too heavy and the real-time NURBS is very difficult for diversified contours. For milling tiny segments, Xiong[9] proposed arc fitting method to realize smooth transfer between adjacent segments, but the fitting distance is short.

Just because the difficulty to realize real-time fitting, present researches focus more on how to deal with velocity after completing curve fitting, but less on transition fitting between long straight segments. To solve these questions mentioned above, considering the heavy computation and amount of fitting error for fitting the whole trace, curve fitting algorithm only for transition segments obtained by transfer processing is proposed in this paper. With this algorithm, transition between adjacent straight lines is better and the maximum allowable speed at transition point is improved.

Principle of curve fitting algorithm

Because curve fitting is applied more in tiny segment interpolation, as well as its fitting result is not good enough for long straight line, conducting curve fitting only at transition segments method is presented in this paper. The recommended fitting length is two steps, which equal to interpolation cycles multiplied by transition point speed. With three control points and shape parameter, quadratic trigonometric Bézier curve [10] has excellent transition property and better approximation than quadratic Bézier curve, as shown in Fig.1, which can ensure the fitting accuracy meeting machining requirements.

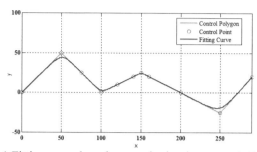

Fig. 1 Fitting curve based on quadratic trigonometric Bézier

Quadratic trigonometric Bézier curve can be defined as follows:

$$P_t = \sum_{i=0}^{i=2} B_{con}(t)P_i \qquad t \in (0, \pi/2) \tag{1}$$

where P_i represent three control points, $P_i = [P_0 \ P_1 \ P_2]^T$. $B_{con}(t)$ is quadratic trigonometric polynomial basis function. $B_{con}(t)$ can be described as follows:

$$B_{con}(t) = \begin{bmatrix} 1 - \alpha\sin t + (\alpha-1)\sin^2 t & 0 & 0 \\ 0 & -\alpha + \alpha\sin t + \alpha\cos t & 0 \\ 0 & 0 & 1 - \alpha\cos t + (\alpha-1)\cos^2 t \end{bmatrix} \tag{2}$$

where α is curve shape parameter. Fitting curves with different shape parameter are shown in Fig.2(a). For simplification, factors $A(t)$, $B(t)$, $C(t)$ are introduced to represent three matrix variables of basis function $B_{con}(t)$, the changing law of $A(t)$, $B(t)$, $C(t)$ are shown in Fig.2(b). As shown in Fig.2(a), fitting curve with $\alpha = 2.5$ is beyond the control polygon and its basis function factors A,B,C are out the regulation range of [0, 1]. Actually, in order to obtain the appropriate fitting curves, the regular range of shape parameter is [0, 2].

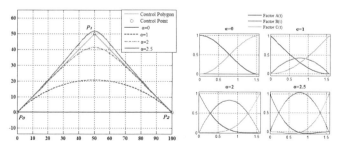

(a)Fitting curves under different parameters   (b) Changing law of factors $A(t)$, $B(t)$, $C(t)$
Fig.2 Fitting curves and basis function variables curves under different parameters

Fitting curve has the following properties: (1) endpoints property: fitting curve must pass through endpoints P_0, P_2, when $t \in [0, \pi/2]$, i.e. $A_{max} = A(0) = P_0$ and $C_{max} = C(\pi/2) = P_2$; (2) symmetry: fitting curves shape the same no matter the control points are P_0, P_1, P_2 or P_2, P_1, P_0 by the law of factors; (3) no-negativity: if the shape parameter ranges from 0 to 2, all values of factors $A(t)$, $B(t)$, $C(t)$ are both positive; (4) weighting property: $A+B+C\equiv1$; (5) monotonic property: when $t\in[0,\pi/2]$, factors $A(t),C(t)$ decrease and $B(t)$ increases as the shape parameter increases.(6) convex hull property: from above properties, fitting curve certainly falls within the convex hull formed of its control points.

Curve fitting in transition segments

According to the above properties of quadratic trigonometric Bézier curve, the maximum error of fitting curve appears at $t = \pi/4$, the maximum error σ is defined as follows:

$$\sigma = P_{\pi/4} - P_1 = (\alpha+1-\sqrt{2}\alpha)H \tag{3}$$

where H designates the linear distance from control point P_2 to the straight line P_1P_3. As shown in Fig.2(a), when $\alpha = 2$ the fitting curve is very close to the original line and the minimum error σ is $0.1716H$. Because the fitting length is only two steps, straight lines P_1P_4, P_4P_3 can be used to replace the fitting curve for NC system, as shown in Fig.3. As a result, the problem of fitting curve solution can be simplified to the solution of the new transition point P_4 with known variables P_1P_3. Instead of fitting curve, straight lines are the real process path, so the shape parameter has a lager value range. $\sigma = 0$ when $\alpha = 2.4142$, so the maximum error σ changes from H to 0 while the shape parameter changes from 0 to 2.4142.

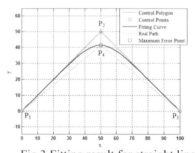

Fig.3 Fitting result for straight lines

For the tuning trace $P_1P_4P_3$, if the theory point should not be reached in the last step, the transition point P_4 requires transfer processing, assuming that transfer processing has been finished at P_4, as shown in Fig4.

When the turning angle is acute, as shown in Fig.4(a), the actual error is P_2P_5 and the error after fitting is the sum of P_2P_4 and P_4P_5. According to the triangle theorem, the actual error must satisfy precision requirement when the fitting error satisfy precision requirement. When the turning angle is obtuse, as shown in Fig.4(b), the actual error is H_2 and the error after fitting is the sum of P_2P_4 and H_1. The actual error satisfy precision requirement as long as the fitting error is satisfied according to the geometry. In conclusion, for the transition point P_4 the machining quality is totally qualified as long as the fitting precision is controlled within the range of machining precision value. In consideration of security, the fitting precision is set to half of the machining precision value in this paper.

In machining process, in order to reduce the impact caused by velocity change, it is important to control the machining velocity size, and the influence of velocity vector change should also be considered. Assuming that curve fitting has been finished and P_4 is the new transition point, as shown in Fig.5. The new turning angle β can be represented as follows:

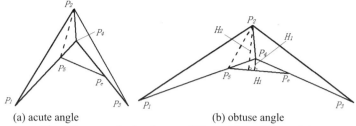

(a) acute angle   (b) obtuse angle

Fig.4 Error analysis of curve fitting at transition points

$$\beta = 2\arctan\left(\frac{\tan(\theta/2)}{(\sqrt{2}-1)\alpha}\right) \tag{4}$$

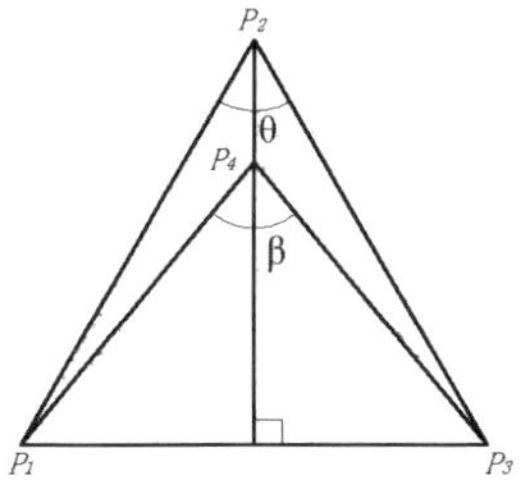

Fig.5 New turning angle obtained by curve fitting

The constraint condition is applied to limit the transition point velocity, and it is relevant to the velocity vector angle(the supplement angle of turning angle). The equation[11] can be expressed as follows:

$$v_i \le \frac{Ta_{max}}{2\sin\left(\dfrac{\pi-\beta}{2}\right)} \qquad i \in [0, N] \tag{5}$$

From Eq.(4) and Eq.(5), when transfer processing is required, under the premise of machining ensure accuracy controlled by fitting precision, the turning angle increases after curve fitting in transition segments, and the velocity vector angle between adjacent segments decreases. Therefore, it is conducive to smooth the transition velocity and improve the maximum allowable speed at the transition point.

Simulation and experiment results

In order to verify the applicability and effectiveness of the proposed algorithm, both simulation and experiment are performed. The machining path is shown in Fig.6, assuming the maximum allowable velocity is 100mm/s, interpolation cycle is 1ms and in conjunction with self-developed advanced S-shape acceleration/deceleration algorithm. Simulation results are shown in Fig.7, where the feed velocity is obtained with completing fitting in transition segments.

The feed velocity of all position points are shown in Table1, including expected velocity and the change range of feed velocity obtained by using the proposed algorithm in this paper and no-using. The simulation result shows that the algorithm proposed above is applicability, the velocity of transition points are increased by curving fitting in transition segments.

This algorithm has been used in self-developed CNC system TDNC - H8. Self-developed system and 6-axis machine tool TDNC-VMC6T410 in our lab is shown in Fig.8. To verify the high reliability of the proposed algorithm in this paper, impeller and bear are processed and the finished parts are shown in Fig.9.

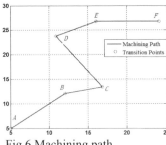
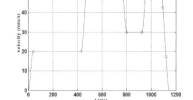

Fig.6 Machining path   Fig.7 Feed velocity curve using proposed algorithm

Table1 Feed velocity at all position points

Position Points	A	B	C	D	E
Expected speed[mm/s]	20	50	60	30	60
Velocity without curve fitting[mm/s]	0-19.126	19.126-47.043	47.043-50.520	50.520-29.385	29.385-60
Velocity with curve fitting	0-20	20-50	50-54.119	54.119-30	30-60

(a) TDNC-VMC6T410 machine tool   (b) TDNC - H8 system

Fig.8 Self-developed machine tool and system

Fig.9 Finished parts

Table 2 Machining time

Machining parts	Time without using curve fitting [min]	Time using curve fitting [min]	Percentage of improvement
Impeller	176	159	9.579%
Bear	283	255	9.734%

With the same processing parameters and machining conditions, the machining time is counted in two cases, including using the proposed algorithm and no-using, the results are shown in Table 2. The experiment result shows that the processing efficiency is promoted nearly 10%.

Summary

To solve transfer processing between long straight segments, this paper presents curve fitting algorithm only for transition segments to avoid heavy computation producing by fitting processing entire trace. With this algorithm, transition between adjacent straight lines is better and the maximum allowable speed at transition point is improved. Simulation and experiment are performed to verify the algorithm applicability, results show that the velocity of transition points is increased and the processing efficiency is promoted.

Acknowledgment

This study is supported by Ministry of Education Doctoral Fund in 2010(No.20100032110006), Fujian Province Science and Technology Project (No.2012H1008) and National Science and Technology Support Program (No.2013BAF06B00).

References

[1] Shen B, Qi DJ, Fan LQ. High-speed adaptive interpolation algorithm based on NURBS fitting for micro sections. China Mechanical Engineering. 15 (2012) 1825~1829.

[2] Liu Y, Wang YZ, Fu HY. Curve interpolation control strategy for 5-axis linkage computer numerical control machine. Computer Integrated Manufacturing Systems. 15(1) (2009) 754~761.

[3] Cao Q, Han M, Xiao YJ. The interpolation algorithm adapting to the trace feature in RPM, China Mechanical Engineering, 8(5)(1997) 56~57.

[4] Shen B, Qi DJ, Fan LQ. Micro sections adaptive interpolation algorithm based on NURBS least square approximation. Manufacturing Technology & Machine Tool, 8 (2012) 54~57.

[5] Ye PQ, Zhao SL, Study on control algorithm for micro-line continuous interpolation. China mechanical engineering, 15(15) (2004) 1354~1356.

[6] You YP, Wang M, Zhu JY, Interpolation algorithm for parametric curve machining. Journal of Nanjing University of Aeronautics & Astronautics, 32 (6) (2000) 667~671.

[7] Huang X, Zeng R,Yue FJ, Application research on NURBS interpolation technique in high speed machining. Journal of Nanjing University of Aeronautics & Astronautics, 34 (1) (2002) 82~85.

[8] Peng FY, He YL, NURBS curve interpolation algorithm with the adaptation to machine tool kinetic character. Journal of Huazhong University of Science and Technology, 33 (7) (2005) 80~83.

[9] Xiong JL, Algorithm of micro-beeline machining for CNC milling. Ordnance Industry Automation, 23(6)(2004) 25~27.

[10]Tang YM, Wu XQ, Han XL, Quadratic trigonometric Bézier curves based on three-points shape parameters, Computer Engineering & Science, 32(3)(2010) 66~67.

[11] Yang L, Zhang CR, Wang K. Speed planning and segment connection in high speed machining. Journal of Shanghai Jiaotong University. Vol.44(1) (2010) 40~45.

Advanced Materials Research Vol. 819 (2013) pp 404-408
© (2013) Trans Tech Publications, Switzerland
doi:10.4028/www.scientific.net/AMR.819.404

Compiling Research and Realization of 3D Cutter Radius Compensation Based on PMAC

Dong Wang[1,a], Shiguang Hu[*1,b], Qingjian Liu[1,c], Dongxiang Chen[1,d], Taiyong Wang[1,e]

[1]Key Laboratory of Mechanism Theory and Equipment Design of Ministry of Education, Tianjin University, Tianjin 300072, China

[a]gailong528m@163.com, [b]sghu@yahoo.cn, [c]lklqj-5759@163.com

[d] dxchen@tju.edu.cn , [e] tywang@189.cn

Keywords：PMAC, 3D cutter radius compensation, CNC

Abstract : This text introduces the compiling research and realization of 3D cutter radius compensation based on PMAC in Five-Axis NC Machines. First, the importance of 3D cutter radius compensation to machining programming of complicated curved surface will be talked about. Then we will present the theory and accomplishment of this function. Finally, based on the train of thought, an experiment will be done on Five-Axis NC Machines. As can be seen from the result, the method, simple and right, can manufacture components with multiple-axis tandem that satisfy complicated curve accuracy class.

Introduction

In recent years, with the development of sophisticated technique, precise or ultra-precise parts play more and more important role at present. Multiple-axis tandem NC machines, especially five-axis or more, are major means to ensure the high -efficiency and high-grade machining of parts with large-scale and abnormal shape. But because of the number of axis increased, the factor and the complexity that influent the working accuracy increased too. The setting of the cutter is one of the important factor, in the aspect of cutter compensation, it only has radius compensation and length compensation in planar. In 3D, especially on five-axis tandem drive NC machine, it still need further study. Only a few well-known foreign manufacturers' High-end CNC systems have equipped with this function in today's market. While our country is still in the research stage, they are not only expensive, but the structures are closed.

In order to simplify NC programming and make the program have nothing to do with cutter size and guarantee the smooth control in velocity, we must equip NC machine with 3D cutter radius compensation. Especially when we cope with spatial complicated curve, such as impeller, this will undoubtedly improve the machining efficiency enormously. So for the purpose of improving the working accuracy of NC machine, the study of five-axis cutter radius compensation in this paper is very useful.

Turbo PMAC provides the capability for performing three-dimensional (3D) cutter (tool) radius compensation on the moves it performs. Unlike the more common two-dimensional (2D) compensation, you can specify independently the offset vector normal to the cutting surface, and the tool orientation vector. The 3D compensation algorithm automatically uses this data to offset the described path of motion, compensating for the size and shape of the tool. This permits you to program the path along the surface of the part, letting Turbo PMAC calculate the path of the center of the end of the tool.

Theory of 3D Cutter Radius Compensation

Generally speaking, 3D cutter radius compensation is aimed at ball-end cutter. However, this paper is directed against any cutter (which is called annular cutter), ball-end cutter (hemispherical tip) and flat-end cutter can be treated as exceptional circumstances.

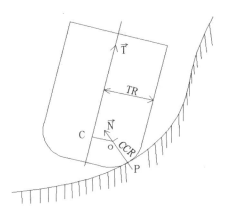

Fig.1 3D Cutter Radius Compensation

Under normal circumstances , cutters(0 <CCR<TR) ,which are called annular cutter, will have a cutter-end radius in between zero and the tool radius, as shown in Fig1(P: Programmed Position; C: Compensated Position; N: Surface-Normal Vector ; T: Tool-Orientation Vector; CCR: Cutter's End Radius ; TR: Tool Shaft Radius) .And also, if the orientation of the cutting tool can change during the compensation, as in five-axis machining, the orientation for purposes of compensation is declared by means of a tool-orientation vector. Note that the tool-orientation vector declared here does not command motion. It merely tells the compensation algorithm the angular orientation that has been commanded of the tool.   Typically the motion for the tool angle has been commanded with A, B, and/or C-axis commands, often processed through an inverse-kinematic subroutine on Turbo PMAC.

As is shown in picture 1, the cutter touches working piece, the coordinate of point P is (x_P, y_P, z_P). The surface-normal vector of part at point P is $\vec{n}(n_x, n_y, n_z)$, obviously,

$$\begin{cases} \overrightarrow{r_{PO}} = \vec{n} * CCR \\ \overrightarrow{r_{OC}} = (TR - CCR) * \vec{e} \\ \overrightarrow{r_C} = \overrightarrow{r_P} + \overrightarrow{r_{PO}} + \overrightarrow{r_{OC}} \end{cases} \qquad (1)$$

According to Eq1, when we get the coordinate (x_P, y_P, z_P) of programmed position and surface-normal vector $\vec{n}(n_x, n_y, n_z)$, the compensated coordinate (NC system actual trajectory) in responding to any cutter radius (TR) and any cutter's end radius can be obtained, as is shown in Eq2 (3,4):

$$x_C = x_P + n_x * CCR + \frac{n_x}{\sqrt{n_x^2 + n_y^2}}(TR - CCR) \tag{2}$$

$$y_C = y_P + n_y * CCR + \frac{n_y}{\sqrt{n_x^2 + n_y^2}}(TR - CCR) \tag{3}$$

$$z_C = z_P + n_z * CCR. \tag{4}$$

Eq2 (3,4) is applied to all kinds of 3D cutters radius compensation. Of course, a flat-end cutter will have a cutter-end radius of zero (CCR=0). A ball-end cutter (hemispherical tip) will have a cutter-end radius equal to the tool(shaft) radius (CCR=TR). In three-axis machining, the orientation of cutting tool is usually declared by the normal vector to the plane of compensation, although the tool-orientation vector may be used.

Realization of 3D Cutter Radius Compensation

To accomplish 3D cutter radius compensation, the author draw support from TDNC-H8 senior opened-system using G45 code. First, this system will do some necessary morphology and grammar check with G45. After passing through (there is nothing wrong with it), NC system will automatically add some information which are essential to achieve other intellectualized subsidiary functions, especially including the sign "CHECKOK". If there is something wrong with G45 code, H8-system will present corresponding error messages, allowing us to modify to check again. Then, H8 will call following relevant compiling sentences against G45 code to accomplish 3D cutter radius compensation. Concrete additional contents mainly include: morphology check sentences, grammar check sentences, compiling sentences.

The interpolation algorithm of 3D cutter radius compensation needs twelve registers (XYZIJKUVWPQ). Register X(Y or Z) deposits the coordinate of programmed position, register I (J or K) deposits the value of surface-normal vector, register U (V or W) deposits the value of tool-orientation vector, register P deposits the value of cutter's end radius, register Q deposits the value of tool shaft radius.

Associating other NC codes, 3D cutter radius compensation can be realized using G45 code which has not been named by international standard. Its basic format is: G01 G45 X_Y_Z_I_J_K_U_V_W_P_Q_. Partial compiling sentences added to cope with G45 based on PMAC are listed as below.

```
N45000 READ (P, Q) IF (Q100&$8000>0) CCR (Q116)
IF (Q100&$10000>0) TR (Q117)
                 CC3
READ (I, J, K) IF (Q100&$100>0) NX(Q109)
IF (Q100&$200>0) NY (Q110)
IF (Q100&$400>0) NZ (Q111)
            READ (U, V, W) IF (Q100&$100000>0) TX (Q121)
IF (Q100&$200000>0) TY (Q122)
IF (Q100&$400000>0) TZ (Q123)
```

Experiment Confirmation

To verify the rationality and accuracy of compiling sentences, we do experiment on TDNC-H8 senior opened-system machine (Fig.3). H8 system can realize five-axis (X, Y, Z, A, C) tandem. For the sake of safety, we adopt plastic material. First, we clamp the $\phi 60$ plastic bar on the rotating

platform and make axis return zero automatically. Also, we must ensure axis A is in horizontal state to do tool setting of X(Y or Z). Then, we preserve NC program containing G45 code in file %0002 and import it into the system to do morphology and grammar check. When nothing is wrong, H8 will present the dialog box "Program is Right！" (Fig.2)and create corresponding compiling ".trc" document. Then we can select executive program and start cycling-button. At last, we manufacture typical five-axis parts (Fig.3) and its precision accuracy meets the requirement with measurement. So we testify the accuracy of 3D cutter radius compensation mentioned above.

Fig.2 Program Check Success

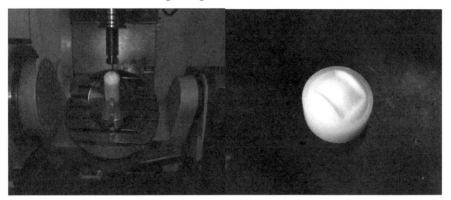

Fig.3 Machining Center and Five-axis Part

Conclusion

3D cutter radius compensation will simplify 3D NC programming to increase NC machining efficiency, regardless of the size of cutter. This paper introduces the theory and accomplishment method of 3D cutter radius compensation in CNC system and put forward universal format. This will not only meet the non-realtime calculating requirement of 3D simulation system, but satisfy the realtime tool setting calculating of CNC system. Also, this will offer significant reference for other complicated codes to accomplish based on pmac. At last, 3D cutter radius compensation will decrease the CNC programming degree of difficulty to make it more flexible.

Acknowledgements

This work is financially supported by the CNC generation of mechanical product innovation demonstration project of Tianjin (2013BAF06B00). The recovery method of weak information and study of algorithm hardened design based on generalized parameter adjustment stochastic resonance (20100032110006) and High-end CNC machining chatter online monitoring and optimization control technology research (12JCQNJC02500).

References

[1] D.N.Moreton, R.Durnford. Three-Dimensional Tool Compensation for a Three-Axis Turning Center[J]. The International Journal of Advanced Manufacturing Technology, 1999,15(9):649-654.

[2] M Kovacic, M Brezocnik, I Pahole. Evolutionary Programming of CNC Machines[J]. Journal of Materials Processing Technology, 2005：1379-1387.

[3] Chen-Hua She & Zhao-Tang Huang. Postprocessor Development of a Five-axis Machine Tool with Nutating Head and Table Configuration[J]. Int J Adv Manuf Technol, 2008(38):728-740.

[4] Liangji Chen. Kinematics Modeling and Post-processing Method of Five-axis CNC Machine[C]. 2009 First International Workshop on Education Technology and Computer Science, 2009: 300-303.

Advanced Materials Research Vol. 819 (2013) pp 409-413
© (2013) Trans Tech Publications, Switzerland
doi:10.4028/www.scientific.net/AMR.819.409

Numerical Simulation of Nonlinear Sloshing in a 2D vertically Moving Container

Zhang Haitao[1, a], Sun Beibei[1, b]

[1]School of Mechanical Engineering, Southeast University, Nanjing, China

[a]zhanghaitao459@163.com, [b]bbsun@seu.edu.cn.

Keywords: Finite difference; Vertical sloshing; Numerical simulation; Parametric instability.

Abstract: Nonlinear liquid sloshing problems in a vertically excited tank are numerically simulated by using a finite difference method. First, the irregular liquid domain is mapped onto a rectangular area by σ-transformation. Then, in the process of time iteration, the free surface is forecasted to estimate the boundary of the next time layer; and some nonlinear terms are approximated to derive linear equations. Free surface elevation and sloshing forces in the vertical sloshing process can be calculated precisely by this method.

Introduction

Sloshing is a complicated phenomenon with strong nonlinearity, which is caused by the movement of the container. The local shock load of the sloshing on the wall would severely affect on the motion stability of the tank, so the system may lose its balance and the structure may be destroyed. In order to avoid these bad consequences, it is necessary to make detailed researches on the sloshing and its impact. In 1960s, because of the demands of aerospace industry, scientists in NASA studied sloshing systematically [1]. The study was mainly on the linear sloshing of small amplitude. With the rapid development of computer technology, the numerical simulation gradually became the main research method. Many numerical methods, such as FEM, FDM, BEM, have been used to calculate violent sloshing in moving containers with complicated shapes.

In the sloshing equations, the free surface boundary are changing over time. Generally speaking, they should be solved by special methods such as MAC, VOF, etc. Frandsen [2,3] and Chern [4] used σ-transformation method to transform the irregular fluid domain into a rectangular area, and then calculated by the pseudo-spectral matrix-element method. In this paper, a new finite difference method has been used to solve the sloshing equations after σ-transformation, in which the free surface elevation is dealt with by a forecast technology, and the nonlinear terms are approximated. 2D sloshing problems in a rectangular tank excited vertically are calculated and analyzed by this method.

Mathematical Model

The fluid is assumed to be incompressible, irrotational and inviscid. Surface tension is neglected. The free surface is assumed to never become overturn or broken during the sloshing process. b is the length of the tank, and h is the still water depth. Two Cartesian coordinate systems are introduced. One is the inertial Cartesian coordinate system XOZ fixed in space. The other is the moving coordinate system xoz connected to the tank, with the origin at the intersection of undisturbed free surface and the left side of the tank. The displacement of the tank in the vertical direction is defined as $Z(t)$. Considered in the coordinate system fixed to the tank, the sloshing equations can be got as:

$$\frac{\partial^2 \varphi}{\partial x^2} + \frac{\partial^2 \varphi}{\partial z^2} = 0, \tag{1}$$

$$\left.\frac{\partial \varphi}{\partial x}\right|_{x=0,b} = 0, \tag{2}$$

$$\left.\frac{\partial\varphi}{\partial z}\right|_{z=-h}=0, \tag{3}$$

$$\frac{\partial\xi}{\partial t}=\left.\frac{\partial\varphi}{\partial z}\right|_{z=\xi}-\left.\frac{\partial\varphi}{\partial x}\right|_{z=\xi}\cdot\frac{\partial\xi}{\partial x}, \tag{4}$$

$$\left.\frac{\partial\varphi}{\partial t}\right|_{z=\xi}=-\frac{1}{2}\left[\left(\frac{\partial\varphi}{\partial x}\right)^{2}+\left(\frac{\partial\varphi}{\partial z}\right)^{2}\right]_{z=\xi}-\left(g+Z''(t)\right)\xi. \tag{5}$$

Eq. 1 is the continuity equation of ideal fluid, φ denotes velocity potential; Eq. 2 and Eq. 3 are rigid wall boundary conditions; Eq. 4 and Eq. 5 are respectively kinematic condition and dynamic condition of free surface, where $\xi(x,t)$ is the free surface elevation in the moving coordinate system.

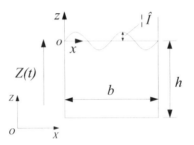

Fig.1 Sloshing model

Numerical Process

First, σ-transformation method is used:

$$\sigma=\frac{z+h}{h+\xi}, \quad \Phi(x,\sigma,t)=\varphi(x,z,t). \tag{6}$$

New equations have been derived by variable substitution:

$$\frac{\partial^{2}\Phi}{\partial x^{2}}-\frac{2\sigma}{\xi+h}\frac{\partial\xi}{\partial x}\frac{\partial^{2}\Phi}{\partial x\partial\sigma}-\left(\frac{1}{\xi+h}\frac{\partial^{2}\xi}{\partial x^{2}}-\frac{2}{(\xi+h)^{2}}\left(\frac{\partial\xi}{\partial x}\right)^{2}\right)\sigma\frac{\partial\Phi}{\partial\sigma}+\frac{1+\sigma^{2}\left(\frac{\partial\xi}{\partial x}\right)^{2}}{(\xi+h)^{2}}\frac{\partial^{2}\Phi}{\partial\sigma^{2}}=0, \tag{7}$$

$$\left.\frac{\partial\Phi}{\partial x}\right|_{x=0,b}-\left.\left(\frac{\partial\Phi}{\partial\sigma}\frac{\sigma}{\xi+h}\frac{\partial\xi}{\partial x}\right)\right|_{x=0,b}=0, \tag{8}$$

$$\left.\frac{\partial\Phi}{\partial\sigma}\right|_{\sigma=0}=0, \tag{9}$$

$$\frac{\partial\xi}{\partial t}=\left.\frac{\partial\Phi}{\partial\sigma}\frac{1}{\xi+h}\right|_{\sigma=1}-\left.\left(\frac{\partial\Phi}{\partial x}-\frac{\partial\Phi}{\partial\sigma}\frac{1}{\xi+h}\frac{\partial\xi}{\partial x}\right)\right|_{\sigma=1}\frac{\partial\xi}{\partial x}, \tag{10}$$

$$\left.\left(\frac{\partial\Phi}{\partial t}-\frac{\partial\Phi}{\partial\sigma}\frac{1}{\xi+h}\frac{\partial\xi}{\partial t}\right)\right|_{\sigma=1}=-\frac{1}{2}\left[\left(\frac{\partial\Phi}{\partial x}-\frac{\partial\Phi}{\partial\sigma}\frac{1}{\xi+h}\frac{\partial\xi}{\partial x}\right)^{2}+\left(\frac{\partial\Phi}{\partial\sigma}\frac{1}{\xi+h}\right)^{2}\right]_{\sigma=1}-\left(g+Z''(t)\right)\xi. \tag{11}$$

So the fluid domain has been changed into a rectangular area.

A finite difference method is used to discretize the governing equation and boundary conditions. Space discretization is not complicated, so the key problem is time discretization. For the integration over time, one of the difficulties is that free surface ξ is an unknown boundary condition; the other is nonlinearity in the equations. In order to solve these two problems, a forecast method and an approximation method are used respectively.

Assume Δt is the time step, consider Eq. 7, Eq. 8 and Eq. 9 on time $(k+1)\Delta t$, that is, on $k+1$ time layer, so ξ should be assigned the value on $k+1$ time layer, which is unfortunately unknown. Explicit scheme of Eq. 10 can be adopted to make predictions of ξ^{k+1}, and then put them into Eq. 7 and Eq. 8. So in the two equations, just Φ^{k+1} are unknowns. Eq. 9 doesn't have ξ, so there's no need to do this.

Consider two boundary conditions Eq. 10 and Eq. 11 on $k+0.5$ time layer. In these two equations, ξ^{k+1} and Φ^{k+1} are set to be unknowns; and they are coupled with nonlinear terms. For example,

$$\left(\frac{\partial \Phi}{\partial x} - \frac{\partial \Phi}{\partial \sigma}\frac{1}{\xi+h}\frac{\partial \xi}{\partial x}\right)\Bigg|_{\sigma=1} \cdot \frac{\partial \xi}{\partial x}, \tag{12}$$

is a nonlinear term, it should have been assigned the value on $k+0.5$ time layer in the calculation process, but here it is assigned the value on k time layer as an approximation. Other nonlinear terms are dealt with by the same idea. Linear terms are discretized by the Crank-Nicolson scheme. Overall, these two boundary conditions are discretized as:

$$\frac{\xi^{k+1}-\xi^k}{\Delta t} = \left(\frac{1}{2}\frac{\partial \Phi^{k+1}}{\partial \sigma} + \frac{1}{2}\frac{\partial \Phi^k}{\partial \sigma}\right)\frac{1}{\xi^k+h}\Bigg|_{\sigma=1} - \left(\frac{\partial \Phi^k}{\partial x} - \frac{\partial \Phi^k}{\partial \sigma}\frac{1}{\xi^k+h}\frac{\partial \xi^k}{\partial x}\right)\Bigg|_{\sigma=1} \cdot \frac{\partial \xi^k}{\partial x}, \tag{13}$$

$$\left(\frac{\Phi^{k+1}-\Phi^k}{\Delta t} - \frac{\partial \Phi^k}{\partial \sigma}\frac{1}{\xi^k+h}\frac{\xi^{k+1}-\xi^k}{\Delta t}\right)\Bigg|_{\sigma=1} = -\frac{1}{2}\left[\left(\frac{\partial \Phi^k}{\partial x} - \frac{\partial \Phi^k}{\partial \sigma}\frac{1}{\xi^k+h}\frac{\partial \xi^k}{\partial x}\right)^2 + \left(\frac{\partial \Phi^k}{\partial \sigma}\frac{1}{\xi^k+h}\right)^2\right]\Bigg|_{\sigma=1} \tag{14}$$

$$-\left(g+Z''\left(t^{k+0.5}\right)\right)\left(\frac{\xi^k+\xi^{k+1}}{2}\right).$$

Then, the governing equation and boundary conditions will be discretized in space, and linear equations are derived. The solutions ξ^{k+1} will be taken as the corrected values of ξ on $k+1$ time layer.

Numerical Results and Discussions

Sloshing induced by containers moving vertically is important in some industries such as oil storage and transportation, aerospace engineering. This vibration phenomenon is caused by the parametric instability, and should be explained by Matheiu function theory [8]. So it is more complicated than transversal sloshing.

A 2D rectangular container model is established, where $b=2$ m, $h=1$ m. The natural frequencies of linear sloshing are

$$\omega_n = \sqrt{g \cdot \frac{n\pi}{b} \cdot Tanh\frac{nh\pi}{b}}, \quad n=1,2,3\cdots. \tag{15}$$

The vertical excitation of the container is assumed to be cosine function:

$$Z(t) = a_v \cos(\omega_v t), \tag{16}$$

The amplitude a_v and the angular frequency ω_v are the parameters which influence the sloshing stability. The initial conditions are defined as:

$$\xi(x,0) = a\cos(\pi x/b), \quad \varphi(x,z,0) = 0, \tag{17}$$

a is the amplitude of the initial wave profile, which influence nonlinearity of the sloshing.

The first example is about stable sloshing. According to Frandsen [2], the parameters satisfy the following conditions:

$$\omega_1/\omega_v = 1.253, \ a_v\omega_v^2/g = 0.5, \ a\omega_1^2/g = 0.288, \tag{18}$$

which means that ω_v=3 rad/s, a_v=0.544 m, a =0.2 m. In the numerical simulation, the time step is 0.005 s, and the space grid size is set to be 40×80. Fig. 2(a) shows the time history of the free surface elevation at the left wall, where the time and the vibration displacement are dealt with using dimensionless method:

$$\xi' = \xi/a, \ t' = \omega_1 \cdot t. \tag{19}$$

The solid line indicates the numerical solutions by the finite difference method in this article; the dotted line indicates the numerical solutions by Frandsen [2]. Good agreement between the two results is achieved. Fig. 2(b) displays wave profiles for a half period from dimensionless time 19.36 to 22.75. Nonlinearity is obvious in the figure.

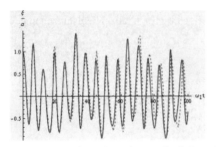
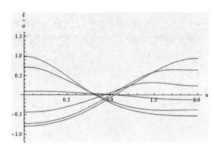

(a) Free surface elevation at the left wall   (b) Wave profiles for a half period

Fig. 2 Stable sloshing

The second example is about unstable sloshing, the parameters satisfy:

$$\omega_1/\omega_v = 0.5, \ a_v\omega_v^2/g = 0.3, \ a\omega_1^2/g = 0.0014. \tag{20}$$

The space grid size is set to be 40×40. Fig. 3(a) shows the time history of the free surface elevation at the left wall. The solutions by the finite difference method are nearly the same as the ones by Frandsen [2]. Assume that the width of the container is d=1 m, then the sloshing forces on the walls of the container can be calculated. Fig. 3(b), (c) show the sloshing forces on the left wall and on the right wall. Fig. 3(d) shows resultant forces on the tank. Dimensionless method is also used:

$$F' = F/\rho ghbd, \ t' = \omega_1 \cdot t. \tag{21}$$

It can be concluded that the forces are becoming greater; eventually the state and the form of the movement will be changed.

5. Conclusions

A new finite difference method is used to analyze sloshing problems in vertically moving containers. Nonlinear fluid equations are transformed to approximate linear equations, and solved by numerical methods. The results show the new method is simple, effective, and has high accuracy. Unstable vertical sloshing will generate great hydrodynamic forces on the wall, and should be avoided.

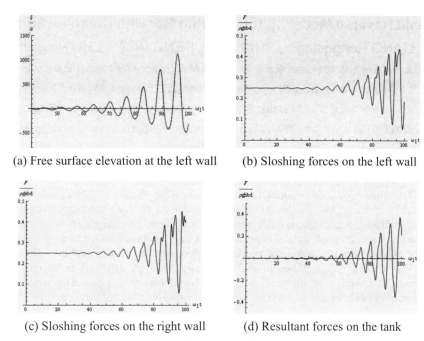

(a) Free surface elevation at the left wall   (b) Sloshing forces on the left wall

(c) Sloshing forces on the right wall   (d) Resultant forces on the tank

Fig. 3 Unstable sloshing

Acknowledgements

This study is supported by the Six Peak Talents Foundation in Jiangsu Province, Prospective and Creative Projects of University-Industry Collaboration under grant number BY2011151.

References

[1]H.N. Abramson, The dynamics of liquids in moving containers, Rep. SP 106, NASA, 1966.

[2]J.B. Frandsen, Sloshing motions in excited tanks, Journal of Computational Physics. 196 (2004) 53-87.

[3]J.B. Frandsen, A.G.L. Borthwick, Simulation of sloshing motions in fixed and vertically excited containers using a 2-D inviscid sigma-transformed finite difference solver, Journal of Fluids and Structures. 18 (2003) 197-214.

[4]M.J. Chern, A.G.L Borthwick, R.E. Taylor, A pseudo-spectral σ-transformation model of 2-D nonlinear waves, Journal of Fluids and Structures. 13 (1999) 607-630.

[5]G.X. Wu, Q.W. Ma, R.E. Taylor, Numerical simulation of sloshing waves in a 3D tank based on a finite element method, Applied Ocean Research. 20 (1998) 337-355.

[6]Yue Baozeng, Zhu Lemei, Yu Dan, Recent advances in liquid-filled tank dynamics and control (in Chinese), Advances in Mechanics. 2011, 41 (1) (2011) 79-92.

[7]T. Ikeda, Nonliear parametric vibrations of an elastic structure with a rectangular liquid tank, Nonlinear Dynamics. 33 (1) (2003) 43-70.

[8]T.B. Benjamin, F. Ursell, The stability of the plane free surface of a liquid in vertical periodic motion, Proceedings of the Royal Society A: Mathematical, Physical, and Engineering Sciences. 225 (1954) 505-515.

Advanced Materials Research Vol. 819 (2013) pp 414-418
© *(2013) Trans Tech Publications, Switzerland*
doi:10.4028/www.scientific.net/AMR.819.414

Rapid Reverse Modeling for a Boring Bar with Complex Surface

LIANG Hongqiang[1,a], YUAN Li[2, b], PENG Wei[2,c], ZHU Pingyu[2, d]

[1]Provincial Key Lab of Hunan Province Health Maintenance for Mechanical Equipment,

Hunan University of Science and Technology, Xiangtan 411201, China

[2]School of mechanical and Electric Engineering, Guangzhou University, 510006, China

[a]leung2013@126.com, [b]bicy2002@126.com, [c]uiopwz129@sina.com,[d]pyzhu@gzhu.edu.cn

Keywords: Coordinate Measuring machine (CMM); reverse engineering; boring bar.

Abstract. For some product development processes, reverse engineering (RE) allows to generate surface models by using the coordinate measuring machine (CMM) scanning technique, for this methodology permits to manufacture different tools in a short development period. A case study on a boring bar with complex surface is present for a brief overview of RE and provide benefits to improve the efficiency of designing and production processes. In order to guarantee the precision of the boring bar model, two methods of software Imageware for rule surface reconstruction and NURBS fitting are used. According to compare experimental modal data with simulation modal data, verified the viability of rapid reverse modeling for damping boring bar.

Introduction

A damping boring bar with complex head surface has admirable effect to reduce chattering quickly, which can improve the vibration damping properties. The advantage of this kind of boring bar due to the cutter head that is complex surface designed [1]. To develop the product, it is attract to know the relationship between the cutter head surface and the effect of reducing vibration. CAD model of the damping boring bar has to be established first. Reverse engineering (RE) allows to generate surface models by using the coordinate measuring machine (CMM) scanning technique without CAD model of the product, and the model can be translated into CAD model so as to modify and optimize the designs[2]. In this paper, a case study on a boring bar with complex surface is the object. Coordinate measuring technology and reverse engineering technology are used to obtain data and reconstruct model for achieving the goal of damping boring bar's rapid reverse modeling.

Data acquisition

On the condition of construct the model effectively, in order to accelerate the speed of data acquisition, we take the structural features of damping boring bar into consideration, select an type of CMM properly to confirm position and quantity of measuring point. As reverse engineering object, one model of Mitsubishi damping boring bar is adopted in this case study. This boring bar consists of cutter head and boring spindle with two symmetrical facets. The boring head has big pits and some chip pockets, which enable chips to evacuate from the hole. This is the key part to scan and make a model with complex surfaces.

This damping boring bar is small in diameter while complex in the structure. There are many curve surfaces in the cutter head part. The portable CMM is ingenious to employ without setting measuring path but with good repeatability and high efficiency, especially, it is competent for measure special narrow curves without caring about dead angle[3]. For the damping boring bar, it is suitable to select pen type probe instead of spherical probe because of it needs to measure some points in the slots.

In general, there are three primary methods to construct curve surfaces: with point-group, with section curves and with existing curve surface[4,5].Combining the latter two methods, measuring positions steps can be as follows:

(1) Divide curve surface of the boring bar into several modules according to its structure feature.

(2) Protocol section curves of curve surface according to shape of curve surface in each module.

(3) Select measuring points with appropriate amount according to the section curves. Fit section curves with measuring points, then construct curve surfaces to finish the reverse modeling of this boring bar.

(4) Points on the joint line among curve surfaces are needs to measure to check out the precision of curve surfaces and rectify the shape of curve surface.

Model reconstruction

After measuring point data with the above methods, the model of this damping boring bar is constructed using the reverse engineering software Imageware. According to the complexity of an boring bar's curve surfaces, the process of reverse modeling can contain two parts: rule surface reconstruction and NURBS fitting[6,7].

The main structure of the damping boring bar includes a lathy cylinder, a thick and short cylinder and a conical named as Cylinder A1, Cylinder A2 and Conical B respectively, as shown in Fig. 1. These surfaces belong to rule curve surfaces, which are the basis of reverse modeling.

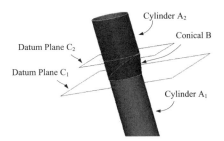

Fig.1 Major structure of boring bar

The main steps of rule surface reconstruction are as follows:

(1) Confirm the data of rule surfaces. The complexity of multiple rule surfaces depends on data points (or data lines, data planes). The precision of data points (or data lines, data planes) directly affects the precision of whole model. The joint face between cutter head and boring spindle is selected as Datum Plane C1. Drawing Data Plane C2 by paralleling Data Plane C1, a point of acrossing the plane can be obtained. This point is on the border between Cylinder A2 and Conical B, which is the nearest to Cylinder A2.

(2) Reconstruct rule surfaces of boring spindle part. Boring spindle part is a lathy cylinder, recorded as Cylinder A1.The diameter D1 and the length H1 can be measured from real boring bar. Secondly, according to point data of Cylinder A1 and the diameter D1, a rounded section curve can be fitted. By projecting this section curve to Data Plane A1, a new rounded curve named as Curve E1 is produced. Finally, Surface F1 of the boring spindle part is reconstructed according to Curve E1 and length H1.

(3) Reconstruct rule surfaces of cutter head part. Cutter head part consists of a thick and short cylinder and a conical that recorded as Cylinder A2 and Conical B. The diameter D2 of Cylinder A2 and the length H2 can be measured from transverse plane of Cylinder A2 to Data Plane C1. Secondly, repeating the same operations as step (2), Curve E2 and Surface F2 of cutter head part are reconstructed. Thirdly, by measuring length H3 of Conical B, a circular conical surface F3 with Curve E1 and Curve E2 is also be constructed. Finally, removing redundant surfaces around the border between Surface F1 and circular conical surface F3,then the whole surfaces of cutter head part is setup.

Through the above operations, most of rule surfaces of this damping boring bar can be reconstructed. It is beneficial to measure structure parameter values of boring bar to reduce point data which need to be measured. And it can save much time to measure point data by CMM.

Moreover, the structure parameter values can be used to correct point data and improve precision of the modeling.

There are two methods of NURBS fitting: (1) Fit NURBS lines to structure curve surfaces[8]; (2) Use a mass of point data to fit NURBS surfaces[9].

Big Pit

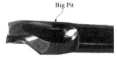

Fig.2 Big pits on cutter head part

The fitting of big pit in cutter head part of the boring bar is important, as shown in Fig.2, this big pit has several irregular surfaces. Hence, it is difficult to describe the feature of this surface directly. It is helpful to measure a lot of point data for big pit to form point cloud by means of CMM. The big pit surfaces with point cloud can then be constructed. But the big pit surfaces belong to non-parametric surface. It is adverse to study structure feature and damping property of damping boring bar. At the same time, it costs much time to measure a mass of point data. It is impossible to achieve the goal of rapid reverse modeling in this way. Based on the above discussions, the method (1) is more appropriate to fit big pit surfaces with section curves. The key steps are as follows.

(1) Machining process judgment. It is possible to find fitting methods for some complex and irregular surfaces with machining process. Through the observation, this big pit is processed by bulb milling tool with certain path. It provides an idea to fit big pit surfaces.

(2) Curve fitting. After confirming the machining process of big pit, digital model of big pits surface with section curves and path curves is constructed. It is simple to fit section curve with point data, which is measured by CMM. But it is complex to fit path curve. First step is to fit a rough curve with point data. Then with decomposing and correcting the rough curve, the path curve is designed, as shown in Fig. 3.

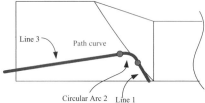

Fig.3 The path curve

(3) Big pit surfaces fitting. It is unable to get whole big pits surface with sweep construction directly because the path curve is complex. Therefore, the path curve should be divided into three segments according to the structure: Line 1, Circular Arc 2 and Line 3. Then three surfaces can be constructed one by one with sweeping construction. Finally, big pit surfaces can be constructed by removing redundant surfaces around the borders of the surfaces.

Through above operations, NURBS fitting is finished. A whole digital model of damping boring bar is obtained, as shown in Fig. 4.

(a) Photo of real boring bar   (b) Constructed digital model

Fig.4 Comparision of real boring bar and constructed model

Comparison validation

The digital model of constructed damping boring bar was transformed into the finite element model using FEM software.Then the first four modal frequencys and vibration modes can be obtained with modal analysis in the condition that boring bar has one fixed end. The results are shown in Fig. 5.

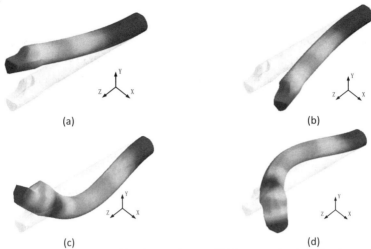

Fig. 5   (a) The first order vibration mode;   (b) The second order vibration mode
(c) The third order vibration mode;   (d) The fourth order vibration mode

From the vibration mode of damping boring bar ,as shown in fig.5,it vibrates in the Y direction at first order vibration, the X direction at second order vibration; the Y direction like "S" at third order vibration, the X direction like "S" at fourth order vibration.

A modal experiment with multipoint implusing input and single-point output has been setup. As the acceleration transducer is only used to detect the Y direction response signal of boring bar. So the First order and second order frequency from modal experiment data are correspond to first order and third order frequency from modal simulation data, the results are shown as Table 1.

Table 1 Modal frequency comparison [Hz]

Data type	First order	Second order	Third order	Fourth order
Simulation data	432.08	458.21	2510.6	2675.1
Experiment data	423	\	2451	\
Relative error	2.15%	\	2.43%	\

From the table 1, one can conclude that the relative errors between simulation data and experiment data are about 2%.It means that the CAD model is almost the same to physical model of damping boring bar, which proved the viability of rapid reverse modeling for damping boring bar.

Conclusion

The process of damping boring bar digital model construction with CMM technology and reverse engineering technology is introduced in details. By comparing analysis and experimental results of the modals, the constructed digital model of the boring bar with complex surfaces is precise. The method of rapid reverse modeling for damping boring bar is effective.

Acknowledgements

The research is supported by the National Natural Science Foundation of China (Grant No. 51105140).

Rreference

[1] L.B. Wang,on Design and Research of Singie-degree-of-freedom High-frequency Damping Boring Bar [D]. Shenyang University of Science and Technology (2011).

[2] H.Y. Chen, D. Liu,Key Techniques and Latest Development in Reverse Engineering of Objects [J].Journal of Machine Design,Vol.23(2006),p.1-4.

[3] KOVACI,FRANKA. Testing and calibration of coordinate measuring arms[J].Precision Engineering,Vol.25(2001),p.90-99.

[4] D.P. Liu, J.J. Chen Reconstruction of surface from measuring points clouds[J]. Machine Tool & Hydraulics,Vol.3(2006),p.32-34.

[5] Y.Y. Fan,Study and Development on Free-Form Surface Reconstruction Technology Based on NURBS [D].Tianjing University(2004).

[6] F.C. Lan, J.Q. Chen and J.G. Lin,Surface Fitting for Scattered Data and Application in Automotive Body Reconstruction [J]. Chinese Journal of Mechanical Engineering, Vol.41(2005),p.213-216.

[7] H. Liu, R.Y. Fan,L. Zeng and L.L. Luo, NURBS Freeform Surface Construction from Points Cloud of Sectional Feature [J].Modern Manufacturing Engineering,Vol.5(2012),p.81-84.

[8] Q.H. Cai, Study on NURBS Surface Modeling and NC Machining Technique of Scattered Data [D]. Shenyang University of Technology(2004).

[9] L. Zhao, Research and Design of Reverse Technology Based on Point Cloud Data [D]. Zhongshan university(2010).

Advanced Materials Research Vol. 819 (2013) pp 419-423
© (2013) Trans Tech Publications, Switzerland
doi:10.4028/www.scientific.net/AMR.819.419

Mechanical Components Design and Research Based on Reliability

Shuying Zhao[1, a], Chensheng Yang [2,b]

[1]College of Sciences, Heilongjiang Institute of Since and Technology Harbin, 150027, China

[2]College of Mathematics and Mechanics, Heilongjiang Institute of Since and Technology Harbin, 150027, China

[a]shuying6354@sina.com, [b]yangchensheng1234@163.com,

Keywords: Reliability, Random Variable, Market Competitiveness

Abstract. In this paper, the reliability design method was used during the process of mechanical parts designing, the design parameters was regarded as random variables, the reliability index of products was involved in the design of the entire process, making the design more in line with the results of the actual situation. The general process of the reliability design was explained by the design of typical part. A good method to effectively improve product quality, reduce part size, save materials and reduce the cost was founded, and the same time the method can make the design of the products more competitive in the market.

Introduction

As far as the traditional mechanical design method of safety coefficient, the design criteria and expression form is simple, intuitive clear, so it is widely used by the vast number of engineering and technical personnel.

Moreover, the trade-off for safety factor has a larger type of experience but because of its stay in participating in the design parameters are deterministic concept, not taking into account the random variable of them teachers, so the design results can not be true reflection of objective reality and blindness, often for security reasons caused unnecessary waste too much access to high-quality materials, or to increase the size of the parts.

Reliability design is random (probability theory and mathematical statistics) analysis the stochastic regularity and reliability of the research system and the part in running state. The basic method of this study parts reliability design, its purpose is to seek to deal more effectively with the problems in the design, improve product quality, reduce part size , material savings, and find a good way to reduce costs.

The basic concepts of reliability and coupling equation of normal distribution stress - strength "interference" and reliability

The ability that products complete regulation function under prescribed conditions and within the time required is called reliability.

The concept of stress within traditional mechanical design is the area ratio of the internal forces generated by the external force in the area of the infinitesimal area. "Stress"of reliability design refers to all the factors that can cause component effectiveness such as temperature, humidity, corrosive effect, particle radiation etc. Strength refers to the magnitude material can withstand the force per unit area. Reliability design " strength " refers to any structure or component failure factors that can prevent it in addition to the mechanical properties of the material (yield limit , ultimate strength , fatigue limit), but also including machining accuracy, surface roughness, surface coating layer etc.

It is reliable that if the consolidated results of the "intensity " factor is greater than the consolidated results of the "stress " when parts or bodies work under the given operating conditions, the contrary is valid.

When the stress random variables S follows a normal distribution, we have

$$f(S) = \frac{1}{\sigma_s\sqrt{2\pi}}\exp\left[-\frac{1}{2}\left(\frac{S-\mu_s}{\sigma_s}\right)^2\right] \quad (0 < S < \infty) \tag{1}$$

Where μ_s is average value of stress, σ_s is standard deviation of stress.

And when the the stress random variables δ follows a normal distribution, we have

$$g(\delta) = \frac{1}{\sigma_\delta\sqrt{2\pi}}\exp\left[-\frac{1}{2}\left(\frac{\delta-\mu_\delta}{\sigma_\delta}\right)^2\right] \quad (0 < \delta < \infty) \tag{2}$$

Where μ_δ is average value of strength, σ_δ is standard deviation of strength.

If random variables S and δ are both normally distributed, then random variables $y = \delta - S$ is also normally distributed. The mean value of y is $\mu_y = \mu_\delta - \mu_s$, its average variance is $\sigma_y = \sqrt{\sigma_\delta^2 + \sigma_S^2}$.

Where probability density function of y can be written as equation followed

$$h(y) = \frac{1}{\sigma_y\sqrt{2\pi}}\exp\left[-\frac{1}{2}\left(\frac{y-\mu_y}{\sigma_y}\right)^2\right]$$

So the reliability of the parts at this time can be expressed as equation (3)

$$R = P(y > 0) = \int_0^\infty \frac{1}{\sigma_y\sqrt{2\pi}}\exp\left[-\frac{1}{2}\left(\frac{y-\mu_y}{\sigma_y}\right)^2\right]dy \tag{3}$$

Formula (3) can be expressed by Fig.1,

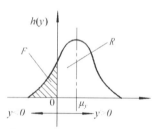

Fig.1 Y distribution of strength between δ and S

The above equation can be regularized, let $z = (y-\mu_y)/\sigma_y$ and t then if we take the differential both sides, because $dz = \frac{1}{\sigma_y}dy$ so $dy = \sigma_y dz$, let $y = 0$, then $z = \frac{-\mu_y}{\sigma_y}$ equation (3) can be transformed into

$$R = P(y > 0) = \int_{-\frac{\mu_y}{\sigma_y}}^\infty \frac{1}{\sigma_y\sqrt{2\pi}}\exp\left[-\frac{z^2}{2}\right]\sigma_y dy = \int_{\frac{\mu_\delta-\mu_S}{\sqrt{\sigma_\delta^2+\sigma_S^2}}}^\infty \frac{1}{\sigma_y\sqrt{2\pi}}\exp\left[-\frac{z^2}{2}\right]dy \tag{4}$$

If $y = 0$, for $\mu_y = \mu_\delta - \mu_S, \sigma_y = \sqrt{\sigma_\delta^2 + \sigma_S^2}$ so

$$z = -\frac{\mu_\delta - \mu_S}{\sqrt{\sigma_\delta^2 + \sigma_S^2}} \tag{5}$$

When $y \to \infty$, the upper limit of z is ∞, so

$$R = \frac{1}{\sigma_y \sqrt{2\pi}} \int_z^{\infty} \exp\left(-\frac{z^2}{2}\right) dz = 1 - \Phi(z)$$

In equation (5), as long as the calculated standard normal variable z, the standard normal probability integral table can check the reliability of the calculated value, it is the characteristic parameters of the structural strength and stress reliability, so it can be called coupling equation.

Design process

In the progress reliability design, you should first establish the mathematical model of the parts work stress and strength, and give the limit state equation. Meanwhile we should also put forward a clear distribution of stress and strength parameters (the mean and standard deviation), the entire design process as shown in Fig.2. General reliability design method will be illustrated in this paper by one typical example of parts reliability design as follow.

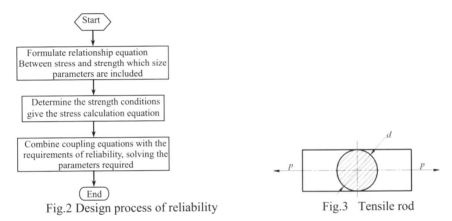

Fig.2 Design process of reliability   Fig.3 Tensile rod

For example, one stretching rod shown in Fig.3. The load acting on the part is random variable P which belongs to normal distribution. The average value and standard deviation of p are $\mu_p = 6 \times 10^4$ and $\sigma_p = 200$ N. Ultimate tensile strength δ of the stretch rod material is also random variable which belongs to normal distribution, and the average value and standard deviation of δ are $\mu_\delta = 500 \times 10^3$ and $\sigma_\delta = 12 \times 10^3$. Seeking the rod diameter to ensure minimum reliability when $R = 0.9_4$.

Solution   Stress $S = \dfrac{P}{A}$

Where P is load (N) overlaps on the rod, A is area of the rod.

With regard to area $A = \pi R^2$, it can be unfolded according to Taylor series at $R = \mu_R$. If the higher order term is negatived, the average area A is

$$\mu_A = \pi \mu_R^2$$

Its standard deviation is $\sigma_A = \sqrt{D(A)} = \sqrt{(2\pi\mu_R)^2 \sigma_R^2} = 2\pi\mu_R\sigma_R$, Assumed processing error of the rod cross section radius R is average value of radius multiplied by one percentage α, According to the characteristics of the normal distribution $3\sigma_R = \alpha\mu_R$, we have $\sigma_R = \alpha\mu_R/3$.

Due to stress

$S = f(P, A) = \dfrac{P}{A}$ So the average value of stress is $\mu_S = \dfrac{\mu_P}{\mu_A} = \dfrac{\mu_P}{\pi\mu_R^2}$

The standard deviation of stress is $\sigma_S^2 = \left(\dfrac{1}{\mu_A}\right)^2 \sigma_P^2 + \left(\dfrac{\mu_P}{\mu_A^2}\right)^2 \sigma_A^2$

Bring μ_A and σ_A into equation above, we can obtain

$$\sigma_S^2 = \left(\frac{1}{\pi\mu_R^2}\right)^2 \sigma_P^2 + \left(\frac{\mu_P}{(\pi\mu_R^2)^2}\right)^2 (2\pi\mu_R\sigma_R)^2 = \frac{\sigma_P^2}{\pi^2\mu_R^4} + \frac{4\pi^2\mu_P^2(\alpha/3)^2\mu_R^4}{\pi^4\mu_R^8} = \frac{\sigma_P^2 + 4/9\alpha^2\mu_P^2}{\pi^2\mu_R^4} \tag{6}$$

Substituted into contact due to the stress S and strength δ are subject to the normal distribution, equation (5) can be obtained

$$z = -\frac{\mu_S - \dfrac{\mu_P}{\pi\mu_R^2}}{\sqrt{\sigma_\delta^2 + \dfrac{\sigma_P^2 + 4/9\alpha^2\mu_P^2}{\pi^2\mu_R^4}}} \tag{7}$$

The further simplification is

$$z = -\frac{\dfrac{\mu_S\pi\mu_R^2 - \mu_P}{\pi\mu_R^2}}{\dfrac{1}{\pi\mu_R^2}\sqrt{\sigma_\delta^2\pi^2\mu_R^4 + (\sigma_P^2 + 4/9\alpha^2\mu_P^2)}} = -\frac{\mu_S\pi\mu_R^2 - \mu_P}{\sqrt{\sigma_\delta^2\pi^2\mu_R^4 + (\sigma_P^2 + 4/9\alpha^2\mu_P^2)}} \tag{8}$$

Both sides of the square are organized into a standard form

$$(\mu_S^2\pi^2 - z^2\sigma_\delta^2\pi^2)\mu_R^4 - 2\mu_S\pi\mu_P\mu_R^2 + \mu_P^2 - z^2(\sigma_P^2 + 4/9\alpha^2\mu_P^2) = 0$$

Reliability by the provisions of Schedule 1 for the standard normal distribution integral table available standard normal random variable value $z = -3.72$, assumptions $\alpha = 0.015$, substituted into the above equation, while the generation of other data using Matlab software programming, the solution of the above equation will get two positive roots:

$\mu_{R_1} = 0.20556$ mm

$\mu_{R_2} = 0.18642$ mm

It is proved, while $\mu_{R_2} = 0.18642$ mm, the parts is reliability $R = 1 - 0.9_4 = 0.0001$. Only 0.20556 mm, reliability $R = 0.9_4$, and thus the results μ_{R_1} is what we seek. The average diameter is:

$\mu_d = 2\mu_R = 2 \times 0.20556\text{mm} = 0.41113\text{mm}$

Because $3\sigma_R = \alpha\mu_R$, $\alpha = 0.015$, so

$3\sigma_R = 0.015 \times 0.20556\,\text{mm} = 0.003083\text{mm}$

$3\sigma_d = 2 \times 0.003083\text{mm} = 0.00617\text{mm}$

The final result obtained is:

$d = (0.41113 \pm 0.00617)\text{mm}$

That is, when the diameter $d = (0.41113 \pm 0.00617)\text{mm}$, under the provisions obtained material and stress levels can be guaranteed reliability $R = 0.9_4$. If $\mu_R = 0.20556$ mm on behalf into equation (8) to change its processing error, you can calculate the standard normal z value of the variable of integration, to a certain structure size, impact of processing errors α on the reliability R, as shown in

Fig.3. Similarly, in the case of a certain structure size (e.g. $\mu_R = 0.20556$ mm) and processing errors (for example $\alpha = 0.015$), changing the dispersion of the material, the degree of influence of the dispersion degree of the material can be obtained by the above method.

Conclusions

As can be seen from the entire part of the design process, its reliability design has the following characteristics:

(1) In order to improve the reliability of the parts: 1.Reduce processing errors, improve processing accuracy can be appropriate in the process, more stringent processing standards and unity; 2. Reduce the dispersion of strength, by strictly controlling the parts timber strength and the process improvement process. 3. Others, such as by improving the strength average, reducing the average stress also can improve the reliability of the parts.

(2) The amount of stress, strength parameters are considered as random variables, more in line with the actual situation, and thus the design results will be closer to actual working parts state.

(3) Reliability index can be throughout the design process, which in turn controls the quality of the parts in the process of selection, material processing technology with the reliability index, so they are in uniform indicators under control, thus contributing to the product further improvement and enhancement of the quality.

(4)Stress - strength "interference" model designed mechanical parts, can give full play to the inherent properties of the material to make parts of the light weight, reasonable structure, low cost which make products more competitive.

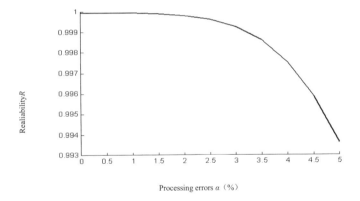

Fig. 3 Impact of processing error α on reliability R

References

[1] Pin Liu: Reliability Engineering Fundamentals. (China Metrology Publishing, China 2008).

[2] Xing Jin, HONG Yanji, Shen Huairong: Reliability data calculation and application. (National Defense Industry Press, China 2003).

[3] Zhili Sun, Liangyu Chen: Practical mechanical reliability design theory and method (Science and Technology Press, China 2003).

Keyword Index

Author Index